Lecture Notes in Computer Sci

T0238527

Commenced Publication in 1973
Founding and Former Series Editors:
Gerhard Goos, Juris Hartmanis, and Jan van Leeuwen

Editorial Board

David Eppstein   Emden R. Gansner (Eds.)

Graph Drawing

17th International Symposium, GD 2009
Chicago, IL, USA, September 22-25, 2009
Revised Papers

 Springer

Volume Editors

David Eppstein
Computer Science Department
Donald Bren School of Information and Computer Sciences
University of California
Irvine, CA, USA
E-mail: eppstein@ics.uci.edu

Emden R. Gansner
AT&T Labs - Research
Florham Park, NJ, USA
E-mail: erg@research.att.com

Library of Congress Control Number: 2010921981

CR Subject Classification (1998): E.1, G.2.2, G.1.6, I.5.3, H.2.8

LNCS Sublibrary: SL 1 – Theoretical Computer Science and General Issues

ISSN   0302-9743
ISBN-10   3-642-11804-6 Springer Berlin Heidelberg New York
ISBN-13   978-3-642-11804-3 Springer Berlin Heidelberg New York

springer.com

© Springer-Verlag Berlin Heidelberg 2010
Printed in Germany

Typesetting: Camera-ready by author, data conversion by Scientific Publishing Services, Chennai, India
Printed on acid-free paper   06/3180   5 4 3 2 1 0

Preface

The 17th International Symposium on Graph Drawing (GD 2009) was held in Chicago, USA, September 22–25, 2009, and was attended by 91 participants from 19 countries.

In response to the call for papers, the Program Committee received 79 submissions. Each submission was reviewed by at least three Program Committee members. Following substantial discussions, the committee accepted 31 long papers and 4 short papers. All authors received detailed reviewers' comments. In a separate submission process, 10 posters were accepted. These were described during the conference, and displayed at the conference site. Each poster was also granted a two-page description in the conference proceedings.

Two invited speakers, János Pach from EPFL Lausanne and Rény Institute, and Martin Wattenberg from IBM Research, gave absorbing talks during the conference. Prof. Pach looked at the class of string graphs, and tantalized us to consider why their properties are so mathematically beautiful. Dr. Wattenberg showed how sometimes twisting the standard rules of graph drawing can illuminate unexpected information contained in graphs.

Keeping with tradition, the symposium hosted the 15th Annual Graph Drawing Contest, including a Graph Drawing Challenge for conference attendees. The contest elicited robust participation from the community with 27 submissions. These proceedings end with a detailed report of the contest.

As always, the success of a conference such as this relies on the help of many people. Our thanks to the Program Committee and all of the external referees who worked so hard to sift for the best among the submitted papers. The Organizing Committee provided us with admirable facilities, a fine banquet, and took care of so many other details that it was hard to believe there were only three members: Jennifer McClelland, Michael J. Pelsmajer and Marcus Schaefer. Also, many thanks to the Chicago-based student volunteers who helped in many ways during the conference. And, of course, special thanks to everyone who submitted a paper or poster, giving us a wealth of raw material from which to build the program.

Despite the uncertain economic times, corporate and institutional sponsors were very generous in their support of GD 2009. AT&T, Tom Sawyer Software and DePaul University served as Gold Sponsors, while IBM ILOG, Illinois Institute of Technology, and the College of Computing and Digital Media at DePaul University were Silver Sponsors.

The 18th International Symposium on Graph Drawing (GD 2010) will be held September 21–24, 2010 in Konstanz, Germany, chaired by Ulrik Brandes.

November 2009

David Eppstein
Emden R. Gansner

Organization

Steering Committee

Franz Josef Brandenburg	University of Passau, Germany
Ulrik Brandes	University of Konstanz, Germany
Giuseppe Di Battista	Università Roma Tre, Italy
Peter Eades	NICTA and University of Sydney, Australia
David Eppstein	University of California, Irvine, USA
Hubert de Fraysseix	CAMS-CNRS, France
Emden R. Gansner	AT&T Labs, USA
Giuseppe Liotta	Università degli Studi di Perugia, Italy
Takao Nishizeki	Tohoku University, Japan
Pierre Rosenstiehl	CAMS-CNRS, France
Roberto Tamassia	Brown University, USA
Ioannis G. Tollis	FORTH-ICS and University of Crete, Greece

Program Committee

Therese Biedl	University of Waterloo, Canada
Franz Josef Brandenburg	University of Passau, Germany
Ulrik Brandes	University of Konstanz, Germany
Erin Chambers	St. Louis University, USA
Sabine Cornelsen	University of Konstanz, Germany
Christian A. Duncan	Lousiana Tech University, USA
Tim Dwyer	Microsoft, USA
David Eppstein	University of California, Irvine (Co-chair), USA
Emden R. Gansner	AT&T Labs (Co-chair), USA
Herman Haverkort	TU Eindhoven, The Netherlands
Patrick Healy	University of Limerick, Ireland
Yifan Hu	AT&T Labs, USA
Michael Lawrence	Fred Hutchinson Cancer Research Center, USA
Giuseppe Liotta	Università degli Studi di Perugia, Italy
Roberto Tamassia	Brown University, USA
Ioannis G. Tollis	FORTH-ICS and University of Crete, Greece
Sue Whitesides	University of Victoria, Canada
Graham Wills	SPSS, USA

Organizing Committee

Jennifer McClelland	DePaul University (Budget Manager), USA
Michael J. Pelsmajer	IIT (Co-chair), USA
Marcus Schaefer	DePaul University (Co-chair), USA

Contest Committee

Christian A. Duncan	Lousiana Tech Univesity, USA
Carsten Gutwenger	Dortmund University of Technology, Germany
Lev Nachmanson	Microsoft, USA
Georg Sander	IBM (Chair), Germany

External Referees

Terry Anderson
Christopher Auer
Christian Bachmaier
Melanie Badent
Michael Baur
Carla Binucci
Wolfgang Brunner
Emilio Di Giacomo
Walter Didimo
Luca Grilli
Martin Harrigan

Andreas Hofmeier
Seok-Hee Hong
Stephen G. Kobourov
Karol Lynch
Martin Mader
Marco Matzeder
Elena Mumford
Nikola Nikolov
Pietro Palladino
Christian Pich
Aimal Rextin

Lesvia Elena Ruiz
 Velázquez
Janet Six
Bettina Speckmann
Michal Stern
Vasilis Tsiaras
Imrich Vrt'o
Michael Wybrow
Hendrik Ziezold

Sponsoring Institutions

Gold Sponsors

Silver Sponsors

Table of Contents

Invited Talks

Papers

Posters

Graph Drawing Contest

Why Are String Graphs So Beautiful?
(Extended Abstract)

János Pach

Chair of Combinatorial Geometry
École Polytechnique Fédérale de Lausanne
janos.pach@epfl.ch

Abstract. String graphs are intersection graphs of continuous simple arcs ("strings") in the plane. They may have a complicated structure, they have no good characterization, the recognition of string graphs is an NP-complete problem. Yet these graphs show remarkably beautiful properties from the point of view of extremal graph theory. What is the explanation for this phenomenon? We do not really know, so we offer three answers. (1) Being a string graph is a hereditary property. (2) String graphs are nicely separable into smaller pieces. (3) As in any geometric picture, one can discover several natural partial orders on a collection of strings.

D. Eppstein and E.R. Gansner (Eds.): GD 2009, LNCS 5849, p. 1, 2010.
© Springer-Verlag Berlin Heidelberg 2010

The Art of Cheating When Drawing a Graph
(Extended Abstract)

Martin Wattenberg

Visual Communication Lab
IBM Research
mwatten@us.ibm.com

Abstract. The prime directive of graph drawing is to depict a network faithfully and accurately. But sometimes it's better to cheat. I will discuss a series of examples - both my own work and that of others - that involve discarding information, distorting the data, encouraging visual clutter, or even adding random noise. The benefits of breaking the rules can range from the scientific to the artistic.

D. Eppstein and E.R. Gansner (Eds.): GD 2009, LNCS 5849, p. 2, 2010.

Drawing Hamiltonian Cycles with No Large Angles

Adrian Dumitrescu[1,*], János Pach[2,**], and Géza Tóth[3,***]

[1] Department of Computer Science, University of Wisconsin-Milwaukee, USA
ad@cs.uwm.edu
[2] Ecole Polytechnique Fédérale de Lausanne and City College, New York
pach@cims.nyu.edu
[3] Alfred Rényi Institute of Mathematics, Budapest, Hungary
geza@renyi.hu

Abstract. Let $n \geq 4$ be even. It is shown that every set S of n points in the plane can be connected by a (possibly self-intersecting) spanning tour (Hamiltonian cycle) consisting of n straight line edges such that the angle between any two consecutive edges is at most $2\pi/3$. For $n = 4$ and 6, this statement is tight. It is also shown that every even-element point set S can be partitioned into at most two subsets, S_1 and S_2, each admitting a spanning tour with no angle larger than $\pi/2$. Fekete and Woeginger conjectured that for sufficiently large even n, every n-element set admits such a spanning tour. We confirm this conjecture for point sets in convex position. A much stronger result holds for large point sets *randomly* and uniformly selected from an open region bounded by finitely many rectifiable curves: for any $\varepsilon > 0$, these sets almost surely admit a spanning tour with no angle larger than ε.

1 Introduction

Consider a set of $n \geq 2$ points. A *spanning tour* is a directed Hamiltonian cycle, drawn with straight line edges; if $n = 2$ the tour consists of the two edges, with opposite orientations, connecting the two points. When three points, p_1, p_2, and p_3, are traversed in this order, their *rotation angle* $\angle p_1 p_2 p_3$ is the angle in $[0, \pi]$ determined by segments $p_1 p_2$ and $p_2 p_3$. If p_3 is on the left (resp. right) side of the oriented line $\overrightarrow{p_1 p_2}$ then we say that the tour, or path makes a *left* (resp. *right*) *turn* at p_2. If a tour (or path) makes only right turns, we call it *pseudo-convex*.

* Supported in part by NSF CAREER grant CCF-0444188, and by the Discrete and Convex Geometry project, in the framework of the European Community's "Structuring the European Research Area" program. Part of the research by this author was done at the Alfred Rényi Institute of Mathematics in Budapest, and at the Ecole Polytechnique Fédérale de Lausanne.
** Research partially supported by NSF grant CCF-08-30272, grants from OTKA, SNF, and PSC-CUNY.
*** Supported by OTKA.

D. Eppstein and E.R. Gansner (Eds.): GD 2009, LNCS 5849, pp. 3–14, 2010.
© Springer-Verlag Berlin Heidelberg 2010

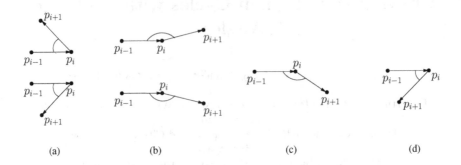

Fig. 1. (a) acute (b) obtuse (c) obtuse and pseudo-convex (d) acute and pseudo-convex

If all of its rotation angles are at most $\pi/2$, we call it an *acute* tour (or path). If all rotation angles are at least $\pi/2$, the tour (or path) is *obtuse*; see Figure 1.

Given a set A of angles, the *angle-restricted tour* (ART) problem is to decide whether a set S of n points in the plane allows a (possibly self-intersecting) spanning tour such that all the n angles between consecutive segments belong to the set A; see [10].

Fekete and Woeginger [10] proved that every finite set of at least *five* points admits a pseudo-convex tour and a non-intersecting pseudo-convex spanning path. They also noticed that every n-element point set S admits an acute spanning path. To see this, start at any point $p_1 \in S$. Assuming that the initial portion $p_1 \ldots p_i$ of such a path has already been defined and $i < n$, let p_{i+1} be an element of $S \setminus \{p_1, \ldots, p_i\}$ farthest away from p_i. It is easy to check that the resulting *path* $p_1 \ldots p_n$ is acute. It is also clear that such a path cannot be always completed to an acute *tour*. Indeed, if all points are on a line and n is *odd*, then along any (spanning) tour, one of the rotation angles must be equal to π.

The question arises: Does every *even*-element point set admit a tour with small rotation angles? More precisely, given an n-element point set S in the plane, where n is even, let $\alpha = \alpha(S) \geq 0$ denote the smallest angle such that S admits a (spanning) tour with the property that all of its rotation angles belong to $[0, \alpha]$. Finally, let $\alpha(n)$ be the maximum of $\alpha(S)$ over all n-element point sets in the plane. Trivially, $\alpha(2) = 0$. The 4-element point set formed by the 3 vertices and the center of an equilateral triangle shows that $\alpha(4) \geq 2\pi/3$. The 6-point configuration depicted in Fig. 2 (left) shows that $\alpha(6) \geq 2\pi/3$.

In this note we show that $\alpha(n) \leq 2\pi/3$, for all even $n \geq 4$.

Theorem 1. *Let $n \geq 4$ be even. Every set of n points in the plane admits a spanning tour such that all of its rotation angles are at most $2\pi/3$. This bound is tight for $n = 4, 6$. Such a tour can be computed in $O(n^{4/3} \log^{1+\varepsilon} n)$ time, for every $\varepsilon > 0$.*

It remains open whether the bound $2\pi/3$ can be replaced by $\pi/2$, for every even $n \geq 8$, as was conjectured in [10]. In other words, every n-element set may admit an acute tour, whenever $n \geq 8$ is even. The point set depicted in Fig. 2 (right)

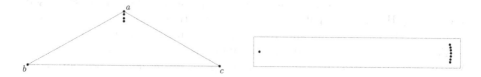

Fig. 2. Left: $\triangle abc$ is an isosceles triangle with $\angle bac = 2\pi/3$. Point a and the 3 points below it are placed on the altitude of the triangle, and very closely inter-spaced. Every tour on these 6 points has a rotation angle of at least $2\pi/3 - \varepsilon$. Right: $n-1$ equidistant points very closely inter-spaced on a small arc of a circle, and one point at the center. Every tour on these n points has a rotation angle of at least $\pi/2 - \varepsilon$.

demonstrates that this statement, if true, cannot be improved. That is, we have $\alpha(n) \geq \pi/2$, for all even $n \geq 8$.

We confirm three weaker versions of this statement. In Section 4, we show that if we enforce acute rotation angles, *two* tours instead of one will certainly suffice.

Theorem 2. *Let $n \geq 8$ be even.*

(i) *Every set of n points in the plane can be partitioned into two even parts, each of which admits an acute spanning tour. Given the n points, the two tours can be computed in $O(n)$ time.*

(ii) *Every set of n points in the plane can be partitioned into two parts of sizes $2\lfloor\frac{n}{4}\rfloor$ and $2\lceil\frac{n}{4}\rceil$, each of which admits an acute spanning tour. Given the n points, the two tours can be computed in $O(n^{4/3}\log^{1+\varepsilon} n)$ time, for every $\varepsilon > 0$.*

In Section 5, we prove the existence of an acute tour in the special case when the points are in convex position.

Theorem 3. *Every even set S of n points in the plane in convex position, with $n \geq 12$, admits an acute spanning tour. Given the n points, such a tour can be computed in $O(n)$ time.*

A much stronger statement holds for *random* point sets, uniformly selected from a not necessarily connected region.

Theorem 4. *Let B be an open region in the plane bounded by finitely many rectifiable Jordan curves and let S be a set of n points, randomly and uniformly selected from B. Then, for any $\varepsilon > 0$, the point set S almost surely admits a spanning tour with no rotation angle larger than ε, as n tends to infinity.*

The last result easily generalizes to higher dimensions.

Related problems and results. Various angle conditions imposed on *geometric graphs* (that is, graphs with straight-line edges) drawn on a fixed vertex set have been studied in [2,3,4,5]. For instance, sharpening an earlier bound of Bárány,

Pór, and Valtr [5], Kynčl [11] proved that any point set admits a (possibly self-intersecting) Hamiltonian path, in which each rotation angle is at least $\pi/6$. This result conjectured by Fekete and Woeginger [10] cannot be improved.

Aichholzer et al. [2] studied similar questions for planar geometric graphs. Among other results, they showed that any point set in general position in the plane admits a non-intersecting Hamiltonian (spanning) path with the property that each rotation angle is at most $3\pi/4$. They also conjectured that this value can be replaced by $\pi/2$. Arkin et al. introduced the notion of *reflexivity* of a point set, as the minimum number of reflex vertices in a polygonalization (i.e., simple polygon) of the set [4]. They gave estimates for the maximum reflexivity of an n-element point set. Recently, Ackerman et al. have made further progress on this problem [1].

2   Balanced Partitions

It is well known (see, e.g. [8], Section 6.6) that every region (every continuous probability measure) in the plane can be cut into *four* parts of equal area (measure) by two orthogonal lines. This statement immediately implies:

Lemma 1. *Given a set S of $n \geq 8$ points in the plane (n even), one can always find two orthogonal lines ℓ_1, ℓ_2 and a partition $S = S_1 \cup S_2 \cup S_3 \cup S_4$ with $|S_1| = |S_3| = \lfloor \frac{n}{4} \rfloor$, $|S_2| = |S_4| = \lceil \frac{n}{4} \rceil$ such that S_1 and S_3 belong to two opposite closed quadrants determined by ℓ_1 and ℓ_2, and S_2 and S_4 belong to the other two opposite quadrants.*

Proof. By a standard compactness argument, it is sufficient to prove this statement for point sets S in general position, in the sense that no 3 points of S are on a line, no 3 determine a right angle, and no two segments spanned by 4 points are orthogonal to each other. Choose a very small $\varepsilon > 0$ and replace each point $p \in S$ by a disk of radius ε around p. Applying the above mentioned result from [8] to the union of these n disks, we obtain two orthogonal lines that meet the requirements of the lemma.   □

Lemma 2. *Given a set S of n points in the plane (n even), there exist three concurrent lines such that the angle between any two of them is $\pi/3$, and there is a partition $S = S_1 \cup \ldots \cup S_6$ with $|S_1| = |S_4|$, $|S_2| = |S_5|$, and $|S_3| = |S_6|$, such that S_i is contained in the i-th closed angular region (wedge) determined by the lines, in counterclockwise order.*

Proof. Just like before, by compactness, it is sufficient to prove the statement for point sets in general position. This time, it is convenient to assume that no 3 points of S determine an angle which is an integer multiple of $\pi/3$, and there are no 2 pairs of points such that the angle between their connecting lines is an integer multiple of $\pi/3$.

Choose again a very small $\varepsilon > 0$ and replace each point $p \in S$ by a disk D_p of radius ε centered at p. Approximate very closely the union of these disks by a continuous measure μ which is strictly positive on every Jordan region in the plane and for which $\mu(\mathbb{R}^2) = n$ and $|\mu(D_p) - 1| < \varepsilon$ for every $p \in S$.

We say that a line ℓ is a *bisecting* line with respect to the continuous measure μ if the measures of both half-planes bounded by ℓ are equal to $n/2$. Clearly, there is a unique bisecting line parallel to every direction, and this line changes continuously as the direction varies. Choose *three* bisecting lines ℓ_1, ℓ_2, ℓ_3 such that the angle between any two of them is $\pi/3$. By changing the direction of ℓ_1, we can achieve that these lines pass through the same point. Indeed, as we turn ℓ_1 by $\pi/3$, the crossing point of the other two lines moves from one side of ℓ_1 to the other. Therefore, there is an intermediate position in which the three lines pass through the same point.

An easy case analysis shows that if ε was sufficiently small, then either no ℓ_i intersects any disk D_p or there is one ℓ_i that intersects two D_p's and the others do not intersect any. In the former case, the lines satisfy the conditions in the lemma, in the latter one, they can be slightly perturbed so as to meet the requirements. □

Given a set S of n points in *general position* in the plane (i.e., no three points are collinear), a line passing through two elements of S is called a *halving line* if there are $\lfloor (n-2)/2 \rfloor$ points on one of its sides and $\lceil (n-2)/2 \rceil$ points on the other [12]. The number of halving lines of an n-element point set in the plane is bounded from above by $O(n^{4/3})$, as was established by Dey [9]. It is also known that the set of halving lines can be computed in $O(n^{4/3} \log^{1+\varepsilon} n)$ time [6], for every $\varepsilon > 0$.

Remark. Starting with an arbitrary halving line ℓ and following the rotation scheme described in [12], one can enumerate all halving lines for S. Using this approach, one obtains algorithmic proofs of Lemmas 1 and 2 that run in $O(n^{4/3} \log^{1+\varepsilon} n)$ time, for every $\varepsilon > 0$.

3   Making a Tour with Rotation Angles at Most $2\pi/3$

In this section, we prove Theorem 1. As we mentioned in the Introduction, for small even values of n, namely for $n = 4$ and $n = 6$, we need to allow rotation angles as large as $2\pi/3$. Here we show that this value suffices for all even n.

Let ℓ_1, ℓ_2, ℓ_3 be three concurrent lines satisfying the conditions of Lemma 2. They divide the plane into six wedges.

Let X, Y, Z, X', Y', Z' denote the six wedges in counterclockwise order, labeled as in Fig. 3. Note that the angle between the x-axis and any edge $p_{i-1}p_i$ of a tour with $p_{i-1} \in X$ and $p_i \in X'$, say, belongs to the interval $[0, \pi/3]$. A piece $p_{i-1}p_ip_{i+1}$ of a tour is of the *form $XX'X$*, say, if $p_{i-1}, p_{i+1} \in X$ and $p_i \in X'$.

Observation 1. *Consider a piece of a tour, which is of the form XQX, where $Q = Y', X'$, or Z. Then the rotation angle at the middle point of this piece, which belongs to Q, is at most $2\pi/3$. The same holds for any other piece consisting of two edges, which starts and ends in the same wedge, and whose middle point belongs to one of the three opposite wedges.*

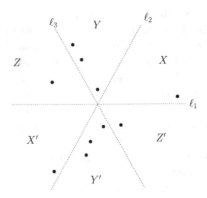

Fig. 3. Three concurrent bisecting lines of S: ℓ_1, ℓ_2, ℓ_3, at angles 0, $\pi/3$, and $2\pi/3$

Observation 2. *Consider a piece of a tour, which is of the form* $XX'Y$ *or* $XX'Z'$. *Then the rotation angle at the middle point of this piece, which belongs to* X', *is at most* $2\pi/3$. *The same holds for any other piece of the form* $X'XZ$, $X'XY'$, $YY'X$, $YY'Z$, $Y'YX'$, $Y'YZ'$, $ZZ'Y$, $ZZ'X'$, $Z'ZX$, $Z'ZY'$.

Proof of Theorem 1. We distinguish two cases:

Case 1. There are at most two nonempty double wedges. If all points are contained in a unique double wedge, say XX' then, by Observation 1, they can be connected by an acute tour of the form $(XX')^*$. The tours starts in X, ends in X', and alternates between the wedges X and X' until all points in $X \cup X'$ are exhausted. Assume now that there are exactly two nonempty double wedges, XX' and YY', say, and refer to Fig. 4. Consider a spanning tour of the form $(XX')^*(YY')^*$, where $(XX')^*$ and $(YY')^*$ are point sequences that alternate between the corresponding opposite wedges until all points in those wedges are

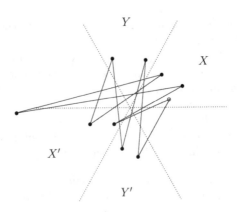

Fig. 4. Case 1: points in two double wedges. A tour of the form $XX'XX'XX'YY'YY'$ is shown; its starting vertex in X is drawn as an empty circle.

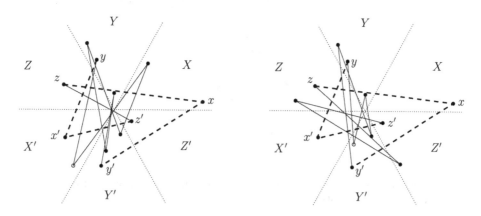

Fig. 5. Case 2: points in three double wedges. Left: a tour of the form $X'XY'YY'Yy'xzz'x'y$ is shown; its starting vertex in X' is drawn as an empty circle. Right: a tour of the form $Y'YY'Yy'xzZ'Zz'x'y$ is shown; its starting vertex in Y' is drawn as an empty circle.

exhausted. By Observations 1 and 2, at each vertex of this tour the rotation angle is at most $2\pi/3$.

Case 2. There are exactly three nonempty double wedges; refer to Fig. 5.

Arbitrarily pick one point from each wedge: $x \in X$, $y \in Y$, $z \in Z$, $x' \in X'$, $y' \in Y'$, $z' \in Z'$. Consider the two triangles $\triangle xzy'$ and $\triangle yx'z'$. The sum of the interior angles of the two triangles is obviously 2π. By averaging, there is one pair of points lying in opposite wedges, say x and x', whose angles sum up to at most $2\pi/3$. Thus, each of these angles is at most $2\pi/3$: $\angle zxy' \le 2\pi/3$, and $\angle yx'z' \le 2\pi/3$.

If $|X \cap S| = |X' \cap S| \ge 2$, consider a spanning tour of the form $(X'X)^+(Y'Y)^+y'xz(Z'Z)^+z'x'y$. Here $(X'X)^+$ denotes a nonempty alternating path between the wedges X' and X, that starts in X', ends in X, and involves all points except x and x'. The notations $(Y'Y)^+$ and $(Z'Z)^+$ are used analogously. An example is depicted in Fig. 5 (left). By Observations 1 and 2, and by our choice of x, y, z, x', y', z', all rotation angles along this tour are at most $2\pi/3$, as required.

If $|X \cap S| = |X' \cap S| = 1$, consider a spanning tour $(Y'Y)^+y'xz(Z'Z)^+z'x'y$; see Fig. 5 (right). The arguments justifying that all rotation angles are at most $2\pi/3$ are the same as before.

The proof of Theorem 1 is now complete.

4   Covering by Two Acute Tours

Proof of Theorem 2. (i) Take a horizontal line ℓ and a partition of our point set $S = S^+ \cup S^-$ into two subsets, each of size $n/2$, such that S^+ and S^- are in the closed half-planes above and below ℓ, respectively. If some points of S lie on

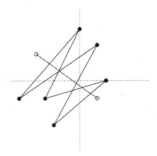

Fig. 6. Even set covered by two tours with 6 and 2 points, respectively; $a = 3$, and $b = 1$ (A double-edge counts as a tour.)

ℓ, we can include them in either of these sets so as to satisfy the condition. Next, take a vertical line ℓ' which gives rise to another equipartition of S. Assume for simplicity that ℓ and ℓ' coincide with the x and y coordinate axes. See Fig. 6, for an illustration.

Thus, we obtain a partition $S = S_1 \cup S_2 \cup S_3 \cup S_4$ such that all points of S_i belong to the i-th closed quadrant determined by the axes (enumerated in the counterclockwise order), $|S_1| = |S_3| = a$, and $|S_2| = |S_4| = b$ for some integers a and b with $a + b = n/2$. Connect now all elements of $S_1 \cup S_3$ by a tour of length $2a$ alternating between S_1 and S_3. Similarly, connect the elements of $S_2 \cup S_4$ by an alternating tour of length $2b$. Obviously, both tours are acute. The above procedure can be performed in linear time, using any linear time selection algorithm [7].

(ii) Find two orthogonal lines and a partition $S = S_1 \cup S_2 \cup S_3 \cup S_4$ satisfying the conditions of Lemma 1. Using the notation of the proof of part (i), now we have $a = \lfloor \frac{n}{4} \rfloor$ and $b = \lceil \frac{n}{4} \rceil$. As above, we obtain two acute tours, of lengths $2\lfloor \frac{n}{4} \rfloor$ and $2\lceil \frac{n}{4} \rceil$, respectively. This completes the the proof of part (ii) of Theorem 1.   □

By keeping only the larger tour, Theorem 2 immediately implies

Corollary 1. *For any even n, every n-element point set in the plane admits an acute even tour covering at least half of its elements.*

5   Acute Tours for Points in Convex Position

Throughout this section, let S denote a set of $n \geq 8$ points in the plane, in convex position and let $S = S_1 \cup S_2 \cup S_3 \cup S_4$ be a partition satisfying the conditions in Lemma 1. A 3-edge path (on 4 points) is called a *hook* if the rotation angles at its two intermediate vertices are acute.

Lemma 3. *Let $P = \{p_1, p_2, p_3, p_4\}$ be the vertex set of a convex quadrilateral, with $p_i \in S_i$, $i = 1, 2, 3, 4$. Then at least one of the following two conditions is satisfied.*

(i) $p_1p_3p_4p_2$ and $p_3p_1p_2p_4$ are hooks, or
(ii) $p_1p_3p_2p_4$ and $p_3p_1p_4p_2$ are hooks.

Proof. At least one of the two angles defined by the diagonals p_1p_3 and p_2p_4 is larger or equal to $\pi/2$. Let x denote the crossing point of these diagonals. If $\angle p_1xp_2 \geq \pi/2$, then the two 3-edge paths $p_1p_3p_4p_2$ and $p_3p_1p_2p_4$ are hooks, while if $\angle p_2xp_3 \geq \pi/2$, then $p_1p_3p_2p_4$ and $p_3p_1p_4p_2$ are hooks.   □

We say that a convex quadrilateral P, as in Lemma 3, is of *type 1* if $\angle p_1xp_2 \geq \pi/2$, and of type 2, otherwise (i.e., if $\angle p_2xp_3 > \pi/2$).

Lemma 4. *Let $P = \{p_1, p_2, p_3, p_4\}$, $Q = \{q_1, q_2, q_3, q_4\}$, and $R = \{r_1, r_2, r_3, r_4\}$ be three vertex-disjoint convex quadrilaterals with $p_i, q_i, r_i \in S_i$, for $i = 1, 2, 3, 4$. Then there exist two hooks induced by two of these quadrilaterals such that the two endpoints of the first one and the two endpoints of the second one lie in different parts of the partition $S_1 \cup S_2 \cup S_3 \cup S_4$. Two such hooks are called opposite. (See Fig. 7 (left).)*

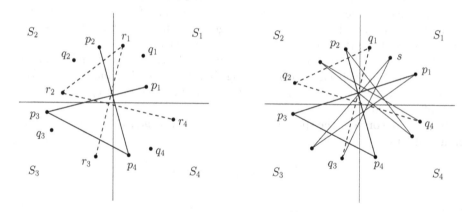

Fig. 7. Left: $p_1p_3p_4p_2$ and $r_3r_1r_2r_4$ are two opposite hooks. Right: an acute tour of S of the form $(S_1S_3)^+p_1p_3p_4p_2(S_4S_2)^+q_4q_2q_1q_3$, starting at $s \in S_1$.

Proof. By the pigeonhole principle, two out of the three quadrilaterals, say P and Q, must have the same type. By Lemma 3, one can find a hook in each of them such that their endpoints are all in different parts of the partition, i.e., two opposite hooks.   □

Proof of Theorem 3. Consider a partition $S = S_1 \cup S_2 \cup S_3 \cup S_4$ satisfying the conditions in Lemma 1. Since $|S| \geq 12$, we have $|S_i| \geq 3$. Pick 3 points from each S_i, and using these points construct three vertex-disjoint convex quadrilaterals, P, Q, and R. By Lemma 4, two of these quadrilaterals, P and Q, say, determine opposite vertex-disjoint hooks. Suppose without loss of generality that P and Q are of type 1, and these two hooks are $p_1p_3p_4p_2$ and $q_4q_2q_1q_3$, where $p_i, q_i \in S_i$, $i = 1, 2, 3, 4$. See Fig. 7(right).

Let $(S_i S_j)^+$ denote a polygonal path starting in S_i, ending in S_j, alternating between S_i and S_j, and exhausting all points of $S_i \cup S_j$, except for p_i, p_j, q_i, q_j. The following tour is acute: $(S_1 S_3)^+ p_1 p_3 p_4 p_2 (S_4 S_2)^+ q_4 q_2 q_1 q_3$, and this completes the proof.   □

6   Random Point Sets

We first verify Theorem 4 for centrally symmetric convex bodies, and then in its full generality.

Lemma 5. *Let B be a centrally symmetric convex body in the plane and let S be a set of n points, randomly and uniformly selected from B. Then, for any $\varepsilon > 0$, S almost surely admits a spanning tour with no rotation angle larger than ε, as n tends to infinity.*

Proof. Let ε be fixed, and let o denote the center of B. Assume without loss of generality that $\text{area}(B) = 1$. Any chord through o divides the area of B into two equal parts. Therefore, there is a positive constant $\delta = \delta(B, \varepsilon)$, depending only on B and ε, such that for every wedge W with angle at most $\pi - \frac{\varepsilon}{2}$ and apex at o, we have that $\text{area}(W \cap B) \leq 1/2 - \delta$. Let $m = \lceil n/2 \rceil$.

Let $p_1, p_2, \ldots p_n$ be n random points, independently and uniformly selected from B, listed in their circular order of visibility from o. The indices are taken modulo n, so that $p_{n+1} = p_1$. Note that almost surely all points p_i are distinct and different from o.

If n is odd, consider the spanning tour $C = p_1 p_{m+1} p_2 p_{m+2} \ldots p_m p_1$. For every i, almost surely we have

$$\pi - \frac{\varepsilon}{2} \leq \angle p_i o p_{m+i-1} \leq \pi + \frac{\varepsilon}{2},$$

and

$$\pi - \frac{\varepsilon}{2} \leq \angle p_i o p_{m+i} \leq \pi + \frac{\varepsilon}{2}.$$

Therefore, we almost surely have $\angle p_{m+i-1} p_i p_{m+i} \leq \varepsilon$, for every i, and the tour C meets the requirements.

If n is even, we choose two odd numbers n_1, n_2 with $n_1 + n_2 = n$ such that $0 \leq n_1 - n_2 \leq 2$. That is, n_1 is m or $m+1$ while n_2 is m or $m-1$. Connect the points p_i by two disjoint cycles, C_1 and C_2, of length n_1 and n_2, with property that (1) in the cyclic order around o, the points $p_1, p_2, \ldots$ belong alternately to C_1 and C_2, as much as possible; and (2) every edge of C_1 and C_2 connects two points, p_i and p_j, with $|j - i - m| \leq 3 \pmod{n}$. We distinguish two cases.

Case 1. $n_1 = n_2 = m$. Let

$$C_1 = p_1 p_{2+m} p_3 p_{4+m} p_5 \cdots p_{n-1} p_m,$$

$$C_2 = p_2 p_{3+m} p_4 p_{5+m} p_6 \cdots p_n p_{1+m}.$$

Switching between these two cycles at two points, we can combine them into a single spanning tour C, as follows.

$$C = p_1 p_{2+m} p_3 p_{4+m} p_5 \cdots p_{n-1} p_m p_2 p_{3+m} p_4 p_{5+m} p_6 \cdots p_n p_{1+m}.$$

It remains true that $|j - i - m| \leq 3 \pmod{n}$ for every edge $p_i p_j$ of C, so that almost surely all rotation angles of C will be smaller than ε.

Case 2. $n_1 = m + 1$, $n_2 = m - 1$. Let

$$C_1 = p_1 p_{2+m} p_3 p_{4+m} p_5 \cdots p_n p_{m+1},$$

$$C_2 = p_2 p_{3+m} p_4 p_{5+m} p_6 \cdots p_{n-1} p_m.$$

We can combine them into a single spanning tour C, as follows.

$$C = p_1 p_{2+m} p_3 p_{4+m} p_5 \cdots p_n p_{m+1} p_2 p_{3+m} p_4 p_{5+m} p_6 \cdots p_{n-1} p_m.$$

It remains true that $|j - i - m| \leq 3 \pmod{n}$ for every edge $p_i p_j$ of C, so that almost surely all rotation angles of C will be smaller than ε. $\qquad\square$

To prove Theorem 4 in its full generality, we need the following technical lemma. Its proof is very similar to that of Lemma 5. The minor modifications are left to the reader.

Lemma 6. *Let B be a centrally symmetric convex set in the plane with nonempty interior. Let o denote the center of B, let $\varepsilon > 0$ be fixed, let s and t be two points of B, and let S' be a set of at most $\varepsilon n/4$ points not belonging to B.*

Then, for any set S of n points randomly and uniformly selected from B, the set $S \cup S'$ almost surely admits a spanning path satisfying the following conditions, as $n \to \infty$:

(i) *all of its turning angles are at most ε;*
(ii) *its first two points are p_1 and p_2 such that $\angle op_1 p_2 \leq \varepsilon/3$, $\angle sop_1 \leq \varepsilon/3$;*
(iii) *its last two points are q_2 and q_1 such that $\angle oq_1 q_2 \leq \varepsilon/3$, $\angle toq_1 \leq \varepsilon/3$.*

Proof of Theorem 4. Assume without loss of generality that area$(B) = 1$. Consider a square lattice of minimum distance δ, for some $\delta > 0$ to be specified later. Let $A = A(\delta)$ denote the total area of all cells (lattice squares of side length δ) completely contained in B, and let $A' = A'(\delta)$ denote the total area of all those cells that intersect B, but are not completely contained in it. Obviously, $A + A' \geq 1$. Since the boundary of B is the union of finitely many rectifiable curves, we have

$$\lim_{\delta \to 0} A = 1, \quad \limsup_{\delta \to 0} \frac{A'}{\delta} < \infty.$$

Therefore, we can choose $\delta > 0$ so that $A' \leq \varepsilon/6$.

Let $X_1, X_2, \ldots, X_m$ denote the cells completely contained in B, in some arbitrary order, and let o_i denote the center of X_i. For any $1 \leq i \leq m$, let s_i be a point on the line $o_i o_{i-1}$ such that o_i belongs to the segment $s_i o_{i-1}$. Analogously, let t_i be a point on the line $o_i o_{i+1}$ such that o_i belongs to the segment $t_i o_{i+1}$. Here the indices are taken modulo m.

Let S be a set of n points in B, selected independently, randomly, and uniformly. Let $S_i = S \cap X_i$, for $1 \le i \le m$, and let $S' = S \setminus \cup_{i=1}^m S_i$. Divide S' into m almost equal parts, $S_1', S_2', \ldots, S_m'$ with $||S_i'| - |S_j'|| \le 1$, for any $i, j = 1, \ldots, m$.

For each $1 \le i \le m$, apply Lemma 6 with S_i, S_i', s_i, and t_i, to obtain a spanning path P_i. The spanning tour $P_1 P_2 \ldots P_m$ obtained by the concatenation of these paths now meets the requirements.   □

References

1. Ackerman, E., Aichholzer, O., Keszegh, B.: Improved upper bounds on the reflexivity of point sets. Computational Geometry: Theory and Applications 42(3), 241–249 (2009)
2. Aichholzer, O., Hackl, T., Hoffmann, M., Huemer, C., Pór, A., Santos, F., Speckman, B., Vogtenhuber, B.: Maximizing maximal angles for plane straight-line graphs. In: Dehne, F., Sack, J.-R., Zeh, N. (eds.) WADS 2007. LNCS, vol. 4619, pp. 458–469. Springer, Heidelberg (2007)
3. Arkin, E.M., Bender, M.A., Demaine, E.D., Fekete, S.P., Mitchell, J.S.B., Sethia, S.: Optimal covering tours with turn costs. SIAM Journal on Computing 35(3), 531–566 (2005)
4. Arkin, E.M., Fekete, S., Hurtado, F., Mitchell, J., Noy, M., Sacristán, V., Sethia, S.: On the reflexivity of point sets. In: Aronov, B., Basu, S., Pach, J., Sharir, M. (eds.) Discrete and Computational Geometry: The Goodman-Pollack Festschrift, pp. 139–156. Springer, Heidelberg (2003)
5. Bárány, I., Pór, A., Valtr, P.: Paths with no small angles. In: Laber, E.S., Bornstein, C., Nogueira, L.T., Faria, L. (eds.) LATIN 2008. LNCS, vol. 4957, pp. 654–663. Springer, Heidelberg (2008)
6. Chan, T.: Remarks on k-level algorithms in the plane, manuscript, Univ. of Waterloo (1999)
7. Cormen, T., Leiserson, C., Rivest, R., Stein, C.: Introduction to Algorithms, 2nd edn. McGraw-Hill, New York (2001)
8. Courant, R., Robbins, H.: What is Mathematics? An Elementary Approach to Ideas and Methods. Oxford University Press, Oxford (1979)
9. Dey, T.K.: Improved bounds on planar k-sets and related problems. Discrete & Computational Geometry 19, 373–382 (1998)
10. Fekete, S.P., Woeginger, G.J.: Angle-restricted tours in the plane. Computational Geometry: Theory and Applications 8(4), 195–218 (1997)
11. Kynčl, J.: Personal communication (2009)
12. Lovász, L.: On the number of halving lines. Ann. Univ. Sci. Budapest, Eötvös, Sec. Math. 14, 107–108 (1971)

Area, Curve Complexity, and Crossing Resolution of Non-planar Graph Drawings

Emilio Di Giacomo[1], Walter Didimo[1], Giuseppe Liotta[1], and Henk Meijer[2]

[1] Dip. di Ingegneria Elettronica e dell'Informazione, Università degli Studi di Perugia
{digiacomo,didimo,liotta}@diei.unipg.it
[2] Roosevelt Academy, The Netherlands
h.meijer@roac.nl

Abstract. In this paper we study non-planar drawings of graphs, and study trade-offs between the crossing resolution (i.e., the minimum angle formed by two crossing segments), the curve complexity (i.e., maximum number of bends per edge), the total number of bends, and the area.

1 Introduction

Planarity is one of the most desirable properties when drawing a graph because planar drawings are more readable and more aesthetically pleasant than non-planar ones. Unfortunately, very few graphs are planar in practice and edge crossings are unavoidable in the vast majority of the application scenarios where relational data are visualized and analyzed by means of graph drawing techniques.

While a large body of literature has been published about the problem of reducing the number of crossings in a non-planar drawing of a graph, by the well-known crossing-lemma this number is quadratic with the number of the edges for dense graphs. This, together with cognitive experiments showing the negative impact of edge crossings on the human understanding of a graph drawing, apparently leads to the conclusion that computing readable node-link visualizations of dense graphs is a hopeless challenge [6,7,8].

However, new cognitive experiments on large drawings of graphs refine the conclusions in [6,7,8] by showing that actually the human's understanding of the relations between the nodes of a network is not bothered by those edge crossings that form large angles [3,4,5]. These experiments suggest a new and fascinating research scenario in which the problem of maximizing the *crossing resolution* of a drawing (i.e. the smallest angle formed by two crossing edges) becomes as important for the non-planar graphs as the problem of avoiding edge crossings is for the planar graphs.

With this motivation, we study the trade-offs between crossing resolution, curve complexity, total number of bends, and area of non-planar drawings of graphs. The area is measured as the number of grid points contained in or on a bounding box of the drawing, i.e., the smallest axis-aligned box enclosing the drawing. The curve complexity is the maximum number of bends along each edge of the drawing.

We recall that the special case where the crossing resolution of a drawing is $\frac{\pi}{2}$ has been studied in [1], where these types of drawings are called *Right Angle Crossing drawings* (*RAC drawings*, for short). The main theme of [1] is to study the trade-off

D. Eppstein and E.R. Gansner (Eds.): GD 2009, LNCS 5849, pp. 15–20, 2010.

between the edge density and the curve complexity in a RAC drawing. In this paper we extend the study of [1] in two ways: we take into account the area requirement and we relax the crossing resolution constraints to angles smaller than $\frac{\pi}{2}$. Our main results can be listed as follows.

- We study the trade-off between area requirement and curve complexity of RAC drawings. We establish an $\Omega(n^2)$ lower bound on the area requirement of RAC drawings and show an $O(n + m)$-time algorithm whose input is a graph G with n vertices and m edges and whose output is a RAC drawing of G having curve complexity 4, total number of bends $O(m)$, and area $O(n^3)$. We observe that the previously known algorithm requires area $O(n^4)$ and curve complexity 3 [1].
- We relax the constraint on the crossing resolution and introduce the *Large Angle Crossing drawings* (*LAC drawings*, for short). In a LAC drawing the edge crossings are allowed to form angles at least $\frac{\pi}{2} - \varepsilon$ for any given $0 < \varepsilon < \frac{\pi}{2}$. For any choice of the constant ε, we show an infinite family of graphs such that any LAC drawing of an n-vertex graph in the family requires curve complexity 1, total number of bends $\Omega(n^2)$, and $\Omega(n^2)$ area.
- We describe an $O(n + m)$-time algorithm whose input is a graph G with n vertices and m edges and a constant $0 < \varepsilon < \frac{\pi}{2}$ and whose output is a LAC drawing of G having crossing resolution $\frac{\pi}{2} - \varepsilon$, curve complexity 1, total number of bends $O(m)$, and area $O(n^2)$, which are worst-case optimal.

The rest of this paper is organized as follows. Preliminary definitions are given in Section 2. Results about RAC drawings and LAC drawings are presented in Section 3 and 4, respectively. Section 5 lists some open problems. Some proofs are sketched or omitted for reasons of space.

2   Preliminaries

Let G be a graph. A *polyline drawing* Γ of G is a geometric representation of G such that each vertex u of G is mapped to a distinct point p_u of the plane, each edge (u, v) of G is drawn as a polyline with end-points p_u and p_v, and two edges can intersect either at shared endvertices or at a finite number of interior points. Each intersection between two or more edges that happens at an interior point is called a *crossing*. Since each edge is drawn as a polyline, each crossing is an intersection of two or more straight-line segments. The *intersection angle* between two crossing straight-line segments is the smallest angle defined by these segments. The *crossing resolution* of a polyline drawing is the minimum intersection angle between any two segments. Each point shared by two consecutive segments of a polyline representing an edge is called a *bend*. The *curve complexity* of a polyline drawing is the maximum number of bends along an edge. A *straight-line drawing* is a polyline drawing with curve complexity zero. A polyline drawing is a *grid* drawing if the points representing vertices and bends have integer coordinates. The *bounding box* of a polyline grid drawing Γ of a graph G is the smallest axis-aligned rectangle containing Γ. If the sides of the bounding box of Γ parallel to the x- and y-axis have lengths $W - 1$ and $H - 1$, respectively, we say that Γ has *width* W, *height* H and *area* $W \cdot H$. A polyline grid drawing with crossing resolution $\frac{\pi}{2}$ is called a *RAC drawing*. A polyline grid drawing with crossing resolution $0 < \alpha < \frac{\pi}{2}$ is called a *LAC_α drawing*. We will write *LAC*-drawing when we are not interested in the value of α.

3 Optimal Crossing Resolution: RAC Drawings

The following results, consequence of the results in [1], establish lower and upper bounds on the curve complexity, on the total number of bends and on the area of *RAC* drawings of graphs.

Theorem 1. *There exists an infinite family of graphs such that any RAC drawing of an n-vertex graph in the family has curve complexity at least 3, total number of bends $\Omega(n^2)$ and area $\Omega(n^2)$.*

Lemma 1. *Every graph with n vertices and m edges admits a RAC drawing with curve complexity 3, total number of bends $O(m)$ and area $O(n^4)$.*

We prove now that the upper bound on the area of a *RAC* drawing can be reduced to $O(n^3)$ at the expense of the curve complexity that increases to 4.

Lemma 2. *The complete graph K_n admits a RAC drawing with curve complexity 4, total number of bends $O(n^2)$, and area $O(n^3)$.*

Sketch of Proof: Refer to Fig. 1 for an illustration of the technique with $n = 6$. Arbitrarily number the vertices of K_n from 0 to $n - 1$. Vertex i with $0 \leq i \leq n - 1$ is placed at point $p_i = (in - 3, 2n)$. For each pair of vertices i and j, with $i < j$, the four bends of edge (i, j) will be placed at the following points (in this order): $a_{i,j} = (in - 2, (j - i) - 1 + 2(n - 1))$, $b_{i,j} = (in, (j - i) - 1)$, $c_{i,j} = (jn - (j - i), 2(j - i) - 1)$, and $d_{i,j} = (jn - (j - i) - 2, 2(j - i) - 1 + 2(n - 1))$.

Clearly the drawing has curve complexity 4 and total number of bends $4\frac{n(n-1)}{2}$. Also, it is easy to see that the crossing resolution is $\frac{\pi}{2}$. We now prove that the area is $O(n^3)$. The point with smallest x-coordinate is $p_0 = (-3, 2n)$, while the point with largest x-coordinate is $c_{n-2,n-1} = (n^2 - n - 1, 1)$ and therefore the width of the drawing is $n^2 - n - 1 + 3 + 1 = n^2 - n + 3 = O(n^2)$. The points with smallest y-coordinates are the points $b_{i,i+1} = (in, 0)$, while the point with largest y-coordinate is $d_{0,n-1} = (n^2 - 2n - 1, 4n - 5)$, and therefore the height of the drawing is $4n - 5 + 1 = 4n - 4 = O(n)$ which gives an $O(n^3)$ area. □

The following theorem summarizes the results about the upper bounds

Theorem 2. *Every graph with n vertices and m edges admits both a RAC drawing with curve complexity 3, total number of bends $O(m)$, and area $O(n^4)$ and a RAC drawing with curve complexity 4, total number of bends $O(m)$ and area $O(n^3)$. Both drawings can be computed in $O(n + m)$ time.*

In the next section we show that a drawing that is worst-case optimal in terms of curve complexity, total number of bends, and area can be computed if one allows a crossing resolution arbitrarily close to the optimal one.

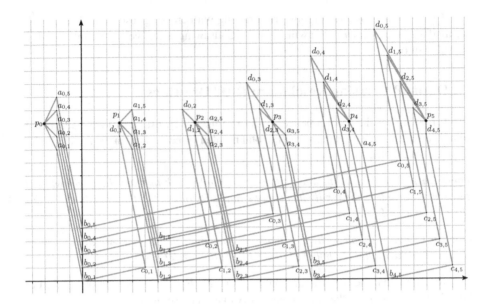

Fig. 1. A RAC drawing of K_6 with curve complexity 4

4 Nearly Optimal Crossing Resolution: LAC Drawings

We start by proving lower bounds on the curve complexity, on the total number of bends, and on the area of $LAC_{\frac{\pi}{2}-\varepsilon}$ drawings for every $0 \leq \varepsilon \leq \frac{\pi}{2}$. The next three lemmas are the technical foundation of our lower bound. The proof of the first one is omitted for reasons of space.

Lemma 3. *In a straight-line drawing of K_n on a convex set of $n > 3$ points there exists an intersection angle $\leq \frac{2\pi}{n}$.*

Lemma 4. *For all $\varepsilon > 0$ there exists a constant N such that for all $n \geq N$, any straight line drawing of K_n has an intersection angle $< \varepsilon$.*

Proof. Let $\varepsilon > 0$. Let $k > 2\pi/\varepsilon$. From the Erdös-Szekeres lemma [2] there exists an N such that any set of $n \geq N$ non-degenerate points (i.e. no three points collinear) contains a set of k convex points. Consider a straight-line drawing of K_n with $n \geq N$. If there are three collinear points in this drawing, there are two edges intersecting at an infinite number of points, i.e., the drawing is not a polyline drawing according to our definition. If there are no three collinear points, then there is a subset of k convex points. From Lemma 3 the drawing of K_n contains an intersection angle $\leq \frac{2\pi}{k} < \varepsilon$.   □

Lemma 5. *For all $\varepsilon > 0$ there exist constants N and c_N such that for all $n > N$, any subdrawing obtained removing at most $n(n-1)/c_N$ edges from a straight-line drawing of K_n has an intersection angle $\leq \varepsilon$.*

Proof. Let $\varepsilon > 0$. From Lemma 4 we know that there is a constant N such that any straight-line drawing of K_N has an intersection angle less than ε. Let $c_N = 2N(N-1)$

and let $n > N$. Consider a straight-line drawing of K_n. Any edge of K_n is in $\binom{n-2}{N-2}$ different subgraphs of K_n isomorph to K_N. So by removing $\leq n(n-1)/c_N$ edges from the drawing we remove at most $\frac{n(n-1)}{c_N}\binom{n-2}{N-2}$ copies of K_N. Therefore, there are at least $\binom{n}{N} - \frac{n(n-1)}{c_N}\binom{n-2}{N-2}$ copies of K_N present in the drawing. Since $\binom{n}{N} - \frac{n(n-1)}{c_N}\binom{n-2}{N-2} = \binom{n}{N} - \frac{n(n-1)(n-2)!}{2N(N-1)(N-2)!(n-N)!} = \binom{n}{N} - \frac{n!}{2N!(n-N)!} = \frac{1}{2}\binom{n}{N} > 1$, there is an intersection angle less than ε. $\square$

Theorem 3. *For every $0 < \varepsilon < \frac{\pi}{2}$ there exists a constant N such that for every $n > N$ there exists a graph G with n vertices such that every $LAC_{\frac{\pi}{2}-\varepsilon}$ drawing of G has curve complexity at least 1, total number of bends $\Omega(n^2)$ and area $\Omega(n^2)$.*

Proof. From Lemma 5 we know that for any value of $\frac{\pi}{2} - \varepsilon$ there are constants N and c_N such that for any straight-line drawing of K_n with $n > N$, even if we remove at most $n(n-1)/c_N$ of the edges of K_n there is an intersection angle less than $\frac{\pi}{2} - \varepsilon$. Thus, a drawing with intersection angles at least $\frac{\pi}{2} - \varepsilon$ has at least $n(n-1)/c_N$ bent edges. Since bends have to be placed on grid points, the result follows. $\square$

We prove now that the lower bounds of Theorem 3 are worst-case optimal.

Lemma 6. *For every $0 < \varepsilon < \frac{\pi}{2}$ and for all $n > 3$ there exists a $LAC_{\frac{\pi}{2}-\varepsilon}$ drawing of K_n with curve complexity 1, total number of bends $O(n^2)$, and area $O(n^2(\cot\frac{\varepsilon}{2})^2)$.*

Sketch of Proof: Let c be an integer such that the angle between a line of slope $c+1$ and a line of slope $\frac{1}{(c+1)}$ is larger than $\frac{\pi}{2} - \varepsilon$. Arbitrarily number the vertices of K_n from 0 to $n-1$. Refer to Fig. 2 for an example with $n = 6$ and $c = 1$. Vertex i with $0 \leq i \leq n-1$ is placed at point $p_i = (ic, (n-i-1)c)$. For each pair of vertices i and j, with $i < j$, the bend of edge (i, j) will be placed at point $a_{i,j} = (jc+1, (n-i-1)c+1)$. Clearly, the curve complexity is 1 and the total number of bends is $\frac{n(n-1)}{2}$. It is easy to see that

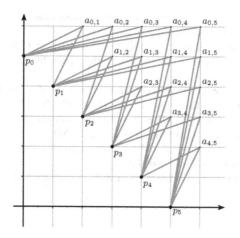

Fig. 2. A LAC_α-drawing of K_6 with curve complexity 1 ($\alpha > 63°$)

the crossing resolution is larger than $\frac{\pi}{2} - \varepsilon$ and that the area is $O(c^2 n^2)$. With simple geometric arguments it can be proved that $c = O(\cot \frac{\varepsilon}{2})$. $\square$

The following theorem summarizes the results of this section.

Theorem 4. *For every* $0 < \varepsilon < \frac{\pi}{2}$, *every graph with* n *vertices admits a* $LAC_{\frac{\pi}{2} - \varepsilon}$ *drawing which is worst-case optimal in terms of curve complexity, total number of bends, and area. Namely, the curve complexity is* 1, *the total number of bends is* $\Theta(m)$, *and the area is* $\Theta(n^2)$. *Furthermore, the drawing can be computed in* $O(n + m)$ *time.*

5   Open Problems

We conclude by listing some open problems that arise from the results of this work. One natural problem is that of closing the gap between the upper and the lower bound on the area of RAC drawings stated by Theorem 1 and Theorem 2.

The lower bounds on the area of RAC and LAC drawings are a consequence of the fact that K_n contains $O(n^2)$ edges. It would be interesting to study whether a $o(n^2)$ area and good crossing resolution can be obtained for graphs with $O(n)$ edges.

A question related to the previous one is the following: Is it possible to obtain straight-line drawings of planar graphs with $o(n^2)$ area if we allow right or large angle crossings? It is worth mentioning that, if no requirement about the crossing resolution exists, then every planar graph admits a non-planar drawing in $O(n)$ area [9].

References

1. Didimo, W., Eades, P., Liotta, G.: Drawing Graphs with Right Angle Crossings. In: Dehne, F., et al. (eds.) WADS 2009. LNCS, vol. 5664, pp. 206–217. Springer, Heidelberg (2009)
2. Erdös, P., Szekeres, G.: A combinatorial problem in geometry. Compositio Math. 2, 463–470 (1935)
3. Huang, W.: Using eye tracking to investigate graph layout effects. In: Hong, S.H., Ma, K.L. (eds.) APVIS, pp. 97–100. IEEE, Los Alamitos (2007)
4. Huang, W.: An eye tracking study into the effects of graph layout. CoRR abs/0810.4431 (2008)
5. Huang, W., Hong, S.H., Eades, P.: Effects of crossing angles. In: PacificVis, pp. 41–46. IEEE, Los Alamitos (2008)
6. Purchase, H.C.: Effective information visualisation: a study of graph drawing aesthetics and algorithms. Interacting with Computers 13(2), 147–162 (2000)
7. Purchase, H.C., Carrington, D.A., Allder, J.A.: Empirical evaluation of aesthetics-based graph layout. Empirical Software Engineering 7(3), 233–255 (2002)
8. Ware, C., Purchase, H.C., Colpoys, L., McGill, M.: Cognitive measurements of graph aesthetics. Information Visualization 1(2), 103–110 (2002)
9. Wood, D.R.: Grid drawings of k-colourable graphs. Computational Geometry 30(1), 25–28 (2005)

On the Perspectives Opened by
Right Angle Crossing Drawings[*]

Patrizio Angelini[1], Luca Cittadini[1], Giuseppe Di Battista[1], Walter Didimo[2],
Fabrizio Frati[1], Michael Kaufmann[3], and Antonios Symvonis[4]

[1] Dipartimento di Informatica e Automazione - Roma Tre University, Italy
{angelini,ratm,gdb,frati}@dia.uniroma3.it
[2] Dip. di Ingegneria Elettronica e dell'Informazione - Perugia University, Italy
walter.didimo@diei.unipg.it
[3] Wilhelm-Schickard-Institut für Informatik - Universität Tübingen, Germany
mk@informatik.uni-tuebingen.de
[4] Department of Mathematics - National Technical University of Athens, Greece
symvonis@math.ntua.gr

Abstract. Right Angle Crossing (RAC) drawings are polyline drawings where
each crossing forms four right angles. RAC drawings have been introduced because cognitive experiments provided evidence that increasing the number of
crossings does not decrease the readability of the drawing if the edges cross at
right angles. We investigate to what extent RAC drawings can help in overcoming
the limitations of widely adopted planar graph drawing conventions, providing
both positive and negative results. First, we prove that there exist acyclic planar
digraphs not admitting any straight-line upward RAC drawing and that the corresponding decision problem is NP-hard. Also, we show digraphs whose straight-
line upward RAC drawings require exponential area. Second, we study if RAC
drawings allow us to draw bounded-degree graphs with lower curve complexity
than the one required by more constrained drawing conventions. We prove that
every graph with vertex-degree at most 6 (at most 3) admits a RAC drawing with
curve complexity 2 (resp. 1) and with quadratic area. Third, we consider a natural
non-planar generalization of planar embedded graphs. Here we give bounds for
curve complexity and area different from the ones known for planar embeddings.

1 Introduction

In Graph Drawing, it is commonly accepted that *crossings* and *bends* can make the
layout difficult to read and experimental results show that the human performance in
path tracing tasks is negatively correlated to the number of edge crossings and to the
number of bends along the edges [16,17,19]. However, further cognitive experiments in
graph visualization show that increasing the number of crossings does not decrease the
readability of the drawing if the edges cross at right angles [10,11]. These results provide evidence for the effectiveness of *orthogonal* drawings (in which edges are chains
of horizontal and vertical segments) with few bends [4,12] and motivate the study of a

---

[*] This work started during the Bertinoro Workshop on Graph Drawing 2009. We acknowledge
Giuseppe Liotta for suggesting the study of upward RAC drawings.

D. Eppstein and E.R. Gansner (Eds.): GD 2009, LNCS 5849, pp. 21–32, 2010.

new class of drawings, called *Right Angle Crossing drawings* (*RAC drawings*) [7]. A RAC drawing of a graph G is a polyline drawing D of G such that any two crossing segments in D are orthogonal. If D has curve complexity 0, D is a *straight-line RAC drawing*, where the *curve complexity* of D is the maximum number of bends along an edge of D.

This paper continues the study of RAC drawings initiated in [7] and investigates RAC drawings with low curve complexity for both directed and undirected graphs. For directed graphs (*digraphs*), a widely studied drawing standard is the *upward drawing* convention, where edges are monotone in the vertical direction. A digraph has an upward planar drawing if and only if it has a straight-line upward planar drawing [5]. However, not all planar digraphs have an upward planar drawing and straight-line upward planar drawings require exponential area for some families of digraphs [6].

We investigate *straight-line upward RAC drawings*, i.e. straight-line upward drawings with right angle crossings. In particular, it is natural to ask if all planar digraphs admit an upward RAC drawing and if all digraphs with an upward RAC drawing admit one with polynomial area. Both these questions have a negative answer (Sect. 3): (i) we prove that there exist acyclic planar digraphs that do not admit any straight-line upward RAC drawing, and that the problem of deciding whether an acyclic planar digraph admits such a drawing is NP-hard; (ii) we show that there exist upward planar digraphs whose straight-line upward RAC drawings require exponential area.

Concerning undirected graphs, it is known [7] that any n-vertex straight-line RAC drawing has at most $4n - 10$ edges, for every $n \geq 4$, and this bound is tight. Further, every graph admits a RAC drawing with at most three bends per edge, and this curve complexity is required in infinitely many cases. Indeed, RAC drawings with curve complexity 1 and 2 have $O(n^{4/3})$ and $O(n^{7/4})$ edges, respectively. Hence, we investigate families of graphs that can be drawn with curve complexity 1 or 2, proving the following results (Sect. 4): (i) every degree-6 graph admits a RAC drawing with curve complexity 2; (ii) every degree-3 graph admits a RAC drawing with curve complexity 1. Both types of drawings can be computed in linear time and require quadratic area. Observe that degree-4 graphs admit orthogonal drawings with curve complexity 2 [14], and that two bends on an edge are sometimes necessary even for degree-3 graphs.

In a *fixed embedding* setting, the input graph G is given with a (non-planar) *embedding*, i.e., a circular ordering of the edges incident to each vertex and an ordering of the crossings along each edge. A RAC drawing algorithm cannot change the embedding of G. For such a setting it has been proved in [7] that any n-vertex graph admits a RAC drawing with $O(kn^2)$ bends per edge, where k is the maximum number of crossings between any two edges. Also, there exist graphs whose RAC drawings require $\Omega(n^2)$ bends along some edges. In Sect. 5 we study the fixed embedding setting, namely: (i) we study non-planar graphs obtained by augmenting a plane triangulation with edges inside pairs of adjacent faces; we call these graphs *kite-triangulations* and prove that one bend per edge is always sufficient and sometimes necessary for a RAC drawing of a kite-triangulation; (ii) we study the area requirement of straight-line RAC drawings of kite-triangulations and prove that cubic area is sometimes necessary. Recall that every embedded planar graph admits a planar drawing with quadratic area [18].

Sect. 6 concludes the paper with some open problems.

2 Preliminaries

We assume familiarity with graph drawing and planarity [4,12]. In the following, unless otherwise specified, all considered graphs are *simple*.

A *Right Angle Crossing drawing* (*RAC drawing*) of a graph G is a polyline drawing D of G such that any two crossing segments in D are orthogonal. The *curve complexity* of D is the maximum number of bends along an edge of D. If a RAC drawing D has curve complexity 0, D is a *straight-line RAC drawing*. A *fan* in a drawing D is a pair of edge segments incident to the same vertex. Two segments s_1 and s_2 crossing the same segment in D are parallel. This leads (Fig. 1) to the following properties (see also [7]).

Property 1. In a straight-line RAC drawing no edge can cross a fan.

Property 2. In a straight-line RAC drawing there cannot be a triangle $\triangle$ and two edges (a, b), (a, c) such that a lies outside $\triangle$ and b, c lie inside $\triangle$.

In an *upward drawing* all the edges are curves monotonically increasing in the upward direction. An *upward planar drawing* of a digraph G is an upward drawing of G without edge crossings. If G admits an upward planar drawing, G is an *upward planar* digraph. An *upward RAC drawing* of a digraph is a RAC drawing that is also upward.

3 Upward RAC Drawings

We now study straight-line upward RAC drawings (in this section, RAC drawings for short). We introduce an upward planar digraph H, shown in Fig. 2(a), that serves as a gadget for proving the main results of this section. The following lemmata show that two copies of H cannot cross each other in any RAC drawing. Let $\mathcal{E}_1$, $\mathcal{E}_2$, and $\mathcal{E}_3$ be the embeddings of H shown in Fig. 2(a), 2(b), and 2(c), respectively.

Lemma 1. *In any RAC drawing of H, its embedding is one of $\mathcal{E}_1$, $\mathcal{E}_2$, or $\mathcal{E}_3$, up to a reversal of the adjacency lists of all the vertices.*

Let G be a digraph containing two copies H' and H'' of H, with vertex sets $\{u', v', w', z'\}$ and $\{u'', v'', w'', z''\}$, respectively, so that one vertex in $\{u', v'\}$ possibly coincides with one in $\{u'', v''\}$, while no other vertex is shared by the two graphs. A vertex of H'' that is coincident with a vertex of a 3-cycle of H' is considered both as internal and as external to the triangle representing the 3-cycle.

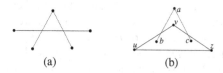

(a)   (b)

Fig. 1. Illustrations for (a) Property 1 and for (b) Property 2

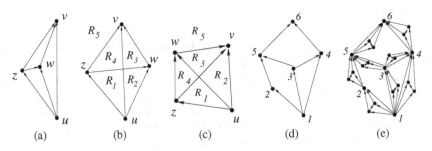

Fig. 2. (a) $\mathcal{E}_1$; (b) $\mathcal{E}_2$; (c) $\mathcal{E}_3$. (d) A planar digraph K. (e) The planar digraph K' obtained by replacing each edge of K with a copy of H.

Lemma 2. *Let D be a RAC drawing of G. For any 3-cycle (a', b', c') of H', which is represented in D by a triangle $\triangle'$, either all the vertices of H'' are inside $\triangle'$ or they are all outside it.*

Lemma 3. *In any RAC drawing D of G, no edge of H' crosses an edge of H''.*

Proof. Let D^* be D restricted to the edges of H' and H''. We show that in D^* there is no crossing among the edges of H' and the edges of H''.

If the embedding of H' in D^* is $\mathcal{E}_1$, one of the four triangular faces of H', say (u', z', v'), is a triangle $\triangle'$ enclosing w' (see Fig. 2(a)). By Lemma 2, either all the vertices of H'' lie outside $\triangle'$ or they all lie inside it. In the former case, if there is a crossing between an edge of H' and an edge of H'', then such an edge of H'' cuts a fan composed of two edges of H', violating Property 1. In the latter case, the vertices of H'' lie in the faces of H' internal to $\triangle'$. By Lemma 2, all the vertices of H'' lie in one of the internal faces of H'. Hence, in both cases, no edge of H' crosses an edge of H''.

If the embedding of H' in D^* is $\mathcal{E}_2$, cycle (u', z', v', w') of H' is a convex quadrilateral with edges (u', v') and (z', w') crossing inside it, since $y(u') < y(z') < y(w') < y(v')$ and since (u', v') is a straight-line segment. Thus, connected regions $R_1, \ldots, R_5$ are created (see Fig. 2(b)). We prove that all the vertices of H'' are inside a region R_i. For every pair of regions R_i and R_j, with $j \neq i$, a 3-cycle (a', b', c') of H', with $a', b', c' \in \{u', z', v', w'\}$, exists containing R_i in its interior and R_j in its exterior, or vice versa. Suppose that vertices a'' and b'' exist such that: (i) $a'', b'' \in \{u'', z'', v'', w''\}$; (ii) a'' is inside R_i and b'' is inside R_j, with $i \neq j$; and (iii) a'' is outside R_j and b'' is outside R_i (notice that R_i and R_j can possibly share a vertex). Then a'' is inside the triangle representing (a', b', c') and b'' is outside such a triangle, or vice versa. However, by Lemma 2, D^* is not a RAC drawing. If all the vertices of H'' are in the same region R_i, with $1 \leq i \leq 4$, no edge of H' crosses an edge of H''. If all the vertices of H'' are in R_5, suppose that a crossing between an edge of H' and an edge of H'' exists. Then, such an edge of H'' cuts a fan composed of two edges of H'.

If the embedding of H' in D^* is $\mathcal{E}_3$, connected regions $R_1, \ldots, R_5$ are created by the edges of H' (see Fig. 2(c)). With the same argument as above, it can be proved that no edge of H' crosses an edge of H''. □

We get the following:

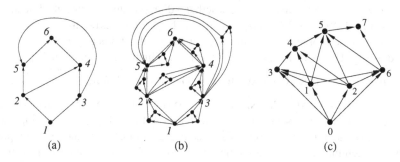

Fig. 3. (a) A planar acyclic digraph G that is not upward planar. (b) The planar acyclic digraph G' obtained by replacing each edge of G with a copy of H. (c) An 8-vertex planar digraph that does not admit any RAC drawing.

Lemma 4. *Consider a planar acyclic digraph K. Replace each edge (a, b) of K with a copy of H (see Figs. 2(d) and 2(e)), by identifying vertices a and b with vertices u and v of H, respectively. Let K' be the resulting planar digraph. Digraph K is upward planar if and only if K' is straight-line upward RAC drawable.*

Proof. First, suppose that K has an upward planar drawing. Then, by the results in [5], K admits a straight-line upward planar drawing D. Consider the drawing D' obtained by drawing each copy of H that replaces an edge (a, b) in such a way that: (i) the drawing of H is upward planar; (ii) the drawing of edge (u, v) of H in D' coincides with the drawing of (a, b) in D; and (iii) the drawing of the rest of H is arbitrarily close to (u, v). Since D is a straight-line upward planar drawing, D' is a straight-line upward planar drawing. Hence, D' is a RAC drawing of K'.

Second, suppose that K' has a RAC drawing D'. By construction, every edge of K' belongs to a copy of H, and, by Lemma 3, no two edges belonging to distinct copies of H cross in D'. Hence, since K is a subgraph of K' such that, for each copy of H belonging to K', only one edge, namely (u, v), belongs to K, the drawing D obtained as D' restricted to the edges of K is a straight-line upward planar drawing of K.   □

We are ready to prove the first theorem of this section.

Theorem 1. *There exist acyclic planar digraphs that do not admit any straight-line upward RAC drawing.*

Proof. Consider any planar acyclic digraph G (as the one of Fig. 3(a)) that is not upward planar. By Lemma 4, the planar acyclic digraph G' obtained by replacing each edge of G with a copy of H is not RAC drawable (see Fig. 3(b)).   □

Note that there exist planar digraphs, as the one in Fig. 3(c), that do not admit any RAC drawing, that are not constructed using gadget H, and whose size is smaller than the one of the digraph in Fig. 3(b). However, proving that they are not RAC drawable could result in a complex case-analysis.

Motivated by the fact that there exist acyclic planar digraphs that do not admit any RAC drawing, we study the time complexity of the corresponding decision problem.

We show that the problem of testing whether a digraph admits a straight-line upward RAC drawing (UPWARD RAC DRAWABILITY TESTING) is NP-hard, by means of a reduction from the problem of testing whether a digraph admits a straight-line upward planar drawing (UPWARD PLANARITY TESTING), which is NP-complete [9].

Theorem 2. UPWARD RAC DRAWABILITY TESTING *is NP-hard.*

Proof. We reduce UPWARD PLANARITY TESTING to UPWARD RAC DRAWABILITY TESTING. Let G be an instance of UPWARD PLANARITY TESTING (see Fig. 2(d)). Replace each edge of G with a copy of H (see Fig. 2(e)). Let G' be the resulting planar digraph. By Lemma 4, G is upward planar if and only if G' admits a RAC drawing. □

As a final contribution of this section we show that there exists a class of planar acyclic digraphs that require exponential area in any RAC drawing.

Consider the class of digraphs G_n [6] which requires $\Omega(2^n)$ area in any straight-line upward planar drawing, under any resolution rule. Let G'_n be the class of digraphs obtained by replacing each edge (a, b) of G_n with a copy of H, so that vertices a and b are identified with vertices u and v of H, respectively.

Theorem 3. *Let G' be a digraph belonging to G'_n. Then, any straight-line upward RAC drawing of G' requires $\Omega(b^n)$ area, under any resolution rule, for some constant $b > 1$.*

Proof. Suppose, for a contradiction, that, for every constant $b > 1$, G' admits a RAC drawing D' with $o(b^n)$ area, under some resolution rule. Consider the digraph $G \in G_n$ corresponding to G'. By construction, G is a subgraph of G' containing only edge (u, v) for each copy of H in G'. By Lemma 3, no two edges belonging to distinct copies of H cross. Hence, the drawing D of G obtained as D' restricted to the edges of G is a straight-line upward planar drawing with $o(b^n)$ area, a contradiction. □

4 RAC-Drawings of Bounded-Degree Graphs

In this section, we present an algorithm for constructing RAC drawings of graphs with degree at most 6. The algorithm is based on the decomposition of a regular multigraph into cycle covers. A *cycle cover* of a directed graph is a spanning subgraph consisting of vertex-disjoint directed cycles. The decomposition into cycle covers follows from a classical result [15] stating that "a regular multigraph G of degree $2k$ has k edge-disjoint factors", where a *factor* is a spanning subgraph consisting of vertex-disjoint cycles (see also [2, pp.227]). A constructive proof of the following theorem was given in [8].

Theorem 4 *(Eades, Symvonis, Whitesides [8]). Let $G = (V, E)$ be an undirected graph of maximum degree Δ and let $d = \lceil \Delta/2 \rceil$. Then, there exists a directed multi-graph $G' = (V, E')$ such that: (i) each vertex of G' has indegree d and outdegree d; (ii) G is a subgraph of the underlying undirected graph of G'; and (iii) the edges of G' can be partitioned into d edge-disjoint cycle covers. Furthermore, for an n-vertex graph G, the directed graph G' and its d cycle covers can be computed in $O(\Delta^2 n)$ time.*

Let u be a vertex placed at a grid point. We say that an edge e exiting u uses the Y-port of u (resp. the $-Y$-port of u) if it exits u along the $+Y$ direction (resp. along the $-Y$ direction). In a similar way, we define the X-port and the $-X$-port.

Theorem 5. *Every n-vertex graph with degree at most 6 admits a RAC drawing with curve complexity 2 in $O(n^2)$ area. Such a drawing can be computed in $O(n)$ time.*

Proof. Let $G = (V, E)$ be a graph of maximum degree 6. Let $G' = (V, E')$ be the directed multigraph obtained from G as in Theorem 4, and let C_1, C_2, and C_3 be the edge-disjoint cycle covers of G'. We show how to obtain a RAC drawing of G'. Then, a RAC drawing of G can be obtained by removing from the drawing all the edges in $E' \setminus E$ and by ignoring the direction of the edges.

The algorithm places the vertices of V on the main diagonal of an $n \times n$ grid, in an order determined by one of the cycle covers, say C_1. Most of the edges of C_1 are drawn as straight lines along the diagonal while the edges of C_2 and C_3 are drawn as 3-segment lines above and below the diagonal, respectively. Finally, the remaining "closing" edges of C_1 (i.e., the edges that cannot be drawn as straight lines on the diagonal) are drawn without creating any overlap with other edges.

We first describe how to place the vertices of the graph along the main diagonal. Arbitrarily name the cycles $c_1, c_2, \ldots, c_k$ of C_1. Consider a cycle c_i, $1 \le i \le k$. If there exists a vertex $u \in c_i$ and an edge $(u, z) \in C_2$ or C_3 such that z belongs to a cycle c_j of C_1 with $j > i$ (note that there could be several of such vertices and edges), then let u be the *topmost vertex* of c_i and let the vertex following u in c_i be the *bottommost vertex* of c_i. Otherwise, if there exists a vertex $v \in c_i$ and an edge $(v, w) \in C_2$ or C_3 such that w belongs to a cycle c_j of C_1 with $j < i$, then let v be the bottommost vertex of c_i and let the vertex preceding v in c_i be the topmost vertex of c_i. If no such vertices exist, all the edges of C_2 and C_3 originating from vertices of c_i are also directed to vertices of c_i. In this case, let an arbitrary vertex w of c_i be the bottommost vertex of c_i and let the vertex preceding w in c_i be the topmost vertex of c_i. Then, for $i = 1, \ldots, k$, place the vertices of c_i in the order they appear as traversing the cycle, with the bottommost vertex placed at the bottommost free grid point.

Fig. 4(a) shows a regular directed multigraph G' of indegree and outdegree 3 and its cycle covers (C_1: thin, C_2: thick, C_3: dashed). C_1 consists of cycles $c_1 : (5, 1, 2, 3, 4, 5)$ and $c_2 : (6, 7, 8, 9, 6)$. We set 4 as the topmost vertex of c_1 since edge $(4, 6)$ of C_2 has vertex 6 of c_2 as its destination. Similarly, we set 6 as the bottommost vertex of c_2 since

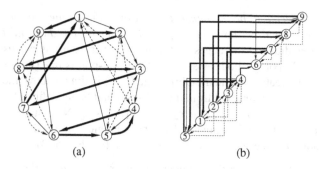

(a)   (b)

Fig. 4. (a) A regular directed multigraph G' with indegree and outdegree equal to 3 and its cycle covers. (b) A RAC drawing of G' with two bends per edge. Vertex 4 is the topmost vertex of c_1 and vertex 6 is the bottommost vertex of c_2.

edge $(6, 5)$ of C_2 has vertex 5 of c_1 as its destination. Fig. 4(b) shows the RAC drawing of G' of curve complexity 2 produced by the algorithm described in this proof.

Having placed the vertices on the grid, we turn our attention to draw the edges of G'. No edge overlaps are allowed and each edge is drawn either as a 1-segment edge along the diagonal, or as a 2/3-segment polyline above or below the diagonal. We draw the edges so that all the crossing line segments are parallel to the axes and, consequently, all the crossings are at right angles. In our drawings, every line segment s that is not parallel to the axes is incident to a vertex v_s of the graph; further, the other end-point of s is confined to a dedicated area within a unit-side rectangle centered at v_s (see Fig. 5(a)).

We first describe how to draw the edges of cycle cover C_2 above the diagonal. Consider an edge (u, v) of C_2 and let u and v be placed at grid points (u_x, u_y) and (v_x, v_y), respectively. If u is placed below v (i.e., $u_y < v_y$), then edge (u, v) is drawn as a 3-segment line exiting vertex u from the Y-port and being defined by bend-points $(u_x, v_y - \frac{3}{8} + \epsilon_2)$ and $(v_x - \frac{3}{8} + \epsilon_1, v_y - \frac{3}{8} + \epsilon_2)$, $0 < \epsilon_1 < \epsilon_2 < \frac{1}{4}$. Note that the second bend-point is located within the lightly-shaded region (above the diagonal) of the south-west quadrant of the square centered at vertex v (see Fig. 5(a)). If u is placed above v (i.e., $u_y > v_y$), then edge (u, v) is drawn as a 3-segment line exiting vertex u from the $-X$-port and being defined by bend-points $(v_x + \frac{1}{8} + \epsilon_1, u_y)$ and $(v_x + \frac{1}{8} + \epsilon_1, v_y + \frac{1}{8} + \epsilon_2)$, $0 < \epsilon_1 < \epsilon_2 < \frac{1}{4}$. Note that, in this case, the second bend-point is located within the lightly-shaded region (above the diagonal) of the north-east quadrant of the square centered at v (see Fig. 5(a)). It is easy to observe that the only line segments that belong to edges of C_2 and that cross other line segments are parallel to the axes, hence they cross at right angles. In a symmetric fashion, the edges of C_3 are drawn below the diagonal. Fig. 4(b) shows the routing of cycle covers C_2 and C_3 for graph G' of Fig. 4(a).

Consider now the edges of C_1. All such edges, except those closing the cycles of C_1, are drawn as straight-line segments along the diagonal. As all the edges of C_2 (resp. C_3) are drawn above (resp. below) the diagonal, these edges are not involved in any edge crossing. To complete the drawing of G', we describe how to draw the edges connecting the topmost vertex to the bottommost vertex of each cycle of C_1. Consider an arbitrary cycle c_i of C_1 and let (u, v) be its closing edge. We consider 3 cases:

Case 1: *u was selected to be the topmost vertex of c_i due to the existence of an edge* (u, z) *of C_2 or C_3 with vertex z being placed higher on the diagonal than u. This* implies that after drawing the edges of C_2 and C_3, vertex u has not used either its $-X$-port or its $-Y$-port, or both. Assume that the $-X$-port is free (the case where the $-Y$-port is free is treated symmetrically). Edge (u, v) is drawn above the diagonal as a 3-segment line exiting vertex u from the $-X$-port and being defined by bend-points $(v_x + \epsilon_1, u_y)$ and $(v_x + \epsilon_1, v_y + \frac{3}{8} + \epsilon_2)$, $0 < \epsilon_1, \epsilon_2 < \frac{1}{8}$. Note that, in this case, the second bend-point is located within the dark-shaded region (above the diagonal) of the north-east quadrant of the square centered at vertex v (see Fig. 5(a)).

Case 2: *v was selected to be the bottommost vertex of c_i due to the existence of an edge* (v, w) *of C_2 or C_3 with vertex w being placed lower on the diagonal than v. This* implies that after drawing the edges of C_2 and C_3, vertex v has not used either its X-

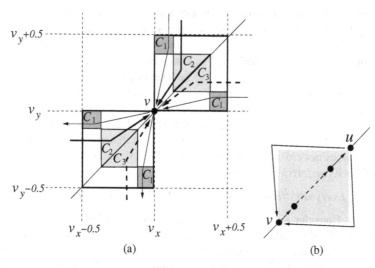

Fig. 5. (a) The area around a vertex. Each specific subarea hosts a single bend-point of an edge belonging to a specific cycle cover. (b) Two alternatives for the drawing of a closing edge of a cycle that has only edges originating and destined for its own vertices.

port or its Y-port, or both. The drawing of the closing edge is done in a way similar to the above case.

Case 3: *None of the above cases applies.* In this case, all the edges of C_2 and C_3 originating from vertices of cycle c_i are also directed to vertices of c_i and the topmost vertex u and the bottommost vertex v of c_i were selected arbitrarily. Then, both the $-X$-port and the $-Y$-port of u and both the X-port and the Y-port of v are already used. Also, the drawing of the edges of C_2 and C_3 connecting vertices of c_i takes place entirely within the square having points (v_x, v_y) and (u_x, u_y) as its opposite corners (the shaded square in Fig. 5(b)). Hence, the closing edge can be easily drawn as a 2-segment line connecting u and v just outside the boundary of the square, either above or below the diagonal (see Fig. 5(b)).

Given the three cycle covers, it is easy to see that the placement of the vertices along the diagonal and the routing of all the edges can be completed in linear time. By Theorem 4, the three cycle covers can be also computed in linear time, resulting in a linear time algorithm. Also, the produced RAC drawing requires $O(n^2)$ area.   ☐

With similar techniques we can prove the following:

Theorem 6. *Every n-vertex graph with degree at most 3 admits a RAC drawing with curve complexity 1 in $O(n^2)$ area. Such a drawing can be computed in $O(n)$ time.*

5   RAC Drawings of Kite-Triangulations

In this section we study the impact of admitting orthogonal crossings on the drawability of the non-planar graphs obtained by adding edges to maximal planar graphs, in a fixed embedding scenario. We show that such graphs always admit RAC drawings with curve complexity 1 and that such a curve complexity is sometimes required.

Let G' be a triangulation and let (u, z, w) and (v, z, w) be two adjacent faces of G' sharing edge (z, w). We say that $[u, v]$ is a *pair of opposite vertices* with respect to (z, w). Let $E^+ = \{[u_i, v_i] | i = 1, 2, \cdots, k\}$ be a set of pairs of opposite vertices of G', where $[u_i, v_i]$ is a pair of opposite vertices with respect to (z_i, w_i) and edge (u_i, v_i) does not belong to G'. Suppose that (z_p, w_p) and (z_q, w_q) do not share a face of G', for any $1 \leq p, q \leq k$ and $p \neq q$. Let G be the embedded non-planar graph obtained by adding an edge (u_i, v_i) to G', for each pair $[u_i, v_i]$ in E^+, so that edge (u_i, v_i) crosses edge (z_i, w_i) and does not cross any other edge of G. We say that G is a *kite-triangulation* and that G' is its *underlying triangulation*.

Theorem 7. *Every kite-triangulation admits a RAC drawing with curve complexity 1.*

Proof sketch. Consider any kite-triangulation G and its underlying triangulation G'. Remove from G' all the edges (z_i, w_i), for $i = 1, \ldots, k$, obtaining a new planar graph G'' whose faces contain at most four vertices. Construct any straight-line planar drawing Γ'' of G''. Construct a RAC drawing Γ of G inserting in Γ'', for each $i = 1, \ldots, k$, edges (u_i, v_i) and (z_i, w_i). Two cases are possible, either face (u_i, w_i, v_i, z_i) is strictly convex in Γ'' (Fig. 6(a)) or it is not (Fig. 6(b)). □

Theorem 8. *There exist kite-triangulations that do not admit straight-line RAC drawings.*

Proof. Consider the graph H in Fig. 6(c). Triangle (u, a, z) and vertices v, x, y create the forbidden structure of Property 2. Hence, every kite-triangulation G containing H as a subgraph requires one bend in any RAC drawing. □

Planar graphs are a proper subset of straight-line RAC drawable graphs. However, while straight-line planar drawings can always be realized on a grid of quadratic size (see, e.g., [3,18]), straight-line RAC drawings may require larger area, as shown in the following.

Theorem 9. *There exists an n-vertex kite-triangulation that requires $\Omega(n^3)$ area in any straight-line grid RAC drawing.*

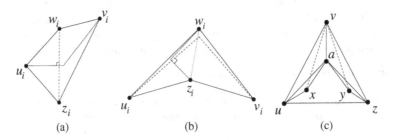

(a)   (b)   (c)

Fig. 6. (a) Drawing (u_i, v_i) and (z_i, w_i) inside (u_i, w_i, v_i, z_i), if (u_i, w_i, v_i, z_i) is strictly convex. (b) Drawing (u_i, v_i) and (z_i, w_i) inside (u_i, w_i, v_i, z_i), if (u_i, w_i, v_i, z_i) is not strictly convex. (c) An embedded graph that is a subgraph of infinitely many kite-triangulations with curve complexity 1 in any RAC drawing.

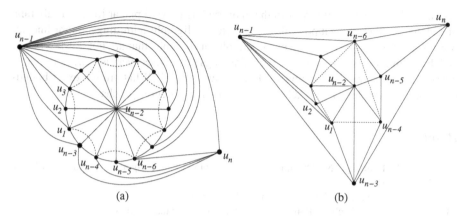

(a)   (b)

Fig. 7. (a) A kite triangulation G requiring $\Omega(n^3)$ area in any straight-line grid RAC drawing. (b) A straight-line RAC drawing of G.

Proof. Consider a triangulation G' defined as follows (see Fig. 7(a)). Let $C = (u_1, u_2, \ldots, u_{n-4}, u_{n-3})$ be a simple cycle, for some odd integer n. Insert a vertex u_{n-2} inside C and connect it to u_i, with $i = 1, 2, \ldots, n-3$. Insert two vertices u_{n-1} and u_n outside C. Connect u_{n-1} to u_i, with $i = 1, 2, \ldots, n-6$ and to u_{n-3}; connect u_n to $u_{n-6}, u_{n-5}, u_{n-4}, u_{n-3}$, and u_{n-1}. Let G be the kite-triangulation obtained from G' by adding edges (u_i, u_{i+2}), for $i = 1, 3, 5, \ldots, n-6$, and edge (u_1, u_{n-4}), so that (u_i, u_{i+2}) crosses edge (u_{i+1}, u_{n-2}) of G', and so that (u_1, u_{n-4}) crosses edge (u_{n-3}, u_{n-2}) of G'.

In the following we prove that, in any straight-line RAC drawing of G, cycle $C' = (u_1, u_3, \ldots, u_{n-6}, u_{n-4}, u_1)$ is a strictly-convex polygon. This claim, together with the observation that G admits a straight-line RAC drawing (see Fig. 7(b)), clearly implies the lemma, since any strictly-convex polygon needs cubic area if its vertices have to be placed on a grid (see, e.g., [1]).

Suppose, for a contradiction, that there exists a straight-line RAC drawing Γ of G with an angle $\widehat{u_i u_{i+2} u_{i+4}} \geq 180°$ inside C'. Then, any two segments orthogonally crossing $\overline{u_i u_{i+2}}$ and $\overline{u_{i+2} u_{i+4}}$, respectively, meet at a point outside C', possibly at infinity, while they should meet at u_{n-2}, which is inside C'. Thus, either $\overline{u_{n-2} u_{i+1}}$ is not orthogonal to $\overline{u_i u_{i+2}}$ or $\overline{u_{n-2} u_{i+3}}$ is not orthogonal to $\overline{u_{i+2} u_{i+4}}$, hence contradicting the assumption that Γ is a RAC drawing. □

6   Conclusions and Open Problems

When a graph G does not admit any planar drawing in some desired drawing convention, requiring that all the crossings form right angles can be considered as an alternative solution for the readability of a drawing of G. In this direction, this paper has shown negative results for digraphs that must be drawn upward with straight-line edges, and positive results for classes of non-planar undirected graphs that must be drawn with curve complexity 1 or 2. We list some open research directions that are related to the results of this paper: (i) It is known that a digraph is upward planar iff it is a subgraph of a planar st-digraph. Is it possible to characterize digraphs admitting straight-line upward RAC drawings? (ii) There exist outerplanar digraphs that are not upward planar and that admit upward straight-line RAC drawings [13]. Does every acyclic outerplanar digraph admit a straight-line upward RAC drawing? (iii) What are the exact bounds for the curve complexity of RAC drawings of bounded degree graphs?

References

1. Bárány, I., Tokushige, N.: The minimum area of convex lattice n-gons. Combinatorica 24(2), 171–185 (2004)
2. Berge, C.: Graphs. North Holland, Amsterdam (1985)
3. de Fraysseix, H., Pach, J., Pollack, R.: How to draw a planar graph on a grid. Combinatorica 10(1), 41–51 (1990)
4. Di Battista, G., Eades, P., Tamassia, R., Tollis, I.G.: Graph Drawing. Prentice Hall, Upper Saddle River (1999)
5. Di Battista, G., Tamassia, R.: Algorithms for plane representations of acyclic digraphs. Theoretical Computer Science 61, 175–198 (1988)
6. Di Battista, G., Tamassia, R., Tollis, I.G.: Area requirement and symmetry display of planar upward drawings. Discrete & Computational Geometry 7, 381–401 (1992)
7. Didimo, W., Eades, P., Liotta, G.: Drawing graphs with right angle crossings. In: WADS 2009. LNCS, vol. 5664, pp. 206–217. Springer, Heidelberg (2009)
8. Eades, P., Symvonis, A., Whitesides, S.: Three-dimensional orthogonal graph drawing algorithms. Discrete Applied Mathematics 103(1-3), 55–87 (2000)
9. Garg, A., Tamassia, R.: On the computational complexity of upward and rectilinear planarity testing. SIAM Journal on Computing 31(2), 601–625 (2001)
10. Huang, W.: An eye tracking study into the effects of graph layout. CoRR, abs/0810.4431 (2008)
11. Huang, W., Hong, S.-H., Eades, P.: Effects of crossing angles. In: PacificVis, pp. 41–46 (2008)
12. Kaufmann, M., Wagner, D. (eds.): Drawing Graphs. Springer, Heidelberg (2001)
13. Papakostas, A.: Upward planarity testing of outerplanar dags. In: Tamassia, R., Tollis, I.G. (eds.) GD 1994. LNCS, vol. 894, pp. 298–306. Springer, Heidelberg (1995)
14. Papakostas, A., Tollis, I.G.: Algorithms for area-efficient orthogonal drawings. Computational Geometry 9(1-2), 83–110 (1998)
15. Petersen, J.: Die theorie der regulären graphen. Acta Mathematicae 15, 193–220 (1891)
16. Purchase, H.C.: Effective information visualisation: a study of graph drawing aesthetics and algorithms. Interacting with Computers 13(2), 147–162 (2000)
17. Purchase, H.C., Carrington, D.A., Allder, J.-A.: Empirical evaluation of aesthetics-based graph layout. Empirical Software Engineering 7(3), 233–255 (2002)
18. Schnyder, W.: Embedding planar graphs on the grid. In: SODA 1990, pp. 138–148 (1990)
19. Ware, C., Purchase, H.C., Colpoys, L., McGill, M.: Cognitive measurements of graph aesthetics. Information Visualization 1(2), 103–110 (2002)

Drawing 3-Polytopes with Good Vertex Resolution

André Schulz[*]

MIT, Computer Science and Artificial Intelligence Laboratory, Cambridge, USA
schulz@csail.mit.edu

Abstract. We study the problem how to obtain a small drawing of a 3-polytope with Euclidean distance between any two points at least 1. The problem can be reduced to a one-dimensional problem, since it is sufficient to guarantee distinct integer x-coordinates. We develop an algorithm that yields an embedding with the desired property such that the polytope is contained in a $2(n-2) \times 1 \times 1$ box. The constructed embedding can be scaled to a grid embedding whose x-coordinates are contained in $[0, 2(n-2)]$. Furthermore, the point set of the embedding has a small spread, which differs from the best possible spread only by a multiplicative constant.

1 Introduction

Let G be a 3-connected planar graph with n vertices $v_1, \ldots, v_n$ and edge set E. Due to Steinitz' seminal theorem [16] we know that G admits a realization as 3-polytope, and every edge graph of a 3-polytope is planar and 3-connected. The question arises how one can obtain a "nice" realization of a 3-polytope when its graph is given. One particular property which is often desired from an aesthetically point of view is that the vertices of the embedding should be evenly distributed. If two vertices lie too close together they are hard to distinguish. Such an embedding may appear as bad "illustration" for the human eye. Of course we can always scale a 3-polytope to increase all of its pointwise distances, but clearly this is not the right solution to create a good drawing. Therefore, we restrict ourselves to an embedding whose vertices have, pairwise, an Euclidean distance of at least 1. We say that in this case the embedding/drawing is under the *vertex resolution rule*. See [1] for a short discussion on resolution rules. Notice, that the resolution rule depends on a particular distance measure. Throughout the paper we use the Euclidean distance, but our results can be easily modified for the L_1 distance.

In 2d, drawings of planar graphs can be realized on an $O(n) \times O(n)$ grid [14,5]. Since the grid is small, these grid embeddings give a good vertex resolution for free. The situation for 3-polytopes is different. The best known algorithm uses a grid of size $O(2^{7.55n})$ [12]. Thus, the induced resolution might be bad for this grid embedding.

---

[*] Supported by the German Research Foundation (DFG) under grant SCHU 2458/1-1.

D. Eppstein and E.R. Gansner (Eds.): GD 2009, LNCS 5849, pp. 33–44, 2010.

Let us briefly discuss some approaches how to realize G as 3-polytope. Steinitz' original proof is based on a transformation of G to the graph of the tetrahedron. The transformation consists of a sequence of local modifications which preserve the realizability of G as 3-polytope. However, the proof does not include a direct method how to construct the 3-polytope given by G. The Koebe-Andreev-Thurston Theorem on circle packings gives a more constructive proof of Steinitz' Theorem (see for example Schramm [15]). This approach relies on non-linear methods which makes many geometric features of the constructed 3-polytope intractable. A third approach uses liftings of planar graphs with equilibrium stresses (known as the Maxwell-Cremona correspondence [18]). This powerful method is used in a series of embedding algorithms [1,6,8,13,12]. A completely different approach is due to Das and Goodrich [4]. It uses an incremental technique which only needs $O(n)$ arithmetic operations for embedding G as 3-polytope, but works only for triangulated planar graphs.

Results. We show how to obtain an embedding of G as 3-polytope inside a $2(n-2) \times 1 \times 1$ box under the vertex resolution rule. It is even possible to make the box arbitrarily flat such that its volume gets arbitrarily small. But for aesthetic reasons we leave the side lengths at least 1. Our algorithm is based on the Maxwell-Cremona approach and extends the ideas of [12]. In contrast to the construction of [12] we can handle more complicated interior edge weights (stresses). Our algorithm creates an embedding with two more interesting properties: (1) it can be scaled to a grid embedding whose x-coordinates are in $[0, 2(n-2)]$, and (2) the point set of the embedding has a good ratio between its largest and shortest pointwise distance. We show that this number differs only by a constant from the best possible ratio. Due to space constraints we omit the proofs for Theorem 3 and 4. The proofs can be found in the full version of the paper, which is available on the website of the author. The full version includes also an example of the algorithm.

Related work. In [1] Chrobak, Goodrich, and Tamassia introduced an algorithm that embeds a 3-polytope with good vertex resolution. However, their result was only published as preliminary version, without giving all details. Moreover, their algorithm is only applicable for polytopes that contain a triangular face. We will reuse some of their ideas for our algorithm, but especially the complicated setting where the polytope does not have a triangular face requires completely new techniques.

2 Preliminaries: Maxwell and Tutte

Since G is 3-connected and planar, the facial structure of G is uniquely determined [19]. Let us pick an arbitrary face f_b, which we call the *boundary face*. We assume further that the vertices are ordered such that $v_1, v_2, \ldots, v_k$ are the vertices of f_b in cyclic order. An edge (vertex) is called *boundary edge (vertex)* if it lies on f_b, otherwise *interior edge (vertex)*. In this paper every embedding is considered as straight-line embedding. A 2d embedding of G is described by giving every vertex v_i a coordinate $\mathbf{p}_i = (x_i, y_i)^T$. We denote the 2d embedding with $G(\mathbf{p})$ and consider only embeddings that realize f_b as convex outer face.

The combination of the Maxwell-Cremona correspondence and barycentric embeddings provides an elegant technique for embedding 3-polytopes. Let us first define the common concept of both methods.

Definition 1 (Stress, Equilibrium). *An assignment* $\omega\colon E \to \mathbb{R}$ *of scalars to the edges of* G *(with* $\omega(i,j) =: \omega_{ij} = \omega_{ji}$*) is called a* stress. *Let* $G(\mathbf{p})$ *be a 2d embedding of* G. *A point* $\mathbf{p}_i$ *is in equilibrium, iff*

$$\sum_{j:(i,j)\in E} \omega_{ij}(\mathbf{p}_i - \mathbf{p}_j) = \mathbf{0}. \tag{1}$$

The embedding of G *is in* equilibrium, *iff all of its points are in equilibrium.*

If $G(\mathbf{p})$ is in equilibrium according to ω, then ω is called *equilibrium stress* for $G(\mathbf{p})$. From special interest are stresses that are positive on every interior edge of G. These stresses are called *positive stresses*.

Suppose we fixed an embedding $G(\mathbf{p})$ in the plane. Let $h_i\colon V \to \mathbb{R}$ be a height assignment for the vertices of G. The function h defines a 3d embedding of G by giving every vertex $\mathbf{p}_i$ the additional z-coordinate $z_i = h(\mathbf{p}_i)$. If in the 3d embedding every face of G lies on a plane the height assignment h is called *lifting*. The so-called Maxwell-Cremona correspondence describes the following.

Theorem 1 (Maxwell [11], Whiteley). *Let* G *be a 3-connected planar graph with 2d embedding* $G(\mathbf{p})$ *and a designated face* $\hat{f}$. *There exists a correspondence between*

A.) *equilibrium stresses* ω *on* $G(\mathbf{p})$,

B.) *liftings of* $G(\mathbf{p})$ *in* $\mathbb{R}^3$, *where face* $\hat{f}$ *lies in the* xy-*plane.*

If the stress is positive, the lifting refers to a convex 3-polytope.

The complete proof of the Maxwell-Cremona correspondence is due to Whiteley [18] (see also [3]). A more constructive proof with detailed rules how o compute the lifting can be found in Richter-Gebert's book [13].

To apply the Maxwell-Cremona correspondence we need a 2d embedding of G with equilibrium stress. Let ω be an arbitrary stress that is zero on the boundary edges and positive on the interior edges. We obtain from the stress ω its *Laplace matrix* $L = \{l_{ij}\}$, which is defined by its entries as

$$l_{ij} := \begin{cases} -\omega_{ij} & \text{if } (i,j) \in E \text{ and } i \neq j, \\ \sum_j \omega_{ij} & \text{if } i = j, \\ 0 & \text{otherwise.} \end{cases}$$

An embedding $G(\mathbf{p})$ is described by the vectors $\mathbf{x} = (x_1, \ldots, x_n)^T$ and $\mathbf{y} = (y_1, \ldots, y_n)^T$. Let $B := \{1, \ldots, k\}$ and $I := \{k+1, \ldots, n\}$ be the index sets of the boundary, respectively interior vertices. We subdivide $\mathbf{x}, \mathbf{y}$, and L by B and I and obtain $\mathbf{x}_B, \mathbf{x}_I, \mathbf{y}_B, \mathbf{y}_I, L_{BB}, L_{II}, L_{BI}$ and L_{IB}. The equilibrium condition

(1) for the interior vertices can be phrased as $L_{IB}\mathbf{x}_B + L_{II}\mathbf{x}_I = \mathbf{0}$. The same holds for the y-coordinates. This implies that we can express the vectors $\mathbf{x}_I$ and $\mathbf{y}_I$ as linear functions in terms of $\mathbf{x}_B$ and $\mathbf{y}_B$, namely by

$$\mathbf{x}_I = -L_{II}^{-1}L_{IB}\mathbf{x}_B, \quad \mathbf{y}_I = -L_{II}^{-1}L_{IB}\mathbf{y}_B. \tag{2}$$

The matrix L_{II} is singular and hence the vectors $\mathbf{x}_I$ and $\mathbf{y}_I$ are properly and uniquely defined (see [13,7]). The embedding is known as (weighted) barycentric embedding. For every positive stress the coordinates give a strictly convex straight-line embedding (see for example [7]). A weighted barycentric embedding has the following nice interpretation: The interior edges of the graph can be considered as springs with spring constants ω_{ij}. The barycentric embedding models the equilibrium state of the system of springs when the boundary vertices are anchored at their fixed positions.

3   Extending an Equilibrium Stress to the Boundary

The interior vertices computed by (2) are in equilibrium by construction. However, the stressed edges of the boundary vertices wont sum up to zero but to

$$\forall i \in B \quad \sum_{j:(i,j)\in E} \omega_{ij}(\mathbf{p}_i - \mathbf{p}_j) =: \mathbf{F}_i. \tag{3}$$

To apply the Maxwell-Cremona correspondence we have to define the (yet unassigned) stresses on the boundary edges such that they cancel the vectors $\mathbf{F}_i$. If the outer face is a triangle this is always possible by solving a linear system with three unknowns [8]. But in general this is only possible for special locations of the boundary face. The problem how to position f_b is challenging, because changing the location of the boundary face will also change the vectors $\mathbf{F}_i$. We use the approach of Ribó, Rote, and Schulz [12] to express this dependence and obtain a formalism that helps us to extend the equilibrium stress to the boundary.

Lemma 1 (Substitution Lemma [12]). *There are positive weights* $\tilde{\omega}_{ij} = \tilde{\omega}_{ji}$, *for* $i, j \in B$, *independent of location of the boundary face such that*

$$\forall i \in B \quad \mathbf{F}_i = \sum_{j\in B: j\neq i} \tilde{\omega}_{ij}(\mathbf{p}_i - \mathbf{p}_j). \tag{4}$$

The weights $\tilde{\omega}_{ij}$ *are the off-diagonal entries of* $-L_{BB} + L_{BI}L_{II}^{-1}L_{IB}$, *which is the Schur Complement of* L_{II} *in* L *multiplied with* -1.

With help of the substitution stresses we can simplify the problem how to locate the boundary face. A feasible position can be found by solving the non-linear system consisting of the $2k$ equations of (4) plus the $2k$ equations

$$\forall i \in B : \omega_{i,suc(i)}(\mathbf{p}_i - \mathbf{p}_{suc(i)}) + \omega_{i,pre(i)}(\mathbf{p}_i - \mathbf{p}_{pre(i)}) = -\mathbf{F}_i, \tag{5}$$

where $suc(i)$ denotes the successor of v_i and $pre(i)$ denotes the predecessor of v_i on f_b in cyclic order. Since the equations are dependent the system is underconstrained. We fix some boundary coordinates to obtain a unique solution.

4   Constructing and Controlling an x-Equilibrium Stress

Let $x_1, \ldots, x_n$ be given as x-coordinates of $G(\mathbf{p})$. We are interested in a positive equilibrium stress that will give these x-coordinates in the barycentric embedding. In particular, we are looking for a positive stress ω such that

$$\forall i \in I \quad \sum_{j:(i,j)\in E} \omega_{ij}(x_i - x_j) = 0. \tag{6}$$

We call a positive stress that fulfills condition (6) a (positive) *x-equilibrium stress*. Since we consider in this paper only positive x-equilibrium stresses we omit the term "positive" in the following. As pointed out in [1], an x-monotone stress exists when every interior edge lies on some x-monotone path, whose endpoints are boundary vertices. Let us assume that we selected the x_i values such that the latter holds. Furthermore, we pick for every edge e some x-monotone path P_e.

We follow the approach of [1] to construct an x-monotone stress. The construction is based on assigning a cost $c_{\{i,j\}} > 0$ to every interior edge (i,j) of G. If the costs guarantee

$$\forall i \in I \quad \sum_{j:(i,j)\in E:x_i<x_j} c_{\{i,j\}} = \sum_{(i,k)\in E:x_i>x_k} c_{\{i,k\}}, \tag{7}$$

we can define an x-equilibrium stress by

$$\omega_{ij} = \frac{c_{\{i,j\}}}{|x_i - x_j|}. \tag{8}$$

We start with $c_{\{i,j\}} \equiv 0$ and increase the costs successively. Let $e = (i,j)$ be an edge with $c_{\{i,j\}} = 0$. We increase the costs of the edges of P_e by 1. In (7) both sides of the equation increase by 1 if v_i lies on P_e, otherwise nothing changes. Hence, (7) still holds. We repeat this procedure until every interior edge is assigned with a positive cost. The total cost for an edge is an integer smaller than $3n - 3$.

We show now how to modify an x-monotone stress to obtain helpful properties for the substitution stresses induced by ω. Our goal is to get a substitution stress that is maximal on an edge we picked. More precisely, let v_s and v_t be two nonincident vertices on the boundary of G and let $\alpha > 1$ be a constant that we fix later. The stress ω should guarantee that for all other pairs of boundary vertices v_i, v_j we have $\tilde{\omega}_{st} > \alpha \tilde{\omega}_{ij}$, unless $ij = st$. The idea how to achieve this is the following: We take an x-monotone path P_{st} from v_s to v_t. Then we take a suitable (large) number K and add K to the costs of every edge which is on P_{st}. The stress induced by the increased costs is still an x-equilibrium stress for our choice of x-coordinates. But if we think of the stresses as spring constants we have increased the force that pushes v_s and v_t away from each other. Our hope is that this will reflect in the substitution stresses and makes $\tilde{\omega}_{st}$ the dominant stress.

First, let us bound all substitution stresses form above and then prove a lower bound for the stress $\tilde{\omega}_{st}$.

Lemma 2. *Let $L = \{l_{ij}\}$ be the Laplace matrix derived from a positive stress ω, and $\tilde{\omega}$ the corresponding substitution stress. For any $i, j \in B$ we have*

$$\tilde{\omega}_{ij} < \min\{l_{ii}, l_{jj}\}.$$

Proof. Since $\mathbf{u}^T L \mathbf{u} = \sum_{(i,j)\in E} \omega_{ij}(u_i - u_j)^2$, which is non-negative for any vector $\mathbf{u}$, the Laplace matrix is positive semidefinite. Let $\tilde{L} = \{\tilde{l}_{ij}\} = L_{BB} - L_{BI}L_{II}^{-1}L_{IB}$ denote the Schur complement of L_{II} in L. Due to [20, page 175] we know that $L_{II} - \tilde{L}$ is positive semidefinite. Therefore, all principal submatrices are positive semidefinite and we have $l_{ii} \geq \tilde{l}_{ii}$. As a consequence of the substitution lemma, we know that $\sum_{j\in B: i\neq j} \tilde{\omega}_{ij} = \tilde{l}_{ii}$, and hence $\sum_{j\in B: i\neq j} \tilde{\omega}_{ij} \leq l_{ii}$. Since each of the $\tilde{\omega}_{ij}$'s is positive, each summand has to be smaller than l_{ii}. By the same argument we can show that $\tilde{\omega}_{ij} \leq l_{jj}$ and the lemma follows.

The next lemma gives us a lower bound for $\tilde{\omega}_{st}$ and tells us how we have to select the number K such that $\tilde{\omega}_{st}$ becomes the dominant stress.

Lemma 3. *Let ω be an x-equilibrium stress obtained from the edge costs $c_{\{i,j\}}$ for $x_1, \ldots, x_n$ that is increased along an x-montone path from v_s to v_t, by incrementing the edge costs for edges on the path by K. The substitution stresses and the matrix L are obtained from ω as usual. Then*

$$\tilde{\omega}_{st} > \frac{K - 3n^2(1 + (k-2)\Delta_x)}{x_t - x_s},$$

where k denotes the number of boundary vertices and Δ_x the largest distance between two x-coordinates. Let $\alpha > 0$ be a parameter that will be fixed later. For $K \geq 3n^2(1 + (\alpha + k - 2)\Delta_x)$ we obtain for any $\tilde{\omega}_{ij}$ that is not $\tilde{\omega}_{st}$

$$\tilde{\omega}_{st} > \alpha\tilde{\omega}_{ij}.$$

Proof. Before bounding $\tilde{\omega}_{st}$ we show an upper bound for the other substitution stresses. Let $\tilde{\omega}_{ij}$ be such a stress. By Lemma 2, $\tilde{\omega}_{ij}$ is less than l_{rr} ($r \in \{i, j\} \cap \{s, t\}$). Since l_{rr} is a diagonal entry of the Laplace matrix it equals $\sum_{k:(r,k)\in E} \omega_{rk}$, which is a sum of at most $n - 1$ summands. We can assume that the path P_{st} uses no boundary edge. In this case each summand in $\sum_{k:(r,k)\in E} \omega_{rk}$ is smaller than $3n - 3$ and we obtain

$$\forall ij \neq st \quad \tilde{\omega}_{ij} < \max_{r\in B\setminus\{s,t\}} \{l_{rr}\} < 3n^2. \tag{9}$$

Let F_t^x be the x-component of $\mathbf{F}_t$. We combine (9) and (4) and obtain as upper bound for F_t^x

$$\tilde{\omega}_{st}(x_t - x_s) + (k-2)3n^2\Delta_x > F_t^x.$$

On the other hand we can express F_t^x by (3). The cost of one edge (k, t), with $x_k < x_t$, was increased by K. All other costs are in total less than $3n^2$. Thus,

$$F_t^x = \sum_{k:(k,t)\in E} \omega_{kt}(x_t - x_k) = \sum_{\substack{k:x_k<x_t \\ (k,t)\in E}} c_{\{t,k\}} - \sum_{\substack{k:x_k>x_t \\ (k,t)\in E}} c_{\{t,k\}} > K - 3n^2.$$

Combining the two bounds for F_t^x leads to the bound for $\tilde{\omega}_{st}$ stated in the lemma. Setting $K \geq 3n^2(1 + (\alpha + k - 2)\Delta_x)$ yields

$$\tilde{\omega}_{st} > \frac{3n^2(\alpha + k - 2)\Delta_x - 3n^2(k - 2)\Delta_x}{x_t - x_s} = \frac{3n^2\alpha\Delta_x}{x_t - x_s} \geq 3n^2\alpha > \alpha\tilde{\omega}_{ij}.$$

As last part of this section we show that we can even enforce a set of substitution stresses to be dominant. Let v_{t_1}, v_{t_2} be two vertices that are both nonincident to v_s and which have the same x-coordinate. We can increment a given x-equilibrium stress by first increasing the edge costs $c_{\{i,j\}}$ along an x-monotone path from v_s to v_{t_1} by K and then do the same for an appropriate path from v_s to v_{t_2}.

Lemma 4. *Assume the same as in Lemma 3, but this time consider two paths: from v_s to v_{t_1} and from v_s to v_{t_2}. Assume further that $x_{t_1} = x_{t_2}$. For $K \geq 3n^2(1 + (\alpha + k - 2)\Delta_x)$ we have for any $\tilde{\omega}_{ij}$ with $\{i,j\} \not\subset \{s, t_1, t_2\}$*

$$\tilde{\omega}_{st_1} > \alpha\tilde{\omega}_{ij}, \quad and \quad \tilde{\omega}_{st_2} > \alpha\tilde{\omega}_{ij}.$$

Proof. Lemma 3 relies on the fact that we can bound $\sum_{j \neq s} \tilde{\omega}_{jt}(x_t - x_j)$ because by (9) the $\tilde{\omega}$'s in this sum are small. We cannot use this bound for $\tilde{\omega}_{t_1 t_2}(x_{t_1} - x_{t_2})$ anymore, but this summand cancels anyway, since $x_{t_1} = x_{t_2}$. Following the steps of the proof of Lemma 3 with first choosing $t = t_1$ and then choosing $t = t_2$ shows the lemma.

5   The Embedding Algorithm

5.1   The Algorithm Template

In this section we present as main result of this paper

Theorem 2. *Let G be a 3-connected planar graph. G admits an embedding as 3-polytope inside a $2(n - 2) \times 1 \times 1$ box under the vertex resolution rule.*

We prove Theorem 2 by introducing an algorithm that constructs the desired embedding. The number of vertices on the boundary face plays an important role. The smaller this number is, the simpler is it to construct an embedding. For this reason we choose as boundary face f_b the smallest face of G. Due to Euler's formula this face has at most 5 vertices. Depending on the size of f_b we obtain three versions of the algorithm which all follow the same basic pattern.

Our goal is to construct a planar embedding of G that has a positive stress and whose x-coordinates are distinct integers. The corresponding lifting of such an embedding fulfills the vertex resolution rule independently of its y and z-coordinates. Hence, we can scale in direction of the y and z-axis without violating the vertex resolution rule. The basic procedure is summarized in Algorithm 5.1.

Steps 1,4, and 5 are mostly independent of the size of f_b and will be discussed first. We start with some strictly convex plane embedding of G – called

---

Algorithm 1. Embedding algorithm as template

1: Choose the x-coordinates.
2: Construct an x-equilibrium stress ω.
3: Choose the boundary y-coordinates.
4: Embed G as barycentric embedding with stress ω in the plane.
5: Lift the plane embedding.

---

pre-embedding. Let $\hat{x}_1, \ldots, \hat{x}_n$ and $\hat{y}_1, \ldots, \hat{y}_n$ be the coordinates of the pre-embedding which we compute as barycentric embedding. We assume that in the pre-embedding all x-coordinates are different. If this is not the case we can perturb the stresses of the (pre)-embedding to achieve this. Since the pre-embedding is strictly convex [17] every edge lies on some x-monotone path. As done in Section 4 we fix for every interior edge e such a path P_e. The x-coordinates of the pre-embedding induce a strict linear order on the vertices of G. We denote with b_i the number of vertices with smaller x-coordinate compared to v_i in the pre-embedding. The x-coordinates of the (final) embedding are defined as $x_i := b_i$. Thus, no two vertices get the same x-coordinate and the largest x-coordinate is less than n. We observe that the paths P_e remain x-monotone. Therefore, they can be used to define an appropriate x-equilibrium stress ω. For technical reasons we might choose the same x-coordinate for some of the boundary vertices. In this case we check the vertex resolution rule for these vertices in the final embedding by hand. Step 4 and 5 can be realized as straight-forward implementations of the barycentric embedding and Maxwell's lifting. Notice that the value of the (extended) stress on the boundary edges is not needed to compute the lifting, because we can place an interior face in the xy-plane and then compute the lifting using only interior edges.

Steps 2 and 3 are the difficult parts of the algorithm. We have to choose the stress ω and the y-coordinates such that an extension of the stress to the boundary is possible. Moreover, the y-coordinates should give a convex boundary face. We continue with the three different cases and discuss step 2 and 3 of the algorithm template for each of them separately.

5.2 Graphs with Triangular Face

The case where f_b is a triangle is the easiest case because we can extend every stress to the boundary for every location of the outer face (see Section 3). This case was already addressed by Chrobak, Goodrich and Tamassia [1]. The discussion in this section will prove the following statement:

Proposition 1. *Let G be a 3-connected planar graph and let G contain a triangular face. G admits an embedding as 3-polytope inside an $(n-1) \times 1 \times 1$ box under the vertex resolution rule.*

Let us compute the pre-embedding with the boundary coordinates $x_1 = 0, x_2 = 1$, and $x_3 = 0$. We use the x-monotone paths P_e to compute a suitable x-equilibrium stress ω as discussed in Section 4. Next we compute the barycentric embedding with boundary coordinates $\mathbf{p}_1 = (0,0)^T, \mathbf{p}_2 = (n-1,0)^T$, and $\mathbf{p}_3 = (0,1)^T$. As result we obtain interior y-coordinates in the interval $(0,1)$. Any two vertices have distance at least 1, since their x-coordinates differ by at least 1. (This is not true for v_1 and v_3, but their distance is $y_3 - y_1 = 1$.)

5.3   Graphs with Quadrilateral Face

Let us assume now that G contains a quadrilateral but no triangular face. In this case it is not always possible to extend an equilibrium stress w to the boundary. The observations of Section 3 help us to overcome this difficulty. We use as boundary coordinates for the pre-embedding $\hat{x}_1 = 0, \hat{x}_2 = 1, \hat{x}_3 = 1$, and $\hat{x}_4 = 0$. The x-coordinates induced by the pre-embedding give $x_2 = x_3 = n - 2$. We redefine the boundary x-coordinates by setting $x_2 = n - 2$ and $x_3 = 2(n-2)$. Notice that this preserves the x-monotonicity of the paths P_e, but makes it easier to extend the stress to the boundary as we will see in the following.

Let w be the x-monotone stress for the obtained x-coordinates. We can express the influence of the stressed edges of G on the boundary with help of the substitution stresses $\tilde{w}$. We modify w with the technique described in Lemma 3 to assure that $\tilde{w}_{13} > \tilde{w}_{24}$. Let us now discuss how to extend the stress w to the boundary. To solve the non-linear system given by (4) and (5) we fix some of the boundary y-coordinates to obtain a unique solution. In particular, we set $y_1 = 0, y_2 = 0$, and $y_4 = 1$. As final coordinate we obtain[1]

$$y_3 = \frac{\tilde{w}_{24}}{2\tilde{w}_{13} - \tilde{w}_{24}}.$$

Since $\tilde{w}_{13} > \tilde{w}_{24}$, we can deduce that $0 < y_3 < 1$. Hence, all y-coordinates are contained in $[0,1]$. Furthermore, we know that the only two vertices with the same x-coordinate, namely v_1 and v_4, have the distance $y_4 - y_1 = 1$. Therefore the vertex resolution rule holds. If we scale the induced lifting such that the largest z-coordinate equals 1 we obtain:

Proposition 2. *Let G be a 3-connected planar graph and let G contain a quadrilateral face. G admits an embedding as 3-polytope inside a $2(n-2) \times 1 \times 1$ box under the vertex resolution rule.*

5.4   The General Case

The most complicated case is the case where we have to use a pentagon as boundary face. However, the basic pattern how to construct the embedding remains the same. We choose as x-coordinates for the pre-embedding $\hat{x}_1 = 0, \hat{x}_2 = 1, \hat{x}_3 = 1, \hat{x}_4 = 0$, and $\hat{x}_5 = -\varepsilon$, for $\varepsilon > 0$ small enough to guarantee that all interior vertices get a positive x-coordinate in the pre-embedding. We change the induced x-coordinates on the boundary without changing the monotonicity of the paths P_e. In particular, we set $x_1 = 0, x_2 = n - 2, x_3 = n - 2, x_4 = 0$, and

---

[1] The solution of the non-linear system was obtained by computer algebra software.

$x_5 = -(n - 2)$. The stress ω is constructed such that the substitution stresses guarantee

$$\tilde{\omega}_{25} > 3\tilde{\omega}_{13}, \tilde{\omega}_{25} > 3\tilde{\omega}_{14}, \tilde{\omega}_{25} > 3\tilde{\omega}_{24}, \tilde{\omega}_{35} > 3\tilde{\omega}_{13}, \tilde{\omega}_{35} > 3\tilde{\omega}_{14}, \tilde{\omega}_{35} > 3\tilde{\omega}_{24}. \quad (10)$$

In other words the substitution stresses $\tilde{\omega}_{25}$ and $\tilde{\omega}_{35}$ dominate all other substitution stresses on interior edges by a factor 3. Since $x_3 = x_2$, we can construct a stress ω that induces a substitution stress that fulfills (10) by the observations of Lemma 4.

Appropriate boundary y-coordinates can be obtained by solving the nonlinear system given by (4) and (5). As done in the previous case we fix some y-coordinates to obtain a unique solution. This time we set $y_1 = -1, y_4 = 1$, and $y_5 = 0$. This yields for the two remaining y-coordinates

$$y_2 = -2 - 2\frac{\tilde{\omega}_{24}\tilde{\omega}_{13} - \tilde{\omega}_{13}^2 - \tilde{\omega}_{35}\tilde{\omega}_{14} - 2\tilde{\omega}_{13}\tilde{\omega}_{35}}{\tilde{\omega}_{24}\tilde{\omega}_{35} + \tilde{\omega}_{25}\tilde{\omega}_{13} + 2\tilde{\omega}_{25}\tilde{\omega}_{35}},$$

$$y_3 = 2 + 2\frac{\tilde{\omega}_{24}\tilde{\omega}_{13} - \tilde{\omega}_{24}^2 - \tilde{\omega}_{14}\tilde{\omega}_{25} - 2\tilde{\omega}_{24}\tilde{\omega}_{25}}{\tilde{\omega}_{24}\tilde{\omega}_{35} + \tilde{\omega}_{25}\tilde{\omega}_{13} + 2\tilde{\omega}_{25}\tilde{\omega}_{35}}.$$

We have to check two things: f_b has to be convex and $y_3 - y_2$ should be large enough to guarantee the vertex resolution rule. First we show that $-2 < y_2$ and $y_3 < 2$, which would imply that f_b is convex if $y_3 > y_2$. The inequalities $-2 < y_2$ and $y_3 < 2$ hold, iff

$$\tilde{\omega}_{24}\tilde{\omega}_{13} - \tilde{\omega}_{13}^2 - \tilde{\omega}_{35}\tilde{\omega}_{14} - 2\tilde{\omega}_{35}\tilde{\omega}_{13} < 0 \text{ and}$$
$$\tilde{\omega}_{24}\tilde{\omega}_{13} - \tilde{\omega}_{24}^2 - \tilde{\omega}_{14}\tilde{\omega}_{25} - 2\tilde{\omega}_{25}\tilde{\omega}_{24} < 0.$$

Both inequalities are true, because as a consequence of (10) the only positive summand $\tilde{\omega}_{24}\tilde{\omega}_{13}$ is smaller than $\tilde{\omega}_{35}\tilde{\omega}_{13}$ and smaller than $\tilde{\omega}_{25}\tilde{\omega}_{24}$. The difference $y_3 - y_2$ equals

$$4 + 2\frac{2\tilde{\omega}_{24}\tilde{\omega}_{13} - \tilde{\omega}_{13}^2 - \tilde{\omega}_{24}^2 - \tilde{\omega}_{35}(\tilde{\omega}_{14} + 2\tilde{\omega}_{13}) - \tilde{\omega}_{25}(\tilde{\omega}_{14} + 2\tilde{\omega}_{24})}{\tilde{\omega}_{24}\tilde{\omega}_{35} + \tilde{\omega}_{25}\tilde{\omega}_{13} + 2\tilde{\omega}_{25}\tilde{\omega}_{35}}.$$

Due to (10) we know that $\tilde{\omega}_{14} + 2\tilde{\omega}_{24} < \tilde{\omega}_{35}$ and $\tilde{\omega}_{14} + 2\tilde{\omega}_{13} < \tilde{\omega}_{25}$. Thus

$$y_3 - y_2 > 4 + 2\frac{2\tilde{\omega}_{24}\tilde{\omega}_{13} - \tilde{\omega}_{13}^2 - \tilde{\omega}_{24}^2 - 2\tilde{\omega}_{35}\tilde{\omega}_{25}}{\tilde{\omega}_{24}\tilde{\omega}_{35} + \tilde{\omega}_{25}\tilde{\omega}_{13} + 2\tilde{\omega}_{25}\tilde{\omega}_{35}} > 2.$$

The estimation holds since (again due to (10)) $\tilde{\omega}_{24}\tilde{\omega}_{35} > \tilde{\omega}_{24}^2$ and $\tilde{\omega}_{25}\tilde{\omega}_{13} > \tilde{\omega}_{13}^2$, and hence the fraction is greater -1.

We multiply all y-coordinates by $1/2$. This yields $y_4 - y_1 = 1$ and $y_3 - y_2 > 1$. The z-coordinates are scaled such that they lie between 0 and 1. Clearly, the vertex resolution rule holds for the computed embedding.

We finish this section with some remarks on the running time of the embedding algorithm. As mentioned in [1] the x-monotone stress can be computed in linear time. The barycentric embedding (which we use twice, once for the pre-embedding and once for the intermediate plane embedding) can be computed by the linear system (2). Since the linear system is based on a planar structure, we can solve it with nested dissections (see [9,10]) based on the planar separator theorem. As a consequence a solution can be computed in $O(M(\sqrt{n}))$ time, where $M(n)$ is the upper bound for multiplying two $n \times n$ matrices. The current record for $M(n)$ is $O(n^{2.325})$, which is due Coppersmith and Winograd [2]. The computation of the lifting can be done in linear time. In total we achieve a running time of $O(n^{1.1625})$.

6   Additional Properties of the Embedding

6.1   Induced Grid Embedding

Besides the small embedding under the vertex resolution rule, the constructed embedding has several other nice properties which we discuss in this section. Since the computed y and z-coordinates are expressed by a linear system of rational numbers, they are rational numbers as well. Thus, we can scale the final embedding to obtain a grid embedding.

Theorem 3. *The embedding computed with Algorithm 5.1 can be scaled to integer coordinates such that*

$$0 \leq x_i \leq 2(n-2),$$
$$0 \leq y_i, z_i \leq 2^{O(n^2 \log n)}.$$

The proof of the theorem can be found in the full version of the paper. Compared to the grid embedding presented in [12], we were able to reduce the size of the x-coordinates (from $2n \cdot 8.107^n$ to $2(n-2)$) at the expense of the y and z-coordinates.

6.2   Spread of the Embedding

The *spread* of a point set is the quotient of the longest pairwise distance (the diameter) and the shortest pairwise distance. The smaller this ratio is, the more densely the point set is packed. A small spread implies that the points are "evenly distributed". We define as *spread of an embedding of a polytope* the spread of its points.

Theorem 4. *The spread of a 3-polytope embedded by Algorithm 5.1 is smaller than n. There are infinitely many polytopes without an embedding with spread smaller than $(n-1)/\pi$.*

The proof of the theorem can be found in the full version of the paper.

References

1. Chrobak, M., Goodrich, M.T., Tamassia, R.: Convex drawings of graphs in two and three dimensions (preliminary version). In: 12th Symposium on Computational Geometry, pp. 319–328 (1996)

2. Coppersmith, D., Winograd, S.: Matrix multiplication via arithmetic progressions. J. Symb. Comput. 9(3), 251–280 (1990)
3. Crapo, H., Whiteley, W.: Plane self stresses and projected polyhedraI: The basic pattern. Structural Topology 20, 55–78 (1993)
4. Das, G., Goodrich, M.T.: On the complexity of optimization problems for 3-dimensional convex polyhedra and decision trees. Comput. Geom. Theory Appl. 8(3), 123–137 (1997)
5. de Fraysseix, H., Pach, J., Pollack, R.: How to draw a planar graph on a grid. Combinatorica 10(1), 41–51 (1990)
6. Eades, P., Garvan, P.: Drawing stressed planar graphs in three dimensions. In: Brandenburg, F.J. (ed.) GD 1995. LNCS, vol. 1027, pp. 212–223. Springer, Heidelberg (1996)
7. Gortler, S.J., Gotsman, C., Thurston, D.: Discrete one-forms on meshes and applications to 3d mesh parameterization. Computer Aided Geometric Design 23(2), 83–112 (2006)
8. Hopcroft, J.E., Kahn, P.J.: A paradigm for robust geometric algorithms. Algorithmica 7(4), 339–380 (1992)
9. Lipton, R.J., Rose, D., Tarjan, R.: Generalized nested dissection. SIAM J. Numer. Anal. 16(2), 346–358 (1979)
10. Lipton, R.J., Tarjan, R.E.: Applications of a planar separator theorem. SIAM J. Comput. 9(3), 615–627 (1980)
11. Maxwell, J.C.: On reciprocal figures and diagrams of forces. Phil. Mag. Ser. 27, 250–261 (1864)
12. Ribó Mor, A., Rote, G., Schulz, A.: Small grid embeddings of 3-polytopes (2009) (submitted for publication), http://arxiv.org/abs/0908.0488
13. Richter-Gebert, J.: Realization Spaces of Polytopes. Lecture Notes in Mathematics, vol. 1643. Springer, Heidelberg (1996)
14. Schnyder, W.: Embedding planar graphs on the grid. In: Proc. 1st ACM-SIAM Sympos. Discrete Algorithms, pp. 138–148 (1990)
15. Schramm, O.: Existence and uniqueness of packings with specified combinatorics. Israel J. Math. 73, 321–341 (1991)
16. Steinitz, E.: Encyclopädie der mathematischen Wissenschaften. In: Polyeder und Raumteilungen, pp. 1–139 (1922)
17. Tutte, W.T.: How to draw a graph. Proceedings London Mathematical Society 13(52), 743–768 (1963)
18. Whiteley, W.: Motion and stresses of projected polyhedra. Structural Topology 7, 13–38 (1982)
19. Whitney, H.: A set of topological invariants for graphs. Amer. J. Math. 55, 235–321 (1933)
20. Zhang, F.: Matrix Theory. Springer, Heidelberg (1999)

Planar Drawings of Higher-Genus Graphs

Christian A. Duncan[1], Michael T. Goodrich[2,*], and Stephen G. Kobourov[3,**]

[1] Dept. of Computer Science, Louisiana Tech Univ.
http://www.latech.edu/~duncan/
[2] Dept. of Computer Science, Univ. of California, Irvine
http://www.ics.uci.edu/~goodrich/
[3] Dept. of Computer Science, University of Arizona
http://www.cs.arizona.edu/~kobourov/

Abstract. In this paper, we give polynomial-time algorithms that can take a graph G with a given combinatorial embedding on an orientable surface $\mathcal{S}$ of genus g and produce a planar drawing of G in $\mathbf{R}^2$, with a bounding face defined by a polygonal schema $\mathcal{P}$ for $\mathcal{S}$. Our drawings are planar, but they allow for multiple copies of vertices and edges on $\mathcal{P}$'s boundary, which is a common way of visualizing higher-genus graphs in the plane. As a side note, we show that it is NP-complete to determine whether a given graph embedded in a genus-g surface has a set of $2g$ fundamental cycles with vertex-disjoint interiors, which would be desirable from a graph-drawing perspective.

1 Introduction

The *classic* way of drawing a graph $G = (V, E)$ in $\mathbf{R}^2$ involves associating each vertex v in V with a unique point (x_v, y_v) and associating with each edge $(v, w) \in E$ an open Jordan curve that has (x_v, y_v) and (x_w, y_w) as its endpoints. If the curves associated with the edges in a classic drawing of G intersect only at their endpoints, then (the embedding of) G is a *plane graph*. Graphs that admit plane graph representations are *planar graphs*, and there has been a voluminous amount of work on algorithms on classic drawings of planar graphs. Most notably, planar graphs can be drawn with vertices assigned to integer coordinates in an $O(n) \times O(n)$ grid, which is often a desired type of classic drawing known as a *grid drawing*. Moreover, there are planar graph drawings that use only straight line segments for edges [2].

The beauty of plane graph drawings is that, by avoiding edge crossings, confusion and clutter in the drawing is minimized. Likewise, straight-line drawings further improve graph visualization by allowing the eye to easily follow connections between adjacent vertices. In addition, grid drawings enforce a natural separation between vertices, which further improves readability. Thus, a "gold standard" in classic drawings is to produce planar straight-line grid drawings

* Work partially supported by NSF grants OCI-0724806, IIS-0713046, CCR-0830403, and ONR MURI N0014-08-1-1015.
** Work partially supported by NSF CAREER grant CCF-0545743.

D. Eppstein and E.R. Gansner (Eds.): GD 2009, LNCS 5849, pp. 45–56, 2010.
© Springer-Verlag Berlin Heidelberg 2010

and, when that is not easily done, to produce planar grid drawings with edges drawn as simple polygonal chains.

Unfortunately, not all graphs are planar. So drawing them in the classic way requires some compromise in the gold standard for plane drawings. In particular, any classic drawing of a non-planar graph must necessarily have edge crossings, and minimizing the number of crossings is NP-hard [6]. One point of hope for improved drawings of non-planar graphs is to draw them crossing-free on surfaces of higher genus, such as toruses, double toruses, or, in general, a surface topologically equivalent to a sphere with g handles, that is, a *genus-g* surface. Such drawings are called *cellular* embeddings or *2-cell* embeddings, since they partition the genus-g surface into a collection of cells that are topologically equivalent to disks. As in classic drawings of planar graphs, these cells are called *faces*, and it is easy to see that such a drawing would avoid edge crossings.

In a fashion analogous to the case with planar graphs, cellular embeddings of graphs in a genus-g surface can be characterized combinatorially. In particular, it is enough if we just have a rotational order of the edges incident on each vertex in a graph G to determine a combinatorial embedding of G on a surface (which has that ordering of associated curves listed counterclockwise around each vertex). Such a set of orderings is called a *rotation system* and, since it gives us a combinatorial description of the set of faces, F, in the embedding, it gives us a way to determine the genus of the (orientable) surface that G is embedded into by using the *Euler characteristic*, $|V| - |E| + |F| = 2 - 2g$, which also implies that $|E|$ is $O(|V| + g)$ [10].

Unfortunately, given a graph G, it is NP-hard to find the smallest g such that G has a combinatorial cellular embedding on a genus-g surface [11]. This challenge need not be a deal-breaker in practice, however, for there are heuristic algorithms for producing such combinatorial embeddings (that is, consistent rotation systems) [1]. Moreover, higher-genus graphs often come together with combinatorial embeddings in practice, as in many computer graphics and mesh generation applications.

In this paper, we assume that we are given a combinatorial embedding of a graph G on an orientable genus-g surface, $\mathcal{S}$, and are asked to produce a geometric drawing of G that respects the given rotation system. Motivated by the gold standard for planar graph drawing and by the fact that computer screens and physical printouts are still primarily two-dimensional display surfaces, the approach we take is to draw G in the plane rather than on some embedding of $\mathcal{S}$ in $\mathbf{R}^3$.

Making this choice of drawing paradigm, of course, requires that we "cut up" the genus-g surface, $\mathcal{S}$, and "unfold" it so that the resulting sheet is topologically equivalent to a disk. The traditional method for performing such a cutting is with a *canonical polygonal schema*, $\mathcal{P}$, which is a set of $2g$ cycles on $\mathcal{S}$ all containing a common point, p, such that cutting $\mathcal{S}$ along these cycles results in a topological disk. These cycles are *fundamental* in that each of them is a continuous closed curve on $\mathcal{S}$ that cannot be retracted continuously to a point. Moreover, these fundamental cycles can be paired up into complementary sets of cycles, (a_i, b_i), one for each handle, so that if we orient the sides of $\mathcal{P}$, then a counterclockwise ordering of the sides of $\mathcal{P}$ can be listed as $a_1 b_1 a_1^{-1} b_1^{-1} a_2 b_2 a_2^{-1} b_2^{-1} \ldots a_g b_g a_g^{-1} b_g^{-1}$, where a_i^{-1} (b_i^{-1}) is a reversely-oriented copy of a_i, so that these two sides of $\mathcal{P}$ are matched in orientation on $\mathcal{S}$. Thus, the canonical polygonal schema for a genus-g surface $\mathcal{S}$ has $4g$ sides that are pairwise identified.

Because we are interested in drawing the graph G and not just the topology of $\mathcal{S}$, it would be preferable if the fundamental cycles are also cycles in G in the graph-theoretical sense. It would be ideal if these cycles form a canonical polygonal schema with no repeated vertices other than the common one. This is not always possible [8] and furthermore, as we show in [3], the problem of finding a set of $2g$ fundamental cycles with vertex-disjoint interiors in a combinatorially embedded genus-g graph is NP-complete. There are two natural choices, both of which we explore in this paper:

- Draw G in a polygon P corresponding to a canonical polygonal schema, $\mathcal{P}$, possibly with repeated vertices and edges on its boundary.
- Draw G in a polygon P corresponding to a polygonal schema, $\mathcal{P}$, that is not canonical.

In either case, the edges and vertices on the boundary of P are repeated (since we "cut" $\mathcal{S}$ along these edges and vertices). Thus, we need labels in our drawing of G to identify the correspondences. Such planar drawings of G inside a polygonal schema $\mathcal{P}$ are called *polygonal-schema* drawings of G. There are three natural aesthetic criteria such drawings should satisfy:

1. *Straight-line edges:* All the edges in a polygonal-schema drawing should be rendered as polygonal chains, or straight-line edges, when possible.
2. *Straight frame:* Each edge of the polygonal schema should be rendered as a straight line segment, with the vertices and edges of the corresponding fundamental cycle, placed along this segment. We refer to such a polygonal-schema drawing as having a *straight frame*.
3. *Polynomial area:* Drawings should have polynomial area when they are normalized to an integer grid.

It is also possible to avoid repeated vertices and instead use a classic graph drawing paradigm, by transforming the fundamental polygon rendering using polygonal-chain edges that run through "overpasses" and "underpasses" as in road networks, so as to illustrate the topological structure of G; see Fig. 1.

Our Contributions. We provide several methods for producing planar polygonal-schema drawings of higher-genus graphs. In particular, we provide four algorithms, one for torodial ($g = 1$) graphs and three for non-toroidal ($g > 1$) graphs. Our algorithm for toroidal graphs simultaneously achieves the three aesthetic criteria for polygonal schema drawings: it uses straight-line edges, a straight frame, and polynomial area. The three algorithms for non-toroidal graphs, *Peel-and-Bend*, *Peel-and-Stretch*, and *Peel-and-Place*, achieve two of the three aesthetic criteria and differ in which criteria they fail to meet.

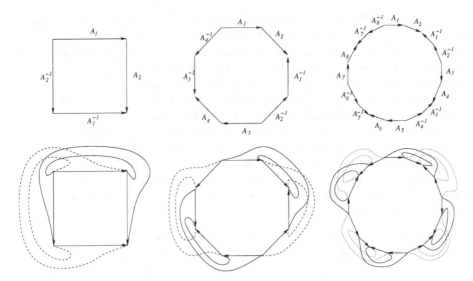

Fig. 1. First row: Canonical polygonal schemas for graphs of genus one, two and four. Second row: Unrolling the high genus graphs with the aid of the overpasses and underpasses.

2   Finding Polygonal Schemas

Suppose we are given a graph G together with its cellular embedding in an (orientable) genus-g surface, $\mathcal{S}$. An important first step in all of our algorithms involves our finding a polygonal schema, $\mathcal{P}$, for G, that is, a set of cycles in G such that cutting $\mathcal{S}$ along these cycles results in a topological disk. We refer to this as the *Peel* step, since it involves cutting the surface $\mathcal{S}$ until it becomes topologically equivalent to a disk. Since these cycles form the sides of the fundamental polygon we will be using as the outer face in our drawing of G, it is desirable that these cycles be as "nice" as possible with respect to drawing aesthetics.

2.1   Trade-Offs for Finding Polygonal Schemas

Unfortunately, some desirable properties are not effectively achievable. As Lazarus *et al.* [8] show, it is not always possible to have a canonical polygonal schema $\mathcal{P}$ such that each fundamental cycle in $\mathcal{P}$ has a distinct set of vertices in its interior (recall that the interior of a fundamental cycle is the set of vertices distinct from the common vertex shared with its complementary fundamental cycle—with this vertex forming a corner of a polygonal schema). In addition, we show in [3] that finding a vertex-disjoint set of fundamental cycles is NP-complete. So, from a practical point of view, we have two choices with respect to methods for finding polygonal schemas.

Finding a Canonical Polygonal Schema. As mentioned above, a canonical polygon schema of a graph G 2-cell embedded in a surface of genus g consists of $4g$ sides, which correspond to $2g$ fundamental cycles all containing a common vertex. Lazarus *et al.* [8] show that one can find such a schema for G in $O(gn)$ time and with total size $O(gn)$, and they show that this bound is within a constant factor of optimal in the worst case, where n is the total combinatorial complexity of G (vertices, edges, and faces), which is $O(|V| + g)$.

Minimizing the Number of Boundary Vertices in a Polygonal Schema. Another optimization would be to minimize the number of vertices in the boundary of a polygonal schema. Erikson and Har-Peled [5] show that this problem is NP-hard, but they provide an $O(\log^2 g)$-approximation algorithm that runs in $O(g^2 n \log n)$ time, and they give an exact algorithm that runs in $O(n^{O(g)})$ time.

In our *Peel* step, we assume that we use one of these two optimization criteria to find a polygonal schema, which either optimizes its number of sides to be $4g$, as in the canonical case, or optimizes the number of vertices on its boundary, which will be $O(gn)$ in the worst case either way. Nevertheless, for the sake of concreteness, we often describe our algorithms assuming we are given a canonical polygonal schema. It is straight-forward to adapt these algorithms for non-canonical schemas.

2.2 Constructing Chord-Free Polygonal Schemas

In all of our algorithms the first step, *Peel*, constructs a polygonal schema of the input graph G. In fact, we need a polygonal schema, $\mathcal{P}$, in which there is no chord connecting two vertices on the same side of $\mathcal{P}$. Here we show how to transform any polygonal schema into a chord-free polygonal schema.

In the *Peel* step, we cut the graph G along a canonical set of $2g$ fundamental cycles getting two copies of the cycle in G^*, the resulting planar graph. For each of the two pairs of every fundamental cycle there may be chords. If the chord connects two vertices that are in different copies of the cycle in G^* then this is a chord that *can* be drawn with a straight-line edge and hence does not create a problem. However, if the chord connects two vertices in the same copy of the cycle in G^*, then we will not be able to place all the vertices of that cycle on a straight-line segment; see Figure 2(a). We show next that a new chord-free polygonal schema can be efficiently determined from the original schema.

Theorem 1. *Given a graph G combinatorially embedded in a genus-g surface and a canonical polygonal schema $\mathcal{P}$ on G with a common vertex p, a chord-free polygonal schema $\mathcal{P}^*$ can be found in $O(gn)$ time.*

Proof. We first use the polygonal schema to cut the embedding of G into a topological disk; see Fig. 2(a). Notice this cutting will cause certain vertices to be split into multiple vertices. For each fundamental cycle in $c_i \in \mathcal{P}$, we stitch the disk graph back together along this cycle forming a topological cylinder. The outer edges (left and right) of the cylinder along this stitch will have two copies of the vertex p, say p_1 and p_2. We perform a shortest path search from p_1 to p_2. This path becomes our new fundamental cycle c_i^*, (since p_1 and p_2 are the same vertex in G). Observe that this cycle must be chord-free or else the path

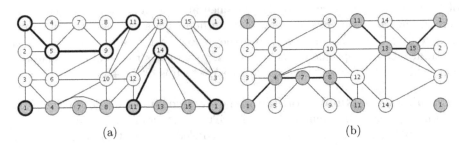

(a)   (b)

Fig. 2. (a) A graph embedded on the torus that has been cut into a topological disk using the cycles $1, 2, 3$ and $1, 4, 7, 8, 11, 13, 15$ with chord $(4, 8)$. The grey nodes correspond to the identical vertices above. The highlighted path represents a shortest path between the two copies of vertex 1. (b) The topological disk *after* cutting along this new fundamental cycle. The grey nodes show the old fundamental cycle.

chosen was not the shortest path; see Fig. 2(b). We then cut the cylinder along c_i^* and proceed to c_{i+1}. The resulting set, $\mathcal{P}^* = \{c_1^*, c_2^*, \ldots, c_{2g}^*\}$, is therefore a collection of chord-free fundamental cycles all sharing the common vertex p. □

It should be noted that, although each cycle c_i^* is at the time of its creation a shortest path from the two copies of p, these cycles are *not* the shortest fundamental cycles possible. For example, a change in the cycle of c_{i+1} could introduce a shorter possible path for c_i^*, but not additional chords.

3   Straight Frame and Polynomial Area

In this section, we describe our algorithms that construct a drawing of G in a straight frame using polynomial area. Here we are given an embedded genus-g graph $G = (V, E)$ along with a chord-free polygonal schema, $\mathcal{P}$, for G from the *Peel* step. We rely on a modified version of the algorithm of de Fraysseix, Pach and Pollack [2] for the drawing. Sections 3.1 and 3.2 describe the details for $g = 1$ and for $g > 1$, respectively. In the latter case we introduce up to $O(k)$ edges with single bends where k is the number of vertices on the fundamental cycles. Thus, we refer to the algorithm for non-toroidal graphs as the *Peel-and-Bend* algorithm.

3.1   Grid Embedding of Toroidal Graphs

For toroidal graphs we are able to achieve all three aesthetic criteria: straight-line edges, straight frame, and polynomial area.

Theorem 2. *Let G^* be an embedded planar graph and $\mathcal{P} = \{P_1, P_2, \ldots, P_{4g}\}$ in G^* be a collection of $4g$ paths such that each path $P_i = \{p_{i,1}, p_{i,2}, \ldots, p_{i,k_i}\}$ is chord-free, the last vertex of each path matches the first vertex of the next path, and when treated as a single cycle, $\mathcal{P}$ forms the external face of G^*. If $g = 1$, we can in linear time draw G^* on an $O(n) \times O(n^2)$ grid with straight-line edges and no crossings in such a way that, for each path P_i on the external face, the vertices on that path form a straight line.*

Proof. For simplicity, we assume that every face is a triangle, except for the outer face (extra edges can be added and later removed). The algorithm of de Fraysseix, Pach and Pollack (dPP) [2] does not directly solve our problem because of the additional requirement for the drawing of the external face. In the case of $g = 1$, the additional requirement is that the graph must be drawn so that the external face forms a rectangle, with P_1 and P_3 as the top and bottom horizontal boundaries and P_2 and P_4 as the right and left boundaries.

Recall that the dPP algorithm computes a canonical labeling of the vertices of the input graph and inserts them one at a time in that order while ensuring that when a new vertex is introduced it can "see" all of its already inserted neighbors. One technical difficulty lies in the proper placement of the top row of vertices. Due to the nature of the canonical order, we cannot force the top row of vertices to all be the last set of vertices inserted, unlike the bottom row which can be the first set inserted. Consequently, we propose an approach similar to that of Miura, Nakano, and Nishizeki [9]. First, we split the graph into two parts (not necessarily of equal size), perform a modified embedding on both pieces, invert one of the two pieces, and stitch the two pieces together.

Lemma 1. *Given an embedded plane graph G that is fully triangulated except for the external face and two edges e_l and e_r on that external face, it is possible in linear time to partition $V(G)$ into two subsets V_1 and V_2 such that*

1. *the subgraphs of G induced by V_1 and V_2, called G_1 and G_2, are both connected subgraphs;*
2. *for edges $e_l = (u_l, v_l)$ and $e_r = (u_r, v_r)$, we have $u_l, u_r \in V_1$ and $v_l, v_r \in V_2$;*
3. *the union U of the set of faces in G that are not in G_1 or G_2 forms an outerplane graph with the property that the external face of U is a cycle with no repeated vertices.*

Proof. First, we compute the dual D of G, where each face in (the primal graph) G is a node in D and there is an arc between two nodes in D if their corresponding primal faces share an edge in common. We ignore the external face in this step. For clarity we shall refer to vertices and edges in the primal and nodes and arcs in the dual; see Fig. 3(a). We further augment the dual by adding an arc between two nodes in D if they also share a vertex in common. Call this augmented dual graph D^*.

Let the source node s be the node corresponding to the edge e_l and the sink node t be the node corresponding to the edge e_r. We then perform a breadth-first shortest-path traversal from s to t on D^*; see Fig. 3(b). Let p^* be a shortest (augmented) path in D^* obtained by this search. We now create a (regular) path p by expanding the augmented arcs added. That is, if there is an arc $(u, v) \in p^*$ such that u and v share a common vertex in G but *not* a common edge in G, i.e. they are part of a fan around the common vertex, we add back the regular arcs from u to v adjacent to this common vertex. The choice of going clockwise or counter-clockwise around the common vertex depends on the previous visited arc; see Fig. 3(c).

All of the steps described above can be easily implemented in linear time. The details of the proof can be found in [3].   □

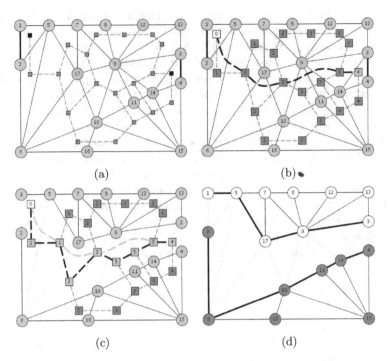

Fig. 3. (a) A graph G and its dual D. The dark edges/nodes represent the sink and source nodes. (b) Each dual node is labeled with its distance (in D^*) from the start node 0. A shortest path p^* is drawn with thick dark arcs. This path includes the augmented arcs of D^*. (c) The path p formed after expanding the augmented arcs. The edges from the primal that are cut by this path are shown faded. (d) The two sets V_1 (light vertices) and V_2 (darker vertices) formed by the removal of path p. The external face of U is defined by the thick edges along with the edges $(1, 2)$ and $(3, 4)$.

Figure 3(d) illustrates the result of one such partition. In some cases we might have to start and end with a set of edges rather than just the two edges e_l and e_r. The following extension of Lemma 1 addresses this issue; the details of the proof can be found in [3].

Lemma 2. *Given an embedded plane graph G that is fully triangulated, except for the external face, and given two vertex-disjoint chord-free paths L and R on that external face, it is possible in linear time to partition $V(G)$ into two subsets V_1 and V_2 such that*

1. *the subgraphs of G induced by V_1 and V_2, called G_1 and G_2, are both connected subgraphs;*
2. *there exists exactly one vertex $v \in V(L)$ (say $v \in V_1$) with neighbors in $V(L) \setminus V_2$ (the opposite vertex set that are not part of $V(L)$), the same holds for $V(R)$; and*

3. *the union U of the set of faces in G that are not in G_1 or G_2 forms an outerplane graph with the property that the external face of U is a cycle with no repeated vertices.*

We can now discuss the steps for the grid drawing of the genus-1 graph G^* with an external face formed by $\mathcal{P}$. Using Lemma 2, with $L = P_4$ and $R = P_2$, divide G^* into two subgraphs G_1 and G_2. We proceed to embed G_1 with G_2 being symmetric. Assume without loss of generality that G_1 contains the bottom path, P_3. Compute a canonical order of G_1 so that the vertices of P_3 are the last vertices removed. Place all of the vertices of P_3 on a horizontal line, $p_{3,k_3}, p_{3,k_3-1}, \cdots, p_{3,1}$ placed consecutively on $y = 0$. This is possible since there are no edges between them (because the path is chord-free). Recall that the standard dPP algorithm [2] maintains the invariant that at the start of each iteration, the current external face consists of the original horizontal line and a set of line segments of slope ± 1 between consecutive vertices. The algorithm also maintains a "shifting set" for each vertex. We modify this condition by requiring that the vertices on the right and left boundary that are part of P_2 and P_4 be aligned vertically and that the current external face might have horizontal slopes corresponding to vertices from P_3; see Fig. 4(a). Upon insertion of a new vertex v, the vertex will have consecutive neighboring vertices on the external face. We label the left and rightmost neighbors x_ℓ and x_r. To achieve our modified invariant, we insert a vertex v into the current drawing depending on its type, 0, 1, or 2, as follows:

Type 0. Vertices not belonging to a path in $\mathcal{P}$ are inserted as with the traditional dPP algorithm. This insertion might require up to two horizontal shifts determined by the shifting sets; see Fig. 4(a).

Type 1. Vertices belonging to P_2, which must be placed vertically along the right boundary, are inserted with a line segment of slope $+1$ between x_ℓ and v and a vertical line segment between v and x_r. Notice that x_r must also be in P_2. And because P_2 is chord-free x_r is the topmost vertex on the right side of the current external face. That is, v can see x_r. By Lemma 2 and the fact that the graph was fully triangulated, we also know that v must have a vertex x_ℓ. This insertion requires only 1 shift, for the visibility of x_ℓ and v. Again the remaining vertices $x_{\ell+1}, \ldots, x_{r-1}$ are connected as usual; see Fig. 4(b).

Type 2. Vertices belonging to P_4, which must be placed vertically along the left boundary, are handled similarly to Type 1.

Because of Lemma 2, after processing both G_1 and G_2, we can proceed to stitch the two portions together. Shift the left wall of the narrower graph sufficiently to match the width of the other graph. For simplicity, refer to the vertices on the external face of each subgraph that are not exclusively part of the wall or bottom row as **upper external vertices**. For each subgraph, consider the point p located at the intersection of the lines of slope ± 1 extending from the left and rightmost external vertices. Flip G_2 vertically placing it so that its point p lies either on or just above (in case of non-integer intersection) G_1's point. Because the edges between the upper external vertices have slope $|m| \leq 1$ and because of the vertical separation of the two subgraphs, every upper external vertex on G_1 can directly

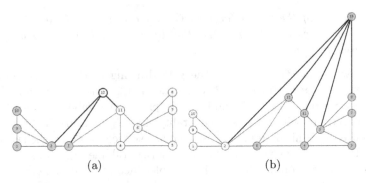

(a)   (b)

Fig. 4. (a) The embedding process after insertion of the first 11 vertices and the subsequent insertion of a Type 0 vertex with $v = 12$, $x_\ell = 2$, and $x_r = 11$. Note the invariant condition allowing the two partial vertical walls $\{8, 7, 5\} \subset P_2$ and $\{1, 9, 10\} \subset P_4$. The light vertices to the right of 12 including x_r have been shifted over one unit. (b) The result of inserting a Type 1 vertex with $v = 13$, $x_\ell = 2$, and $x_r = 8$. Note, the light vertices to the left of and including $x_\ell = 2$ are shifted over one unit.

see every upper external vertex on G_2. By Lemma 2, we know that the set of edges removed in the separation along with the edges connecting the upper external vertices forms an outerplanar graph. Therefore, we can reconnect the removed edges, joining the two subgraphs, without introducing any crossings.

We claim that the area of this grid is $O(n) \times O(n^2)$. First, let us analyze the width. From our discussion, we have accounted for each insertion step using shifts. Since the maximum amount of shifting of 2 units is done with Type 0 vertices, we know that each of the two subgraphs has width at most $2n$. In addition, the stitching stage only required a shifting of the smaller width subgraph. Therefore, the width of our drawing is at most $2n$. The stitching stage for example only adds at most $W \leq 2n$ units to the final height. After the insertion of each wall vertex we know that the height increases by at most W. Therefore, we know that the height is at most Wn or $2n^2$ and consequently we have a correct drawing using a grid of size $O(n) \times O(n^2)$. Ideally, the height of our drawing would also match the width bound.   □

3.2 The Peel-and-Bend Algorithm

The case for $g > 1$ is similar but involves a few alterations. First, we use $n = |V|$ unlike prior sections which used $n = |V| + g$. However, the main difference is that having chord-free fundamental cycles is insufficient to allow rendering the outer face as a rectangle unless edge bends are allowed. The following theorem describes our resulting drawing method, called the *Peel-and-Bend* algorithm.

Theorem 3. *Let G^* be an embedded planar graph and $\mathcal{P} = \{P_1, P_2, \ldots, P_{4g}\}$ in G^* be a collection of $4g$ paths such that each path $P_i = \{p_{i,1}, p_{i,2}, \ldots, p_{i,k_i}\}$ is chord-free, the last vertex of each path matches the first vertex of the next path, and when treated as a single cycle, $\mathcal{P}$ forms the external face of G^*. Let*

$k = \sum_{i=1}^{4g}(k_i - 1)$ be the number of vertices on the external cycle. We can draw G^* on an $O(n) \times O(n^2)$ grid with straight-line edges and no crossings and at most $k - 3$ single-bend edges in such a way that for each path P_i on the external face the vertices on that path form a straight line.

Proof. First, let us assume that the entire external face, represented by $\mathcal{P}$, is completely chord-free. That is, if two vertices on the external cycle share an edge then they are adjacent on the cycle. In this case we can create a new set of 4 paths, $\mathcal{P}' = \{P_1, \cup_{i=2,\ldots,2g}P_i, P_{2g+1}, \cup_{i=2g+2,4g}P_i\}$. We can then use Theorem 2 to prove our claim using no bends.

If, however, there exist chords on the external face, embedding the graph with straight-lines becomes problematic, and in fact impossible to do using a rectangular outer face. By introducing a temporary bend vertex for each chord and retriangulating the two neighboring faces, we can make the external face chord-free. Clearly this addition can be done in linear time. Since there are at most k vertices on the external face and since the graph is planar, there are no more than $k - 3$ such bend points to add. We then proceed as before using Theorem 2, subsequently replacing inserted vertices with a bend point.   □

4 Algorithms for Non-toroidal Graphs

In this section, we describe two more algorithms for producing a planar polygonal-schema drawing of a non-toroidal graph G, which is given together with its combinatorial embedding in an (orientable) genus-g surface, $\mathcal{S}$, where $g > 1$. As mentioned above, these algorithms provide alternative trade-offs with respect to the three primary aesthetic criteria we desire for polygonal-schema drawings. For the sake of space, we describe these algorithms at a very high level and leave their details and full analysis to the full version of this paper [3].

The Peel-and-Stretch Algorithm. In the Peel-and-Stretch Algorithm, we find a chord-free polygonal schema $\mathcal{P}$ for G and cut G along these edges to form a planar graph G^*. We then layout the sides of $\mathcal{P}$ in a straight-frame manner as a regular convex polygon, with the vertices along each boundary edge spaced as evenly as possible. We then fix this as the outer face of G^* and apply Tutte's algorithm [12,13] to construct a straight-line drawing of the rest of G^*. This algorithm therefore achieves a drawing with straight-line edges in a (regular) straight frame, but it may require exponential area when normalized to an integer grid, since Tutte's drawing algorithm may generate vertices with coordinates that require $\Theta(n \log n)$ bits to represent.

The Peel-and-Place Algorithm. For this method, we start by finding a polygonal schema $\mathcal{P}$ for G and cut G along these edges to form a planar graph G^*, as in all our algorithms. We then create a new triangular face, T, place G^* in the interior of T, and fully triangulate this graph. We then apply the dPP algorithm [2] to construct a drawing of this graph in an $O(n) \times O(n)$ integer grid with straight-line edges. Finally, we remove all extra edges to produce a polygonal schema drawing of G. The result will be a polygonal-schema drawing with straight-line edges having polynomial area, but there is no guarantee that it is a straight-frame drawing, since the dPP algorithm makes no collinear guarantees for vertices adjacent to the vertices on the bounding triangle.

5   Conclusion and Future Work

In this paper, we present several algorithms for polygonal-schema drawings of higher-genus graphs. Our method for toroidal graphs achieves drawings that simultaneously use straight-line edges in a straight frame and polynomial area. Previous algorithms for the torus were restricted to special cases or did not always produce polygonal-schema renderings [4,7,14]. Our methods for non-toroidal graphs can achieve any two of these three criteria. It is an open problem whether it is possible to achieve all three of these aesthetic criteria for non-toroidal graphs. To our knowledge, previous algorithms for general graphs in genus-g surfaces were restricted to those with "nice" polygonal schemas [15].

References

1. Chen, J., Kanchi, S.P., Kanevsky, A.: A note on approximating graph genus. Information Processing Letters 61(6), 317–322 (1997)
2. de Fraysseix, H., Pach, J., Pollack, R.: How to draw a planar graph on a grid. Combinatorica 10(1), 41–51 (1990)
3. Duncan, C.A., Goodrich, M.T., Kobourov, S.G.: Planar drawings of higher-genus graphs. Technical report (August 2009), http://arxiv.org/abs/0908.1608
4. Eppstein, D.: The topology of bendless three-dimensional orthogonal graph drawing. In: Tollis, I.G., Patrignani, M. (eds.) GD 2008. LNCS, vol. 5417, pp. 78–89. Springer, Heidelberg (2009)
5. Erickson, J., Har-Peled, S.: Optimally cutting a surface into a disk. In: Proc. of the 18th ACM Symp. on Computational Geometry (SCG), pp. 244–253 (2002)
6. Garey, M.R., Johnson, D.S.: Crossing number is NP-complete. SIAM J. Algebraic Discrete Methods 4(3), 312–316 (1983)
7. Kocay, W., Neilson, D., Szypowski, R.: Drawing graphs on the torus. Ars Combinatoria 59, 259–277 (2001)
8. Lazarus, F., Pocchiola, M., Vegter, G., Verroust, A.: Computing a canonical polygonal schema of an orientable triangulated surface. In: Proc. of the 17th ACM Symp. on Computational Geometry (SCG), pp. 80–89 (2001)
9. Miura, K., Nakano, S.-I., Nishizeki, T.: Grid drawings of 4-connected plane graphs. Discrete and Computational Geometry 26(1), 73–87 (2001)
10. Mohar, B., Thomassen, C.: Graphs on Surfaces. Johns Hopkins U. Press, Baltimore (2001)
11. Thomassen, C.: The graph genus problem is NP-complete. J. Algorithms 10(4), 568–576 (1989)
12. Tutte, W.T.: Convex representations of graphs. Proceedings London Mathematical Society 10(38), 304–320 (1960)
13. Tutte, W.T.: How to draw a graph. Proc. Lon. Math. Soc. 13(52), 743–768 (1963)
14. Vodopivec, A.: On embeddings of snarks in the torus. Discrete Mathematics 308(10), 1847–1849 (2008)
15. Zitnik, A.: Drawing graphs on surfaces. SIAM J. Disc. Math. 7(4), 593–597 (1994)

Splitting Clusters to Get C-Planarity[*]

Patrizio Angelini, Fabrizio Frati, and Maurizio Patrignani

Università Roma Tre
{angelini,frati,patrigna}@dia.uniroma3.it

Abstract. In this paper we introduce a generalization of the c-planarity testing problem for clustered graphs. Namely, given a clustered graph, the goal of the SPLIT-C-PLANARITY problem is to split as few clusters as possible in order to make the graph c-planar. Determining whether zero splits are enough coincides with testing c-planarity. We show that SPLIT-C-PLANARITY is NP-complete for c-connected clustered triangulations and for non-c-connected clustered paths and cycles. On the other hand, we present a polynomial-time algorithm for flat c-connected clustered graphs whose underlying graph is a biconnected series-parallel graph, both in the fixed and in the variable embedding setting, when the splits are assumed to maintain the c-connectivity of the clusters.

1 Introduction

Let $C(G,T)$ be a clustered graph and suppose that a c-planar drawing of C is impossible (or very difficult) to find. A natural question is whether C admits a drawing where each cluster is represented by a small set of connected regions instead of a single connected region of the plane. We formalize this concept by introducing the *split* operation, that replaces a cluster μ of T with two clusters μ_1 and μ_2 with the same parent as μ, and distributes the children of μ between μ_1 and μ_2. We search for the minimum number of splits turning C into a c-planar clustered graph. Formally, the corresponding decision problem is as follows:

Problem: SPLIT-C-PLANARITY
Instance: A clustered graph $C = (G,T)$ and an integer $k \geq 0$.
Question: Can $C(G,T)$ be turned into a c-planar clustered graph $C(G,T')$
by performing at most k split operations?

SPLIT-C-PLANARITY is motivated not only by the practical need of drawing non-c-planar clustered graphs, but also by its implications on the c-planarity theory. In fact, the long-standing problem of testing c-planarity [8] is a particular case of SPLIT-C-PLANARITY, where zero splits are allowed. Therefore, SPLIT-C-PLANARITY extends the c-planarity testing problem to a more general setting, where we are able to show the NP-hardness even for flat clustered graphs whose underlying graphs are paths or cycles.

Hence, following a strategy that is analogous to the one used in the literature for the c-planarity testing problem, we focus on peculiar classes of clustered graphs.

---

[*] This work is partially supported by the Italian Ministry of Research, Grant number RBIP06BZW8, FIRB project "Advanced tracking system in intermodal freight transportation".

D. Eppstein and E.R. Gansner (Eds.): GD 2009, LNCS 5849, pp. 57–68, 2010.
© Springer-Verlag Berlin Heidelberg 2010

Table 1. Time complexity of SPLIT-C-PLANARITY for non-c-connected graphs

Graph Family	Fixed Embedding Setting	Variable Embedding Setting
Paths, cycles, trees, & outerplanar graphs	NP-hard (Th. 5)	NP-hard (Th. 5)
Series-parallel graphs	NP-hard (Th. 5)	NP-hard (Th. 4)
General graphs	NP-hard (Th. 1)	NP-hard (Th. 1)

Table 2. Time complexity of SPLIT-C-PLANARITY for c-connected graphs

Graph Family	Fixed Embedding Setting	Variable Embedding Setting
Paths, cycles, & trees	$\Theta(1)$ (trivial)	$\Theta(1)$ (trivial)
Outerplanar graphs	?	$\Theta(1)$ (trivial)
Series-parallel graphs*	Polynomial (Th. 2)	Polynomial (Th. 3)
Series-parallel graphs	?	?
General graphs	NP-hard (Th. 1)	NP-hard (Th. 1)

*Flat hierarchy, biconnected underlying graph, c-connectivity preserved.

Restrictions on the c-planarity testing problem that have been considered in the literature include: (i) assuming that each cluster induces a small number of connected components [8,4,11,10,1,2,12] (in particular, the case in which the graph is *c-connected*, that is, each cluster induces one connected component, has been deeply investigated); (ii) considering only *flat* hierarchies, where all clusters different from the root of T are children of the root [3,6]; (iii) focusing on particular families of underlying graphs [3,13]; and (iv) fixing the embedding of the underlying graph [6,12].

We show that SPLIT-C-PLANARITY is NP-hard even for flat c-connected clustered graphs whose underlying graph is triconnected (hence even for flat c-connected embedded clustered graphs). On the other hand, we show that SPLIT-C-PLANARITY is polynomial-time solvable for flat c-connected clustered graphs whose underlying graph is a biconnected series-parallel graph (both if the underlying graph has fixed or variable embedding) if the splits are assumed to preserve the c-connectivity of the graph.

Tables 1 and 2 summarize the time complexity of SPLIT-C-PLANARITY. Observe that, being acyclic, every c-connected clustered tree is trivially c-planar. Also, in an outerplanar embedding of any outerplanar graph no cycle contains a vertex in its interior. Therefore, every c-connected clustered outerplanar graph is c-planar.

The rest of the paper is organized as follows. In Sect. 2 we introduce some preliminaries; in Sect. 3 we prove the NP-hardness of SPLIT-C-PLANARITY for flat c-connected clustered triangulations; in Sect. 4 we show a polynomial-time algorithm for SPLIT-C-PLANARITY on flat c-connected biconnected clustered series-parallel graphs; in Sect. 5 we show the NP-hardness of SPLIT-C-PLANARITY for flat non-c-connected clustered paths and cycles; in Sect. 6 we conclude and present some open problems.

2   Background

We refer to [5] for basic definitions about graphs and embeddings, and to [8,4,11,3,10, 1,6,2,13,12] for basic definitions about clustered graphs and c-planar drawings.

A *series-parallel graph* is inductively defined as follows. An edge (u, v) is a series-parallel graph with *poles* u and v. Denote by u_i and v_i the poles of a series-parallel graph G_i. A *series composition* of a sequence $G_1, \ldots, G_k$ of series-parallel graphs, with $k \geq 2$, is a series-parallel graph with poles $u = u_1$ and $v = v_k$ such that v_i and u_{i+1} have been identified, for each $i = 1, \ldots, k - 1$. A *parallel composition* of a set $G_1, \ldots, G_k$ of series-parallel graphs, with $k \geq 2$, is a series-parallel graph with poles $u = u_1 = \cdots = u_k$ and $v = v_1 = \cdots = v_k$. The SPQ-tree of a series-parallel graph G is the tree representing the series and parallel compositions of G. Let G be a series-parallel graph with poles u and v and with a fixed plane embedding $\mathcal{E}_o$. The *leftmost path* (resp. *rightmost path*) of G is the path $(w_1 = u, w_2, \ldots, w_k = v)$ (resp. $(z_1 = u, z_2, \ldots, z_h = v)$) such that: (i) w_2 follows w_1 (resp. z_2 precedes z_1) in the counter-clockwise order of the vertices incident to the outer face of $\mathcal{E}_o$; (ii) edge (w_i, w_{i+1}) follows (w_{i-1}, w_i) (resp. (z_i, z_{i+1}) precedes (z_{i-1}, z_i)) in the counter-clockwise order of the edges incident to w_i (resp. incident to z_i). The leftmost and rightmost paths of G are also called *extreme paths* of G.

3   General C-Connected Clustered Graphs

We show the NP-hardness of SPLIT-C-PLANARITY for flat c-connected clustered graphs whose underlying graph is triconnected. This is done by means of a reduction from HAMILTONIAN-CIRCUIT [9], which takes as an input a triconnected, planar, and cubic graph $G(V, E)$ and asks whether a simple cycle exists in G traversing each node $v \in V$ exactly once. Given an instance of HAMILTONIAN-CIRCUIT, consider a planar drawing of it and the dual graph G' of G (see Fig. 1(a)). Observe that, since G is cubic, G' is a triangulation. Construct an instance $\langle C(G'', T), k \rangle$ of SPLIT-C-PLANARITY as follows. Graph G'' is obtained by adding to G' a node v_i in each face f_i and by connecting v_i to the three vertices incident to f_i (see Fig. 1(b)). Tree T has height two and has a cluster μ_i for each added vertex v_i and a cluster μ_0 containing all the vertices of G'. The value of k is set to one. We make use of the following result appeared in [7]:

Lemma 1. *(Feng [7]) Let $C(G, T)$ be a clustered graph where G is a triangulation. Then C is c-planar only if C is c-connected.*

Lemma 2. *Instance G of* HAMILTONIAN-CIRCUIT *admits a solution if and only if the corresponding instance $\langle C(G'', T), 1 \rangle$ of* SPLIT-C-PLANARITY *does.*

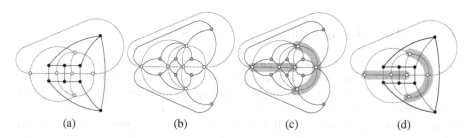

    (a)   (b)   (c)   (d)

Fig. 1. (a) A planar graph G (black vertices) and its dual graph G' (white vertices). (b) Graph G'' (the vertices added to G' are drawn gray). (c) A split of cluster μ_0 turning G'' into a c-planar clustered graph. (d) The corresponding Hamiltonian circuit on G (thick edges).

Proof: Suppose $\langle C(G'', T), 1\rangle$ admits a solution. Since G'' is triconnected, in any planar drawing of G'' each vertex v_i inserted into an internal face f_i of G' is inside a cycle of vertices belonging to cluster μ_0. Hence, $C(G'', T)$ is not c-planar, and at least one split of cluster μ_0 has to be performed in order to turn $C(G'', T)$ into a c-planar graph. Suppose that a split of cluster μ_0 into two clusters μ_a and μ_b exists such that the obtained clustered graph $C(G'', T')$ is c-planar (see Fig. 1(c)). The split is a bipartition of the vertices of G' into V_a and V_b. By Lemma 1, the two graphs induced by V_a and V_b are connected. Hence, the edges between V_a and V_b form a cutset. A cutset in G' corresponds to a cycle C in G [14, pg. 16]. Since $C(G'', T')$ is c-planar, each vertex v_i inserted into a face f_i of G' is adjacent both to a vertex in μ_a and to a vertex in μ_b. This is equivalent to saying that C traverses each vertex of G exactly once (see Fig. 1(d)).

Suppose that a Hamiltonian circuit C exists in G. Split μ_0 so that nodes internal to C belong to μ_a and nodes external to C belong to μ_b. The obtained graph $C(G'', T')$ is c-planar. In fact, C determines a cutset in G', hence μ_a and μ_b induce connected graphs. Further, since C is Hamiltonian, the graphs induced by μ_a and μ_b are acyclic. □

Since $\langle C(G'', T), 1\rangle$ can be constructed in polynomial time and since the problem is easily seen to be in NP, the following holds.

Theorem 1. SPLIT-C-PLANARITY *is NP-complete when the input graph is a flat c-connected clustered graph and $k = 1$.*

4   Series-Parallel C-Connected Clustered Graphs

In this section, we show that SPLIT-C-PLANARITY is polynomial-time solvable if: (i) the input graph is a flat c-connected clustered graph whose underlying graph is a biconnected series-parallel graph, and (ii) the splits have to maintain the c-connectivity of the input graph. Observe that the reduction shown in Sect. 3 proves that SPLIT-C-PLANARITY is NP-complete if: (i) the input graph is a flat c-connected clustered graph, and (ii) the splits have to maintain the c-connectivity of the input graph (namely, such a condition is always met when splitting clusters of a clustered triangulation). Throughout this section, we assume that every set of splits turning a c-connected clustered graph into a c-planar clustered graph maintains the c-connectivity of the graph.

4.1   Series-Parallel Graphs with Fixed Embedding

We show a polynomial-time algorithm that, given a flat c-connected clustered graph $C(G, T)$, where G is a biconnected series-parallel graph with fixed planar embedding $\mathcal{E}$, computes the minimum number of splits turning C into a c-planar clustered graph. The algorithm performs a bottom-up visit of the SPQ-tree T of G, rooted at any P-node corresponding to a parallel composition of two series-parallel graphs B_1 and B_2, where B_1 is an edge e and B_2 is the rest of the graph. Topologically, such a choice corresponds to assuming that e is on the outer face of a plane embedding $\mathcal{E}_o$ corresponding to the planar embedding $\mathcal{E}$. However, there are $O(n)$ ways of making such a choice, hence the test is repeated a linear number of times. Throughout this subsection, we assume

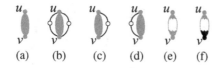

Fig. 2. Representation of a node t of $\mathcal{T}$ satisfying (a) Condition A, (b) Condition B, (c) Condition C, (d) Condition D, (e) Condition E, and (f) Condition F

that $\mathcal{E}$ is fixed and that e is on the outer face of $\mathcal{E}_o$. We denote by $\mu(u)$ the only cluster different from the root of T containing vertex u.

For each node t of $\mathcal{T}$ corresponding to a series-parallel graph B with poles u and v, the algorithm computes six labels $\alpha(t), \beta(t), \gamma(t), \delta(t), \epsilon(t)$, and $\phi(t)$. Such labels represent the minimum number of splits on C turning $(B, T'[B])$ (that is, the clustered graph whose cluster hierarchy is the tree obtained from T by performing the splits on C and by restricting to the clusters containing vertices of B) into a c-planar clustered graph satisfying, respectively, the following conditions (see Fig. 2):

- *Condition A:* all the vertices of B belong to $\mu(u) = \mu(v)$;
- *Condition B:* $\mu(u) = \mu(v)$, there exists a path between u and v whose vertices all belong to $\mu(u)$, and $p_r(B)$ and $p_l(B)$ contain vertices not belonging to $\mu(u)$;
- *Condition C:* $\mu(u) = \mu(v)$, there exists a path between u and v whose vertices all belong to $\mu(u)$, $p_r(B)$ contains vertices not belonging to $\mu(u)$, and all the vertices of $p_l(B)$ belong to $\mu(u)$;
- *Condition D:* $\mu(u) = \mu(v)$, there exists a path between u and v whose vertices all belong to $\mu(u)$, all the vertices of $p_r(B)$ belong to $\mu(u)$, and $p_l(B)$ contains vertices not belonging to $\mu(u)$;
- *Condition E:* $\mu(u) = \mu(v)$ and there exists no path between u and v whose vertices all belong to $\mu(u)$; and,
- *Condition F:* $\mu(u) \neq \mu(v)$.

When $(B, T'[B])$ satisfies a certain condition, we equivalently say that t satisfies the same condition. In general, it could be not possible to make t satisfy a certain condition with any set of splits. For example, if $\mu(u) \neq \mu(v)$, no set of splits makes u and v belong to the same cluster, hence labels $\alpha(t), \beta(t), \gamma(t), \delta(t)$, and $\epsilon(t)$ have no meaning for t. In such cases, we set the corresponding labels to ∞.

We observe the following lemmata:

Lemma 3. *Consider any set of splits turning $C(G, T)$ into a c-planar clustered graph $C'(G, T')$. Then, $(B, T'[B])$ satisfies exactly one of Conditions A, B, C, D, E, and F.*

Lemma 4. *If $(B, T'[B])$ satisfies Condition A, B, C, D, or F, then $(B, T'[B])$ is a c-connected clustered graph. Also, if $(B, T'[B])$ satisfies Condition E, then each cluster in $T'[B]$ induces one connected component in B, except for $\mu(u)$, which induces two connected components, one containing u, and the other containing v.*

We now sketch how to compute $\alpha(t), \beta(t), \gamma(t), \delta(t), \epsilon(t)$, and $\phi(t)$. In the base case, t is an edge (u, v) and the six labels can be easily computed. Namely, if u and v belong

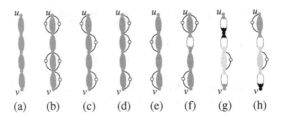

(a)   (b)   (c)   (d)   (e)   (f)   (g)   (h)

Fig. 3. Constraints on the children of t, if t is an S-node satisfying Condition x. If $x = A$, then all the t_i satisfy Condition A (a). If $x = B$, then either there exists t_i satisfying Condition B and all other t_j satisfy Condition A, B, C, or D (b), or there exist t_i satisfying Condition C, t_j satisfying Condition D, and all other t_l satisfy Condition A, C, or D (c). If $x = C$, then there exists t_i satisfying Condition C and all other t_j satisfy Condition A or C (d). If $x = D$, then there exists t_i satisfying Condition D and all other t_j satisfy Condition A or D (e). If $x = E$, then u and v belong to the same cluster in the input clustered graph and either there exists t_i satisfying Condition E and all other t_j satisfy Condition A, B, C, or D (f), or there exist t_i and t_j satisfying Condition F and all other t_l satisfy Condition A, B, C, D, or F (g). If $x = F$, then there exists t_i satisfying Condition F and all other t_j satisfy Condition A, B, C, D, or F (h).

to distinct clusters, then $\alpha(t) = \beta(t) = \gamma(t) = \delta(t) = \phi(t) = \infty$, and $\epsilon(t) = 0$. If u and v belong to the same cluster, then $\alpha(t) = 0$, $\beta(t) = \gamma(t) = \delta(t) = \epsilon(t) = \infty$, and $\phi(t) = 1$.

Consider a node t of $\mathcal{T}$ corresponding to a series-parallel graph B. Let $t_1, \ldots, t_k$ be the children of t, corresponding to series-parallel graphs $B_1, \ldots, B_k$. Let u_i and v_i be the poles of B_i. Inductively suppose that the labels of $t_1, \ldots, t_k$ have been computed.

The main idea is that if a set S of splits makes $(B, T'[B])$ satisfy Condition A, B, C, D, E, or F, then several constraints on the conditions that are satisfied by the children of t can be deduced, also based on whether t is an S-node or a P-node.

As an example, if t is a P-node satisfying Condition C, then either t_i exists satisfying Condition C or not. If such a t_i exists, then all the t_j with $j < i$ satisfy Condition A and all the t_j with $j > i$ satisfy Condition E; namely, if any t_j with $j < i$ satisfies Condition B or C, then $(B, T'[B])$ is not c-planar, as it contains a cycle, whose vertices belong to the same cluster, enclosing a vertex not belonging to such a cluster; if any t_j with $j < i$ satisfies Condition D or E, then either $(B, T'[B])$ is not c-planar or t does not satisfy Condition C; no t_j satisfies Condition F because $\mu(u) = \mu(v)$; finally, if any t_j with $j > i$ satisfies Condition A, B, C, or D, then $(B, T'[B])$ is not c-planar. If no t_i satisfies Condition C, then a sequence of consecutive t_j, including t_1, satisfy Condition A, and all other t_j, including t_k, satisfy Condition E. See Figs. 3 and 4.

As a result of the above argumentations, a set of k-tuples is associated to Condition x, where $x \in \{A, B, C, D, E, F\}$, for each node t of $\mathcal{T}$ with k children. Each tuple is such that if t_i satisfies the condition indicated at the i-th item of the tuple, for each i, then t satisfies Condition x. Then, the minimum number of splits turning $(B, T'[B])$ into a c-planar clustered graph satisfying Condition x is the minimum among the values associated with the tuples, where the value associated with each tuple is obtained by summing up the labels corresponding to the conditions of the tuple, paying attention to those splits counted more than once in different nodes t_i. We get the following:

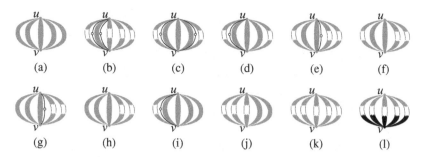

Fig. 4. Constraints on the children of t, if t is a P-node satisfying Condition x. If $x = A$, then all the t_i satisfy Condition A (a). If $x = B$, then either there exists t_i satisfying Condition B and all other t_j satisfy Condition E (b), or there exists t_i satisfying Condition D, t_j satisfying Condition C, with $j > i$, and all the t_l satisfy Condition E, if $l < i$ and $l > j$, or Condition A, if $i < l < j$ (c), or there exists t_i satisfying Condition D, all the t_l satisfy Condition E, if $l < i$ and if $l > y$, for some $i \le y < k$, and all the t_l satisfy Condition A, if $i < l \le y$ (d), or there exists t_i satisfying Condition C, all the t_l satisfy Condition E, if $l > i$ and if $l < x$, for some $1 < x \le i$, and all the t_l satisfy Condition A, if $x \le l < i$ (e), or all the t_l satisfy Condition E, if $l < x$ and if $l > y$, for some $1 < x \le y < k$, and all the t_l satisfy Condition A, if $x \le l \le y$ (f). If $x = C$, then either there exists t_i satisfying Condition C, all the t_j with $j > i$ satisfy Condition E, and all the t_j with $j < i$ satisfy Condition A (g), or all the t_j satisfy Condition A, with $1 \le j \le y$ for some $1 \le y < k$, and all the t_j satisfy Condition E, with $j > y$ (h). If $x = D$, then either there exists t_i satisfying Condition D, all the t_j with $j < i$ satisfy Condition E, and all the t_j with $j > i$ satisfy Condition A (i), or all the t_j satisfy Condition A, with $x \le j \le k$ for some $1 < x \le k$, and all the t_j satisfy Condition E, with $j < x$ (j). If $x = E$, then all the t_i satisfy Condition E (k). If $x = F$, then all the t_i satisfy Condition F (l).

Theorem 2. *Let $C(G, T)$ be a flat c-connected clustered graph whose underlying graph G is an n-vertex biconnected series-parallel graph with a fixed planar embedding $\mathcal{E}$. The minimum number of splits turning C into a c-planar clustered graph while maintaining the c-connectivity of every cluster can be computed in $O(n^4)$ time.*

4.2 Series-Parallel Graphs with Variable Embedding

We sketch how to extend the result of Sect. 4.1 to the variable embedding scenario.

As in the fixed embedding case, we perform a bottom-up visit of the rooted SPQ-tree $\mathcal{T}$ of G, while computing some labels for each node t of $\mathcal{T}$. However, in this case, we have to determine some embeddings of the series-parallel graph B corresponding to t. For each node t of $\mathcal{T}$, we compute *five* labels $\alpha(t), \beta(t), \gamma\delta(t), \epsilon(t),$ and $\phi(t)$. Labels $\alpha(t), \epsilon(t),$ and $\phi(t)$ have the same meaning as in the fixed embedding case. Label $\beta(t)$ represents the minimum number of splits turning $(B, T[B])$ into a c-planar clustered graph $(B, T'[B])$ containing a path between u and v whose vertices all belong to $\mu(u)$, and having vertices not belonging to $\mu(u)$ on both extreme paths of a computed planar embedding. Label $\gamma\delta(t)$ represents the minimum number of splits turning $(B, T[B])$ into a c-planar clustered graph $(B, T'[B])$ containing a path between u and v whose vertices all belong to $\mu(u)$ and having vertices not belonging to $\mu(u)$ on exactly one extreme path of a computed planar embedding. Observe that labels $\beta(t)$ and $\gamma\delta(t)$ replace

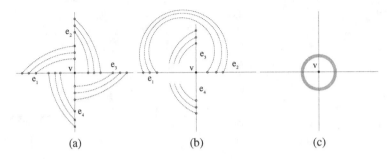

Fig. 5. (a) A pinwheel gadget of size three. Dashed lines join vertices of the same cluster. (b) An illustration for the proof of Lemma 5. (c) A symbolic representation of the pinwheel gadget.

labels $\beta(t)$, $\gamma(t)$, and $\delta(t)$ of the fixed embedding scenario, as in the variable embedding setting it is not known *a priori* which are the rightmost and the leftmost path of t.

Theorem 3. *Let $C(G,T)$ be a flat c-connected clustered graph whose underlying graph G is an n-vertex biconnected series-parallel graph. The minimum number of splits turning C into a c-planar clustered graph while maintaining the c-connectivity of every cluster at each split can be computed in $O(n^4)$ time.*

5   Non-C-Connected Clustered Graphs

We open this section by showing that, given a flat non-c-connected clustered graph $C(G,T)$, where G is a biconnected series-parallel graph, it is NP-hard to find the minimum number of splits turning C into a c-planar clustered graph. Namely, we perform a reduction from NAE3SAT [9], which takes in input a collection of clauses, each consisting of three literals, and asks whether a truth assignment to the variables exists such that each clause has at least one true literal and at least one false literal.

Given a clustered graph $C(G,T)$ and a vertex v of G with four incident edges e_1, e_2, e_3, and e_4, we introduce a gadget that forces such edges to appear in this circular order around v in any c-planar drawing of any clustered graph obtained from C with less than σ splits. We construct around v a *pinwheel gadget* of size σ by inserting, in each edge e_i, 2σ vertices $v_{i,j}$, with $j = 1, \ldots, 2\sigma$. For each pair (e_i, e_{i+1}) we add σ child-clusters to the root of T and assign $v_{i,j}$ and $v_{i+1,j+\sigma}$ to the same cluster, for $j = 1, \ldots, \sigma$. Figure 5 provides an example for $\sigma = 3$.

Lemma 5. *Let $C(G,T)$ be a clustered graph containing a pinwheel gadget of size σ around a vertex v. Any c-planar drawing of a clustered graph obtained from C with less than σ splits preserves the circular order of the edges around v, up to a reversal.*

Proof: Suppose, for a contradiction, that there exists a c-planar drawing of a clustered graph obtained from C with less than σ splits such that the order of the edges around v is e_1, e_3, e_2, and e_4, the other cases being analogous. Consider the σ clusters involving vertices of both e_1 and e_2. Since less than σ splits are allowed, at least one of such

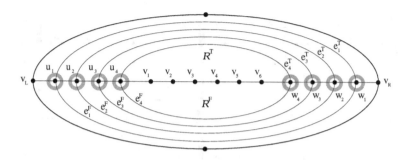

Fig. 6. An illustration for the construction of instance $\langle C(G_\varphi, T_\varphi), k_\varphi \rangle$ of SPLIT-C-PLANARITY corresponding to an instance φ of NAE3SAT with four variables and six clauses

clusters is not split. Hence, the region of the plane delimited by the border of such a cluster, by e_1, and by e_2 either encloses vertices $v_{3,j}$, with $j = 1, \ldots, 2\sigma$, and does not enclose vertices $v_{4,j}$, with $j = 1, \ldots, 2\sigma$, or vice versa. It follows that all the σ clusters involving vertices of both e_3 and e_4 are split, contradicting the hypothesis.   □

Given an instance φ of NAE3SAT with n variables and c clauses we construct the corresponding instance $\langle C_\varphi(G_\varphi, T_\varphi), 2c \rangle$ of SPLIT-C-PLANARITY as follows. Graph G_φ contains a cycle C with two notable vertices v_L and v_R (see Fig. 6), and a path $(v_L, u_1, u_2, \ldots, u_n, v_1, v_2, \ldots, v_c, w_n, w_{n-1}, \ldots, w_1, v_R)$. Observe that, in any planar embedding of G_φ, such a path, together with C, determines two regions (both inside or both outside C) that we arbitrarily denote by $\mathcal{R}^T$ and $\mathcal{R}^F$. G_φ also contains two edges $e_i^T = (u_i, w_i)$ and $e_i^F = (u_i, w_i)$, for each $i = 1, \ldots, n$. Denote $u_0 = v_L$, $u_{n+1} = v_1$, $w_0 = v_R$, and $w_{n+1} = v_c$. For $i = 1, \ldots, n$, two pinwheel gadgets of size $2c + 1$ are inserted around u_i and w_i so that the circular order of the edges around u_i and w_i is (u_{i-1}, u_i), e_i^T, (u_i, u_{i+1}), e_i^F, and (w_{i-1}, w_i), e_i^F, (w_i, w_{i+1}), e_i^T, respectively. Figure 6 shows an example with $n = 4$ and $c = 6$. The insertion of the pinwheel gadgets turns e_i^T and e_i^F into two paths, that we denote by p_i^T and p_i^F, respectively. Observe that, by Lemma 5, in any c-planar embedding of a clustered graph obtained from C_φ with less than $2c + 1$ splits, if p_i^T lies into $\mathcal{R}^T$ ($\mathcal{R}^F$), then p_i^F lies into $\mathcal{R}^F$ ($\mathcal{R}^T$).

For each clause j, we introduce two clusters $\nu_{j,1}$ and $\nu_{j,2}$. Also, we define two *literal gadgets* $l^{\notin}(j)$ and $l^{\in}(j)$ as follows. Gadget $l^{\notin}(j)$ is a sequence of three vertices v_a, v_b, and v_c belonging to clusters $\nu_{j,1}$, $\nu_{j,2}$, and $\nu_{j,1}$, respectively (see variable x_1 of Fig. 7(a)). Gadget $l^{\in}(j)$ contains a sequence of four vertices v_d, v_e, v_f, and v_g, plus two additional vertices v_h and v_i attached to both v_d and v_e. While v_d and v_e are assigned to the root of T_φ, v_f belongs to $\nu_{j,2}$ and v_g, v_h, and v_i belong to $\nu_{j,1}$. Finally, two pinwheel gadgets of size $2c + 1$ are inserted around v_d and v_e so that, in any c-planar drawing of a clustered graph obtained from C_φ with less than $2c + 1$ splits, v_h and v_i are on opposite sides with respect to edge (v_d, v_e) (see variable x_2 of Fig. 7(a)).

For each variable x_i, with $i = 1, \ldots, n$, and for each clause c_j, with $j = 1, \ldots, c$, we insert into p_i^T (p_i^F) gadget $l^{\in}(j)$ if x_i ($\overline{x}_i$, respectively) is a literal of c_j and gadget $l^{\notin}(j)$ otherwise, in such a way that the gadgets for clauses $c_1, c_2, \ldots, c_c$ appear in this order from u_i to w_i in p_i^T and p_i^F.

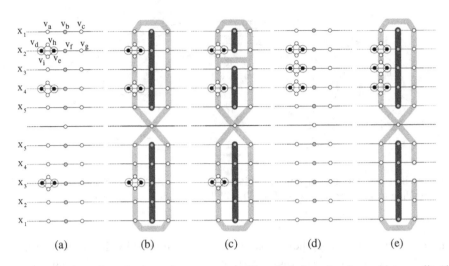

Fig. 7. (a) Configuration of a clause $(x_2 \vee \overline{x}_3 \vee x_4)$. (b) and (c) show drawings with two split. (d) A configuration of a clause with all true literals. Any c-planar drawing of it needs three split (e).

We assign to the root of T_φ vertices v_L, v_R, u_i and w_i, with $i = 1, \ldots, n$. Vertex v_j, for $j = 1, \ldots, c$, is assigned to $\nu_{j,1}$.

Lemma 6. *Instance φ of* NAE3SAT, *with n variables and c clauses, admits a solution if and only if instance $\langle C_\varphi(G_\varphi, T_\varphi), 2c \rangle$ of* SPLIT-C-PLANARITY *admits a solution.*

Proof sketch: Suppose φ admits a solution and consider an assignment of truth values to the variables that satisfies φ. If variable x_i is TRUE (FALSE), then draw p_i^T into $\mathcal{R}^T$ ($\mathcal{R}^F$) and p_i^F into $\mathcal{R}^F$ ($\mathcal{R}^T$). Observe that, for each clause c_j no three $l^\in(j)$ are in the same region. Figure 7(b) shows a portion of a c-planar drawing of a clustered graph obtained from C_φ with two splits per clause. Hence, $\langle C_\varphi, 2c \rangle$ admits a solution.

Suppose $\langle C_\varphi, 2c \rangle$ admits a solution. In order to obtain a c-planar clustered graph from C_φ, at least two splits are needed for $\nu_{j,1}$ and $\nu_{j,2}$ as a whole (see Figs 7(b) and 7(c)); further, if the literal gadgets $l^\in(j)$ of clause c_j are all three in the same region, then at least three splits are needed for $\nu_{j,1}$ and $\nu_{j,2}$ as a whole (see Fig. 7(e)). It follows that, since only $2c$ splits turn C_φ into a c-planar graph, there exists a truth assignment such that each clause has a TRUE and a FALSE literal.   □

Since $\langle C_\varphi, 2c \rangle$ can be constructed in polynomial time and since the problem is easily seen to be in NP, the following holds.

Theorem 4. SPLIT-C-PLANARITY *is NP-complete when the input is a flat non-c-connected clustered series-parallel graph.*

By modifying the above reduction, it is possible to show that SPLIT-C-PLANARITY is NP-complete even for a non c-connected clustered tree, path, or cycle.

Namely, we introduce the *open pinwheel gadget* of size σ, whose vertices have degree at most two, to replace a pinwheel gadget of size σ in the reduction from

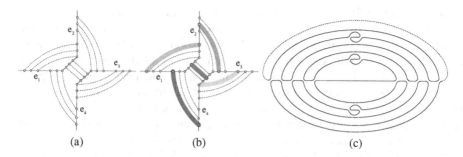

Fig. 8. (a) An open pinwheel gadget. (b) A picture for the proof of Lemma 7. (c) A picture for the proof of Theorem 5 (the literal gadgets of only one cluster are shown).

NAE3SAT. Such a gadget is obtained from a pinwheel gadget around vertex v by removing v and joining edges e_1 and e_2 and edges e_3 and e_4 (or edges e_1 and e_4 and edges e_2 and e_3) with a path of σ vertices belonging to clusters $\mu_1, \ldots, \mu_\sigma$ (see Fig. 8).

Lemma 7. *Let $C^*(G^*, T^*)$ be the clustered graph obtained from $C_\varphi(G_\varphi, T_\varphi)$ by replacing each pinwheel gadget of size σ with an open pinwheel gadget of the same size. Then, C^* can be turned into a c-planar clustered graph with less than σ splits if and only if C_φ can be turned into a c-planar clustered graph with less than σ splits.*

Theorem 5. *Problem* SPLIT-C-PLANARITY *is NP-complete when the input graph is a non-c-connected cycle or path.*

Proof sketch: Construct instance $\langle C_\varphi(G_\varphi, T_\varphi), 2c \rangle$ corresponding to the instance φ of NAE3SAT with c clauses as in the reduction used in Theorem 4. Add an edge connecting v_L with v_R and add two pinwheel gadgets around v_L and v_R. Observe that all the vertices have degree two or four and that all the vertices of degree four have a pinwheel gadget around them. Replace each pinwheel gadget with an open pinwheel gadget of the same size in such a way that the underlying graph G^* of the obtained clustered graph $C^*(G^*, T^*)$ is a cycle, as shown in Fig. 8(c). By Lemma 7, any c-planar drawing of a clustered graph obtained from C^* with less than $2c$ splits corresponds to a c-planar drawing of a clustered graph obtained from C_φ with less than $2c$ splits, and vice versa. Lemma 6 ensures that instance φ admits a solution if and only if instance $\langle C^*(G^*, T^*), 2c \rangle$ of SPLIT-C-PLANARITY admits a solution.

To prove that the problem is NP-complete also for paths it suffices to turn G^* into a path by "opening" edge (v_L, v_R) (dashed edge of Fig. 8(c)).   □

Theorem 5 implies that SPLIT-C-PLANARITY is NP-complete when the input is a non-c-connected tree both in the fixed and in the variable embedding setting.

6   Conclusions

In this paper we introduced the SPLIT-C-PLANARITY problem, which takes as an input a clustered graph $C(G, T)$ and an integer $k \geq 0$ and asks whether C can be turned into a c-planar clustered graph $C'(G, T')$ by performing at most k cluster splits.

We proved that SPLIT-C-PLANARITY is NP-hard, even for non-c-connected clustered paths and cycles, and for c-connected clustered triangulations. Further, SPLIT-C-PLANARITY is not fixed-parameter tractable with respect to k, as it is NP-hard even with $k = 1$. However, it could still be the case that SPLIT-C-PLANARITY is fixed-parameter tractable with respect to k, when the underlying graph of the input clustered graph is a path, a cycle, or a graph in a similarly simple graph family. Namely, the reduction we presented in the c-connected case uses a constant k, but deals with triconnected graphs, while the reduction we presented for the non-c-connected case deals with paths and cycles, but uses a k which is function of the size of the problem.

We proved that for flat clustered graphs whose underlying graph is a biconnected series-parallel graph SPLIT-C-PLANARITY is polynomial-time solvable, if the splits are assumed to maintain the c-connectivity of the clusters. We believe the following extensions of such a result to be interesting: (i) non-flat clustered graphs; (ii) simply-connected series-parallel graphs; (iii) splits not maintaining the c-connectivity.

References

1. Cornelsen, S., Wagner, D.: Completely connected clustered graphs. J. Discrete Algorithms 4(2), 313–323 (2006)
2. Cortese, P.F., Di Battista, G., Frati, F., Patrignani, M., Pizzonia, M.: C-planarity of c-connected clustered graphs. J. Graph Alg. Appl. 12(2), 225–262 (2008)
3. Cortese, P.F., Di Battista, G., Patrignani, M., Pizzonia, M.: Clustering cycles into cycles of clusters. J. Graph Alg. Appl. 9(3), 391–413 (2005)
4. Dahlhaus, E.: A linear time algorithm to recognize clustered graphs and its parallelization. In: Lucchesi, C.L., Moura, A.V. (eds.) LATIN 1998. LNCS, vol. 1380, pp. 239–248. Springer, Heidelberg (1998)
5. Di Battista, G., Eades, P., Tamassia, R., Tollis, I.G.: Graph Drawing: Algorithms for the Visualization of Graphs. Prentice-Hall, Englewood Cliffs (1999)
6. Di Battista, G., Frati, F.: Efficient c-planarity testing for embedded flat clustered graphs with small faces. In: Hong, S.-H., Nishizeki, T., Quan, W. (eds.) GD 2007. LNCS, vol. 4875, pp. 291–302. Springer, Heidelberg (2008)
7. Feng, Q.: Algorithms for Drawing Clustered Graphs. Ph.D. thesis, The University of Newcastle, Australia (1997)
8. Feng, Q., Cohen, R.F., Eades, P.: Planarity for clustered graphs. In: Spirakis, P.G. (ed.) ESA 1995. LNCS, vol. 979, pp. 213–226. Springer, Heidelberg (1995)
9. Garey, M.R., Johnson, D.S.: Computers and Intractability: A Guide to the Theory of NP-Completeness. W.H. Freeman, New York (1979)
10. Goodrich, M.T., Lueker, G.S., Sun, J.Z.: C-planarity of extrovert clustered graphs. In: Healy, P., Nikolov, N.S. (eds.) GD 2005. LNCS, vol. 3843, pp. 211–222. Springer, Heidelberg (2006)
11. Gutwenger, C., Jünger, M., Leipert, S., Mutzel, P., Percan, M., Weiskircher, R.: Advances in c-planarity testing of clustered graphs. In: Goodrich, M.T., Kobourov, S.G. (eds.) GD 2002. LNCS, vol. 2528, pp. 220–235. Springer, Heidelberg (2002)
12. Jelinek, V., Jelinkova, E., Kratochvil, J., Lidicky, B.: Clustered planarity: Embedded clustered graphs with two-component clusters. In: Tollis, I.G., Patrignani, M. (eds.) GD 2008. LNCS, vol. 5417, pp. 121–132. Springer, Heidelberg (2009)
13. Jelinkova, E., Kara, J., Kratochvil, J., Pergel, M., Suchy, O., Vyskocil, T.: Clustered planarity: Small clusters in eulerian graphs. In: Hong, S.-H., Nishizeki, T., Quan, W. (eds.) GD 2007. LNCS, vol. 4875, pp. 303–314. Springer, Heidelberg (2008)
14. Nishizeki, T., Chiba, N.: Planar Graphs: Theory and Algorithms. Ann. Discrete Math., vol. 32. North-Holland, Amsterdam (1988)

On the Characterization of Level Planar Trees by Minimal Patterns*

Alejandro Estrella-Balderrama, J. Joseph Fowler, and Stephen G. Kobourov

Department of Computer Science, University of Arizona
{alexeb,fowler}@email.arizona.edu,
kobourov@cs.arizona.edu

Abstract. We consider characterizations of level planar trees. Healy *et al.* [8] characterized the set of trees that are level planar in terms of two minimal level non-planar (MLNP) patterns. Fowler and Kobourov [7] later proved that the set of patterns was incomplete and added two additional patterns. In this paper, we show that the characterization is still incomplete by providing new MLNP patterns not included in the previous characterizations. Moreover, we introduce an iterative method to create an arbitrary number of MLNP patterns, thus proving that the set of minimal patterns that characterizes level planar trees is infinite.

1 Introduction

An important application of automatic graph drawing can be found in the layout of graphs that represent hierarchical relationships. When drawing graphs in the xy-plane, this translates to a restricted form of planarity where the y-coordinate of a vertex is given and the drawing algorithm only has the freedom to choose the x-coordinate. This restricted form of planarity is called *level planarity*, and each given y-coordinate corresponds to a *level*.

Jünger, Leipert, and Mutzel [13] provide a linear-time recognition algorithm for level planar graphs. This algorithm is based on the level planarity test given by Heath and Pemmaraju [9,10]. The algorithm by Heath and Pemmaraju is based on the more restricted PQ-tree level planarity testing algorithm of *hierarchies* (level graphs of directed acyclic graphs in which all edges are between adjacent levels and all the source vertices are on the uppermost level) given by Di Battista and Nardelli in [3]. In the paper, the authors also characterize such hierarchies in terms of level non-planar (LNP) patterns. Jünger and Leipert [12] provide a linear-time level planar embedding algorithm that outputs a set of linear orderings in the x-direction for the vertices on each level. However, to obtain a straight-line planar drawing one needs to subsequently run an $O(|V|)$ algorithm given by Eades *et al.* [4] who demonstrate that every level planar embedding has a straight-line drawing, though it may require exponential area.

Healy *et al.* [8] use LNP patterns to provide a set of *minimal level non-planar* (MLNP) subgraph patterns that characterize level planar graphs. This is the counterpart for level graphs to the characterization of planar graphs by Kuratowski [14] in terms of forbidden subdivisions of K_5 and $K_{3,3}$. Two new MLNP tree patterns were added in [7]

---

* This work was supported in part by NSF grant CCF-0545743.

D. Eppstein and E.R. Gansner (Eds.): GD 2009, LNCS 5849, pp. 69–80, 2010.
© Springer-Verlag Berlin Heidelberg 2010

by Fowler and Kobourov to the previous set of patterns given by Healy *et al.* In this paper, we show that the characterization remains incomplete by providing new MLNP patterns not included in the previous characterizations. Moreover, we introduce an iterative method to create an arbitrary number of MLNP patterns, thus proving that the set of minimal patterns that characterizes level planar trees is infinite.

The study of MLNP patterns is motivated in part by the problem of visualizing hierarchical structures. Sugiyama *et al.* [15] described what has become the standard framework for drawing directed acyclic graphs. In this framework vertices are assigned to levels and then on each level vertices are ordered, with the overall goal of minimizing the number of edge crossings. There exists good heuristics and some exact methods based upon integer linear programs (ILPs) to find good orders within levels [11]. However, typically the assignment of vertices to levels is done with the help of greedy local optimizations [2]. Understanding the underlying obstructions to level planarity (such as MLNP patterns) could lead to better solutions to the level assignment step.

Level planarity is also related to simultaneous embedding [1]. In general, a set of restrictions on the layout of one graph may help in the layout of a second graph on the same vertex set. Specifically, when embedding a path with a planar graph, if the graph can be drawn on horizontal levels, then the path can be drawn in a y-monotone fashion without crossings. Estrella-Balderrama *et al.* [6] characterized the set of unlabeled level planar (ULP) trees on n vertices that are level planar over all possible labelings of the vertices in terms of two forbidden trees: T_8 and T_9. A level non-planar labeling of T_9 was used to obtain MLNP patterns P_3 and P_4 in [7]; see Fig. 3.

2 Preliminaries

A *k-level graph* $G(V, E, \phi)$ on n vertices is a directed graph $G(V, E)$ with a level assignment $\phi : V \rightarrow \{1, \ldots, k\}$ such that the induced partial order is strict: $\phi(u) < \phi(v)$ for every $(u, v) \in E$. A k-level graph is a k-partite graph in which ϕ partitions V into k independent sets $V_1, V_2, \ldots, V_k$, which form the k *levels* of G. A *level-j* vertex v is on the jth level V_j if $\phi(v) = j$ (*i.e.* $v \in V_j$). In a level graph, an edge (u, v) is *short* if $\phi(v) = \phi(u) + 1$ while edges spanning multiple levels are *long*. A *proper level graph* has only short edges. Any level graph can be made proper by subdividing long edges into short edges. In this paper, a level graph is proper unless stated otherwise.

A level graph G has a *level drawing* if there exists a drawing such that every vertex in V_j is placed along the horizontal line $\ell_j = \{(x, j) \mid x \in \mathbb{R}\}$ and the edges are drawn as strictly y-monotone polylines. The order that the vertices of V_j are placed along each ℓ_j in a level drawing of a proper graph induces a family of linear orders along the x-direction, which form a *linear embedding* of G. A level drawing, and consequently its level embedding, is *level planar* if it can be drawn without edge crossings. A level graph G is level planar if it admits a level planar embedding. The definition of level drawings allowing only straight-line segments for edges is equivalent, given that Eades *et al.* [4] have shown that every level planar graph has a straight-line planar drawing.

A *path* is a non-repeating ordered sequence of vertices $(v_1, v_2, \ldots, v_n)$ for $n \geq 1$. A *star* with n vertices is a tree with one vertex of degree $n - 1$, called the *root*, and $n - 1$ vertices of degree 1. A *spider* is an arbitrarily subdivided star, where *subdividing* an

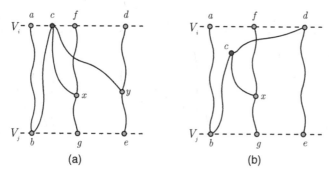

Fig. 1. Original MLNP patterns P_1 in (a) and P_2 in (b) proposed by Healy *et al*

edge (u, v) replaces the edge with a new vertex w and new edges (u, w) and (w, v). In a *degree-k spider*, the root has degree k.

A *chain-link*, denoted $u \rightsquigarrow v$, is a path from vertex u to vertex v with $u \neq v$ such that each internal vertex w that lies along the path has degree 2. Let $\phi(u \rightsquigarrow v)$ denote the set of levels of the internal vertices where $i \leq \phi(u \rightsquigarrow v) \leq j$ is a short-hand for saying that $i \leq \phi(w) \leq j$ for each internal vertex w of the chain-link $u \rightsquigarrow v$. Unless stated otherwise we assume that $\phi(u) \leq \phi(u \rightsquigarrow v) \leq \phi(v)$ for each chain-link $u \rightsquigarrow v$. A *linking chain*, or simply a *chain*, is a sequence of one or more chain-links. Notice that a vertex in the intersection of two chains is not considered a crossing between the chains. In all figures, a curve connecting two vertices, represents a chain.

In a level non-planar graph, a *pattern* is an obstructing subgraph with a level assignment that forces a crossing. Since here we define particular patterns in terms of chains, they represent a set of graphs with similar properties in terms of leveling. A level non-planar pattern is *minimal* if the removal of an arbitrary edge makes the pattern level planar. All the patterns described here (with the exception of a few that are symmetrical) have a corresponding horizontally flipped version.

3 Previous Work

3.1 Characterization of Level Planar Trees by Healy *et al.*

Healy *et al.* [8] defined MLNP patterns as follows: Let i and j be the minimum and maximum level, respectively, of any vertex in the pattern. Let x be a vertex of degree 3 with three subtrees with the following properties: (i) each subtree has at least one vertex on both extreme levels; (ii) a subtree is either a chain or it has two subtrees that are chains; (iii) all leaves are located on extreme levels (and each leaf is the only vertex in its subtree on the extreme level); and (iv) the subtrees that are chains and have non-leaf vertices on one extreme level, also have at least one leaf vertex on the opposite extreme level.

Then they distinguish two patterns; P_1 with x on an extreme level and P_2 with x on a non-extreme level (Healy *et al.* denote them T1 and T2). Figure 1 shows P_1 and P_2. Notice that these patterns are defined in terms of subtrees. This implies, for example,

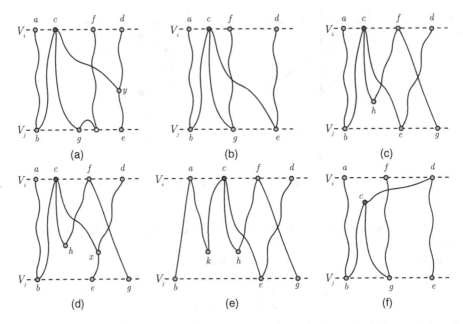

Fig. 2. (a-e) Five variations of pattern P_1 in addition to the one in Fig. 1(a); (f) One variation of pattern P_2 in addition to the one in Fig. 1(b)

that a subtree with a vertex of degree 3 may be replaced by a path. Fowler and Kobourov, on the other hand, defined the patterns in terms of paths. Hence, to properly compare the set of patterns we need to consider the different cases, or *variations*, of the subtrees in P_1 and P_2. Hence, P_1 leads to variations $P_1^A, \ldots, P_1^F$ and P_2 leads to variations P_2^A and P_2^B; see Fig. 2. Notice that when a chain reaches an extreme level with a degree-2 vertex, more degree-2 vertices of the chain can also be on the extreme level. This is illustrated in Fig. 2(a) for the chain $c \rightsquigarrow g \rightsquigarrow f$ with a second degree-2 vertex. Healy *et al.* [8] showed that both of these patterns are minimal level non-planar.

3.2   Characterization of Level Planar Trees by Fowler and Kobourov

The two trees T_8 and T_9 were shown to be the only obstructions in the context of un-labeled level planarity for trees [6]. However, as the tree T_9 does not match any of the MLNP patterns by Healy *et al.* [8], a new pattern P_3 was proposed [7]; see Fig. 3(b). Note that matching T_9 with either of the earlier patterns P_1 or P_2 would be impossible as both P_1 and P_2 are based on a central vertex of degree 3 (vertex x in Fig. 1), while T_9 and its matching pattern P_3 have a central vertex of degree 4 (vertex x in Fig. 3(b)).

Yet another pattern P_4 can be obtained from P_3 by "splitting" vertex x of degree 4 such that $i < l \le \phi(x) \le m < j$ into two vertices of degree 3 connected by a path. In Fig. 3(b) vertex x is replaced by a chain $x \rightsquigarrow y$ such that $l \le \phi(x \rightsquigarrow y) \le m$ in Fig. 3(c). Patterns P_3 and P_4 were added to the previous set of two patterns (eight variations) to obtain a new characterization consisting of four patterns (ten variations). A sketch of a

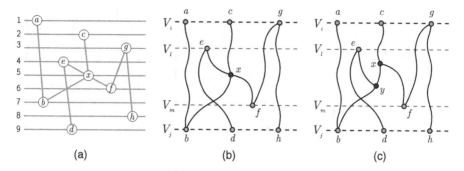

Fig. 3. Fowler and Kobourov generalized the forbidden ULP tree T_9 in (a) to produce the MLNP patterns P_3 in (b) and P_4 in (c)

proof for the claim that this new characterization is complete was made in [7], but in the next section we show that the characterization remains incomplete.

4 New Minimal Level Non-planar Patterns

In this section, we show that the characterization of level planar trees by minimal patterns is still incomplete. In Sect. 4.1, we show that there are variations of P_3 and P_4 that were not considered. Then in Sect. 4.2, we describe a new pattern previously not considered as it has a vertex of degree 5, whereas, all of the previously known MLNP patterns have maximum degree 4.

4.1 Variations of Patterns P_3 and P_4

The previous characterization introduces the new patterns P_3 and P_4. Just as with the variations of P_1 and P_2, different variations of P_3 and P_4 can be produced by replacing some chains with degree-3 spiders. We describe these variations next.

- *Pattern P_3^A.* This is the original pattern P_3; see Fig. 3(b).
- *Pattern P_3^B.* This pattern is similar to P_3^A but replaces the chain $x \rightsquigarrow f \rightsquigarrow g$ such that $l \leq \phi(x) \leq m$, $\phi(f) = m$, $\phi(g) = i$, and $i \leq \phi(f \rightsquigarrow g) \leq m$, with a degree-3 spider rooted at f' and leaves f, g, and x such that $l \leq \phi(f') \leq m$, $\phi(x) = \phi(f) = m$, $\phi(g) = i$, and $l \leq \phi(f \rightsquigarrow f') \leq m$; see Fig. 4(a).
- *Pattern P_3^C.* This pattern is similar to P_3^A but replaces the chain $x \rightsquigarrow e \rightsquigarrow d$, such that $l \leq \phi(x) \leq m$, $\phi(e) = l$, $\phi(d) = j$, and $l \leq \phi(x \rightsquigarrow e) \leq m$ with a degree-3 spider rooted at e' and leaves e, d, and x such that $l \leq \phi(e') \leq m$, $\phi(x) = \phi(e) = l$, $\phi(d) = j$, and $l \leq \phi(e \rightsquigarrow e') \leq m$; see Fig. 4(b).
- *Pattern P_3^D.* This pattern makes both replacements made by patterns P_3^B and P_3^C on P_3^A such that $\phi(e) = \phi(f') = l$, $\phi(e') = \phi(f) = m$, $l \leq \phi(x) \leq m$, $i \leq \phi(x \rightsquigarrow g) \leq m$, and $l \leq \phi(x \rightsquigarrow h) \leq j$; see Fig. 4(c).
- *Pattern P_4^A.* This is the original pattern P_4; see Fig. 3(c).
- *Patterns P_4^B, P_4^C, and P_4^D.* These patterns make analogous replacements on P_4^A as those made by P_3^B, P_3^C, and P_3^D on P_3^A; see Fig. 4(d-f).

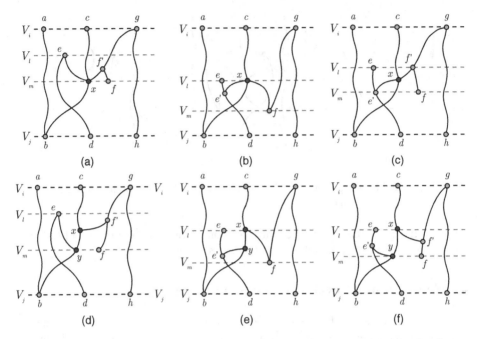

Fig. 4. (a-c) Variations of pattern P_3 (P_3^B, P_3^C, and P_3^D); (d-f) Variations of pattern P_4 (P_4^B, P_4^C, and P_4^D)

The importance of the new variations of P_3 and P_4 is that they break the fundamental assumption made in the early attempts at characterizations, namely that in any minimal level non-planar pattern, leaves must lie on extreme levels i or j. All of the new patterns have leaves on non-extreme levels. We omit the proofs for the variations of P_3 and P_4 as in the next section we formally show that a new pattern, P_5 with non-extreme leaves is MLNP. Moreover, in Sect. 5, we show that the set of MLNP patterns for trees is not just missing a few more patterns but is actually infinite.

4.2   New Pattern P_5

In this section, we describe a new pattern P_5 and its variations. The main characteristic of this pattern is the presence of a vertex x with degree 5.

- *Pattern P_5^A.* This pattern is a degree-5 spider, rooted at x, with two levels l and m between the extreme levels i and j such that $i < l < \phi(x) \le m < j$. There is a chain $x \rightsquigarrow c$ such that $\phi(c) = i$, a chain $x \rightsquigarrow d$ such that $\phi(d) = j$; a chain $x \rightsquigarrow p \rightsquigarrow q$ such that $\phi(p) = m$ and $\phi(q) = l$; a chain $x \rightsquigarrow e \rightsquigarrow f \rightsquigarrow g \rightsquigarrow h$ such that $\phi(e) = l$, $\phi(f) = m$, $\phi(g) = i$, and $\phi(h) = j$; and a chain $x \rightsquigarrow k \rightsquigarrow b \rightsquigarrow a$ such that $l < \phi(k) < \phi(x)$, $\phi(b) = j$, $\phi(a) = i$ and $l < \phi(x \rightsquigarrow k \rightsquigarrow b) \le j$; see Fig. 5(a).
- *Pattern P_5^B.* Similar to P_5^A but replaces the chain $x \rightsquigarrow e \rightsquigarrow f \rightsquigarrow g \rightsquigarrow h$ with a degree-3 spider rooted at f' such that $l < \phi(f') < m$, with x, g, and f such that $l < \phi(x) \le m$, $\phi(e) = l$, $\phi(g) = i$, $\phi(f) = m$, and there is a chain $x \rightsquigarrow e \rightsquigarrow f'$

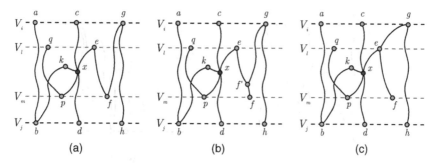

Fig. 5. Patterns P_5^A, P_5^B, and P_5^C

such that $\phi(e) = l$ where $l \leq \phi(x \rightsquigarrow e \rightsquigarrow f') \leq m$, $l \leq \phi(f \rightsquigarrow f') \leq m$, and $i \leq \phi(f' \rightsquigarrow g) \leq m$; see Fig. 5(b).

- Pattern P_5^C. Similar to P_5^A but replaces the chain $x \rightsquigarrow e \rightsquigarrow f \rightsquigarrow g \rightsquigarrow h$ with a degree-3 spider rooted at e such that $\phi(e) = l$ with leaves g, x, and f; see Fig. 5(c).

In the following two lemmas we show that this new pattern is MLNP.

Lemma 1. *Pattern P_5 is level non-planar.*

Proof. We show that P_5^A is level non-planar (the cases for P_5^B and P_5^C are similar). First notice that to avoid a crossing with chain $c \rightsquigarrow x \rightsquigarrow d$, all the vertices of the chain $x \rightsquigarrow e \rightsquigarrow f \rightsquigarrow g \rightsquigarrow h$ must lie to the right of the chain $c \rightsquigarrow x \rightsquigarrow d$ while all the vertices of the chain $x \rightsquigarrow k \rightsquigarrow b \rightsquigarrow a$ must lie to the left, or vice versa; see Fig. 5(a). Assume w.l.o.g. that $x \rightsquigarrow k \rightsquigarrow b \rightsquigarrow a$ lies to the left and $x \rightsquigarrow e \rightsquigarrow f \rightsquigarrow g \rightsquigarrow h$ lies to right of chain $c \rightsquigarrow x \rightsquigarrow d$ (as in Fig. 5(a)). Now observe that in order to avoid a crossing of chain $x \rightsquigarrow p \rightsquigarrow q$ with chains $a \rightsquigarrow b$, $c \rightsquigarrow x \rightsquigarrow d$ or $g \rightsquigarrow h$, the chain $x \rightsquigarrow p \rightsquigarrow q$ must lie between chains $a \rightsquigarrow b$ and $c \rightsquigarrow x \rightsquigarrow d$ or lie between chains $c \rightsquigarrow x \rightsquigarrow d$ and $g \rightsquigarrow h$. However, in the first case a crossing will occur with chain $x \rightsquigarrow k \rightsquigarrow b$ (since $\phi(k) < \phi(x)$ and $\phi(x) \leq \phi(x \rightsquigarrow p) \leq m$) and in the later case a crossing will occur with chain $x \rightsquigarrow e \rightsquigarrow f \rightsquigarrow g$. $\qquad\square$

Lemma 2. *The removal of any edge in pattern P_5 makes it level planar.*

Proof. We consider the different cases of edge removal from the chains in P_5^A (P_5^B and P_5^C are similar):

case 1) If any edge is removed from chain $x \rightsquigarrow p \rightsquigarrow q$, then the crossing with chain $x \rightsquigarrow e \rightsquigarrow f$ is avoided when $x \rightsquigarrow p \rightsquigarrow q$ is to the right of $c \rightsquigarrow x \rightsquigarrow d$ as in Fig. 6(a).

case 2) If any edge is removed from chains $x \rightsquigarrow k \rightsquigarrow b \rightsquigarrow a$ or $x \rightsquigarrow e \rightsquigarrow f \rightsquigarrow g \rightsquigarrow h$, then all the vertices in the chain (except x) can be to the left or to the right of chain

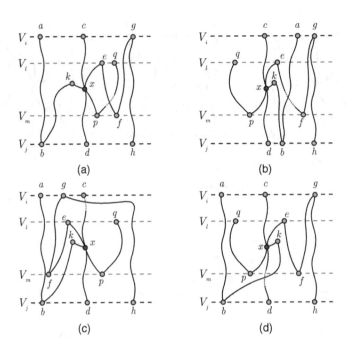

Fig. 6. Different cases of removing an edge (dotted) from pattern P_5^A

$c \leadsto x \leadsto d$ where chain $x \leadsto p \leadsto q$ can be on the other side avoiding the crossing as in Fig. 6(b).

case 3) If any edge is removed from chain $c \leadsto x$, then chains $x \leadsto k \leadsto b \leadsto a$ and $x \leadsto e \leadsto f \leadsto g$ can be on the same side with respect to $c \leadsto x \leadsto d$. Thus avoiding the crossing with chain $x \leadsto p \leadsto q$; see Fig. 6(c).

case 4) If any edge is removed from chain $x \leadsto d$, then chain $x \leadsto k \leadsto b$ can lie to the right of chain $x \leadsto p \leadsto q$ as in Fig. 6(d). $\square$

We now use Lemmas 1 and 2 to show that P_5 is indeed MLNP.

Theorem 1. *P_5 is a minimal level non-planar pattern for trees.*

Proof. By Lemma 1, P_5 is level non-planar and by Lemma 2, P_5 is minimal. Minimality also implies that P_5 does not contain any MLNP pattern as a subgraph. Moreover, pattern P_5 does not match any of the previous patterns given that vertex x has degree 5, while all of the previously known patterns have maximum degree 4. $\square$

In this section, we have shown that a new pattern P_5 is MLNP. However, P_5 is not the only pattern missing from earlier characterizations. New patterns $P_6, \ldots, P_{11}$ are shown along with their variations in [5]. The proofs of level non-planarity and minimality of these patterns are similar to the one given for P_5. Thus, instead of proving that each of these patterns is MLNP, we describe a constructive method for generating an infinite number of distinct MLNP patterns in the next section.

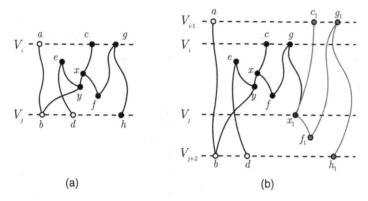

Fig. 7. Construction of a new pattern $(P_4^A)_1$ in (b) from pattern P_4^A in (a)

5 Infinite Minimal Level Non-planar Patterns

Our approach for creating new MLNP patterns is to take a known pattern as a base and then repeat a subgraph of the pattern making modifications on the leveling such that the new pattern does not strictly contain the previous one. Here we use P_4^A but the method applies to other patterns as well.

The first step is to make a copy of the path $p_0 = c \leadsto x \leadsto f \leadsto g \leadsto h$ such that $\phi(c) = \phi(g) = i < \phi(x) < \phi(f) < \phi(h) = j$ as in Fig. 7(a) in order to get a new path $p_1 = c_1 \leadsto x_1 \leadsto f_1 \leadsto g_1 \leadsto h_1$ such that $\phi(c_1) = \phi(g_1) = i - 1$, $\phi(x_1) = j$, $\phi(f_1) = j + 1$, and $\phi(h_1) = j + 2$ as in Fig. 7(b). The second step is to add p_1 to P_4^A by merging vertices x_1 and h creating a new vertex of degree 3 that takes the place of h. This new level assignment creates two new extreme levels $i - 1$ and $j + 2$. We complete the construction of the new pattern by moving vertices a, b, and d to the new extreme levels, specifically, we set $\phi(a) = i - 1$ and $\phi(b) = \phi(d) = j + 2$.

We now generalize the previous construction to an arbitrary number of iterations. We denote the pattern created at iteration t from pattern P as $(P)_t$. Thus, the original P_4^A is $(P_4^A)_0$ and the pattern created in Fig. 7(b) is $(P_4^A)_1$. The vertices in the pattern are labeled in the same way, for example $x_0 = x$. Therefore, in order to create a new pattern $(P_4^A)_{t+1}$ from pattern $(P_4^A)_t$, we first copy the path $p_t = c_t \leadsto x_t \leadsto f_t \leadsto g_t \leadsto h_t$ to get a new path $p_{t+1} = c_{t+1} \leadsto x_{t+1} \leadsto f_{t+1} \leadsto g_{t+1} \leadsto h_{t+1}$ such that $\phi(c_{t+1}) = \phi(g_{t+1}) = i - t - 1$, $\phi(x_{t+1}) = j + 2t$, $\phi(f_{t+1}) = j + 2t + 1$, and $\phi(h_{t+1}) = j + 2t + 2$. We then merge x_{t+1} with h_t to obtain the new x_{t+1}. Finally, we set the levels as $\phi(a) = i - t - 1$, and $\phi(b) = \phi(d) = j + 2t + 2$; see Fig. 8.

In the next lemma we show that a pattern, $(P_4^A)_t$, generated with the previous method is level non-planar.

Lemma 3. *Pattern* $(P_4^A)_t$ *for* $t \geq 0$, *is level non-planar.*

Proof. We use induction on t, the number of iterations in the generation method. The base case is $t = 0$; this is the original pattern P_4 which is proven to be level non-planar in the characterization by Fowler and Kobourov [7]. We now assume that $(P_4^A)_t$ is level

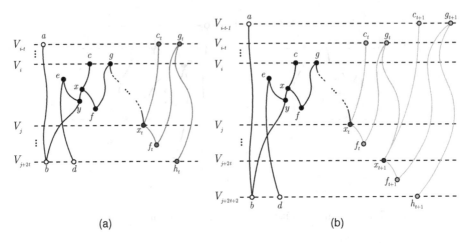

Fig. 8. Construction of a new pattern $(P_4^A)_{t+1}$ in (b) from pattern $(P_4^A)_t$ in (a)

non-planar in order to prove that $(P_4^A)_{t+1}$ is level non-planar. That is, we show that the modifications made to $(P_4^A)_t$ to obtain $(P_4^A)_{t+1}$ do not affect the level non-planarity of the new pattern.

Clearly, the addition of vertices and edges cannot affect the level non-planarity of a tree, hence the addition of the path p_{t+1} does not make the pattern level planar. Moreover, since the chains $a \rightsquigarrow b$ and $e \rightsquigarrow d$ in $(P_4^A)_t$ are contained in the chains $a \rightsquigarrow b$ and $e \rightsquigarrow d$ of $(P_4^A)_{t+1}$, the change on the levels of a and d are simply addition of vertices and edges that cannot affect the level non-planarity of the pattern. Finally, we consider the change of level of vertex b. Notice that the crossing between the chain $a \rightsquigarrow b \rightsquigarrow y \rightsquigarrow x \rightsquigarrow f \rightsquigarrow g$ and the chain $y \rightsquigarrow e \rightsquigarrow d$ in $(P_4^A)_t$ cannot be avoided in $(P_4^A)_{t+1}$ with the change of level of b. This is because as d is moved to the level of b the chain $f \rightsquigarrow g \rightsquigarrow \cdots \rightsquigarrow x_{t+1} \rightsquigarrow f_{t+1} \rightsquigarrow h_{t+1}$ plays an analogous role in the pattern $(P_4^A)_{t+1}$ that the chain $f \rightsquigarrow \cdots \rightsquigarrow h_t$ plays in the pattern $(P_4^A)_t$. That is, the addition of the chain $c_{t+1} \rightsquigarrow h_{t+1}$ to the pattern $(P_4^A)_{t+1}$ prevents the switch of side of the chain $a \rightsquigarrow b$ in order to avoid the crossing with $y \rightsquigarrow e \rightsquigarrow d$ as this will produce a crossing with the chain $c_{t+1} \rightsquigarrow x_{t+1}$ (as in Fig. 9(d)). Therefore, by induction the pattern $(P_4^A)_t$ is level non-planar for all non-negative integers $t \geq 0$.   □

We next show the minimality of the patterns generated with the method above.

Lemma 4. *The removal of any edge in* $(P_4^A)_t$ *for any* $t \geq 0$, *makes it level planar.*

Proof. We consider the cases of edge removal in $(P_4^A)_t$.

case 1) If any edge is removed from the chain $a \rightsquigarrow b \rightsquigarrow y \rightsquigarrow e \rightsquigarrow d$, then the self-intersection is avoided as in Fig. 9(a).

case 2) If any edge is removed from the chain $x \rightsquigarrow y$, then the chain $e \rightsquigarrow d$ can use the gap to avoid the crossing as in Fig. 9(b).

case 3) If any edge is removed from the chain $x_\alpha \rightsquigarrow f_\alpha$ or $g_\alpha \rightsquigarrow h_\alpha$ for any $\alpha = 0, \ldots, t$, then chain $a \rightsquigarrow b \rightsquigarrow y$ can use the gap to be drawn between the chains $c_\alpha \rightsquigarrow x_\alpha$ and $f_\alpha \rightsquigarrow g_\alpha$ as in Fig. 9(c) or between g_α and h_α.

case 4) If any edge is removed from the chains $c_\alpha \rightsquigarrow x_\alpha$ or $f_\alpha \rightsquigarrow g_\alpha$ for any $\alpha = 0, \ldots, t$, then the chain $a \rightsquigarrow b$ can interchange sides with the chain $h_\alpha \rightsquigarrow g_\alpha$ if $\alpha = t$ as in Fig. 9(d). When $\alpha < t$, all the chains $c_\beta \rightsquigarrow x_\beta \rightsquigarrow f_\beta \rightsquigarrow g_\beta \rightsquigarrow h_\beta$ for $\beta = \alpha + 1, \ldots, t$ are moved along with the chain $h_\alpha \rightsquigarrow g_\alpha$. □

With the last two lemmas we now show that a pattern generated with the iterative method described in this section is MLNP.

Theorem 2. *Pattern $(P_4^A)_t$ for $t \geq 0$, is a minimal level non-planar pattern for trees.*

Proof. By Lemma 3, $(P_4^A)_t$ is level non-planar and by Lemma 4, $(P_4^A)_t$ is minimal. Minimality implies that $(P_4^A)_t$ does not contain any MLNP pattern as a subgraph. In particular, $(P_4^A)_t$ does not contain the previous pattern $(P_4^A)_{t-1}$. To see this in Fig. 8(b), observe that in the subgraph between levels i and j, the chain $a \rightsquigarrow b \rightsquigarrow y$ is separated by level j into two disjoint chains. Moreover, pattern $(P_4^A)_t$ does not match any of the previous patterns $(P_4^A)_\alpha$ for $\alpha = 0, \ldots, t - 1$ since $(P_4^A)_t$ contains an additional vertex of degree 3, x_t. □

Theorem 2 implies that we can generate an arbitrary number of different MLNP patterns. This gives our main result.

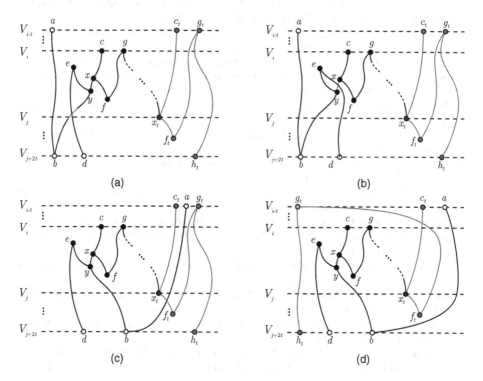

Fig. 9. Different cases of removing an edge (dotted) from pattern $(P_4^A)_t$

Theorem 3. *The set of minimal level non-planar patterns for trees is infinite.*

6   Conclusions and Future Work

In this paper, we showed why two earlier attempts to characterize the set of level non-planar trees in terms of minimal level non-planar patterns failed. In both cases, there was an implicit assumption that the set of different MLNP patterns is small and finite. However, it turns out that there are infinitely many different MLNP patterns, and an altogether different approach might be needed for a complete characterization.

References

1. Braß, P., Cenek, E., Duncan, C.A., Efrat, A., Erten, C., Ismailescu, D., Kobourov, S.G., Lubiw, A., Mitchell, J.S.B.: On simultaneous planar graph embeddings. Computational Geometry: Theory and Applications 36(2), 117–130 (2007)
2. Chimani, M., Gutwenger, C., Mutzel, P., Wong, H.-M.: Layer-free upward crossing minimization. In: McGeoch, C.C. (ed.) WEA 2008. LNCS, vol. 5038, pp. 55–68. Springer, Heidelberg (2008)
3. Di Battista, G., Nardelli, E.: Hierarchies and planarity theory. IEEE Transactions on Systems, Man, and Cybernetics 18(6), 1035–1046 (1989)
4. Eades, P., Feng, Q., Lin, X., Nagamochi, H.: Straight-line drawing algorithms for hierarchical graphs and clustered graphs. Algorithmica 44(1), 1–32 (2006)
5. Estrella-Balderrama, A.: Simultaneous Embedding and Level Planarity. PhD thesis, Department of Computer Science, University of Arizona (2009)
6. Estrella-Balderrama, A., Fowler, J.J., Kobourov, S.G.: Characterization of unlabeled level planar trees. Computational Geometry: Theory and Applications 42(7), 704–721 (2009)
7. Fowler, J.J., Kobourov, S.G.: Minimum level nonplanar patterns for trees. In: Hong, S.-H., Nishizeki, T., Quan, W. (eds.) GD 2007. LNCS, vol. 4875, pp. 69–75. Springer, Heidelberg (2008)
8. Healy, P., Kuusik, A., Leipert, S.: A characterization of level planar graphs. Discrete Mathematics 280(1-3), 51–63 (2004)
9. Heath, L.S., Pemmaraju, S.V.: Stack and queue layouts of directed acyclic graphs. II. SIAM Journal on Computing 28(5), 1588–1626 (1999)
10. Heath, L.S., Pemmaraju, S.V., Trenk, A.N.: Stack and queue layouts of directed acyclic graphs. I. SIAM Journal on Computing 28(4), 1510–1539 (1999)
11. Jünger, M., Lee, E.K., Mutzel, P., Odenthal, T.: A polyhedral approach to the multi-layer crossing minimization problem. In: DiBattista, G. (ed.) GD 1997. LNCS, vol. 1353, pp. 13–24. Springer, Heidelberg (1997)
12. Jünger, M., Leipert, S.: Level planar embedding in linear time. Journal of Graph Algorithms and Applications 6(1), 67–113 (2002)
13. Jünger, M., Leipert, S., Mutzel, P.: Level planarity testing in linear time. In: Whitesides, S.H. (ed.) GD 1998. LNCS, vol. 1547, pp. 224–237. Springer, Heidelberg (1999)
14. Kuratowski, C.: Sur les problèmes des courbes gauches en Topologie. Fundamenta Mathematicae 15, 271–283 (1930)
15. Sugiyama, K., Tagawa, S., Toda, M.: Methods for visual understanding of hierarchical system structures. IEEE Transactions on Systems, Man, and Cybernetics 11(2), 109–125 (1981)

Characterization of
Unlabeled Radial Level Planar Graphs
(Extended Abstract)

J. Joseph Fowler

Department of Computer Science, University of Arizona
jfowler@cs.arizona.edu

Abstract. Suppose that an n-vertex graph has a distinct labeling with the integers $\{1, \ldots, n\}$. Such a graph is *radial level planar* if it admits a crossings-free drawing under two constraints. First, each vertex lies on a concentric circle such that the radius of the circle equals the label of the vertex. Second, each edge is drawn with a radially monotone curve. We characterize the set of *unlabeled radial level planar* (URLP) graphs that are radial level planar in terms of 7 and 15 forbidden subdivisions depending on whether the graph is disconnected or connected, respectively. We also provide linear-time drawing algorithms for any URLP graph.

1   Introduction

Visualizing social networks with respect to centrality, the relative importance that actors hold within a relational structure, yields a graphical representation that conveys domain-specific hierarchical information that aids in policy network analysis [2]. When possible, planar layouts are preferred where actors are placed at distances from the origin based upon their level of importance. Thus, vertices are constrained to lie along concentric circles, called *rings*, with radii proportional to the centrality of the respective actors. Radially monotone curves can be used for edges, which denote relationships between actors. Each curve lies between the rings of the endpoints of the edge. Such layouts are radial level planar if the drawing is crossings free. If straight-line edges are used, edges may cross rings interior to their endpoints, which can decrease readability; see Fig. 1.

Spiral edges have the added advantage that the radii used for the rings can vary uniformly. This is not the case with straight-line edges. For instance, an outerplanar graph admits a straight-line planar drawing in which the vertices lie on a unit circle. If each vertex is perturbed to lie on a circle of different radii, planarity may be lost unless extremely minute perturbations are used [4].

Some social networks have a structure that are conducive to the dynamic visualization of actors whose centrality changes over time. The most dynamic networks are radial level planar regardless of the centrality of the actors. In representing a social network, we take the corresponding graph and label each vertex with an integer denoting its degree of importance. This then becomes the distance that each actor is placed from the origin.

D. Eppstein and E.R. Gansner (Eds.): GD 2009, LNCS 5849, pp. 81–93, 2010.

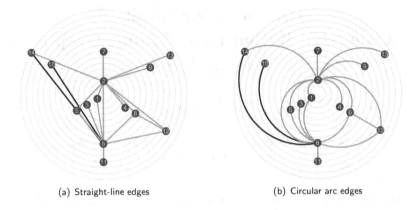

(a) Straight-line edges   (b) Circular arc edges

Fig. 1. Drawing a radial graph with different edge types

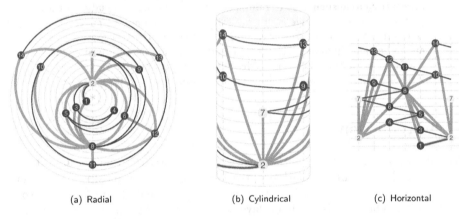

(a) Radial   (b) Cylindrical   (c) Horizontal

Fig. 2. Three equivalent representations of radial level graphs

This leads us to consider the set of *unlabeled radial level planar* (URLP) graphs that have a radial level planar layout for any *distinct labeling*. The term "unlabeled" implies that the underlying structure of the graph determines whether radial level planarity is always possible, independent of any labeling scheme. Alternatively, URLP graphs can be defined as the set of graphs that can always be drawn simultaneously without crossings with a radially monotone path. Labels are given by the order that the vertices occur along the path; see Fig. 2(a). This is related to the problem of *simultaneous embedding* in which multiple planar graphs are drawn simultaneously on the same vertex set [3].

As an equivalent graphical representation, a radial level graph can be drawn on a cylinder in which the rings are circles of equal radii at different heights along the cylinder; see Fig. 2(b). The radial drawing may wrap around the cylinder. A third alternate representation comes from cutting the cylinder along a vertical line (a ray from the origin in Fig. 2(a)) and flattening the cut surface onto a plane. The rings then become horizontal lines, and the edges becomes straight-lines that can "wrap" from the right to the left side of the drawing; see Fig. 2(c).

If no edges wrap in this third representation, then the graph is *level planar*. Analogous to URLP graphs, graphs that are level planar for any distinct labeling are *unlabeled level planar* (ULP). We use this last representation when drawing radial graphs given its compactness and relationship to ULP graphs.

1.1   Related Previous Work

Radial level planar graphs can be recognized and embedded in linear time [1]. A linear-time straight-line drawing algorithm exists for level planar graphs [5,11],

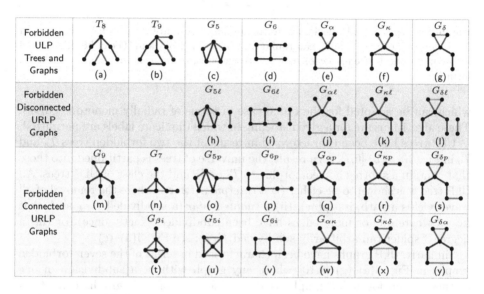

Fig. 3. Forbidden ULP and URLP trees and graphs

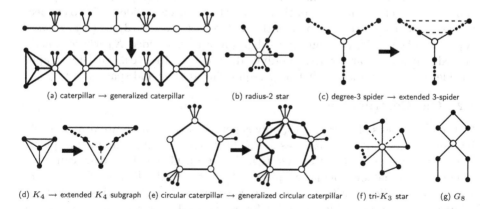

(a) caterpillar → generalized caterpillar   (b) radius-2 star   (c) degree-3 spider → extended 3-spider

(d) K_4 → extended K_4 subgraph   (e) circular caterpillar → generalized circular caterpillar   (f) tri-K_3 star   (g) G_8

Fig. 4. Classes of ULP trees and graphs in (a)–(d) and classes of additional URLP graphs in (e)–(g) where dashed edges are optional and white vertices are cut vertices

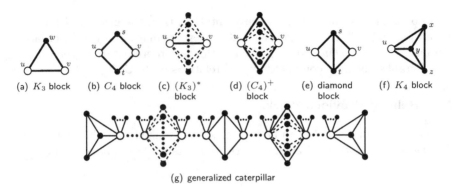

(a) K_3 block   (b) C_4 block   (c) $(K_3)^*$ block   (d) $(C_4)^+$ block   (e) diamond block   (f) K_4 block

(g) generalized caterpillar

Fig. 5. Six types of ULP blocks where the joining blocks in (a)–(e) can be substituted for an internal edge (u, v) of a caterpillar and the ending block in (f) can be substituted for at most one leaf edge (u, ℓ) incident to each endpoint u of the spine in order to form (g)

which can be adapted for the radial case in terms of radially monotone curves. These algorithms are for a given labeling in which duplicate labels are permitted.

ULP trees have been characterized in terms of the two forbidden trees T_8 and T_9 in Fig. 3(a)–(b) [6,7]. As a result, the universe of trees is partitioned into those that contain a subtree homeomorphic to T_8 or T_9 and the class of ULP trees. An ULP tree was shown to be either (i) a *caterpillar* (a tree where the removal of all leaves yields a path, its *spine*), (ii) a *radius-2 star* (a subdivided $K_{1,k}$ such that $k \geq 3$ where one or more edges have been subdivided exactly once), or (iii) a degree-3 spider (an arbitrarily subdivided $K_{1,3}$); see Fig. 4(a)–(c).

Similarly, ULP graphs have been characterized in terms of the seven forbidden graphs in Fig. 3(a)–(g) [9,10] where any graph without a subdivision of one of these seven forbidden graphs is ULP. The class of ULP graphs consists of (i) *generalized caterpillars* (GCs) (formed by substituting edges of a caterpillar for ULP blocks as described in Fig. 5), (ii) *rstars* (R2Ss), (iii) *extended 3-spiders* (E3Ss) (formed by adding two optional edges to a degree-3 spider so as to connect two leaf vertices or two neighbors of the root vertex of degree 3), or (iv) *extended K_4 subgraphs* (EK4s) (a connected subgraph of a subdivided K_4 where exactly one edge has been arbitrarily subdivided); see Fig. 4(a)–(d). All ULP graphs have level planar drawings that only require linear time and space; see Fig. 6.

1.2 Our Contribution

Of the four classes of ULP graphs, only generalized caterpillars are drawn block-by-block proceeding left to right as in Fig. 6(a). This is unlike extended 3-spiders that can have spirals when drawn as in Fig. 6(c). Given that edges can wrap around in a radial level planar graph, this leads to the URLP class of *generalized circular caterpillars* (GCCs). These are constructed by substituting any of the joining ULP blocks in Fig. 5(a)–(e) for a cycle edge of a circular caterpillar (a graph where the removal of all endpoints yields a cycle); see Fig. 4(e). Observe that extended K_4 subgraphs are a subclass of generalized (circular) caterpillars.

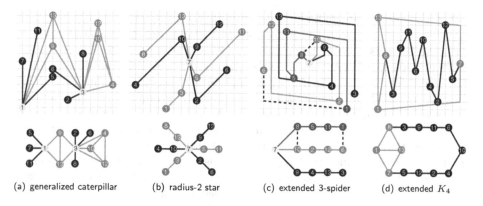

(a) generalized caterpillar   (b) radius-2 star   (c) extended 3-spider   (d) extended K_4

Fig. 6. Level planar drawings of the four classes of ULP graphs

A *tri-K_3 star* (TK3S) is formed by adding up to three edges to a radius-2 star (with at most three subdivided edges) that connect one, two, or three leaves to the root vertex as in Fig. 4(f). The graph G_8 in Fig. 4(g) is a subdivision of G_δ in Fig. 3(g), a tri-K_3 star, where the 3-cycle has been subdivided to form a 4-cycle. Both classes of graphs admit radial level planar drawings where an edge of each cycle can wrap around so as to avoid a crossing.

We extend the ULP characterization to all URLP graphs as follows:

1. First, we show that a disconnected graph is URLP if and only if it neither contains a subgraph homeomorphic to either tree in Fig. 3(a) or (b) nor to one of the disconnected graphs in Fig. 3(h)–(l) for a total of 7 forbidden subdivisions. We prove that this is equivalent to each component being ULP.
2. Second, we show that a connected graph is URLP if and only if it neither contains a subgraph homeomorphic to either of the trees in Fig. 3(a)–(b) nor to one of the connected graphs in Fig. 3(m)–(y) for a total of 15 forbidden subdivisions. We prove that this is equivalent to the graph either being ULP or belonging to one of the three additional classes of URLP connected graphs, which are generalized circular caterpillars, tri-K_3 stars, and graphs isomorphic to G_8.
3. Third, we provide $O(n)$-time drawing algorithms for each new class of n-vertex URLP graphs that are drawn on $O(n) \times n$ integer grids.

2   Preliminaries

Track ℓ_j is the horizontal line $\{(x, j) : x \in \mathbb{R}\}$. *Ring* c_j is the circle $\{(x, y) : x^2 + y^2 = j^2, (x, y) \in \mathbb{R}^2\}$ in Cartesian coordinates or $\{(j, \theta) : j \in \mathbb{R}, 0 \le \theta \le 2\pi\}$ in radial coordinates. Curve $L = \{(x(s), y(s)) : s_1 \le s \le s_2\} = \{(r(s), \theta(s)) : s_1 \le s \le s_2\}$ in parametrized Cartesian and radial coordinates, respectively, is *y-monotone* if $y(s) < y(s')$ when $s < s'$ and *r-monotone* if $r(s) < r(s')$ when $s < s'$. A $m \times n$ *circular grid* consists of n concentric rings $\{(r', \theta) : 0 \le \theta \le 2\pi\}$ for $r' \in \{1, \dots, n\}$ and m rays $\{(r, \theta') : r \in \mathbb{R}\}$ for $\theta' \in \{\frac{2k\pi}{m} : k \in \{1, \dots, m\}\}$.

Let $G(V, E)$ be an undirected graph with vertex set V and edge set $E = \{(u, v) : u, v \in V, u \neq v\}$ with no isolated vertices. Let ϕ be a *labeling* on V with the integers 1 to k, i.e. $\phi : V \mapsto \{1, \ldots, k\}$ where $k \leq |V|$. A *level* is a set of vertices with the same label where the jth *level* is the vertex subset $V_j = \{v \in V : \phi(v) = j\}$. If $\phi(u) \neq \phi(v)$ for any $(u, v) \in E$, then $G(V, E, \phi)$ forms a *(radial) level graph* on k levels. Radial level graph $G(V, E, \phi)$ is *(radial) level planar* if a drawing can be *realized* such that (i) each vertex of V_j is placed along track ℓ_j (or ring c_j), (ii) each edge (u, v) is drawn with a y-monotone (or r-monotone) curve (that can wrap), and (iii) each pair of curves can only intersect at their endpoints. Graph $G(V, E)$ is *unlabeled (radial) level planar* if (radial) level graph $G(V, E, \phi)$ is (radial) level planar for every bijective labeling ϕ. A *(radial) level planar drawing* is also called a *realization*.

3   Drawing Unlabeled Radial Level Planar Graphs

Any graph that has a level planar drawing has a radial level planar realization as seen by the equivalence of the layouts in Fig. 2. Hence, the drawing algorithms for ULP graphs given in [7,9,10] naturally extend to the radial setting. The drawing algorithm for generalized caterpillars detailed in [9,10] draws each ULP block proceeding left to right; see Fig. 6(a). Extending this algorithm to draw each ULP block of a generalized circular caterpillar proceeding clockwise in a radial setting gives the next lemma. See [8] for a more detailed proof.

Lemma 1. *An n-vertex generalized (circular) caterpillar with m ULP blocks can be realized on a $4m \times n$ (circular) grid in $O(n)$ time for any distinct labeling.*

We can also directly extend the drawing algorithms from [7,9,10] for radius-2 stars and extended 3-spiders to give the next two lemmas, respectively.

Lemma 2. *An n-vertex radius-2 star can be realized in $O(n)$ time on a (circular) $(2n + 1) \times n$ grid for any distinct labeling.*

Lemma 3. *An n-vertex extended 3-spider can be realized in $O(n)$ time on an $(n + 1) \times n$ (circular) grid for any distinct labeling.*

Next, we show how to realize the remaining two classes of URLP graphs in the subsequent two lemmas; the full proofs of each can be found in [8].

Lemma 4. *An n-vertex tri-K_3 star G can be realized on a $5 \times n$ circular grid in $O(n)$ time for any distinct labeling.*

Proof Sketch: Figure 7 illustrates how to handle the four distinct cases of drawing a tri-K_3 star with the three 3-cycles $r-s-t-r$, $r-v-w-r$, and $r-x-y-r$ where $\phi(s) > \phi(t)$, $\phi(v) > \phi(w)$, $\phi(x) > \phi(y)$, and $\phi(s) > \phi(v) > \phi(x)$.

Initially, s, t, x, and y are placed two units to the right of r, while v, w are placed to two units to the left of r. In the first case $\phi(t) > \phi(x)$ so that no crossings will occur. In the remaining three cases where $\phi(x) > \phi(t)$, the vertices t, w, and y have the minimum labels of any 3-cycle, respectively, and vertices switch sides as depicted in Fig. 7(b)–(d). Finally, leaf edges are either drawn one unit to the left or to the right of r so as to avoid any edge overlap.   □

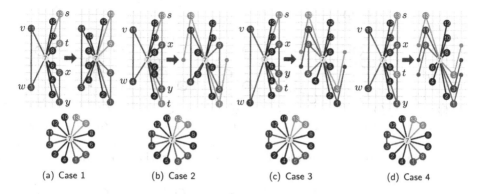

Fig. 7. Four cases of how to avoid edge overlaps for tri-K_3 stars

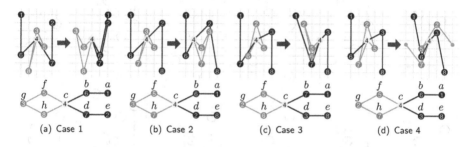

Fig. 8. Four cases of how to avoid edge overlaps for G_8

Lemma 5. *The graph G_8 can be realized on a 5×8 circular grid in $O(1)$ time for any distinct labeling.*

Proof Sketch: If any chain of length 2 of G_8 is radially monotone, then drawing G_8 is equivalent to drawing the generalized caterpillar or tri-K_3 star obtained by replacing the chain with a single edge where either the algorithms of Lemmas 1 or 4 can be used. Otherwise, Fig. 8 depicts the four remaining cases of how to draw G_8 if this is not the case.   □

4   Forbidden Unlabeled Radial Level Planar Graphs

We define $\mathcal{F}_{\mathsf{ULP}} := \{T_8, T_9, G_5, G_6, G_\alpha, G_\kappa, G_\delta\}$ in Fig. 3(a)–(g), $\mathcal{T}_{\mathsf{URLP}} := \{T_8, T_9\}$ (forbidden URLP trees) in Fig. 3(a)–(b), $\mathcal{D}_{\mathsf{URLP}} := \{G_{5\ell}, G_{6\ell}, G_{\alpha\ell}, G_{\kappa\ell}, G_{\delta\ell}\}$ (disconnected forbidden URLP graphs) in Fig. 3(h)–(l), $\mathcal{C}_{\mathsf{URLP}} := \{G_9, G_7, G_{5p}, G_{6p}, G_{\alpha p}, G_{\kappa p}, G_{\delta p}, G_{\beta i}, G_{5i}, G_{6i}, G_{\alpha\kappa}, G_{\kappa\delta}, G_{\delta\alpha}\}$ (connected forbidden URLP graphs) in Fig. 3(m)–(y), and $\mathcal{F}_{\mathsf{URLP}} := \mathcal{T}_{\mathsf{URLP}} \cup \mathcal{D}_{\mathsf{URLP}} \cup \mathcal{C}_{\mathsf{URLP}}$[1].

The labelings in Fig. 9 of $\mathcal{F}_{\mathsf{ULP}}$ were shown to be level non-planar in [9,10]. In this section, we prove that of these, only the two trees, T_8 and T_9, in $\mathcal{T}_{\mathsf{URLP}}$ are also radial level non-planar. We show how to add an extra edge to each of

---

[1] Subscripts ℓ, p, and i stand for "lone", "pendant", and "internal" edges, respectively.

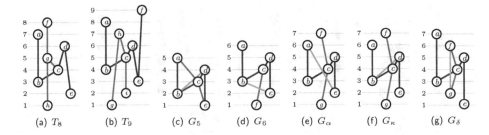

Fig. 9. Level non-planar labelings of the seven forbidden ULP graphs

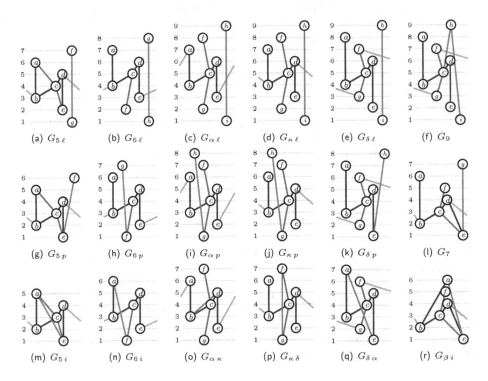

Fig. 10. Radial level non-planar labelings of the 20 cyclic forbidden URLP graphs

the five cyclic graphs of $\mathcal{F}_{\mathsf{ULP}}$ in three different ways so as to produce 15 of the graphs of $\mathcal{F}_{\mathsf{URLP}}$ in Fig. 3(h)–(l), (o)–(s), and (u)–(y) that have radial level non-planar labelings in Fig. 10(a)–(e), (g)–(k), and (m)–(q), respectively. For the three remaining graphs of $\mathcal{F}_{\mathsf{URLP}}$ in Fig. 3(m),(n), and (t), we show that the labelings in Fig. 10(f),(l), and (r), respectively, are also radial level non-planar.

Full proofs for omitted or shortened proofs in this section can be found in [8]. First, we observe a property common to all the embeddings in Figs. 9 and 10.

Observation 6. *Suppose a radial level graph G has a labeling ϕ and a path a–b–c–d–e such that $\phi(a) > \phi(d) > \phi(c) > \phi(b) > \phi(e)$. For G to be radial level* *planar, G must have an embedding where the path does not wrap such that the edge $a{-}b$ intersects the tracks ℓ_c and ℓ_d to the left of c and d (with respect to the ray that cuts the cylinder as in Fig. 2(b)–(c)), respectively, and the edge $d{-}e$ intersects the tracks ℓ_b and ℓ_c to the right of b and c, respectively.*

We next see that only T_8 and T_9 of $\mathcal{F}_{\mathsf{ULP}}$ are also radial level non-planar.

Lemma 7. *Of the seven forbidden ULP graphs in $\mathcal{F}_{\mathsf{ULP}}$, only T_8 and T_9 are also radial level non-planar with distinct labels.*

Proof Sketch: Using the labeling in Fig. 9(a) for T_8, Observation 6 implies that $a{-}b{-}c{-}d{-}e$ proceeds left to right. Since the vertex c lies between the edges $a{-}b$ and $d{-}e$, so must the edge $c{-}g$. As a result, either the edge $f{-}g$ or $g{-}h$ must cross some edge of $a{-}b{-}c{-}d{-}e$. A similar argument can be given for T_9 using the labeling in Fig. 9(b). Finally, each remaining graph is either a generalized circular caterpillar or a tri-K_3 star, which are URLP by Lemmas 1 and 4.   □

Next we show that all of the graphs in $\mathcal{F}_{\mathsf{URLP}}$ are radial level non-planar.

Lemma 8. *There exist distinct labelings preventing each of the cyclic forbidden URLP graphs in $\mathcal{F}_{\mathsf{URLP}}$ from being radial level planar.*

Proof Sketch: Each of the 16 graphs in Fig. 10(a)–(k), (m)–(q) contain a subgraph of one of the five cyclic forbidden graphs of $\mathcal{F}_{\mathsf{ULP}}$ using the same relative labeling as in Fig. 9(c)–(g). This means that in order for any of these graphs to be radial level planar, an edge e must wrap. However, in each case (except for $G_{\alpha\kappa}$), the graph has an extra edge with extreme labels that prevents e from wrapping. In the case of $G_{\alpha\kappa}$ in Fig. 10(o), if the edge $a{-}e$ wraps, then the extra edge $b{-}d$ must cross the path $f{-}c{-}g$ whose endpoints have extreme labels. In the cases of G_7 and $G_{\beta i}$ in Fig. 10(l) and (r), the labeling is the same as the labelings of $G_{\alpha p}$ and $G_{\delta\alpha}$ in Fig. 10(k) and (q) where vertices e and g have been merged into vertex e with an extreme label that forces crossing in each case.   □

We can extend any radial level non-planar labeling with the next lemma.

Lemma 9. *If a graph G contains a subgraph homeomorphic to a graph $\tilde{G}$ with a radial level non-planar labeling, then G also has a radial level non-planar labeling.*

Lemma 9 allows us to generalize Lemmas 7 and 8 to the following lemma:

Lemma 10. *If a graph contains a subgraph homeomorphic to any of the graphs in $\mathcal{F}_{\mathsf{URLP}}$, then it cannot be URLP with distinct labels.*

5   Characterizing Unlabeled Radial Level Planar Graphs

In this section, we characterize disconnected and connected URLP graphs separately. The omitted proofs in this section can be found in [8].

5.1   Disconnected Unlabeled Radial Level Planar Graphs

We characterize the class of disconnected URLP graphs in terms of the forbidden graphs $\mathcal{T}_{\mathsf{URLP}} \cup \mathcal{D}_{\mathsf{URLP}}$. First, we consider minimality.

Lemma 11. *Each forbidden graph in $\mathcal{T}_{\mathsf{URLP}}$ and $\mathcal{D}_{\mathsf{URLP}}$ is minimal in that the removal of any edge yields one or more URLP graphs.*

In a disconnected graph G, each component (discounting isolated vertices) contains at least one edge. Since one of the components of each graph in $\mathcal{D}_{\mathsf{URLP}}$ is an isolated edge, that edge prevents any other component of G from having a subdivision of a cyclic graph from $\mathcal{F}_{\mathsf{ULP}}$. Hence, we can extend Theorem 8.3.13 of [9] that characterizes ULP graphs in terms of $\mathcal{F}_{\mathsf{ULP}}$ to the following theorem:

Theorem 12. *For a disconnected graph G, with no isolated vertices, the following three statements are equivalent:*

1. *G does not contain a subgraph homeomorphic to T_8, T_9, $G_{5\ell}$, $G_{6\ell}$, $G_{\alpha\ell}$, $G_{\kappa\ell}$, or $G_{\delta\ell}$.*
2. *Each component of G is either a generalized caterpillar, a radius-2 star, an extended 3-spider, or an extended K_4 subgraph.*
3. *G is URLP with distinct labels.*

5.2   Connected Unlabeled Radial Level Planar Graphs

Here we characterize the class of connected URLP graphs in terms of the forbidden graphs $\mathcal{T}_{\mathsf{URLP}} \cup \mathcal{C}_{\mathsf{URLP}}$. We start by considering minimality.

Lemma 13. *Each connected forbidden graph in $\mathcal{C}_{\mathsf{URLP}}$ is minimal in that the removal of any edge that does not disconnect the graph yields an URLP graph.*

Lemma 8.3.7 of [9] characterizes generalized caterpillars in terms of the four forbidden graphs in Fig. 11(a)–(d) as follows:

Lemma 14. *A connected graph G is a generalized caterpillar if and only if G does not have a subgraph homeomorphic to G_6, C_5, G_ω, or T_7.*

This requires first characterizing graphs in which the maximum length of any cycle is 4, which is Corollary 8.3.2 from [9]:

Lemma 15. *Every block of a connected graph G is isomorphic to a K_4, $(K_3)^*$, or $(C_4)^+$ block, or G contains a C_5 subdivision.*

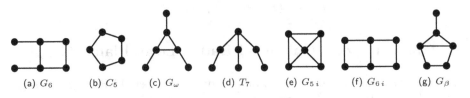

| (a) G_6 | (b) C_5 | (c) G_ω | (d) T_7 | (e) G_{5i} | (f) G_{6i} | (g) G_β |

Fig. 11. Forbidden graphs for generalized caterpillars in (a)–(d) and for generalized circular caterpillars in (d)–(g)

We extend this characterization to include generalized circular caterpillars in terms of the four forbidden graphs in Fig. 11(d)–(g) as follows:

Lemma 16. *A connected graph G is either a generalized caterpillar or a generalized circular caterpillar if and only if G does not contain a subgraph homeomorphic to $T_7, G_{5\,i}, G_{6\,i}$, or G_β.*

Two of the four forbidden graphs of generalized (circular) caterpillars are contained in $\mathcal{C}_{URLP}$. For the other two graphs T_7 and G_β, Figs. 12 and 13 consider all possible ways to add an edge to either obtain an URLP graph or another forbidden graph in $\mathcal{T}_{URLP} \cup \mathcal{C}_{URLP}$. This gives the following two lemmas:

Lemma 17. *If G is a connected graph that contains a subgraph homeomorphic to T_7, but does not contain a subgraph homeomorphic to $G_{6\,p}$, $G_{\alpha\,p}$, $G_{\kappa\,p}$, $G_{\delta\,p}$, $G_{6\,i}$, $G_{\kappa\,\delta}$, $G_{\delta\,\alpha}$, G_7, G_9, or T_8, then G is an extended 3-spider, a radius-2 star, an tri-K_3 star, or is isomorphic to G_8.*

Lemma 18. *If G is a connected graph that contains a subgraph homeomorphic to G_β, but does not contain a subgraph homeomorphic to $G_{5\,p}$, $G_{6\,p}$, $G_{6\,i}$, $G_{\alpha\,\kappa}$, $G_{\beta\,i}$, or T_8, then G is an extended 3-spider.*

We next show that $\mathcal{T}_{URLP} \cup \mathcal{C}_{URLP}$ forms a set of forbidden URLP graphs with the subsequent lemma; the full proof of which can be found in [8].

Lemma 19. *The class of connected URLP graphs with distinct labels does not contain a subgraph homeomorphic to any of graph in $\mathcal{T}_{URLP} \cup \mathcal{C}_{URLP}$.*

Proof Sketch: By applying Lemma 16 and comparing degree sequences, cycle lengths, and the number of cycles of each graph $G \in \mathcal{T}_{URLP} \cup \mathcal{C}_{URLP}$ to each class of URLP graphs, we see that in each case G is a forbidden graph. □

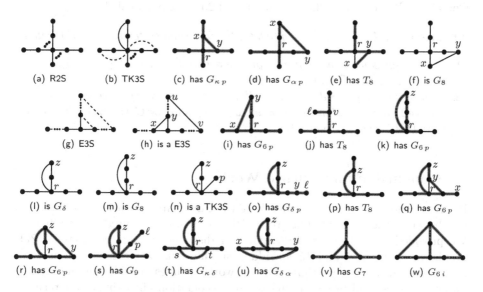

Fig. 12. Cases for adding one or more edges to T_7

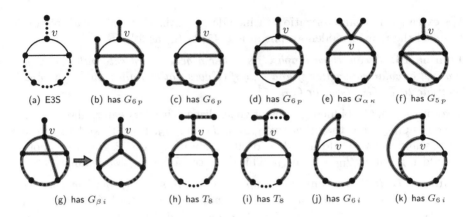

Fig. 13. Cases for adding one or more edges to G_β

Lemma 20. *If G does not contain a subgraph homeomorphic to a graph in $\mathcal{T}_{\text{URLP}} \cup \mathcal{C}_{\text{URLP}}$, then G is a generalized (circular) caterpillar, a radius-2 star, an extended 3-spider, a tri-K_3 star, or is isomorphic to G_8.*

Proof. If G is not a generalized (circular) caterpillar, then by Lemma 14 G must contains a subgraph homeomorphic to T_7, G_{5i}, G_{6i}, or G_β. However, since both G_{5i} and G_{6i} are in $\mathcal{C}_{\text{URLP}}$, G must either contain a subgraph homeomorphic to T_7 or G_β. In the first case, G must be an extended 3-spider, a radius-2 star, tri-K_3 star, or is isomorphic to G_8 by Lemma 17. In the second case, G can only be a extended 3-spider by Lemma 18.   □

Combining Lemmas 1, 2, 3, 4, 5, 10 19, and 20 gives our final theorem.

Theorem 21. *For a connected graph G, the following statements are equivalent:*

1. *G does not contain a subgraph homeomorphic to T_8, T_9, G_7, G_9, $G_{\beta i}$, G_{5p}, G_{6p}, $G_{\alpha p}$, $G_{\kappa p}$, $G_{\delta p}$, G_{5i}, G_{6i}, $G_{\alpha \kappa}$, $G_{\kappa \delta}$, or $G_{\delta \alpha}$.*
2. *G is either a generalized caterpillar, a generalized circular caterpillar, a radius-2 star, an extended 3-spider, a tri-K_3 star, or G is isomorphic to G_8.*
3. *G is URLP with distinct labels.*

6   Conclusion and Future Work

In this extended abstract, we generalized the ULP characterization to radial level graphs with URLP graphs that are radial level planar for any distinct labeling. We provided separate characterizations for disconnected and connected URLP graphs in terms of 7 and 15 forbidden subdivisions, respectively, along with linear-time drawing algorithms. Future work includes considering the case of duplicate labels and providing linear-time recognition algorithms to determine whether a graph is URLP.

References

1. Bachmaier, C., Brandenburg, F.J., Forster, M.: Radial level planarity testing and embedding in linear time. Journal of Graph Algorithms and Applications 9(1), 53–97 (2005)
2. Brandes, U., Kenis, P., Wagner, D.: Communicating centrality in policy network drawings. IEEE Transactions on Visualization and Computer Graphics 09(2), 241–253 (2003)
3. Brass, P., Cenek, E., Duncan, C.A., Efrat, A., Erten, C., Ismailescu, D., Kobourov, S.G., Lubiw, A., Mitchell, J.S.B.: On simultaneous graph embedding. Computational Geometry: Theory and Applications 36(2), 117–130 (2007)
4. Cappos, J., Estrella-Balderrama, A., Fowler, J.J., Kobourov, S.G.: Simultaneous graph embedding with bends and circular arcs. Computational Geometry: Theory and Applications 42(2), 173–182 (2009)
5. Eades, P., Feng, Q., Lin, X., Nagamochi, H.: Straight-line drawing algorithms for hierarchical graphs and clustered graphs. Algorithmica 44(1), 1–32 (2006)
6. Estrella-Balderrama, A., Fowler, J.J., Kobourov, S.G.: Characterization of unlabeled level planar trees. In: Kaufmann, M., Wagner, D. (eds.) GD 2006. LNCS, vol. 4372, pp. 367–379. Springer, Heidelberg (2007)
7. Estrella-Balderrama, A., Fowler, J.J., Kobourov, S.G.: Characterization of unlabeled level planar trees. Computational Geometry: Theory and Applications 42(7), 704–721 (2009)
8. Fowler, J.J.: Characterization of unlabeled radial level planar graphs. Technical Report TR09-03, University of Arizona (2009),
http://ulp.cs.arizona.edu/TR09-03.pdf
9. Fowler, J.J.: Unlabeled Level Planarity. PhD thesis, University of Arizona (2009),
http://ulp.cs.arizona.edu/ulp-thesis.pdf
10. Fowler, J.J., Kobourov, S.G.: Characterization of unlabeled planar graphs. In: Hong, S.-H., Nishizeki, T., Quan, W. (eds.) GD 2007. LNCS, vol. 4875, pp. 37–49. Springer, Heidelberg (2008)
11. Jünger, M., Leipert, S.: Level planar embedding in linear time. Journal of Graph Algorithms and Applications 6(1), 67–113 (2002)

Upward Planarization Layout

Markus Chimani, Carsten Gutwenger, Petra Mutzel, and Hoi-Ming Wong*

Chair for Algorithm Engineering, TU Dortmund, Germany

Abstract. Recently, we presented a new practical method for upward crossing minimization [6], which clearly outperformed existing approaches for drawing hierarchical graphs in that respect. The outcome of this method is an *upward planar representation (UPR)*, a planarly embedded graph in which crossings are represented by dummy vertices. However, straight-forward approaches for drawing such UPRs lead to quite unsatisfactory results. In this paper, we present a new algorithm for drawing UPRs that greatly improves the layout quality, leading to good hierarchal drawings with few crossings. We analyze its performance on well-known benchmark graphs and compare it with alternative approaches.

1 Introduction

The visualization of hierarchical graphs representing some natural flow of information is one of the key topics in graph drawing. It has numerous practical applications and received a lot of scientific attention since the very beginning of graph drawing. Formally, we are given a directed acyclic graph (DAG) G and we want to find an *upward drawing* of G, i.e., a drawing of G in which all arcs are drawn as curves monotonically increasing in the vertical direction.

In 1981, Sugiyama et al. [18] proposed their well-known three-phase framework for creating such drawings, which is still widely used:

1. **Layer assignment:** Assign nodes to layers such that arcs point from lower to higher layers; split long arcs spanning several layers by creating *dummy nodes*.
2. **Node Ordering/Crossing reduction:** Order nodes on the layer to reduce the number of arc crossings.
3. **Coordinate assignment:** Assign coordinates to original nodes and dummy nodes (bend points) such that we get only few bend points and short arcs.

A vast amount of modifications and alternatives for the individual steps have been proposed, e.g., Gansner et al. [11] give a LP-based formulations for layer and coordinate assignment. The layer assignment computes a layering which minimize the sum of the vertical edge lengths (i.e., the number of layers an edge spans). The coordinate assignment minimizes the objective function $\sum_{e=(v,w)\in A} w(e) \cdot |X(v) - X(w)|$ where $w(e)$ gives the priority for drawing e vertically and A is the arc set after splitting long arcs. Brandes and Köpf [3] propose an approach which is simpler and faster than [11], but

---

* Hoi-Ming Wong was supported by the German Research Foundation (DFG), priority project (SPP) 1307 "Algorithm Engineering", subproject "Planarization Practices in Automatic Graph Drawing".

D. Eppstein and E.R. Gansner (Eds.): GD 2009, LNCS 5849, pp. 94–106, 2010.

nevertheless it compute coordinate assignment with similar quality. Brankes et al. [4] investigate the computation complexity of the *width-restricted graph layering problem*. They proved that width-restricted graph layering is NP-hard when taking the dummy nodes into account. Healy and Nikolov give an experimental analysis of existing layering algorithms for DAGs. They also give an ILP formulation which computes a layering with minimum number of dummy nodes with a given upper bound on the width and height of the layering and an branch and cut algorithm to solve it [14,13].

However, a major drawback of Sugiyama's framework could not be solved by any of these modifications: Since layer assignment and crossing reduction are realized as independent steps, the resulting drawing might have many unnecessary crossings caused by an unfortunate layer assignment. A main challenge is to perform crossing reduction *independent* of a layer assignment. First steps to adapt the planarization approach for undirected graphs [1,12] have been presented in [2,7]; Eiglsperger et al. [10] presented the more advanced *mixed upward planarization* approach. However, even the latter approach still needs some kind of layering. Experimental results suggested that this approach produces considerably less crossings than Sugiyama's algorithm. In [6], we presented a novel approach for upward planarization that does not require any layering. We could experimentally show that this new approach clearly outperforms Eiglsperger's mixed upward planarization and Sugiyama's algorithm with respect to crossing reduction.

The output of an upward planarization procedure is an *upward planar representation*, i.e., a representation of the original digraph, in which crossings are replaced by dummy vertices (*crossing dummies*), with a planar embedding and designated external face. In our case, the upward planar representation will always be a single-source digraph; if the input digraph contains multiple sources we introduce a super-source $\hat{s}$ connected to all sources and do not count crossings with arcs incident to $\hat{s}$.

A simple method to draw a DAG by applying upward planarization consists of using Sugiyama's coordinate assignment phase for drawing the upward planar representation, where we use a straight-forwardly obtained layering and the ordering of the nodes on each layer implied by the upward-planar representation and embedding. However, this method produces quite unsatisfactory drawings with too many layers and much too long arcs. The main objective of this paper is to significantly improve on this simple method, by enhancing the computation of layers and node orderings, taking into account the

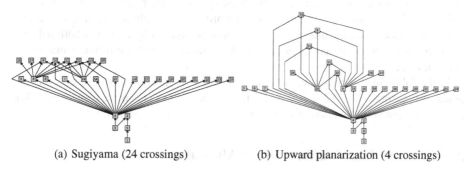

(a) Sugiyama (24 crossings)   (b) Upward planarization (4 crossings)

Fig. 1. Instance *g.29.16* (North DAGs) with 29 nodes and 38 arcs

special roles of crossing dummies. This will allow us to reduce the heights of the draw-ings and lengths of the arcs substantially, resulting in much more pleasant drawings.

Since upward planarization yields an upward planar representation, we can alter-natively also use drawing methods for upward planar digraphs to draw it, cf. [7]. We consider two such algorithms in our experimental study:

Dominance drawings: The linear-time algorithm by Di Battista et al. [9] draws planar st-digraphs with small area on a grid. We apply this algorithm by augmenting the upward planar representation to a planar st-digraph and omitting the augmenting arcs in the final drawing.

Visibility representations: We use the algorithm by Rosenstiehl and Tarjan [17] for computing a visibility representation on the grid. This algorithm is based on bipo-lar orientations implied by an st-numbering. By augmenting the upward planar representation to an st-planar digraph, we obtain a bipolar orientation such that the resulting drawing is upward planar. Again, we omit augmenting edges in the final drawing.

Fig. 1 shows a relatively small digraph, where the benefits of the new upward planariza-tion approach can be easily seen: While the classical Sugiyama's approach leads to few layers, our approach can expand the layout of the subgraph that looks very congested otherwise.

Upward Planarization. We briefly sketch the planarization approach proposed in [6]. It can be divided into two phases: the feasible subgraph computation and the reinsertion phase. In the first phase the input DAG G is transformed into a single source digraph G' by adding an artificial super source $\hat{s}$ and connecting it to the sources of G. Then we compute a spanning tree T of G' and iteratively try to insert the remaining arcs into T. Thereby, we perform a subgraph feasibility test after each inserted arc e: we do not only test upward planarity but also if all remaining edges can potentially still be inserted (with crossings) in an upward fashion. If the resulting digraph is not feasible in this sense, we add e to a set of deleted arcs B instead. By these operations, we obtain an embedded feasible upward planar subgraph U.

In the second phase, the arcs of B are reinserted into U one after another such that few crossings arise. Thereby, the crossings caused by the reinsertion are replaced by crossing dummies. As a result, we obtain an upward planar representation R of G' (cf. Fig. 2). Note that R is an embedded upward planar single-source digraph. It can easily be augmented to a single-source, single-sink digraph by adding additional arcs (*face-arcs*) and an additional super sink $\hat{t}$ that is connected with the former sinks on the external face.

In the following, R will always be an embedded single-source, single-sink upward planar representation of G. Let v and e be a node and an arc in G, respectively. We denote the corresponding node and arc in R by v_R and e_R, respectively.

2   Upward Planarization Layout Algorithm

The crucial starting point of our algorithm can be stated as follows:

Proposition 1. Given an upward planar representation R of a DAG G, there exists a layering of the nodes of G, a node order per layer, and a node placement including bend points, such that the thereby induced drawing of G *realizes* R, i.e., the crossings arising in G are the ones modeled by R.

We observe that a realizing drawing of G hence follows Sugiyama's framework, but the individual steps do not simply optimize their respective objectives, but follow the overall goal of simulating R. Our algorithm hence divides naturally into the three steps known from Sugiyama's framework, whereby the first two steps are closely related.

As sketched in Sect. 1 (and investigated in the experimental comparison, Sect. 3.1), it is easy to find *some* solution that realizes R. Yet, even if G is only of moderate size, R can become much larger due to the crossing dummies and long arcs. This causes weak runtime performance, many layers, and overall unsatisfactory drawings. Hence our algorithm aims at minimizing the required layers, thereby also reducing the necessary dummy nodes from splitting long arcs.

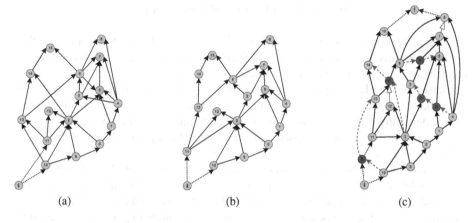

Fig. 2. Illustration of the upward planarization approach by Chimani et al. [6]. (a) input DAG G augmented to G' via the artificial super source $\hat{s}$; (b) embedded feasible subgraph U of G' obtained by deleting the arcs $(10, 13)$, $(2, 14)$, $(4, 3)$, $(6, 5)$; (c) upward planar representation R of G after reinserting the deleted arcs (dashed line). R contains five crossing dummies. By adding face-arcs (drawn with hollow arrow heads) and the super sink $\hat{t}$, R becomes a single-source, single-sink digraph.

2.1 Layer Assignment and Node Ordering

Let H be a copy of G that we will use to obtain a valid layering for G, cf. Fig. 3. For any two nodes $u, v \in V(G)$, we add an auxiliary arc (u, v) to H if: (a) there exists no directed path from u to v in G, but (b) there exists a directed path from u_R to v_R in R. Part (a) prohibits the unnecessary generations of transitive arcs. Part (b), in conjunction with the face-arcs and the single-source, single-sink property of R, ensures that the hierarchical order of R is mapped to H. Since G and R are DAGs, H is also acyclic, and we can use any existing layering algorithm on H. Let $\mathcal{L} = \langle L_1, L_2, ..., L_\ell \rangle$ be the

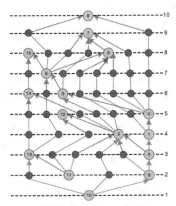

(a) The auxiliary digraph H with respect to R from Fig. 2. Arcs in H but not in G are drawn as dotted lines.

(b) Layering of H. Long-arc dummies are drawn as smaller circles.

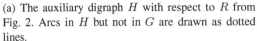

Fig. 3. Layer Assignment and Node Ordering

layering of H, and therefore also for G; i.e., $\bigcup_{1 \le i \le \ell} L_i = V(G)$. We can extend this layering naturally, by splitting any arc that spans more than one layer into a chain of arc segments by introducing dummy nodes (*long-arc dummies*).

Considering the layering $\mathcal{L}$, we now have to arrange the nodes on each layer according to the order induced by R. For this purpose, we consider the order of the arcs around each node, as given by R. In particular, we can recognize the *left incoming* arc for any node v, which is the embedding-wise left-most arc with target v. Note that this arc is defined for each node except for the super source.

Now consider any two disjoint nodes u and v on the same layer. We can decide their correct order using the following strategy: we construct a *left path* p_u from $\hat{s}$ to u_R from back to front, i.e., starting at u_R we select its left incoming arc e as the end of p_u and proceed from the source node of e, choosing its left incoming arc as the second to last arc in p_u, and so on. The construction of p_u ends when we reach the super source, which will always happen as R is a single-sink DAG. Analogously, let p_v be the left path from $\hat{s}$ to v_R.

The paths p_u and p_v may share a common subpath starting at $\hat{s}$; let c_R be the last common node of p_u and p_v, and let e_u and e_v be the first disjoint path elements, respectively. We determine the ordering of u and v directly by the order of e_u and e_v at c_R. For example in Fig. 3, if u_R and v_R are the nodes '4' and '12' respectively (layer 5), then node '0' is the last common node of their left paths, and hence v_R is left of u_R.

Algorithmically, we can consider each layer independently. Introducing an auxiliary digraph A, the above relationship between two nodes on the same layer can be modeled as an arc between these two nodes in A. We can construct correct ordering for the layer, by computing the topological order in A. Note that therefore we do not have to compute the arc direction for all node pairs, but only for the ones that are not already "solved" by other arcs through transitivity.

The above approach already gives a good layering and ordering realizing R. Yet, in order to further improve the solution we introduce two postprocessing strategies:

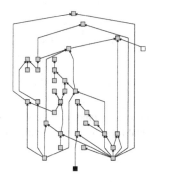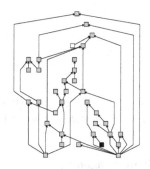

Fig. 4. A drawing of graph *grafo2379.35* (Rome graphs): (left) without postprocessing, (right) after applying source repositioning (white node) and long-arc dummy reduction (black node)

Long-Arc Dummy Reduction. A *dominated subgraph* of G w.r.t. a node s is the subgraph induced by the nodes v for which G contains a directed path from s to v. Most layering algorithms—in particular also the optimal LP-based approach [11]—will put the nodes on the lowest possible layer. While this is generally a good idea, this approach can be counter-productive in the context of the super source node that will be removed from the final drawing: Since every source node s in G is attached to the super source node $\hat{s}$ (which is on the lowest layer), s may end up very low in the drawing, even though most of its dominated subgraph requires higher layers, hence introducing long arcs.

We tackle this problem using an approach similarly to the *promotion node method* by Nikolov and Tarassov [15] by re-layering parts of dominated subgraphs after the removal of $\hat{s}$ incrementally, without modifying the hierarchical order induced by R. Layers that become empty by these operations can be removed afterwards:

1. **For Each** source s in G (in decreasing order of their layer index j):
   (a) Mark the subgraph dominated by s. Let M_i be the marked nodes on layer L_i ($1 \leq i \leq \ell$).
   (b) **For** $i = j + 1$ **To** ℓ:
      If M_i are all long-arc dummies **Then**
         1. Remove the nodes M_i and lift the marked subgraph on the layers below L_i by one layer.
         2. **If** the new layering causes more edge crossings **Or** more long-arc dummies **Then Undo** step 1. and **Return**

Repositioning the Sources. Since our upward planarization algorithm considers G augmented with $\hat{s}$ and inserts additional arcs considering a fixed embedding, the final upward planar representation may contain artifacts in form of seemingly unnecessary crossings, see, e.g., node '5' in Fig. 4. To overcome this, we sift each source s through all possible positions on its layer and choose the position where it causes the fewest crossings.

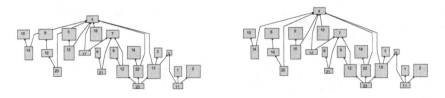

Fig. 5. A drawing of graph *grafo159.24* (Rome graphs) with random node sizes: without (left) and with (right) our bending arcs method and individual layer distance assignment

2.2 Coordinate Assignment

After the previous steps we get a correct layering and node ordering, realizing R. Conceptually, we can use *any* coordinate assignment strategy (e.g., [11,5]) known for Sugiyama's layout; it will always maintain the given number of crossings. Such strategies assign horizontal coordinates to the nodes, while maintaining the given ordering. The aim is to generate drawings such that the subdivided long arcs are drawn as vertical straight lines for their most part.

Yet, when considering the hard-to-measure "beauty" or "readability" of the resulting drawings, we realize that we can improve on traditional coordinate assignment strategies as they usually do not accommodate for the following two drawing problems:

- *node-arc crossings*: A line segment connecting nodes or bend-points between layer L_i and L_{i+1} may cross through some nodes of these two layers. This can easily happen when node sizes are relatively large compared to the layer distance.
- *long-line segments*: The general direction of upward drawings should naturally be along the vertical direction. Yet, there can be arc segments between some layers L_i and L_{i+1} which are very long since they span a large horizontal distance. Such arcs can make Sugiyama-style drawings hard to read.

Fig. 5 shows the benefit of the two strategies described below. Note that these strategies are not only applicable to our layout algorithm, but to any Sugiyama-style layout.

Vertical Coordinates. Usually, the vertical coordinates for the nodes on layer L_i are simply given by $\delta \cdot i$, where δ is the minimal layer distance. Yet, often we may prefer larger distances between layers in order to improve readability: larger distances counter both above problems, but in our context we are in particular interested in long-line segments—we will discuss how to tackle node-arc crossings in the paragraph hereafter.

Buchheim et al. [5] propose a solution for variable layer distance computing which dependents on the gradient of the line segments. However, our experimental results show that drawing DAGs using upward planarization tends to produce drawings with large height. Therefore we use a different approach which limits the maximal layer distance to 3δ.

Let σ_i be the number of arcs between L_i and L_{i+1} whose horizontal dimension is at least 3δ. Then we set the vertical distance between these two layers to $(1 + \min\{\sigma_i/4, 2\})\delta$.

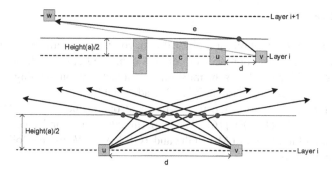

Fig. 6. (top) Avoiding node line-overlapping by introducing new bend-point into an arc e. (bottom) The horizontal coordinates of the bend-points must be all distinct, even if all involved arcs require a bend.

Bending Arcs. While enlarging the layer distance also helps to prevent node-line crossings, the necessary height increase is usually not worth it, from the readability perspective. We therefore propose a strategy that allows trading additional bend-points for layer distance. The strategy can be parameterized to find one's favorite trade-off between these two measures.

Let $X(v)$ and $Y(v)$ denote the horizontal and vertical coordinates of a node v, respectively. An arc (or line segment) $e = (v, w)$ is pointing upward from left to right (right to left) if $X(v) < X(w)$ ($X(v) > X(w)$, resp.). Since purely vertical line segments cannot cross through nodes, we distinguish four cases: e is pointing upward from right to left (or left to right) and v (or w) is a node on layer i. In all these cases, e has to bend if it overlaps some nodes of L_i. However, bending e might cause additional crossings. To avoid this, we also have to bend the line segments that cross the just bended line. W.l.o.g., we only discuss the case $X(v) > X(w)$ with $v \in L_i$. The other cases can be solved analogously.

Let width(v) and height(v) denote the width and height of the bounding box of a node v. Let a be the node on layer L_i with the highest bounding box, and let $\alpha :=$ height$(a)/2$. If v is a bend-point, we do not need to introduce an additional bend. Instead we move v upwards by α. It can happen that v was already shifted downwards before, due to one of our other cases. Then, we bend e by introducing a new bend-point b and set $X(b) := X(v)$ and $Y(b) := Y(v) + 2\alpha$.

Assume v is not a bend-point. Then we have to introduce a bend point along e. Thereby we have to consider that other arcs might also get rerouted, and accommodate enough space for them as well such that no two bend points may coincide. In particular, it might be that the arcs leaving v's left neighbor to the right might also require additional bend points (cf. Fig. 6). Let u be the left neighbor of v on L_i and $d := X(v) - X(u) - $ width$(v)/2 - $ width$(u)/2$ their inner distance. Let r be the number of line segments adjacent to v and pointing from right to left; among these, assume that e is the j-th segment when counting from left to right. Let q be the number of line segments adjacent to u and pointing from left to right. Then, $\Delta := \left(\frac{d}{q+r+1}\right)$ gives the distances between the potential bend points and the coordinates of the new bend-point b are given by $X(b) := X(u) + \frac{\text{width}(u)}{2} + \Delta \cdot (j + \min\{q, j-1\})$ and $Y(b) := Y(v) + \alpha$.

3   Experiments

We investigate the quality of our new algorithm in comparison with known algorithms. We first compare different approaches to draw a computed upward planar representation, i.e., if the crossing number is the most important factor in our drawing. Afterwards, we also compare our approach to Sugiyama's traditional framework.

All algorithms are implemented in the free and open-source (GPL) *Open Graph Drawing Framework (OGDF)* [16]. The experiments were conducted on an IBM Thinkpad with an Intel Pentium M 1.7Ghz and 1GB RAM. We use the following well-known benchmark sets:

Rome Graphs: The Rome graphs [8] are a widely used benchmark set in graph drawing, obtained from a basic set of 112 real-world graphs. The benchmark contains 11528 instances with 10–100 nodes and 9–158 edges. Although the graphs are originally undirected, they have been used as directed graphs by artificially directing the edges according to the node order given in the input files, see, e.g., [10,6].

North DAGs: The North DAGs[1] have been introduced in an experimental comparison of algorithms for drawing DAGs [7]. The set consists of 1277 DAGs collected by Stephen North. The digraphs are grouped into 9 sets, where the first set contains digraphs with 10 to 20 nodes and the i-th set contains $10i + 1$ to $10(i + 1)$ nodes for $i = 2, \ldots, 9$.

3.1   Planarization Layouts

As outlined in the introduction, there are various other possibilities to draw an upward planar representation R of a digraph. Therefore, we use R as the input for the following algorithms. After computing the drawing, we can replace the dummy nodes by usual arc crossings and remove the face-arcs and the super source/sink. By this approach we guarantee that the resulting drawing realizes the specified representation.

We denote our new algorithm by Upward Planarization Layout (*UPL*). We compare it to the *Dominance* drawing style [9], the *Visibility* Representation drawing style [17], and to a straight-forward application of Sugiyama's framework (*UPSugiyama*). For the latter, we use Optimal Ranking [11] for layering, extract the node orders directly from the upward planar representation, and use Fast Hierarchy [5] for coordinate assignment.

Fig. 7(a)–(d) give the height and width of the resulting drawings, respectively, averaged over the digraphs with the same number of nodes. For a fair comparison between the approaches, disregarding any differences due to spacing parameters, we have: The *height* of a drawing is the number of required layers, in case of *UPL* and *UPSugiyama*, and the number of vertical grid coordinates in case of *Dominance* and *Visibility*. The *width* of a drawing is the maximum number of elements per layer, or per horizontal grid line, respectively.

For a fair runtime comparison (Fig. 7(e) and (f), time in seconds), we use the same coordinate assignment algorithm for *UPL* as for *UPSugiyama*. This choice is due to the fact that the alternative ILP approach [11] would require too much time for *UPSugiyama*, which has to consider very large digraphs due to the crossing and long-arc

---

[1] www.graphdrawing.org/data/index.html

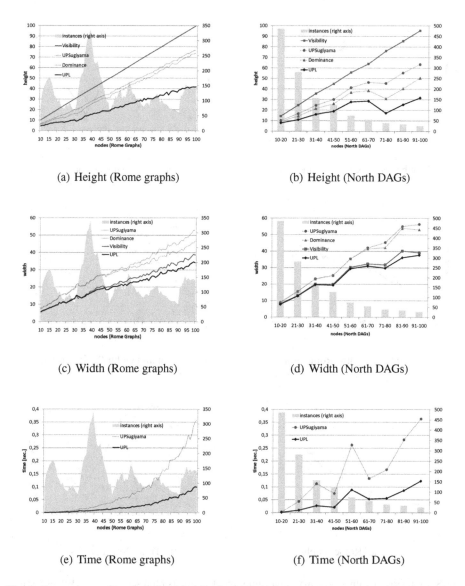

(a) Height (Rome graphs)

(b) Height (North DAGs)

(c) Width (Rome graphs)

(d) Width (North DAGs)

(e) Time (Rome graphs)

(f) Time (North DAGs)

Fig. 7. Comparison with other algorithms to draw upward planar representations; the data points represents average values for the corresponding node groups

dummies. Note that the above height and width measures are invariant under the choice of coordinate assignment. We omit the runtime figures for the linear-time algorithms *Dominance* and *Visibility*, as they are usually below any measurable threshold.

We see that our new approach clearly outperforms all other approaches on the geometry measures, independent of the benchmark set. Also, the runtime comparison shows *UPL* as the clear winner compared to *UPSugiyama*.

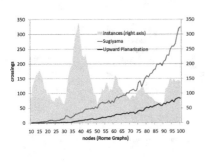

(a) Number of crossings (Rome graphs)

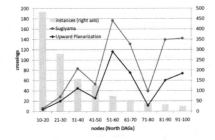

(b) Number of crossings (North DAGs)

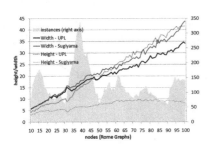

(c) Height and width (Rome graphs)

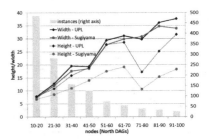

(d) Height and width (North DAGs)

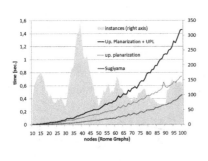

(e) Time (Rome graphs)

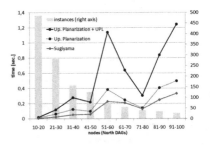

(f) Time (North DAGs)

Fig. 8. Comparison with Sugiyama's algorithm; the data points represents average values for the corresponding node groups

3.2 Comparison with Traditional Sugiyama

Finally, we can investigate how much the requirement of having a drawing with few crossings costs in terms of other quality measures. Therefore we compare *UPL* to a traditional Sugiyama approach that is not bound to a specific upward planar representation. For the latter we use the experimentally most competitive combination by layering via Optimal Ranking [11], using the barycenter heuristic for the crossing reduction step (best of 5 randomized runs), and assigning the coordinates via the exact LP approach [11]. For a fair comparison, *UPL* also uses the latter coordinate assignment algorithm. This time, the runtime of *UPL* furthermore includes the computation of the planarization, as this step is not necessary for Sugiyama. Note that the implementation of the planarization was vastly improved compared to the performance given in [6].

Clearly, the number of crossings in the pure Sugiyama approach are much higher (Fig. 8(a) and (b)), which is consistent with the findings in [6]. As shown in Fig. 8(c) and (d), our *UPL* drawings are of course higher than Sugiyama's, by construction. Yet we observe that the difference is not as large as one might have expected, and *UPL* seems to be a good fit also with respect to this measure, if a small number of crossings is an important issue. We can also see that *UPL* has certain advantages over Sugiyama's approach: A strong packing into few layers will usually require a wider drawing than our planarization. Furthermore, such few layers can in fact be counterproductive from the point of readability, see, e.g., Fig. 1. Overall, we can observe that *UPL* obtains a more balanced aspect ratio than Sugiyama's approach.

In terms of running time (Fig. 8(e),(f)), we see that while Sugiyama's approach is generally faster, *UPL* is not too slow either, requiring below 1.5 seconds for the large instances.

4   Conclusion

Traditional methods of drawing DAGs consider the number of crossings only as a second order priority. If it is of highest priority, one has to use algorithms to draw upward planar representations. Our algorithm constitutes the first such algorithm that takes the special crossing nodes into account. As our experiments show, it generates drawings that are preferable over the other known methods to draw a given upward planar representation. Furthermore, the drawing is also comparable to Sugiyama's approach with respect to other quality measures, while offering a smaller number of crossings.

References

1. Batini, C., Talamo, M., Tamassia, R.: Computer aided layout of entity relationship diagrams. J. Syst. Software 4, 163–173 (1984)
2. Di Battista, G., Pietrosanti, E., Tamassia, R., Tollis, I.G.: Automatic layout of PERT diagrams with X-PERT. In: Proc. IEEE Workshop on Visual Languages, pp. 171–176 (1989)
3. Brandes, U., Köpf, B.: Fast and simple horizontal coordinate assignment. In: Mutzel, P., Jünger, M., Leipert, S. (eds.) GD 2001. LNCS, vol. 2265, pp. 31–44. Springer, Heidelberg (2002)
4. Branke, J., Leppert, S., Middendorf, M., Eades, P.: Width-restricted layering of acyclic digraphs with consideration of dummy nodes. Inf. Process. Lett. 81(2), 59–63 (2002)
5. Buchheim, C., Jünger, M., Leipert, S.: A fast layout algorithm for k-level graphs. In: Marks, J. (ed.) GD 2000. LNCS, vol. 1984, pp. 229–240. Springer, Heidelberg (2001)
6. Chimani, M., Gutwenger, C., Mutzel, P., Wong, H.-M.: Layer-free upward crossing minimization. In: McGeoch, C.C. (ed.) WEA 2008. LNCS, vol. 5038, pp. 55–68. Springer, Heidelberg (2008)

7. Di Battista, G., Garg, A., Liotta, G., Parise, A., Tamassia, R., Tassinari, E., Vargiu, F., Vismara, L.: Drawing directed acyclic graphs: An experimental study. Int. J. Comput. Geom. Appl. 10(6), 623–648 (2000)

8. Di Battista, G., Garg, A., Liotta, G., Tamassia, R., Tassinari, E., Vargiu, F.: An experimental comparison of four graph drawing algorithms. Comput. Geom. Theory Appl. 7(5-6), 303–325 (1997)

9. Di Battista, G., Tamassia, R., Tollis, I.G.: Area requirement and symmetry display of planar upward drawings. Discrete Comput. Geom. 7(4), 381–401 (1992)

10. Eiglsperger, M., Kaufmann, M., Eppinger, F.: An approach for mixed upward planarization. J. Graph Algorithms Appl. 7(2), 203–220 (2003)

11. Gansner, E., Koutsofios, E., North, S., Vo, K.-P.: A technique for drawing directed graphs. Software Pract. Exper. 19(3), 214–229 (1993)

12. Gutwenger, C., Mutzel, P.: An experimental study of crossing minimization heuristics. In: Liotta, G. (ed.) GD 2003. LNCS, vol. 2912, pp. 13–24. Springer, Heidelberg (2004)

13. Healy, P., Nikolov, N.S.: A branch-and-cut approach to the directed acyclic graph layering problem. In: Goodrich, M.T., Kobourov, S.G. (eds.) GD 2002. LNCS, vol. 2528, pp. 98–109. Springer, Heidelberg (2002)

14. Healy, P., Nikolov, N.S.: How to layer a directed acyclic graph. In: Mutzel, P., Jünger, M., Leipert, S. (eds.) GD 2001. LNCS, vol. 2265, pp. 16–30. Springer, Heidelberg (2002)

15. Nikolov, N.S., Tarassov, A.: Graph layering by promotion of nodes. Discrete Applied Mathematics 154(5), 848–860 (2006)

16. OGDF – the Open Graph Drawing Framework. Technical University of Dortmund, Chair of Algorithm Engineering, http://www.ogdf.net

17. Rosenstiehl, P., Tarjan, R.E.: Rectilinear planar layouts and bipolar orientations of planar graphs. Discrete Comput. Geom. 1(1), 343–353 (1986)

18. Sugiyama, K., Tagawa, S., Toda, M.: Methods for visual understanding of hierarchical system structures. IEEE Trans. Sys. Man. Cyb. 11(2), 109–125 (1981)

More Flexible Radial Layout

Ulrik Brandes[1] and Christian Pich[2]

[1] Department of Computer & Information Science, University of Konstanz
Ulrik.Brandes@uni-konstanz.de
[2] Chair of Systems Design, ETH Zürich*
cpich@ethz.ch

Abstract. We describe an algorithm for radial layout of undirected graphs, in which nodes are constrained to the circumferences of a set of concentric circles around the origin. Such constraints frequently occur in the layout of social or policy networks, when structural centrality is mapped to geometric centrality, or when the primary intention of the layout is the display of the vicinity of a distinguished node. We extend stress majorization by a weighting scheme which imposes radial constraints on the layout but also tries to preserve as much information about the graph structure as possible.

1 Introduction

In radial graph layout the nodes are constrained to be located on a set of concentric rings; for some or all nodes in a graph a radius is given, which typically encodes the results of a preceding analysis step. Such drawings date back to the 1940s and are called *ring* or *target diagrams* [18]. The interpretation of these rings is specific to the particular application at hand. The overall goals which guide the design of radial layouts can be expressed as two criteria and are possibly contradictory:

- *Representation of distances:* The Euclidean distance between two nodes in the drawing should correspond to their graph-theoretical distance. This is a general objective common to all "organic" layout styles.
- *Radial constraints:* Nodes are associated with the radius of a circle centered in the origin, and are constrained to be placed on the circumference of this circle.

Radial layout occurs as a task in several applications. It is used for the exploration of large hierarchies in [19]; the hierarchy is laid out radially as a tree, followed by an incremental force-based placement. This approach was later modified for dynamic real-time exploration of a filesharing network in [20], where users interactively select a node to be moved into the center, while the current immediate surrounding of that node is updated.

---

* Part of this work was done while the author was at the University of Konstanz, Department of Computer & Information Science.

D. Eppstein and E.R. Gansner (Eds.): GD 2009, LNCS 5849, pp. 107–118, 2010.

A different approach is to extend the definition of level planarity to discrete radial levels [1]; the traditional Sugiyama framework is enhanced accordingly for linear-time embedding of level-planar graphs.

In the case of continuous radii representing some kind of substance, unary constraints are imposed on the drawing for mapping centrality scores to visual centrality [4]. The layout is done by simulated annealing, which allows for penalty costs, e.g. for edge crossings, and is very flexible but also computationally demanding; this prohibits interactivity even for moderately sized graphs.

In the following, the two essential goals above are explicitly formulated as objective functions, which measure how far a layout is from meeting the criteria, and which are sought to be minimized. While the first objective is captured by the traditional energy or stress measures, we try to fulfill the second objective by introducing radial constraints into the energy-based layout model and using a linear combination of the two objectives.

Quite recently, extensions of the stress term have been used for drawing graphs with explicitly formulated aesthetic criteria, such as the uniform scattering of the nodes in a graph over a unit disk [16], penalizing node overlaps [11], or preserving a given topology [10].

All these approaches modify the *distances* themselves in one form or another, while the approach presented in this contribution is based on engineering the *weights* used in the stress minimization model. The weights are coefficients of error terms involved in the quality criteria to be minimized. If chosen carefully, the weights can be used to influence the configuration resulting from optimizing the stress function modified by these weights; see Fig. 1 for an example. We are not aware of previous work which makes systematic use of such a weighting scheme to take up a particular perspective on a data set.

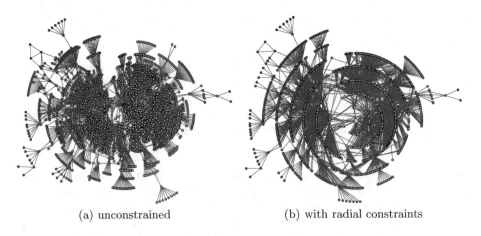

     (a) unconstrained   (b) with radial constraints

Fig. 1. Layouts of a social network (2075 nodes, 4769 edges), consisting of two known clusters. The brightness of node colors is proportional to the graph-theoretical distances from a distinguished focal node, which also defines the radii used in the constrained layout, in which the two clusters are still visible.

2   Preliminaries

Let $G = (V, E)$ be an undirected graph, $E \subseteq \binom{V}{2}$. We will denote the cardinalities of the node and edge sets by $n = |V|$ and $m = |E|$, respectively; it will be convenient to index nodes by numbers, $V = \{1, \ldots, n\}$. The graph-theoretical distance between two nodes i, j is the number of edges on a shortest path between i and j and is denoted $d_{i,j}$ or, when there is no danger of confusion, d_{ij}. The matrix $D = (d_{ij})_{ij} \in \mathbb{R}^{n \times n}$ contains the distances between every two nodes in G; the diameter of G is the maximum distance between any two nodes in G, $\text{diam}(G) = \max_{i,j \in V} d_{ij}$. All graphs are assumed to be connected; otherwise, all connected components are just considered individually.

Two-dimensional node positions are denoted $p(i) = (x_i, y_i)$; a layout of n nodes is captured by two column vectors $x = [x_1, \ldots, x_n]^T, y = [y_1, \ldots, y_n]^T \in \mathbb{R}^n$. The Euclidean distance between two nodes in a given layout p is defined as $\|p(i) - p(j)\| = \left((x_i - x_j)^2 + (y_i - y_j)^2\right)^{1/2}$.

3   Stress, Weights, and Constraints

3.1   Stress

The foundation of the method presented in the following is multidimensional scaling (MDS) [2,7]; originating in psychometrics and the social sciences, MDS has been established and widely used in the graph drawing community for more than three decades, as energy-based placement [15]. While there is a wide range of variants and extensions, we will concentrate on the *stress minimization* approach [12] in this contribution.

Given a set of desired distances among a set of n objects, the overall goal is to place these objects in a low-dimensional Euclidean space in such a way that the resulting distances fit the desired ones as well as possible. In the graph drawing literature, the desired distances are usually graph-theoretical distances d_{ij}, and the goal is to find two-dimensional positions $p(1), \ldots, p(n)$ with

$$\|p(i) - p(j)\| \approx d_{ij} \tag{1}$$

attained as closely as possible for all pairs i, j. When the configuration is not required to satisfy any further constraints, the objective function, called *stress*, is the sum of squared residuals

$$\sigma(p) = \sum_{i<j} w_{ij}\left(d_{ij} - \|p(i) - p(j)\|\right)^2 \tag{2}$$

over all the $n(n-1)/2$ pairs of nodes, where $w_{ij} \geq 0$ is a weight for the contribution of the particular error term $(d_{ij} - \|p(i) - p(j)\|)^2$ for pair i, j to the stress.

There is a wide consensus that configurations with a small stress value tend to be aesthetically pleasing. The state-of-the-art approach to finding such layouts is *stress majorization* [8,12]; starting from an initial configuration, it generates a sequence of improving layouts. When no geometric coordinates are at hand, the iteration may be initialized at random; however, more favorable and robust strategies are available for initial layouts, such as classical MDS [5].

In an iterative process, new coordinates $\hat{x} = [\hat{x}_1, \ldots, \hat{x}_n]^T, \hat{y} = [\hat{y}_1, \ldots, \hat{y}_n]^T \in \mathbb{R}^n$ are computed from the current ones with the update

$$\hat{x}_i \leftarrow \frac{\sum_{j \neq i} w_{ij} (x_j + d_{ij} \cdot (x_i - x_j) \cdot b_{ij})}{\sum_{j \neq i} w_{ij}} \tag{3}$$

$$\hat{y}_i \leftarrow \frac{\sum_{j \neq i} w_{ij} (y_j + d_{ij} \cdot (y_i - y_j) \cdot b_{ij})}{\sum_{j \neq i} w_{ij}} \tag{4}$$

where

$$b_{ij} = \begin{cases} \frac{1}{\|p(i) - p(j)\|} & \text{if } \|p(i) - p(j)\| > 0, \\ 0 & \text{otherwise.} \end{cases} \tag{5}$$

This is repeated until the relative change in the configuration is below some threshold value, or after a predefined number of steps. The sequence of layouts generated in this way can be shown to have non-increasing stress and to converge towards a local minimum [9].

3.2   Weights for Constraints

In early applications of MDS, each pair i, j of objects was assigned the same unit weight by setting $w_{ij} = 1$ in (2); when a desired distance was not known for a pair, this pair was simply ignored by using a zero weight for its contribution to the stress.

In graph drawing it is a de-facto standard to set $w_{ij} = d_{ij}^{-2}$ to emphasize the quality of the fit of local distances, i.e., the contribution of pairs i, j with smaller target distances is increased compared to pairs with larger distances. This weighting scheme was introduced in *elastic scaling* by McGee [17], and is equal to the one used by Kamada and Kawai [15]. Instead of fitting absolute values by minimizing absolute residual error terms $(d_{ij} - \|p(i) - p(j)\|)^2$, the goal is to achieve a fit of the distance magnitudes, expressed by relative error terms $(1 - \|p(i) - p(j)\|/d_{ij})^2$. Summing these over all pairs gives the sum

$$\sum_{i<j} \left(1 - \frac{\|p(i) - p(j)\|}{d_{ij}}\right)^2 = \sum_{i<j} \frac{1}{d_{ij}^2} (d_{ij} - \|p(i) - p(j)\|)^2. \tag{6}$$

In this sum the impact of larger distances in the unweighted stress (2) is lessened, which is due to the square in the error term.

A reason for the favorable aesthetic properties of low-stress layouts is that no node is preferred over others because minimizing the objective function tries to achieve a balance in the fit of the desired distances. In most scenarios this is appropriate and tends to give the drawing a balanced appearance.

In some cases, users want to put more emphasis on some nodes, while other nodes are regarded less important, by centering the view on a node and visualizing this node's neighborhood more prominently. This can be done by introducing suitable constraints on the configuration; when these constraints can be formulated as desired distances, choosing the weights in a suitable way allows for imposing them on the resulting layout.

What follows is a general framework for constraining drawings; while the range of possible applications is wide, our contribution will concentrate on the radial scenario. To avoid confusion, the objective function (2) will be termed *distance stress* and denoted $\sigma_W(p)$, with the subscript indicating that the stress is modified by a weight matrix $W = (w_{ij})_{ij} \in \mathbb{R}^{n \times n}$. This stress model is enhanced by a second set of weights $Z = (z_{ij})_{ij}$ used in the *constraint stress*

$$\sigma_Z(p) = \sum_{i<j} z_{ij} \left(d_{ij} - \|p(i) - p(j)\| \right)^2 \tag{7}$$

whose minimization tries to fit the same distances and hence aims at representing the same information, but highlights different aspects.

3.3 Interpolated Weights

A straightforward approach to imposing the constraints expressed in a weight matrix is to directly minimize (7), but the resulting solutions tend to be trivial; for example, consider a linear layout in which $x_i = r_i, y_i = 0$ for all $i \in V$. Instead, it is more effective to combine distance and constraint stress into a joint majorization process, operating on a linear combination of the stress measures $\sigma_W(p)$ and $\sigma_Z(p)$.

Initially, the nodes are allowed to move freely without considering the constraints at all, by minimizing just $\sigma_W(x, y)$. Then, the constraints are granted more and more control over the appearance of the drawing by dynamically changing the coefficients in this combination, and the bias is shifted from one to the other criterion [3]. The influences of the distance and the radial components are determined by the coefficients in the linear combination

$$\sigma^t(p) = (1 - t) \cdot \sigma_W(p) + t \cdot \sigma_Z(p) \tag{8}$$

and the update terms for the majorization process (3) and (4) become

$$\hat{x}_i \leftarrow \frac{\sum_{j \neq i} \left((1 - t) \cdot w_{ij} + t \cdot z_{ij} \right) \cdot \left(x_j + d_{ij} \cdot (x_i - x_j) \cdot b_{ij} \right)}{\sum_{j \neq i} \left((1 - t) \cdot w_{ij} + t \cdot z_{ij} \right)}, \tag{9}$$

$$\hat{y}_i \leftarrow \frac{\sum_{j \neq i} \left((1 - t) \cdot w_{ij} + t \cdot z_{ij} \right) \cdot \left(y_j + d_{ij} \cdot (y_i - y_j) \cdot b_{ij} \right)}{\sum_{j \neq i} \left((1 - t) \cdot w_{ij} + t \cdot z_{ij} \right)}. \tag{10}$$

In the majorization, the radial constraints are not directly and immediately enforced; rather, the main visual features of the initial configuration are preserved. Then the bias is shifted from the distance component towards the radial component by gradually increasing t from 0 to 1. When the number of iteration steps

k is predefined, a linear interpolation gives values $t = 0, \frac{1}{k}, \frac{2}{k}, \ldots, \frac{k-1}{k}, 1$; otherwise, the iterative process may be simply be repeated with a sequence of values for k converging to 1 from below until the layout is sufficiently stable. Using either variant, in each step, a slightly different objective function is sought to be minimized, and the current iterate preconditions the next step, thus keeping the series of iterates continuous.

In the multidimensional scaling literature, occasionally a distinction is made between *weakly* and *strongly constrained* MDS problems [13]. In the former, the solutions are allowed to deviate from the given constraints, and this deviation is penalized by additional stress; in the latter, only solutions which exactly satisfy the constraints are feasible. In a way, a strongly constrained MDS problem can be thought of as a special case of a weakly constrained problem, in which the deviation penalty is zero.

In this terminology, setting $t = 1$ in (8) turns a weakly constrained problem into a strongly constrained one, provided that the set of constraints is realizable, i.e., a solution with zero constraint stress exists. In all other cases, it should be noted that, even though the distance component vanishes when $t \to 1$, minimizing $\sigma^t(p)$ is *not* the same as minimizing $\sigma_Z(p)$ because of the running preconditioning described above.

4   Radial Layout

4.1   Target Diagrams

The *focus* is put on a node by emphasizing the visual display of its vicinity, constraining all others to attain Euclidean distances corresponding to their graph-theoretical distances, i.e., relative to the focused node, all structural distance-k neighborhoods are mapped to a geometric k-neighborhood.

The constraint weight matrix takes only pairs of nodes into account in which the focused node is involved, while reducing all other weights to zero. Without loss of generality, let n be the index of the node to be focused. D and W are defined as above, and the constraint weight matrix $Z = (z_{ij})_{ij}$ has only zero entries except for the n-th row and column, which contains weights

$$
z_{ij} = \begin{cases} w_{ij} & \text{if } i = n \text{ or } j = n \\ 0 & \text{otherwise} \end{cases} \tag{11}
$$

derived from the distances to the focal node, so that interpolating from W to Z gradually increases the focal node's relative impact on the configuration.

A famous social network was studied by Zachary and, subsequently, many other sociologists. It describes the friendship relations among 34 members in a karate club in a US university in the 1970s [21]. Over the course of a two-year study, the network breaks apart into two clubs because of disagreements between the administrator and the instructor, who leaves the club and takes about half of the members with him. Fig. 2 shows how the same initial layout is modified to a radial layout focused on the instructor (a) and the administrator (b).

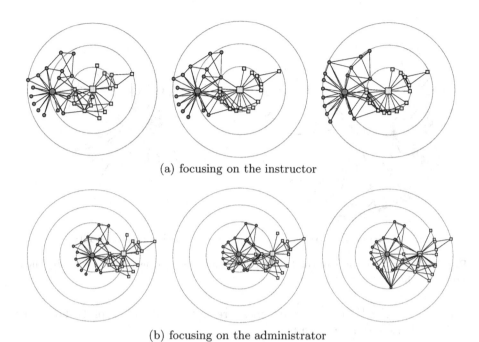

(a) focusing on the instructor

(b) focusing on the administrator

Fig. 2. Radial layouts of Zachary's karate club network ($n = 34, m = 77$), by weight interpolation, for $t \in \{0, 0.9, 1\}$. Members leaving with the instructor are shown as yellow squares, members staying with the administrator as red circles.

4.2 Centrality Drawings

When the radial constraints do not directly correspond to one of the columns in the distance matrix, we assume that this additional input is given as a vector $r = [r_1, \dots, r_n]^T \in \mathbb{R}^n$, with $r_i \geq 0$ for all $i \in \{1, \dots, n\}$. The radial constraints can be formulated in terms distances to the center, added to the distance matrix; since node i is located on a circle with radius r_i if its Euclidean distance to the center is equal to r_i, and the center has coordinates $(0, 0)$, this is equivalent to

$$\|p(i)\| \approx r_i. \tag{12}$$

The origin is treated as an additional *dummy* node indexed with $n + 1$. The stress majorization procedure is applied to a layout problem of $n + 1$ objects; in [3] such a dummy is used to enforce a circular configuration by using the same radius for all objects. The distance and weight matrices involved in (8) are

$$D = \begin{bmatrix} d_{11} & \cdots & d_{1n} & r_1 \\ \vdots & \ddots & \vdots & \vdots \\ d_{n1} & \cdots & d_{nn} & r_n \\ r_1 & \cdots & r_n & 0 \end{bmatrix}, \; W = \begin{bmatrix} d_{11}^{-2} & \cdots & d_{1n}^{-2} & 0 \\ \vdots & \ddots & \vdots & \vdots \\ d_{n1}^{-2} & \cdots & d_{nn}^{-2} & 0 \\ 0 & \cdots & 0 & 0 \end{bmatrix}, \; Z = \begin{bmatrix} 0 & \cdots & 0 & r_1^{-2} \\ \vdots & \ddots & \vdots & \vdots \\ 0 & \cdots & 0 & r_n^{-2} \\ r_1^{-2} & \cdots & r_n^{-2} & 0 \end{bmatrix} \tag{13}$$

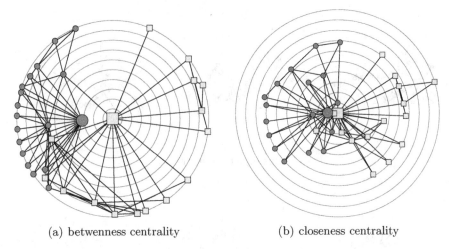

(a) betwenness centrality   (b) closeness centrality

Fig. 3. Centrality layouts of the karate club social network, using two centrality measures to define the radii of nodes

Let $c = [c_1, \ldots, c_n]^T$ be a centrality measure on the nodes of graph G. For every node $i \in V$ its radius is given by

$$r_i = \frac{\text{diam}(G)}{2} \cdot \left(1 - \frac{c_i - \min\limits_{j \in V} c_j}{\max\limits_{j \in V} c_j - \min\limits_{j \in V} c_j + c(G)} \right), \qquad (14)$$

where multiplying with half the diameter serves to keep distances and the radial constraints on the same scale, and $c(G)$ is an offset parameter that prevents more than one maximally central nodes from coinciding in the center [4]. Simplified pseudo-code, which is targeted at radial constraints, is given in Algorithm 1. The majorization is realized as a local variant, in which the coordinates are updated node-by-node immediately.

For dynamic visualization scenarios, an inherently smooth transition between layouts with different foci is obtained by simply using the intermediate layouts given by the steps in the majorization process. In the transition from one focus to the other, it is advantageous to not directly interpolate between the two corresponding constraint weight matrices, but to take a detour via the original weight matrix having entries d_{ij}^{-2}, so as to re-introduce all the shortest-path distances to remove artifacts potentially introduced after focusing on the first node.

4.3   Travel Time Maps

When traveling with a public transportation system, schematic maps are essential for many users. Such maps depict lines, stations, zones, and connections to other traffic systems. Since the primary use for such a map is travel planning within this network, usability and readability are more important criteria than

Algorithm 1. Radial layout

Input: connected undirected graph $G = (\{1, \ldots, n\}, E)$, radii $r_1, \ldots, r_n \in \mathbb{R}_+$
number of iterations $k \in \mathbb{N}$
Output: coordinates $x, y \in \mathbb{R}^n$ with $\sqrt{x_i^2 + y_i^2} = r_i$ for all $i \in \{1, \ldots, n\}$

$D \leftarrow$ matrix of shortest path distances d_{ij}
$W \leftarrow$ matrix of inverse squared distances d_{ij}^{-2}
$(x, y) \leftarrow$ layout with low $\sigma_W(x, y)$
for $t = 0, \frac{1}{k}, \frac{2}{k}, \ldots, \frac{k-1}{k}, 1$ **do**

   $a_i \leftarrow \|p(i)\|^{-1}$ if $\|p(i)\| > 0, 0$ otherwise
   set all b_{ij} as in (5)
   foreach $i \in \{1, \ldots, n\}$ **do**

$$x_i \leftarrow \frac{\sum_{j \neq i}(1-t) \cdot w_{ij}\big(x_j + d_{ij} \cdot (x_i - x_j) \cdot b_{ij}\big) + t \cdot r_i^{-2}(r_i x_i a_i)}{(1-t)\sum_{j \neq i} w_{ij} + t \cdot r_i^{-2}}$$

$$y_i \leftarrow \frac{\sum_{j \neq i}(1-t) \cdot w_{ij}\big(y_j + d_{ij} \cdot (y_i - y_j) \cdot b_{ij}\big) + t \cdot r_i^{-2}(r_i y_i a_i)}{(1-t)\sum_{j \neq i} w_{ij} + t \cdot r_i^{-2}}$$

the accurate representation of actual geographic positions. In the graph drawing literature, this drawing style is called *metro map layout* [14].

One of the most prominent schematic maps is Beck's famous London tube map; it has been and is still being reworked and improved and has inspired similar maps for systems of public transportation all over the world. While schematic maps are widely perceived as very useful, a potential drawback is that they tend to distort a user's perception of closeness, thus compromising the decisions made in the travel planning process, e.g., because stations are displayed as more proximate than they actually are.

If the starting and ending stations of a planned journey are known, the radial constraints can be used to highlight the time needed for traveling between them, by focusing only on one, as described above. In addition, shortest paths between the two stations can be highlighted by putting the focus on both of them at the same time.

Let the nodes in the focus be $n - 1$ and n, without loss of generality. Again, $D, W \in \mathbb{R}^{n \times n}$ are defined as the matrices of shortest-path distances and their inverse squares, respectively. The weight matrix used is $Z = (z_{ij})_{ij} \in \mathbb{R}^{n \times n}$ with

$$z_{ij} = \begin{cases} w_{ij} & \text{if } i \in \{n-1, n\} \text{ or } j \in \{n-1, n\} \\ 0 & \text{otherwise} \end{cases} \tag{15}$$

and contains a $(n - 2) \times (n - 2)$ submatrix with zero entries.

We use the connection graph of the London tube with estimated minimal travel times obtained from the Transport for London web site, or derived from the geographic distance, where estimates are not available. For the sake of simplicity, we did not consider the time needed to walk from one track to the other when changing lines (see also [6]).

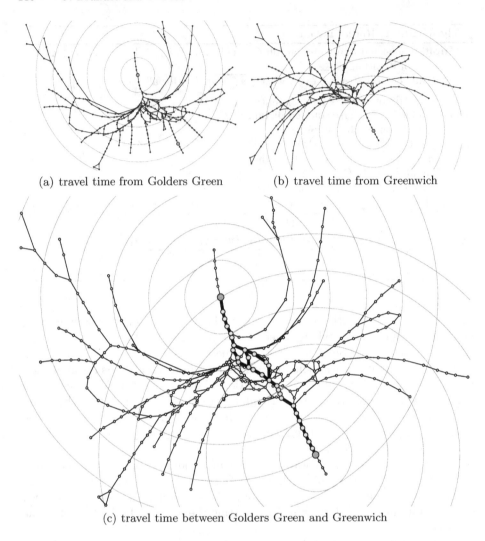

(a) travel time from Golders Green   (b) travel time from Greenwich

(c) travel time between Golders Green and Greenwich

Fig. 4. Radial layouts of the London Tube graph using estimated travel times. The concentric circles indicate travel times in multiples of 10 minutes. The stations are constrained to be at distance equal to their minimum travel times.

Radial layouts are given in Fig. 4, where stations are placed at a distance from the center proportional to their estimated minimum travel times from (a) Golders Green and (b) Greenwich independently and (c) from both stations at the same time.

5   Conclusion

Radial constraints fit well into the framework of multidimensional scaling by stress majorization because the radii can be expressed in terms of desired

Euclidean distances, which requires only minor modifications of available implementations of the stress minimization.

Motivated by the results obtained from the relatively simple radial constraints, and other experiments, we feel that they deserve more attention, because they allow the aesthetic goals of the visualization results to be explicitly formulated and quantified, and can be easily plugged into existing algorithms.

We think that, with the careful choice of a weighting scheme, the ideas presented above are easily carried over to layout tasks with more general constraints, such as the display of grouping, the computation of dynamic layouts, and the visualization of edge strength, certainty, or probability.

References

1. Bachmaier, C., Brandenburg, F.J., Forster, M.: Radial level planarity testing and embedding in linear time. Journal of Graph Algorithms and Applications 9(1), 53–97 (2005)
2. Borg, I., Groenen, P.: Modern Multidimensional Scaling. Springer, Heidelberg (2005)
3. Borg, I., Lingoes, J.: A model and algorithm for multidimensional scaling with external constraints on the distances. Psychometrika 45(1), 25–38 (1980)
4. Brandes, U., Kenis, P., Wagner, D.: Communicating centrality in policy network drawings. IEEE Transactions on Visualization and Computer Graphics 9(2), 241–253 (2003)
5. Brandes, U., Pich, C.: An experimental study on distance-based graph drawing. In: Tollis, I.G., Patrignani, M. (eds.) GD 2008. LNCS, vol. 5417, pp. 218–229. Springer, Heidelberg (2009)
6. Carden, T.: Travel time tube map,
   http://www.tom-carden.co.uk/p5/tube_map_travel_times/applet/
7. Cox, T.F., Cox, M.A.A.: Multidimensional Scaling. CRC/Chapman and Hall, Boca Raton (2001)
8. de Leeuw, J.: Applications of convex analysis to multidimensional scaling. In: Barra, J.R., Brodeau, F., Romier, G., van Cutsem, B. (eds.) Recent Developments in Statistics, pp. 133–145. North-Holland, Amsterdam (1977)
9. de Leeuw, J.: Convergence of the majorization method for multidimensional scaling. Journal of Classification 5(2), 163–180 (1988)
10. Dwyer, T., Marriott, K., Wybrow, M.: Topology preserving constrained graph layout. In: Tollis, I.G., Patrignani, M. (eds.) GD 2008. LNCS, vol. 5417, pp. 230–241. Springer, Heidelberg (2009)
11. Gansner, E.R., Hu, Y.: Efficient node overlap removal using a proximity stress model. In: Tollis, I.G., Patrignani, M. (eds.) GD 2008. LNCS, vol. 5417, pp. 206–217. Springer, Heidelberg (2009)
12. Gansner, E.R., Koren, Y., North, S.: Graph drawing by stress majorization. In: Pach, J. (ed.) GD 2004. LNCS, vol. 3383, pp. 239–250. Springer, Heidelberg (2005)
13. Heiser, W.J., Meulman, J.: Constrained multidimensional scaling, including confirmation. Applied Psychological Measurement 7(4), 381–404 (1983)
14. Hong, S.-H., Merrick, D., do Nascimiento, H.A.D.: The metro map layout problem. In: Proceedings of the 2004 Australasian symposium on Information Visualisation. ACM International Conference Proceeding Series, pp. 91–100 (2004)

15. Kamada, T., Kawai, S.: An algorithm for drawing general undirected graphs. Information Processing Letters 31, 7–15 (1989)
16. Koren, Y., Çivril, A.: The binary stress model for graph drawing. In: Tollis, I.G., Patrignani, M. (eds.) GD 2008. LNCS, vol. 5417, pp. 193–205. Springer, Heidelberg (2009)
17. McGee, V.E.: The multidimensional scaling of "elastic" distances. The British Journal of Mathematical and Statistical Psychology 19, 181–196 (1966)
18. Northway, M.L.: A method for depicting social relationships obtained by sociometric testing. Sociometrics 3, 144–150 (1940)
19. Wills, G.J.: NicheWorks – interactive visualization of very large graphs. In: Di-Battista, G. (ed.) GD 1997. LNCS, vol. 1353, pp. 403–414. Springer, Heidelberg (1997)
20. Yee, K.-P., Fisher, D., Dhamija, R., Hearst, M.: Animated exploration of dynamic graphs with radial layout. In: Proc. InfoVis, pp. 43–50 (2001)
21. Zachary, W.W.: An information flow model for conflict and fission in small groups. Journal of Anthropological Research 33, 452–473 (1977)

WiGis: A Framework for
Scalable Web-Based Interactive Graph Visualizations

Brynjar Gretarsson, Svetlin Bostandjiev, John O'Donovan, and Tobias Höllerer

Department of Computer Science
University of California, Santa Barbara
{brynjar,alex,jod,holl}@cs.ucsb.edu

Abstract. Traditional network visualization tools inherently suffer from scalability problems, particularly when such tools are interactive and web-based. In this paper we introduce *WiGis* –Web-based Interactive Graph Visualizations. *WiGis*[1] exemplify a fully web-based framework for visualizing large-scale graphs natively in a user's browser at interactive frame rates with no discernible associated startup costs. We demonstrate fast, interactive graph animations for up to hundreds of thousands of nodes in a browser through the use of asynchronous data and image transfer. Empirical evaluations show that our system outperforms traditional web-based graph visualization tools by at least an order of magnitude in terms of scalability, while maintaining fast, high-quality interaction.

1 Introduction

This paper presents a novel visualization framework which supports user interactions with large graphs in a web browser without the need for plug-ins or special-purpose runtime systems. Our framework follows the common visual information browsing principle: "Overview first, zoom and filter, then details on demand"[15] for exploration of large information spaces. The *WiGis* framework supports information discovery in two phases. Firstly, by enabling users to visualize and interact with large scale network data, our framework provides a "big picture" of the information space. Secondly, interaction is used to mold large scale data into the user's own mental model, which serves as a useful starting point for more fine-grained analysis.

Many tools for visualizing graphs have previously been developed. Some of these tools run in a web-browser [6, 16, 17] while others are scalable for up to hundreds of thousands of nodes [2, 14]. However, to our knowledge, no web-based tools exist which are capable of interactive visualization of graphs at such scale [13]. We have developed an extensible framework which enables interactive visualization of hundreds of thousands of nodes natively in a web browser.

Based on our analysis of existing interactive web-based graph visualization systems, we find that their main scalability limitation is due to the fact that most of them implement some form of a thick client solution and subsequently need to load the entire graph onto the client machine. In addition to large startup costs, this limits the maximum

---

[1] www.wigis.net

D. Eppstein and E.R. Gansner (Eds.): GD 2009, LNCS 5849, pp. 119–134, 2010.

size of visualized graphs due to memory limitations of the browser or the client computer. We circumvent these limitations by leveraging a novel technique for storing and processing graph data on a powerful remote server. The server continuously produces bitmap images and asynchronously sends them to the client's browser. This provides a smooth interactive animation natively in the browser. For example, we achieve more than 10 frames per second for graphs of 10,000 nodes and 20,000 edges, with minimal requirements for memory and processing power on the client machine. We believe that this is an important contribution which supports interactive visualization of large graphs even on machines with limited resources, such as mobile devices.

We use the term *WiGis*, or Web-based Interactive Graph Visualizations, for our framework, which is a new addition to the set of currently available tools providing "visualization as a service" [6]. This paper describes and evaluates the new framework with a range of test datasets with focus on performance in terms of speed and scalability. Furthermore, we present a comparative experiment in which our system exhibits similar results to the best performing desktop applications, while supporting visualization of over one million nodes, albeit at lower frame rates. None of the tested systems could visualize graphs of this size. In comparison with other web-based systems, *WiGis* improved on the next best performer by an order of magnitude in terms of scale while also performing at least as well or better in terms of speed.

The remainder of this paper is organized as follows: A critical review of current relevant work in the area of large graph visualizations with a focus on web-based approaches is presented in Section 2. Section 3 describes the architecture of *WiGis* in terms of design and implementation. Section 4 discusses an empirical evaluation of our visualization tool (and its component parts) in terms of scalability and speed with respect to popular graph visualization tools. Section 5 contains a brief discussion of the benefits and limitations of our technique as well as various deployments of the system. The paper concludes with a summary of the main contributions.

2 Background

Much research has been conducted on large scale graph visualizations, e.g. [2, 8, 10, 14]. Traditionally, graph visualization applications have been desktop based. For example, Cytoscape [14], Pajek [4], Tulip [2], and some implementations of Tom Sawyer Visualization [16]. Over the past few years, increased web-accessibility and bandwidth improvements have triggered a general shift towards rich internet applications (RIAs) capable of providing interactive and responsive interfaces through a web browser. This shift has a potential benefit for resource-intensive graph visualization, and applications which take advantage of the rich-internet paradigm are beginning to emerge for visualization of graph and network data. Examples of such applications include Touchgraph [17], Tom Sawyer Visualization [16] and IBM's Many Eyes [6]. In general these applications either do not scale past thousands of nodes or are not fully interactive.

Thick v/s Thin Clients. RIAs can be loosely classified into thick and thin clients. A thick client typically provides rich functionality that is largely independent of a central server with the majority of processing done on the client, whereas a thin client requires constant communication with a server to provide functionality. Client-based visualization [20, 1] can be considered a thick-client solution since data is downloaded from a server and the visualization and rendering are done at the client side. The popular graph visualization tool Touchgraph Navigator [17] is a good example of a client-based tool, since it processes graph interactions locally in Java. A thick client solution needs to initially download the entire graph data, which may be on the order of GBs for large graphs, onto the client machine. This severely limits the size of the largest graph a thick client application can handle and poses significant startup times. Once the graph data is obtained there is no guarantee that the client has enough memory available to handle the data. Visualizations can be created on a remote server and passed across a network to the client. This is referred to by Wood et al. [20] as "server-based" visualization, and is an example of a thin client solution. *WiGis* uses both client-based and server-based visualization techniques. An important innovation is the way the system can automatically and transparently switch between the two modes while allowing smooth interaction in both.

Plug-in v/s Native Applications. RIAs can be further classified based on the manner in which they are deployed. Many RIAs are implemented using some form of browser plug-in, for example Java Applets, Adobe Flash or Microsoft Silverlight. The majority of graph visualization tools available on the web are plug-in based, e.g. [17, 6]. There are some fundamental drawbacks with the plug-in approach however. Firstly from a scalability perspective, plug-in based RIAs are limited to the capabilities of the plug-in itself. For example, the default memory limit for Java Applets is usually around 60-90 MB. Furthermore, from a security perspective third party plug-ins are usually not open source and need to access client resources, making them a potential security threat. Many large organizations with sensitive data require every line of third party code to be checked for malicious behaviour before deployment on an analysts machine. The alternative approach to plug-in based RIA implementation is to provide functionality natively in the browser through a combination of DHTML and AJAX. Since this approach does not require any access to client resources outside of the browser it is much more secure with respect to integrity of the client machine. Examples of native RIA's include Google Maps and the JSP and ASP.net implementations of Tom Saywer Visualization [16]. Since we are concerned about scalability we opted to design our framework as a native RIA to avoid the inherent limitations and security drawbacks of browser plug-ins.

While plug-ins such as Flash and Applets provide rich functionality such as object support and dynamic components, they are not directly suited to solve the problem of large scale graph visualization on the web. In order to fully utilize these rich features for visualization of large graphs the entire graph model would have to exist on the client machine. For large graphs this is not a feasible solution due to potential memory limitations on the client machine. The only feasible solution is to store the graph data on a server and pass bitmap images of the graph across the network. While such images can be displayed inside a plug-in, the visualization framework would not benefit from the rich functionality of the plug-in. A simpler approach is to display the image natively in a browser, thus eliminating other drawbacks of the plug-ins, such as memory limitations, start-up cost, and security issues.

3   Architecture

To reiterate, the main contribution of this work is a scalable web-based technique for providing smooth interaction with very large graphs in a user's web-browser. Figure 1 describes the novel, lightweight and flexible architecture we use to achieve this goal. This architecture does not rely heavily on a client's resources, and requires only a basic browser with no external plugins. Two modes of operation are supported: client-mode and server-mode. In server-mode the client's browser only contains a single bitmap image of the graph. All layout and interaction algorithms are run on the server and the bitmap image in the browser is updated on the fly, giving a very smooth interaction experience. In client-mode the browser represents each node as an image and each edge as an SVG line. Layout and interaction algorithms run locally in the browser. The server always maintains a model of the entire graph,

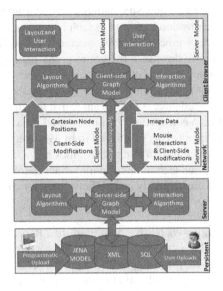

Fig. 1. A scalable web-based architecture for interactive graph visualization, as used in *WiGis*

while the browser can have anywhere from the whole graph model (client-mode with a small graph) down to no graph model at all (server-mode). The system automatically switches between these two modes of operation based on the size of the graph being displayed at a given time.

At the outset of this work, a primary concern was the refresh rate that could be achieved with this type of design. When considering rendering time, network delay, and other processing overheads, one is tempted to picture a slow, jumpy interaction experience. However, as we prove in our evaluations and in our online demonstration[2], this design does achieve fast, smooth graph interactions for up to hundreds of thousands of nodes, even when these potential bottlenecks are considered.

3.1   Visualization Modes

Client-Mode. When a graph in the viewing window (shown in Figure 2) is sufficiently small, all layouts, interactions and renderings are performed in the client browser. This can be either the whole graph or a focused area of a larger graph. The top layer in Figure 1 represents the browser, which contains a model of the graph, referred to as the client-side model. As this model is updated by JavaScript layout and interactions, its state is asynchronously transferred to a remote server, which updates a server-side graph model accordingly (shown in the third layer from top in Figure 1). Rendering is

---

[2] www.wigis.net

performed by JavaScript using SVG for edges (replicable in VML for Internet Explorer) and HTML image tags for nodes. This combination was chosen because it exhibited the best performance out of a multitude of rendering options over a combination of metrics in our preliminary tests. In client-mode, *WiGis* can still make use of rich server-side functionalities such as clustering for instance. The client simply calls the remote function on the server through AJAX, the server runs a process, updates its model and passes it to the client.

The following list outlines the motivations for, and benefits of using client-side graph processing:

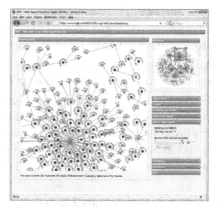

Very Smooth Interaction - Client side computations provide fast interactions for small graphs because there is no direct network overhead.

Network Independent - Client-side processing does not need a fast network to function well, and can even function in an off-line state.

Easy on Server Resources - With a potentially large user base, *WiGis* can be heavy on server resources. Utilizing client resources wherever possible eases the load on a centralized server or server set. We note that in client-mode, the remote server still holds a model of full the graph, so only CPU load is reduced, as opposed to memory.

Fig. 2. A screenshot of *WiGis* displaying results expanded from the seed query "Graph Visualization" on the Citeseer dataset. Graph shows 1104 author and article nodes, with 1125 associations. An overview window of the entire graph is shown in the top right corner with the detail view highlighted by the zoom box.

Server-Mode. For large graphs, *WiGis* automatically switches into server-mode. In this mode, all computations for both layout and interaction are processed on the remote server. Instead of passing a graph model back to the client browser for reconstruction, the server generates a bitmap image of the updated graph. This image is passed across the network and rendered in the browser. Swapping from client to server mode is a *seamless transition* for the end-user, with no jumpiness or image differences, as shown in Figure 3. While in server-mode, interaction is facilitated by capturing mouse interactions on the image of the graph using JavaScript. Mouse interactions are passed asynchronously to the server and the interaction/layout algorithms are triggered on the server-side graph model based on the new input. The server computes an updated graph, renders it, and sends an image of the rendered graph back to the client. The key success of our tool lies in the fact that this entire process occurs at interactive speeds giving very smooth desktop-like interactions with very large graphs.

Our system achieves update rates of 10 frames per second for graphs up to the order of 10K nodes, while graphs of the order of 100K nodes are rendered at approximately 1 to 2 frames per second (c.f. results in Section 4.6). Theoretically, with sufficient hardware resources on the server-side, the upper bound for the number of nodes *WiGis* can usefully display in an interactive fashion approaches the pixel resolution of the client display.

Server-side operation of *WiGis* can loosely be compared to a Google-Maps interface with the difference that transmitted images are not static or pre-defined. Instead, images are computed on-the-fly based on a combination of user input and the existing graph state. The following list shows the benefits and drawbacks of using the server-side approach for large graph computation.

Scalability - Client side graph visualizations generally fail as the graph size approaches thousands of nodes and edges. Using our server-side technique we can interactively visualize graphs of up to 1 millon nodes natively in the browser.

Remote Resources - Server-side processing extends the power of the browser well beyond the resources of the local machine by using a thin client implementation.

Bandwidth Limitation - Server side graph processing relies heavily on network resources, and can perform poorly on slow networks. While many universities operate very fast connections, home and wireless broadband connections typically range from 64 kb/s to about 1 Mb/s. Our evaluations show that network overhead for the image transfer becomes negligible for graphs of over 100 thousand nodes.

3.2 System Architecture Layers

Following is a description of the architecture based around the four layers in Figure 1 from top to bottom. These layers represent physical locations or communications between them, as opposed to the previously discussed client-mode and server-mode, which are modes of *operation* spanning across all layers, and are depicted by the vertical data flows in Figure 1.

Client Browser Layer. The top layer in Figure 1 represents a web-browser running on a client machine. Depending on the mode of operation, the browser holds either a JavaScript model and an SVG/HTML visualization of the graph (client-mode) or a single bitmap image of the graph in its current state (server-mode). The browser contains a JavaScript implementation of a selected layout algorithm and a selected interaction algorithm, *both of which are scripted "replicas" of server-side algorithms*. Depending on the current operation mode (client or server), the browser layer communicates either graph model data or mouse interaction data across the network to the remote server.

Server Layer. The server *layer* is the "powerhouse" of *WiGis*, where most of the heavy processing occurs for large graphs. The server holds a model of the full graph (in memory if possible), a set of graph layout, clustering, and interaction algorithms (currently implemented in Java, but extensible to any language). The key concept of the architecture is that the client layer mirrors the server graph model to the capacity of its available resource. Again, depending on the scale of the visible part of the graph and resources available on the client, the server either accepts mouse interaction data (server-mode) or an updated graph model (client-mode) from the client browser. In return, the server communicates either graph model data or GIF images back to the browser depending on the current mode of operation. The graph model on the server is always kept in synch with the client model through AJAX updates.

Network Layer. The network layer in Figure 1 represents the communication between the server and client layer. Depending on the mode of operation, image data and inter-action data (server-mode) or graph model data (client-mode) is sent across the network to maintain synchronization between the client and server layers.

Persistent Layer. Graph data can be uploaded to the system through a web interface by users or programmatically by other systems to add interactive visualization capabil-ities to them. User uploads are provided in several common formats including XML, GraphML and a simple CSV representation. Regardless of the original source, all data is converted to an XML representation and read into the graph server. The persistent layer of *WiGis* is kept modular to allow data from a broad range of sources to be plugged in easily. For instance, current data sources include citation data from a publication search tool and dynamically generated topic models from New York Times articles.

3.3   Client/Server Implementations

Each algorithm in the client browser is coded in JavaScript to exactly mimic the correspond-ing server-side java ver-sion. The algorithms are designed to be identi-cal with one exception: the client side algorithm operates only on a sub-graph containing all vis-ible nodes and their immediate   neighbors.

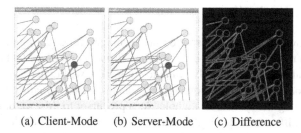

(a) Client-Mode   (b) Server-Mode   (c) Difference

Fig. 3. Seamless transition between client and server renderings. (a.) Client-Mode. Rendering and layout done with DHTML and SVG. (b.) Server-Mode. Rendering and layout done in Java on a remote machine. (c.) Difference image between a and b.

This constraint is necessary because of the scalability limitations of JavaScript and the potentially limited resources on the client machine. Since we must use different platforms and implementation languages, there are also small differences between the resultant graph layouts. However, in most cases these differences are too small to be discerned visually by the end-user. Figure 3 depicts a sample graph visualized in (a.) client-mode and (b.) server-mode. It is clear from Figure 3 that the two representations are very similar, although they might be misaligned by one pixel as a result of float-ing point errors in the conversion between different coordinate systems. Other minor differences exist in the anti-aliasing of the lines and circles. The system automatically switches from server-mode to client-mode when zoomed into a sufficiently small part of the graph and back to server-mode when zoomed out to a larger portion of the graph.

3.4   Layout

The *WiGis* architecture is modular and provides an interface for plugging in multiple different layout algorithms which can then be selected through the user interface. The focus of this paper is on scalability with regard to interaction, and accordingly, details of various layout algorithms are not included here. For the purpose of our analysis we use an efficient implementation of a simple force-directed graph layout algorithm [7] [9].

3.5 Interaction

Interaction with large graphs is not as straightforward as interaction with a small number of nodes, since moving one node at a time can be very time consuming when molding a layout of thousands of nodes. Moreover, the commonly used rectangular area selection of multiple nodes that happen to lie close to each other is not ideal when interacting with large graphs because the selected nodes do not necessarily have meaningful associations with each other apart from proximity as computed by a layout algorithm. For our framework we have chosen an interaction algorithm, which we refer to as the interpolation method, originally developed by Trethewey and Höllerer for use in a desktop application [18]. This method gives great performance for large graphs in terms of speed while also making it easy for users to mold a large graph.

4 Evaluation

Now that we have described our technique for enabling interactions with large graphs on the web, we focus on an empirical evaluation of the technique in terms of speed and scalability. To properly evaluate our system we break down the interactive visualization process into its component pieces and present evaluations of each individually, before testing the system as a whole. As a precursor to this, we define a test dataset of graphs at different scales and discuss their properties. All of the following experiments use the same test data. For the purpose of this evaluation we define three steps (potential bottlenecks) in our interactive visualization process:

Step 1: Rendering - Drawing graphs after a change has been made.

Step 2: Interaction - Capturing user input and calculating modifications to the graph.

Step 3: Network - Passing graph and/or image data across the network from a remote server.

After evaluating each step in isolation, we combine our results to produce our estimated overall time for the variety of graphs in our test suite. As a sanity-check this is then compared against recorded times for interaction with the system as a whole. A discussion of the relative impact of each step is presented. Our evaluation concludes with a comparison against three popular interactive graph visualization systems.

4.1 Setup

All experiments were performed on a 64 Bit Dell Inspiron 530 with an Intel Q9300 2.5GHz quad core processor, 8GB of RAM, an ATI Radeon HD 3650 video card and a serial ATA hard drive with 7200 rpm. The operating system was Windows Vista SP1. No other heavy processes were allowed to run during experiments. A Dell 24" Ultra-Sharp flatscreen monitor with a refresh rate of 60Hz was connected with a DVI cable. Screen resolution was constant at 1920x1200 pixels for all experiments. Graph window sizes were kept constant at 600x600 pixels. This value was chosen because it will fit in most browser windows with 1024x768 resolution. Frame rates were recorded either by *WiGis* or by FRAPS[3]. On our multi core machine, FRAPS did not introduce significant delays in any renderings. All web-based experiments were performed in Mozilla Firefox 3.0.3 and in Google Chrome 1.0.1 with no plug-ins or add-ons running. *WiGis* is Java-based and hosted on a JBoss 4.2.2 server running on the same machine as the client browser. This was done to control network overhead which allowed us to determine the theoretical network overhead of all connection speeds and provide a more reliable and consistent result set based on the the exact size of the data which is passed across the network. It should be noted that we have also successfully tested the system across a real network, with similar results. Moreover, *WiGis* is currently accessible on the world wide web at www.wigis.net.

4.2 Description of Test Data

There are many possible approaches for testing a web-based system for large-graph layout and interaction. Our

Table 1. Description of generated small-world data

Graph	G1	G2	G3	G4	G5	G6
Nodes	10	100	1K	10K	100K	1M
Edges	20	200	2K	20K	200K	2M

system works well with real world data, for example citation networks, computational provenance graphs and topical relations among newspaper articles. We have also applied our system successfully to specific graph types such as meshes, trees, highly connected and highly sparse data. For this paper we have chosen to perform our tests on "small world" data [3, 19], because it is abundant in the social web, financial, biological and many other naturally occurring networks [12]. Small world networks are connected graphs in which most nodes are not direct neighbors but can be reached in a small number of hops from most points in the graph. Our test data was generated using the Barabási-Albert (BA) Model [3] for creation of small world networks. The BA model uses preferential attachment [3] for the addition of new nodes. Table 1 describes our test graphs $G1, ..., G6$. Graph size is increased exponentially from $G1$ (10 nodes, 20 edges) to $G6$ (1M nodes, 2M edges). To confirm the small world nature of our test data, the degree of connectivity versus number of nodes for all graphs was plotted on a log-log scale. This test produced linear trends in logarithmic space for all graphs, exhibiting the trademark power-law distribution of small world networks [3]. Our test data and graph analysis are available for download online[4].

4.3 Rendering

This section describes the procedure and results of an experiment to test the first of three potential bottlenecks for our system. This bottleneck occurs while re-rendering the graph after it has been modified. Modifications can happen either on client or server side, directly by the user, or by interaction and layout algorithms. To evaluate rendering speeds in both client and server modes, each of our test graphs $G1, ..., G6$ was rendered and the average time per frame was recorded. Graphs were redrawn at every frame and all nodes and edges were constrained to the viewing window for the entire test.

---

[3] www.fraps.com
[4] www.wigis.net

As described in Section 3.1, client-side rendering was performed by JavaScript using SVG for edges and HTML image tags for nodes. In server-mode, graphs were rendered into GIF images using Sun Microsystems Java2D graphics library from Java 6.0 and those images were passed to the browser. Our standard setup (outlined in Section 4.1) was applied. For this test, edge width was kept constant at 1 pixel and node size was a constant 4x4 pixels.

Figure 4 shows the results of the rendering experiment. Since our test graphs increase exponentially in size, results are presented on a log-log scale. Note that on this scale, *small differences at upper parts of the graphs represent significantly larger differences on a linear scale*. The four plots in Figure 4 represent the

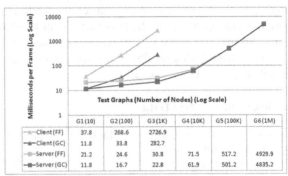

Fig. 4. Results of the rendering experiment showing milliseconds per frame at various graph sizes

client and server side frame rates in ms/frame for rendering of test graphs $G1, ..., G6$ in both Mozilla Firefox (FF) and Google Chrome (GC). The graph shows that both client and server methods eventually scale approximately linearly with number of nodes, however the client side method is notably slower than the server side, because the browser takes longer to update position data in the document object model as graph size is increased. The smaller graphs create a curve effect because of various overheads, but importantly, the larger graphs $G4, ..., G6$ show a linear trend.

In fact, the server side method performs better than linear, as the overhead of loading the image into the browser and displaying it is almost constant for all graph sizes. At $G1$ the methods have identical performance, while at $G3$ the difference is 260 ms for GC and approaches 3 seconds in FF. For small graph sizes, typically less than 100 nodes, client side rendering takes about the same time as server side rendering, in the range of 0 to 260 ms/frame. The client-side approach could not scale past test graph $G3$ (1K nodes). This occurs either because the max number of SVG lines the browser is capable of rendering is exceeded and the page fails to load (FF), or because the browser slows to an unusable state (GC). Irrespective of these failures however, there is stronger motivation for using the server side method for large graphs: at about 1000 nodes it simply becomes more efficient to pass an image of the rendered graph across the network as opposed to sending raw node/edge data and rendering it on the client.

Client side rendering consists of two parts: drawing of SVG lines for edges and HTML image tags for nodes. In addition to the results in Figure 4, these two parts were tested individually. Averaged across all test graphs, line drawing took 53% and image repositioning took 47% of the total rendering time.

Clear performance differences were exhibited between browsers, Chrome was significantly faster than Firefox, which took an average of 6 and 1.3 times longer for client and server methods respectively across all test graphs. This result is as expected because more work is being done by the browser while in client-mode, and Google Chrome's V8 JavaScript engine is faster than the Firefox 3.0.X engine. Looking forward, improvements to JavaScript engines are underway in most major browsers, and we expect that our technique will perform better as faster engines are released.

This experiment shows that our technique is capable of rendering graphs in a browser in the order of hundreds of thousands of nodes and edges in a fraction of a second. This is clearly indicated by the results in Figure 4, which show that G5 (100K nodes) takes about half a second while G6 (1M nodes) takes less than 5 seconds. It is important to note that although 5 seconds may not seem fast, it is a significant improvement over the other systems which could not load G6. Also, our experiment represents a worst-case scenario where all geometric primitives are redrawn. This happens only when the user is working at the overview level that contains the entire graph. If the user is in a zoomed-in view where a smaller number of primitives need to be redrawn for each frame, rendering may be on the order of milliseconds.

4.4 Interaction

The second potential bottleneck in our framework is the simple but effective interaction algorithm we used. However, after evaluation it became clear that the relative time spent on the interaction algorithm was very little as shown in Table 3. For this reason, and due to space restrictions this evaluation is not discussed here.

4.5 Network

The third and final factor in the component analysis of our technique is a look at various network delays that occur as

Table 2. Average image size in kB for each test graph

Graph	G1	G2	G3	G4	G5	G6
Avg. Img Size	6.5	20.5	34	32	34	29

we pass data asynchronously between the client and server. In client-mode, network delay is minimal since we are only passing initial layout data for graphs of the order of G3 or less. Additionally, rendering and interactions are computed locally, so there is no network delay for interaction. However, updates from the client model are passed across the network to maintain synchronicity between client and server graph models. This allows us to switch to server-mode at any time. In server-mode, network capacity has a severe impact on system performance since images of rendered graphs are constantly passed from server to client side. Table 2 shows the average image size in kB for all graphs in our test set. Assuming a network speed of 1000 kB/s (which is common for university campuses) the values in Table 2 are also equivalent to the transfer time in milliseconds for each image. Table 3 presents a breakdown of the total interactive visualization process with network delays included. For graphs of about 1 million nodes network delay represents less than 1% of the total processing time. For smaller graphs, e.g G3, the delay can account for about 59% of the entire process since the size of an image of the rendered graph is relatively stable across graphs G3, ..., G6. We also evaluated the amount of delay introduced by a slower connection of 1000 kb/s which is a common connection speed for residential areas in the USA. This would obviously introduce eight times more network overhead, resulting in about 330 ms per frame for

$G4$ and about 890 ms per frame for $G5$. Interaction with $G6$ would still be under 6 seconds per frame. Again, we note that no other system tested was able to load graphs of the order of $G6$.

4.6 Putting It All Together

Up to this point, we have focused on analysis of the various components of *WiGis* at an individual level. Now we put them all together to evaluate the performance of the system as a whole. This evaluation is performed in two parts, firstly an analysis of speed and scalability is presented based on our test data. Secondly, we present a comparison of our technique against three popular graph visualization systems with respect to speed and scalability.

Scalability Test. Figure 5 shows the time in milliseconds for the full interactive visualization process on Graphs $G1$,..., $G6$, which includes rendering, interaction and network delays. These results are for interaction with the entire graph, i.e: the effect parameter was set to maximum value, making interactions effect every node. This represents the worst case

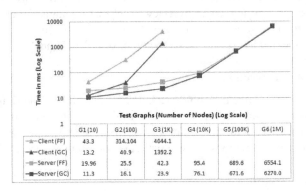

Fig. 5. Results of interactive visualization experiment, showing average times per frame for the worst case scenario, where every node is repositioned in every frame

	G1 (10)	G2 (100)	G3 (1K)	G4 (10K)	G5 (100K)	G6 (1M)
Client (FF)	43.3	314.104	4044.1			
Client (GC)	13.2	40.9	1392.2			
Server (FF)	19.96	25.5	42.3	95.4	689.6	6554.1
Server (GC)	11.3	16.1	23.9	76.1	671.6	6270.0

scenario for our system since every node is repositioned in each frame. The test was run in Firefox (FF) and Google Chrome (GC) browsers in client and server modes. There is an obvious difference in scalability between client and server modes. At $G1$ (10 nodes) there is only a few milliseconds difference between them, but at $G3$ (1000 nodes) the client process is taking 96 times longer than the server (4044 ms compared with 42.3 ms). Again in this test we can see that for the client side process, FF is far slower than GC, taking 4.6 times longer on average. The surprising result in this test is that our technique for computing graphs remotely (i.e: the server side method) is actually faster than JavaScript for large *and* small graphs. (in GC, 1.2 times faster for $G1$, 2.5 times faster for $G2$, and 58 times faster for $G3$). The test was also performed with single node interaction and a similar trend was revealed. For full graph interaction in GC, the million node graph ($G6$) took about 6.3 s, while the single node interaction took 5.7 s. It is important to notice that the rendering of $G6$ took 77% of the total time and, as noted earlier, if the user is working at a zoomed-in level instead of at the overview level the rendering time may be significantly smaller. Thus, interaction with the 1 million node graph may take as little as 1 second per frame.

Delay Breakdown. To gain an understanding of the delays caused by each part of the online interactive visualization process we computed a percentage analysis for each step over all of our test graphs. Table 3 outlines the results for graphs $G1, ..., G6$. For $G1$ and $G2$, client-mode was used because this is the system default for small graphs and gives the best performance in most cases. The table shows the percentage time for rendering, interaction, and the expected network costs. The total column shows an empirically tested value for the entire process over each graph. The difference between the total and the sum of component pieces is shown as "Other". We suspect that this value is due to various system processes, browser overheads, other unmeasured parts of our system and other performance inhibiting overheads.

Image size is influential for the performance of our tool when operating in server-mode. For our evaluations, the graph window was maintained 600x600 pixels to fit in the browser at most screen resolutions, producing for example, an average im-

Table 3. Percentage breakdown of the online interactive visualization process in Google Chrome for our test graphs

Graph	$G1$	$G2$	$G3$	$G4$	$G5$	$G6$
Mode	Client	Client	Server	Server	Server	Server
Rendering	89%	83%	39%	57%	71%	77%
Interaction	0.04%	0.04%	0.4%	3.9%	6.5%	7.1%
Network	0%	0%	59%	30%	5%	0.5%
Other	11%	17%	2%	9%	18%	16%
Total ms	13.2	40.9	57.9	108.6	705.6	6299.5

age size of 34kB for $G5$. However, since we are interested in potentially huge graphs, which may require more screen estate to display adequately, we also considered the impact of bigger window sizes. When we increase the window size to 1200x1200 pixels (4 times the original area), the average image size becomes 154kB. Running the system in server-mode with 600x600 pixel screen size takes 648 ms per frame while the 1200x1200 size takes 1051 ms per frame over a 1000 kB/s network. This is due to network overhead, and increased rendering times since graphical primitives contain more pixels. To summarize, most of the delay in our web-based graph visualization framework can be attributed to rendering while other delays account for only about 20% of the total.

4.7 Comparison

To conclude the evaluation of our system, we now discuss a comparative test against three popular graph visualization systems: Touchgraph Navigator [17] (A Java Applet), IBM Many Eyes [6] (Java Applet), and Cytoscape [14] (a desktop application). A direct comparison with the plug-in free web-based version of Tom Sawyer Visualization was desired, but since this discussion focuses on scalable interaction, a direct comparison became infeasible because we were unable to interact with graphs in that system when more than a few hundred nodes are displayed. Our test dataset from Table 1 was converted to appropriate formats for each system and interactions timings were recorded for each using FRAPS, while keeping all graph elements in the viewing window. We note that the primary focus for these applications is not necessarily on scalability as they have many rich data exploration features for a variety of specific tasks, but this experiment does highlight that some of these systems are quite limited in scale. Since the other systems did not support interaction with the full graph based on single node movements, we restricted our system to movement of one node only. However, we note that in the worst-case, when the entire graph is repositioned based on the interaction algorithm, the timings for *WiGis* increase only by a very small amount. Since our system runs natively in the browser, FRAPS could not record timings. A JavaScript test harness was written to emulate a real user interacting with the graph. (Note: manual tests were also performed and similar results were achieved.) A click was simulated on a random node and it was moved to a random position in the view window, thus triggering selection and movement processes. The movement step was repeated 500 times and an average time was recorded. The experiment was repeated for graphs $G1, ..., G6$. Our system was tested with the browser running on the same machine as the server, and then network overhead with a connection of 1000kB/s was projected based on the image sizes. The fastest mode was used, which was server-side for all except when network overhead was included on $G1$ and $G2$.

Figure 6 shows the results of the interaction experiment in Google Chrome. For graphs of size $G3$ or less, all the systems completed the test in less than 100 ms per frame on average, except Touchgraph which took 265 ms per frame. Our system showed a time increase with respect to graph size that is slightly less than linear. This occurs because overheads such as network time

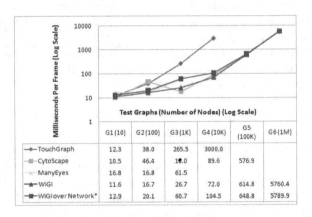

Test Graphs (Number of Nodes) (Log Scale)					
	G1 (10)	G2 (100)	G3 (1K)	G4 (10K)	G5 (100K)	G6 (1M)
TouchGraph	12.3	38.0	265.5	3000.0		
CytoScape	10.5	46.4	18.0	89.6	576.9	
ManyEyes	16.8	16.8	61.5			
WiGi	11.6	16.7	26.7	72.0	614.8	5760.4
WiGi over Network*	12.9	20.1	60.7	104.5	648.8	5789.9

Fig. 6. Results of scalability and speed comparison against other systems. *Network delay over a 1000 kB/s network connection is estimated based on average image size.

take up a smaller percentage of the overall process as graph size increases. The best performer from the other systems was Cytoscape, which took 570 ms for $G5$, which was 37.9 ms (6%) faster than our tool. An interesting trend in the graph occurs between $G2$ and $G3$ on the Cytoscape plot, where time per frame is reduced by 28 ms despite the increase in graph size. This occurs because Cytoscape renders nodes as squares instead of circles for graphs above a certain size. *WiGis* completed the test for $G6$ in an average of 5.7 seconds. These results show that the server side technique used in our system is more efficient than current graph visualization standards on the web.

The blank spaces in the table of Figure 6 represent failed attempts to load data. Under the setup described in Section 4.1, the largest of our test graphs we could load in Many Eyes was $G3$. TouchGraph failed at $G5$, while Cytoscape failed at $G6$. *WiGis* was the only system to successfully load the million node graph $G6$. Furthermore, we were unable to find another web-based graph visualization tool that could display graphs of the order of $G5$ or higher.

Load times for each system were also noted, as they contribute greatly to the overall user experience. *WiGis* outperformed all other systems for every graph. $G1$ and $G2$ were loaded by all systems in less than 1 second. *WiGis*, Touchgraph and Cytoscape loaded

$G3$ in less than one second, while Many Eyes took 5 s. Only *WiGis* and Cytoscape loaded $G5$, taking 2.6 and 4 seconds respectively, making *WiGis* 1.5 times faster.

5   Discussion and Conclusion

The main contribution of this paper to the graph drawing community is a framework for interactive visualization of large graphs over the web. We have presented an argument for our choice of a native browser implementation over a plug-in based approach. The framework provides user interaction with hundreds of thousands of nodes through the use of bitmap graph representations streamed from a remote server. Another novel contribution is the automatic switching between client and server graph models to maximize use of available resources. This is done in a manner which is transparent to the end user.

The approach used in the *WiGis* framework has several limitations. Firstly, since we are transferring data across a network there is a potential security risk and potential for data-loss. This can be mitigated somewhat by the use of SSL communications. Secondly, since we have chosen to display graphs natively in the browser, the current implementation cannot make use of rich functionality provided by plug-ins such as Java and Flash. Thirdly, as shown in our evaluations, the biggest bottleneck in our system occurs during rendering. There are a few possible avenues to address this issue. For example, use of a more powerful rendering technique, such as GPU rendering. Another possible improvement is to keep track of the nodes that will be re-rendered each frame and render those on top of an image of all the static nodes.

We have presented a detailed breakdown of the various components of the system in terms of speed and scalability. We compared *WiGis* against three popular systems and showed that our framework outperforms the best performing web-based system we could find by an order of magnitude in terms of scalability and achieves similar scale to the best performing desktop-based systems. In addition to the scalability advantages of our system, the fact that it is fully web-based (i.e: native) gives it the flexibility and ease-of-use to easily be applied to solve real-world graph visualization problems where users need to access graph data quickly and easily. For example, the tool is currently deployed by the U.S government in Blackbook- a data integration and search system used for counter-terrorism [11]. In this tool, *WiGis* visualize interconnections between artifacts from a variety of diverse datasets, such as security reports or financial information. At the University of California, Irvine, *WiGis* have been deployed for visualization of a topic detection system [5] for newspaper articles. With a view to gathering useful and informative feedback on our visualization and interaction techniques from a large number of users, we are currently deploying a *WiGis* application on Facebook to visualize networks of friends and their various tastes in music, movies, etc. Due to the flexible nature of the framework it is easily adaptable to this task and we hope to report results of user evaluations in a future publication.

Acknowledgements

This work was partially supported by NSF grant IIS-0635492 through funds from the ITIC/IARPA KDD program, by NSF grants CNS-0722075 and IIS-0808772, as well as an ARO MURI award for proposal #56142-CS-MUR.

References

[1] Abello, J., Korn, J.: Mgv: A system for visualizing massive multi-digraphs. IEEE Transactions on Visualization and Computer Graphics 8, 21–38 (2002)

[2] Auber, D.: Tulip. In: Mutzel, P., Jünger, M., Leipert, S. (eds.) GD 2001. LNCS, vol. 2265, pp. 335–337. Springer, Heidelberg (2002)

[3] Barabasi, A.-L., Albert, R.: Emergence of scaling in random networks. Science 286, 509 (1999)

[4] Batagelj, V., Mrvar, A.: Pajek - program for large network analysis. Connections 21, 47–57 (1998)

[5] Chemudugunta, C., Smyth, P., Steyvers, M.: Text modeling using unsupervised topic models and concept hierarchies. CoRR, abs/0808.0973 (2008)

[6] Danis, C.M., Viegas, F.B., Wattenberg, M., Kriss, J.: Your place or mine?: visualization as a community component. In: CHI 2008: Proceeding of the twenty-sixth annual SIGCHI conference on Human factors in computing systems, pp. 275–284. ACM, New York (2008)

[7] Eades, P.: A heuristic for graph drawing. Congressus Numerantium 42, 149–160 (1984)

[8] Eades, P., Huang, M.: Navigating clustered graphs using force-directed methods (2000)

[9] Fruchterman, T.M.J., Reingold, E.M.: Graph drawing by force-directed placement. Softw. Pract. Exper. 21(11), 1129–1164 (1991)

[10] Herman, G.M., Marshall, M.S.: Graph visualization and navigation in information visualization: A survey. IEEE Transactions on Visualization and Computer Graphics 6(1), 24–43 (2000)

[11] Intelligence Technology Innovation Center (ITIC). Blackbook prototype framework for the knowledge discovery and dissemination (kdd) program. McLean, VA, USA, October 3–4 (2006)

[12] Milgram, S.: The small world problem. Psychology Today (2), 60–67 (1967)

[13] Pinaud, B., Kuntz, P., Picarougne, F.: The website for graph visualization software references (GVSR). In: Kaufmann, M., Wagner, D. (eds.) GD 2006. LNCS, vol. 4372, pp. 440–441. Springer, Heidelberg (2007),
http://gvsr.polytech.univ-nantes.fr/GVSR/

[14] Shannon, P., Markiel, A., Ozier, O., Baliga, N.S., Wang, J.T., Ramage, D., Amin, N., Schwikowski, B., Ideker, T.: Cytoscape: a software environment for integrated models of biomolecular interaction networks. Genome Res. 13(11), 2498–2504 (2003)

[15] Shneiderman, B.: The eyes have it: A task by data type taxonomy for information visualizations (1996)

[16] Tom Sawyer Software. Tom sawyer visualization (2009)

[17] Touchgraph. Touchgraph navigator. Proprietary online application, Touchgraph inc.,
http://www.touchgraph.com

[18] Trethewey, P., Höllerer, T.: Interactive manipulation of large graph layouts. Technical report, Department of Computer Science, University of California, Santa Barbara (2009)

[19] Watts, D.J., Strogatz, S.H.: Collective dynamics of 'small-world' networks. Nature (393), 440–442 (1998)

[20] Wood, J., Brodlie, K., Wright, H.: Visualization over the world wide web and its application to environmental data. In: VIS 1996: Proceedings of the 7th conference on Visualization 1996, p. 81. IEEE Computer Society Press, Los Alamitos (1996)

Port Constraints in Hierarchical Layout of Data Flow Diagrams

Miro Spönemann[1], Hauke Fuhrmann[1],
Reinhard von Hanxleden[1], and Petra Mutzel[2]

[1] Real-Time and Embedded Systems Group, Christian-Albrechts-Universität zu Kiel
{msp,haf,rvh}@informatik.uni-kiel.de
[2] Chair of Algorithm Engineering, Technische Universität Dortmund
petra.mutzel@tu-dortmund.de

Abstract. We present a new application for graph drawing in the context of graphical model-based system design, where manual placing of graphical items is still state-of-the-practice. The KIELER framework aims at improving this by offering novel user interaction techniques, enabled by automatic layout of the diagrams. In this paper we present extensions of the well-known hierarchical layout approach, originally suggested by Sugiyama et al. [22], to support port constraints, hyperedges, and compound graphs in order to layout diagrams of data flow languages. A case study and experimental results show that our algorithm is well suited for application in interactive user interfaces.

1 Introduction

Graphical modeling languages have evolved to appealing and convenient instruments for the development and documentation of systems, both in hardware and in software. There are various examples for graphical modeling frameworks that have become an important part of modern development processes. An important class of modeling diagrams are *data flow diagrams*, which are graphical representations of *data flow models* for design of complex systems. Applications of data flow diagrams can be found in modern software and hardware development tools. Some of these, such as Simulink (The MathWorks, Inc.), LabVIEW (National Instruments Corporation), and ASCET (ETAS Inc.), are mainly used for model-based design and simulation of embedded systems and digital or analog hardware, while others, such as SCADE (Esterel Technologies, Inc.), are optimized for automatic code generation from high-level system models. The Ptolemy project [8] features data flow diagrams for *actor-oriented design*. All these examples feature a graphical editor for data flow diagrams, so that users can create diagrams in drag-and-drop manner.

Typical graphical modeling tools do not support the developer with automatic diagram layout, or do so only in a rudimentary fashion. This leads to unnecessarily high development times, as the developer has to manually adapt the layout after each structural change of the model. In this paper we present methods to apply the hierarchical layout approach [22] to data flow diagrams.

D. Eppstein and E.R. Gansner (Eds.): GD 2009, LNCS 5849, pp. 135–146, 2010.

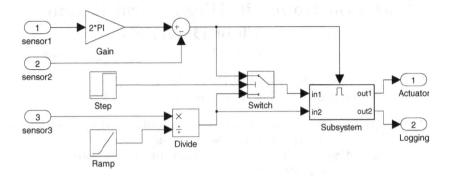

Fig. 1. A data flow diagram from Simulink

We describe constraints which are imposed by such diagrams and show how to extend existing methods to satisfy these constraints. This includes methods for crossing reduction with port constraints and routing of directed hyperedges.

A data flow model is described by a directed graph where the vertices represent *operators* that compute data and the edges represent data paths [6]. Such a data path has a specified *source port* where data is created and a *target port* where data is consumed. A source port may be connected with multiple target ports, thus forming a *hyperedge*. Furthermore, the edges of data flow diagrams are required to be drawn orthogonally. A diagram from Simulink is shown in Fig. 1, which demonstrates the use of ports and hyperedges.

We will proceed as follows. Port constraints, hyperedges and other specialties of data flow diagrams are presented in Section 2. Here, we also introduce four scenarios of port constraints that appear frequently in our applications. Related work is discussed in Section 3. Section 4 describes our methods to handle the special requirements of data flow diagrams within the hierarchical approach. Results of our implementation are shown in Section 5, and we conclude in Section 6. A much more in-depth presentation covering the full hierarchical layout algorithm and details on its implementation can be found on-line [19,20].

2   Port Constraints and Hyperedges

A *port based graph* is a directed graph $G = (V, E)$ together with a finite set P of *ports*. For each $v \in V$ we write $P(v)$ for the subset of ports that belong to v, and we require $P(u) \cap P(v) = \emptyset$ for $u \neq v$. Each edge $e = (u, v) \in E$ has a specified *source port* $p_s(e) \in P(u)$ and a *target port* $p_t(e) \in P(v)$. We write $v(p)$ for the vertex u for which $p \in P(u)$.

A *drawing* of a port based graph G is a mapping Γ of the vertices, edges, and ports of G to subsets of the plane $\mathbb{R}^2$. In general graph drawing it is sufficient that the drawing of each edge $e = (u, v)$ contacts the drawings of u and v anywhere at their border. For port based graphs the drawing of each port $p \in P(v)$ has a specific position on the border of $\Gamma(v)$, and the edges that have p as source

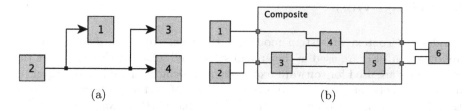

Fig. 2. (a) A hyperedge that connects four vertices (b) The vertex Composite contains connections to external ports, which are shown as small dark boxes on its border

or target port may touch $\Gamma(v)$ only at that position. We consider four different scenarios for the positions of the ports $P(v)$ on a vertex v:

FREEPORTS. Ports may be drawn at arbitrary positions on the border of $\Gamma(v)$.
FIXEDSIDES. The side of $\Gamma(v)$ is prescribed for each port, i. e. the top, bottom, left, or right border, but the order of ports is free on each side.
FIXEDPORTORDER. The side is fixed for each port, and the order of ports is fixed for each side.
FIXEDPORTS. The exact position is fixed for each port.

Mixed-case scenarios, in which some ports of a single vertex have fixed positions and others are free, are not yet covered in our approach, because they require very complex handling and are not needed in our applications.

A *hyperedge* has an arbitrary number of endpoints, thus it may connect more than two vertices. Although there are approaches to directly handle hyperedges [10,17], we split all hyperedges into sets of plain edges in order to simplify the algorithms. For this reason we consider all edges that are incident at the same port of a vertex as parts of a single hyperedge. For example, the hyperedge shown in Fig. 2(a) would be represented by the edges $(2, 1)$, $(2, 3)$, and $(2, 4)$. Such splitting of hyperedges is not unique if the hyperedge has multiple sources and multiple sinks, but many data flow languages do not allow multiple sources for hyperedges.

In data flow diagrams, each vertex may contain a nested diagram; in this context we have to extend our notion of a graph. A *compound graph* or *clustered graph* $G = (V, H, E)$ consists of a set of vertices V, a set of *inclusion edges* H, and a set of *adjacency edges* E [21]. The *inclusion graph* (V, H) must form a tree, hence for each vertex v we can write $V_{\mathrm{ch}}(v)$ for the set of children of v and $v_{\mathrm{par}}(v)$ for the parent of v. For data flow diagrams the adjacency edges are only allowed to connect vertices that have the same parent in the inclusion tree. However, we treat the ports $P(v)$ of each vertex v as *external ports* of the diagram contained in v, and the children $V_{\mathrm{ch}}(v)$ may be connected to the ports $P(v)$ (see Fig. 2(b)). We employ special edge routing mechanisms to properly connect the ports of a node v with its children.

3   Related Work

Besides the context of system modeling, the term *data flow diagram* and its abbreviation DFD are used in the area of structured software analysis [5]. In this sense DFDs are used for software requirements specification and modeling of the interaction between processes and data. Layout of DFDs has been covered by Batini et al. [3] and Doorley et al. [7]. As DFDs have little in common with data flow diagrams for system modeling, these layout algorithms cannot be applied to our specific problem.

The main specialties that make layout of data flow diagrams for system modeling more difficult than layout of general graphs are ports, hyperedges, orthogonal edge routing, and compound graphs. Previous work on layout with port constraints includes that of Gansner et al. [11] and Sander [14], who gave extensions of the hierarchical approach to consider attachment points of edges. These methods are mainly designed for the special case of displaying data structures and are not suited for the more general constraints of data flow diagrams. A more flexible approach is chosen in the commercial graph layout library yFiles (yWorks GmbH), which supports two models of port constraints and hyperedge routing for the hierarchical approach[1], but no details on the algorithm have been published [23]. Either a *weak* port constraints model (corresponding to FIXED-SIDES) or a *strong* port constraints model (corresponding to FIXEDPORTS) can be chosen in yFiles. Other unpublished solutions to drawing with port constraints include ILOG JViews [18] and Tom Sawyer Visualization[2]. Handling of hyperedges in hierarchical layout has been covered by Eschbach et al. [10] and Sander [17]. Sugiyama et al. [21] and Sander [16] showed how to draw general compound graphs, but due to the presence of external ports (see Section 2), our requirements for compound graphs are different. We adapt the orthogonal edge routing approach suggested by Eschbach et al. [10]; alternative approaches have been given by Sander [15,17] and Baburin [2].

The topic of visualization of hardware schematics is quite related to drawing of data flow diagrams. While traditional approaches for layout of schematic diagrams follow the general *place and route* technique from VLSI design [1,12], more recent work includes some concepts from the area of graph drawing [9]. However, these concepts are not sufficient for the needs of our application, since they do not address our scenarios for port constraints, but concentrate on partitioning and placement for large schematics and hyperedge routing.

4   Extensions of the Hierarchical Layout Approach

The hierarchical layout method is well suited for laying out directed graphs and aims at emphasizing the direction of flow, thus expressing the hierarchy of vertices in the graph. It was proposed by Sugiyama, Tagawa, and Toda [22] and

---

[1] yFiles Developer's Guide, http://www.yworks.com/
[2] Tom Sawyer Software, http://www.tomsawyer.com/

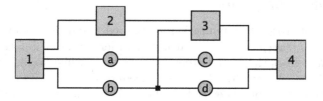

Fig. 3. A layered graph with four layers and three long edges, two of which are part of a hyperedge; the circular vertices are dummy vertices used to split the long edges

has been extensively studied and improved afterwards. We chose the methods of Di Battista et al. [4] and Sander [15] as a base for our implementation.

Handling of port constraints, hyperedges, orthogonal edge routing, and compound graphs is not addressed in the basic versions of the hierarchical layout algorithm. The following sections will depict our approaches to handle these problems. In hierarchical drawings the directed edges are arranged either horizontally or vertically, but only horizontal layout direction is discussed here, as both variants are symmetric.

4.1 Assignment of Dummy Vertices

In the layer assignment phase we compute layers $L_1, \ldots, L_k$ for the vertices of the acyclic graph G using any standard method. A layering is called *proper* if all edges e connect only vertices from subsequent layers. As illustrated in Fig. 3, a proper layering is constructed from a general layering by splitting *long edges* using dummy vertices. We use *linear segments* to organize the dummy vertices: each vertex v in the layered graph is contained in exactly one linear segment $S(v)$, and a linear segment contains either a single regular vertex or all dummy vertices created for a long edge. These linear segments are used in the vertex placement phase to arrange the dummy vertices of each long edge in a straight line adapting Sander's methods [15]. In Fig. 3, the linear segment of the dummy vertex a is $S(a) = \{a, c\}$.

We customized the linear segments approach to support hyperedges which span multiple layers. In this case, care must be taken to merge the dummy vertices of their corresponding point-to-point edges. For this reason we split long edges by processing them iteratively and associating the linear segment of their dummy vertices with their source and target port. If for any long edge there is already a linear segment associated with its source or target port, the dummy nodes of this linear segment are reused. An example is shown in Fig. 3, where the long edges $(1, 3)$ and $(1, 4)$ share the dummy vertex b.

If the diagram contains external ports, they are also added to the layered graph: input ports, which have only outgoing connections, are assigned to the first layer, while output ports, which have only incoming connections, are assigned to the last layer. In this way the external ports can be treated as normal vertices in the following phases of the algorithm.

4.2 Crossing Minimization

The problem of crossing minimization for layered graphs is usually solved with a *layer-by-layer sweep*: choose an arbitrary order for layer L_1, then for each $i \in \{1, \ldots, k-1\}$ optimize the order for layer L_{i+1} while keeping the vertices of layer L_i fixed. Afterwards the same procedure is applied backwards, and it can then be repeated for a specified number of iterations. We will only cover the forward sweep here, because the backward case is symmetric.

Since the standard layer-by-layer sweep is only applied to vertex positions, we will now look at our extensions for port positions. When ports are used to determine the source and target point of each edge, the number of crossings does not only depend on the order of vertices, but also on the order of ports for each vertex. For each vertex v we define *port ranks* for the ordered ports $P(v) = \{p_1, \ldots, p_m\}$ as $r(p_i) = i$. Furthermore we define extended vertex ranks so that for each $v \in L_i$ and $p \in P(v)$ the sum of the rank of v and the rank of p is unique. The *rank width* of a vertex $v \in L_i$ is $w(v) := 1$ if v was created for a dummy vertex of a long edge or for an external port, and $w(v) := |P(v)|$ otherwise. The extended vertex ranks of the ordered vertices in the layer $L_i = \{v_1, \ldots, v_h\}$ are defined as $r(v_j) := \sum_{g<j} w(v_g)$ for all $j \leq h$.

We implemented the *Barycenter* method for the two-layer crossing problem: first calculate values $a(v) \in \mathbb{R}$ for each $v \in L_{i+1}$, then sort the vertices in L_{i+1} according to these values. Let $E_i(v)$ be the set of incoming edges of v. In our approach, the $a(v)$ values are determined as the average of the combined vertex and port ranks for all source ports of incoming edges of v:

$$a(v) := \frac{1}{|E_i(v)|} \sum_{(u,v) \in E_i(v)} (r(u) + r(p_s(u,v))) \ . \tag{1}$$

Vertices v_j that have no incoming edges should be assigned values $a(v)$ that respect the previous order of vertices, thus we define $a(v_j) := \frac{1}{2}(a(v_{j-1}) + a(v_{j+1}))$ if $E_i(v_{j+1}) \neq \emptyset$ and $a(v_j) := a(v_{j-1})$ otherwise. By setting $a(v_0) := 0$ and calculating the missing $a(v_j)$ values with increasing j we can assure that $a(v_{j-1})$ is always defined.

For vertices with FIXEDSIDES or FREEPORTS port constraints we have the additional task of finding an order of ports for each vertex that minimizes the number of crossings. We extend the method described above as follows: instead of calculating values $a(v)$ to order the vertices, calculate values $a(p)$ to order the ports first, then calculate

$$a(v) := \frac{1}{|P(v)|} \sum_{p \in P(v)} a(p) \ . \tag{2}$$

For each port p let $E_i(p)$ be the set of edges which are incoming at that port. Analogously to Equation 1 we define

$$a(p) := \frac{1}{|E_i(p)|} \sum_{(u,v) \in E_i(p)} (r(u) + r(p_s(u,v))) \ . \tag{3}$$

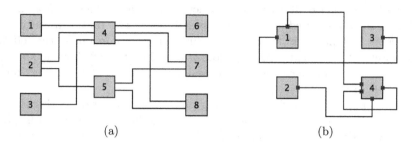

(a)   (b)

Fig. 4. (a) Routing between layers using vertical line segments (b) Routing around vertices due to prescribed port positions

If there are long hyperedges that share common dummy vertices, as described in Section 4.1, crossing reduction must be adapted to avoid inconsistencies in the following phases. If, for example, backwards crossing reduction is performed for the second layer of the graph in Fig. 3 while keeping the vertices of the third layer fixed as $(3, \mathsf{c}, \mathsf{d})$, it can happen that the dummy vertex b is placed above a because of its outgoing connection to vertex 3. This would lead to a crossing of the edges (a, c) and (b, d), thus the corresponding linear segments $\{\mathsf{a}, \mathsf{c}\}$ and $\{\mathsf{b}, \mathsf{d}\}$ could not be drawn as straight horizontal lines.

To resolve this problem, two new rules must be added for each long edge that is split into dummy vertices $v_1, \dots, v_k$:

1. For each dummy vertex v_i, $i \in \{2, \dots, k\}$, only one incoming connection may be considered for crossing reduction, namely (v_{i-1}, v_i).
2. For each dummy vertex v_i, $i \in \{1, \dots, k-1\}$, only one outgoing connection may be considered for crossing reduction, namely (v_i, v_{i+1}).

4.3   Orthogonal Edge Routing

In order to achieve orthogonal edge routing, each edge that cannot be represented by a single horizontal line needs a vertical line segment (see Fig. 4(a)). A proper order of vertical line segments is important to avoid additional edge crossings, and grouping of hyperedges must be considered. To accomplish this, each port p of a vertex in layer L_i that contains outgoing connections to layer L_{i+1} is assigned a *routing slot* $s(p)$. The resulting routing slots are sorted and given appropriate horizontal positions, and each edge that has p as source port is given two bend points with the respective horizontal position of $s(p)$.

We employ the basic sorting of routing slots as depicted by Eschbach et al. [10], but have to extend it to support the different scenarios of port positions. An additional difficulty arises when the source port of an edge is not on the right side of the source vertex, or the target port is not on the left side of the target vertex. In these cases additional bend points are needed to route the edge around the vertex, as seen in Fig. 4(b). For this purpose routing slots must be assigned on each side of a vertex, similarly to layer-to-layer edge routing. This is done in an additional phase after crossing reduction; all edges which need additional bend points are processed here, as well as self-loops. The *rank* of a routing slot indicates its distance from the corresponding vertex. For example, the self-loop $(4, 4)$ in Fig. 4(b) is assigned routing slots of rank 1 on the left, bottom and right side of vertex 4, while the edge $(2, 4)$ is assigned a routing slot of rank 2 on the bottom side of vertex 4.

As an output of this additional routing phase, the number of routing slots for the top and the bottom side of each vertex v, together with the given height of v, determines the amount of space that is needed to place v inside its layer. This information is passed to the vertex placement phase, so that the free space that is left around each vertex suffices for its assigned routing slots.

4.4 Compound Graphs with External Ports

For general compound graphs $G = (V, H, E)$, the adjacency edges E are allowed to connect vertices from different levels of the inclusion tree (V, H). As this is not the case for data flow diagrams, we do not need to employ the special versions of the hierarchical layout method for compound graphs [21,16], but can follow a simpler approach, which consists of executing the layout algorithm recursively, starting with the leaves of the inclusion tree.

However, the presence of external ports (see Fig. 2(b)) leads to the additional problem that edges of the nested graph may be connected to these ports, which may be subject to any of the four scenarios of port constraints described in Section 2. During edge routing such connections must be specially handled, in particular if there are input ports which are not on the left side of the nested diagram, or output ports which are not on the right side. These cases require additional bend points, and if there are multiple edges which need to be routed along the top or bottom side of the nested diagram, the order of these edges must be adjusted to minimize the number of crossings. We achieve this through similar techniques as those used for layer-to-layer edge routing.

5   Implementation and Results

An implementation of our layout algorithm is part of the Kiel Integrated Environment for Layout for the Eclipse RichClientPlatform (KIELER)[3]. KIELER is a platform for experimental approaches to graphical model-based design and for combination of different aspects of graphical modeling, such as methods of model editing, visualization of simulation, and automatic layout. Unlike its preceding project, the Kiel Integrated Environment for Layout (KIEL), which was developed as a stand-alone Java application [13], KIELER builds on Eclipse, an extensible platform comprised of various integrated development environments. Our Eclipse interface enables the layout functionality for editors of the Eclipse

---

[3] http://www.informatik.uni-kiel.de/rtsys/kieler/

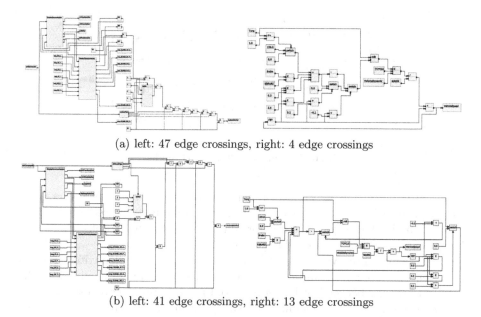

(a) left: 47 edge crossings, right: 4 edge crossings

(b) left: 41 edge crossings, right: 13 edge crossings

Fig. 5. Comparison with yFiles: (a) our layout method (b) layout with yFiles

Graphical Modeling Framework (GMF)[4] and hence for a wide variety of graphical editors. Since the algorithm is written in Java, it can also be used as a plain class library outside of Eclipse.

Fig. 5(a) shows results of automatic layout in the FIXEDPORTS scenario for port positions. Here we see the effectiveness of our method of crossing minimization, as the order of vertices in each layer is adapted to the fixed port positions. Figure 5(b) shows the same diagrams with layouts created in yEd, a free graph editor of yWorks GmbH which includes the yFiles layout library. The results demonstrate that our layout method is comparable with the commercial library yFiles with regard to layout with port constraints.

To test our layout algorithm in an existing modeling framework, it was integrated into *Vergil*, the editor for Ptolemy II developed at UC Berkeley by Lee et al. [8]. Ptolemy II is a graphical modeling tool for exploration of the semantics of different models of computation of formalisms for embedded software design. Its heterogeneous nature enables to mix it with models of physical phenomena to result in full system models including the software controller and its physical environment.

Graphical representations of Ptolemy models can be mapped almost directly to the layout problem described in this paper. Ptolemy *actors* are the interconnected software components represented by nodes which consume and produce data at dedicated ports. Connections can be joined by *relation vertices* to obtain hyperedges that share a common data source. However, Ptolemy does not

---

[4] http://www.eclipse.org/modeling/gmf/

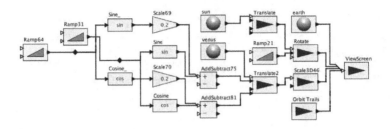

Fig. 6. Layout of a Ptolemy II model

yet support the setting of connection bendpoints, but dynamically routes them internally. Additionally, the data flow of ports is sometimes bidirectional, which results in undirected edges. As our algorithm requires directed graphs, a heuristic chooses explicit directions for all edges first.

The Ptolemy editor Vergil is not based on Eclipse but implemented in plain Java. Hence we used our stand-alone algorithm library and interfaced it with the graphical drawing backend of Ptolemy. Initial results produce diagrams such as those depicted in Fig. 6 and show that the algorithm is applicable for an important set of real-world system modeling tools. More details about the Ptolemy integration can be found elsewhere [20].

Measurement data for the execution time of the hierarchical layout method are shown in Fig. 7. For graphs with about 25 000 vertices and the same number of edges the algorithm takes less than a second, which proves its suitability for automatic layout in a user interface environment. However, the execution time highly depends on the average vertex degree, since layout for a graph with 2 000 vertices and 2 000 edges is 8 times faster than layout for 100 vertices and 2 000 edges. One reason for this is that for vertices with a lot of incident edges the number of long edges that stretch over multiple layers is likely to be high, so that dummy vertices must be inserted to obtain a proper layering. The consequence is that the problem size rises with regard to the total number of vertices.

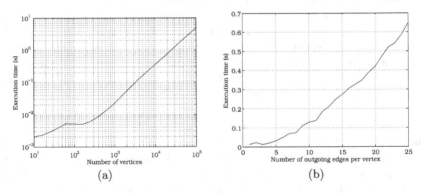

Fig. 7. Execution time for (a) varying number of vertices and one outgoing edge per vertex (b) 100 vertices and varying number of outgoing edges per vertex

6 Conclusion

We introduced four scenarios of port constraints for graph drawing and presented methods to extend the hierarchical layout approach to handle ports, hyperedges, orthogonal edge routing, and compound graphs. These methods are implemented in KIELER, an Eclipse based framework for research on the pragmatics of graphical modeling. The results of our implementation and the low execution times demonstrate its suitability to enhance graphical modeling tools by automatic layout of data flow diagrams. Further work can be done to improve the layout quality:

- Additional support for layout of edge labels.
- Direct support of directed hyperedges with multiple sources and multiple targets.
- Some data flow languages such as SCADE allow to integrate *Statecharts* in their data flow diagrams. A layout algorithm should be able to handle this, i. e. arbitrarily mix nodes with and without port constraints and hyperedges.
- Some vertices in data flow diagrams are very large, thus forcing their respective layer to be large. This could be improved by possibly stretching large vertices over multiple layers.

References

1. Arya, A., Kumar, A., Swaminathan, V.V., Misra, A.: Automatic generation of digital system schematic diagrams. In: DAC 1985: Proceedings of the 22nd ACM/IEEE Conference on Design Automation, pp. 388–395. ACM, New York (1985)
2. Baburin, D.E.: Using graph based representations in reengineering. In: Proceedings of the Sixth European Conference on Software Maintenance and Reengineering, pp. 203–206 (2002)
3. Batini, C., Nardelli, E., Tamassia, R.: A layout algorithm for data flow diagrams. IEEE Transactions on Software Engineering 12(4), 538–546 (1986)
4. Di Battista, G., Eades, P., Tamassia, R., Tollis, I.G.: Graph Drawing: Algorithms for the Visualization of Graphs. Prentice Hall, Englewood Cliffs (1999)
5. Chapin, N.: Some structured analysis techniques. SIGMIS Database 10(3), 16–23 (1978)
6. Davis, A.L., Keller, R.M.: Data flow program graphs. Computer 15(2), 26–41 (1982)
7. Doorley, M., Cahill, A.: Experiences in automatic leveling of data flow diagrams. In: WPC 1996: Proceedings of the 4th International Workshop on Program Comprehension, pp. 218–229. IEEE Computer Society, Los Alamitos (1996)
8. Eker, J., Janneck, J.W., Lee, E.A., Liu, J., Liu, X., Ludvig, J., Neuendorffer, S., Sachs, S., Xiong, Y.: Taming heterogeneity—the Ptolemy approach. Proceedings of the IEEE 91(1), 127–144 (2003)
9. Eschbach, T.: Visualisierungen im Schaltkreisentwurf. PhD thesis, Institut für Informatik, Albert-Ludwigs-Universität Freiburg (June 2008)
10. Eschbach, T., Guenther, W., Becker, B.: Orthogonal hypergraph drawing for improved visibility. Journal of Graph Algorithms and Applications 10(2), 141–157 (2006)

11. Gansner, E.R., Koutsofios, E., North, S.C., Vo, K.-P.: A technique for drawing directed graphs. Software Engineering 19(3), 214–230 (1993)
12. Lageweg, C.R.: Designing an automatic schematic generator for a netlist description. Technical Report 1-68340-44(1998)03, Laboratory of Computer Architecture and Digital Techniques (CARDIT), Delft University of Technology, Faculty of Information Technology and Systems (1998)
13. Prochnow, S., von Hanxleden, R.: Statechart development beyond WYSIWYG. In: Engels, G., Opdyke, B., Schmidt, D.C., Weil, F. (eds.) MODELS 2007. LNCS, vol. 4735, pp. 635–649. Springer, Heidelberg (2007)
14. Sander, G.: Graph layout through the VCG tool. Technical Report A03/94, Universität des Saarlandes, FB 14 Informatik, 66041 Saarbrücken (October 1994)
15. Sander, G.: A fast heuristic for hierarchical Manhattan layout. In: Brandenburg, F.J. (ed.) GD 1995. LNCS, vol. 1027, pp. 447–458. Springer, Heidelberg (1996)
16. Sander, G.: Layout of compound directed graphs. Technical Report A/03/96, Universität des Saarlandes, FB 14 Informatik, 66041 Saarbrücken (June 1996)
17. Sander, G.: Layout of directed hypergraphs with orthogonal hyperedges. In: Liotta, G. (ed.) GD 2003. LNCS, vol. 2912, pp. 381–386. Springer, Heidelberg (2004)
18. Sander, G., Vasiliu, A.: The ILOG JViews graph layout module. In: Mutzel, P., Jünger, M., Leipert, S. (eds.) GD 2001. LNCS, vol. 2265, pp. 438–439. Springer, Heidelberg (2002)
19. Spönemann, M.: On the automatic layout of data flow diagrams. Diploma thesis, Christian-Albrechts-Universität zu Kiel, Department of Computer Science (March 2009),
    http://rtsys.informatik.uni-kiel.de/~biblio/downloads/theses/msp-dt.pdf
20. Spönemann, M., Fuhrmann, H., von Hanxleden, R.: Automatic layout of data flow diagrams in KIELER and Ptolemy II. Technical Report 0914, Christian-Albrechts-Universität zu Kiel, Department of Computer Science (July 2009),
    http://rtsys.informatik.uni-kiel.de/~biblio/downloads/papers/report-0914.pdf
21. Sugiyama, K., Misue, K.: Visualization of structural information: automatic drawing of compound digraphs. IEEE Transactions on Systems, Man and Cybernetics 21(4), 876–892 (1991)
22. Sugiyama, K., Tagawa, S., Toda, M.: Methods for visual understanding of hierarchical system structures. IEEE Transactions on Systems, Man and Cybernetics 11(2), 109–125 (1981)
23. Wiese, R., Eiglsperger, M., Kaufmann, M.: yFiles: Visualization and automatic layout of graphs. In: Mutzel, P., Jünger, M., Leipert, S. (eds.) GD 2001. LNCS, vol. 2265, pp. 588–590. Springer, Heidelberg (2002)

Fast Edge-Routing for Large Graphs

Tim Dwyer and Lev Nachmanson

Microsoft Research,
Redmond, USA
{t-tdwyer,levnach}@microsoft.com

Abstract. To produce high quality drawings of graphs with nodes drawn as shapes it is important to find routes for the edges which do not intersect node boundaries. Recent work in this area involves finding shortest paths in a tangent-visibility graph. However, construction of the full tangent-visibility graph is expensive, at least quadratic time in the number of nodes. In this paper we explore two ideas for achieving faster edge routing using approximate shortest-path techniques.

1   Introduction

Most graphs that people need to visualize have nodes with associated textual or graphical content. For example, in UML class diagrams the nodes are drawn as boxes with textual content describing the class attributes or methods. In metabolic pathway diagrams nodes representing chemical compounds may have long textual labels or a graphic representation of the molecular structure. If edges that are not directly connected to a particular node are drawn over that node then the label may be obscured. Alternately, if edges are drawn behind a node then the reader may erroneously assume a connection to the node. Routing edges *around* nodes can avoid this ambiguity.

Some layout algorithms, such as the level scheme for directed graphs or the topology-shape-metrics approach for orthogonal graph drawing (see [1]) consider edge routing as an integral step in the layout process. However, the popular force-directed family of layout algorithms for general undirected graphs do not usually consider routing edges around node hulls; except perhaps as a post-processing step (e.g. Gansner and North [9]). Recent work such as [6,7] has proposed force-directed methods which are able to preserve the topology of a given edge routing, but a feasible initial routing must still be found using a standard routing algorithm. As described in Section 2, for graphs with hundreds of nodes, the quadratic (in the number of nodes) or worse cost of constructing the visibility graph can be too slow, especially for interactive applications where the layout is changing significantly from iteration to iteration.

In this paper we present two approaches to achieve faster routing using *approximate* shortest paths. The first approach uses a spatial decomposition of the nodes, moving them slightly to obtain strictly disjoint convex hulls around groups of nodes, and then computing visibility graphs over these composite hulls

D. Eppstein and E.R. Gansner (Eds.): GD 2009, LNCS 5849, pp. 147–158, 2010.

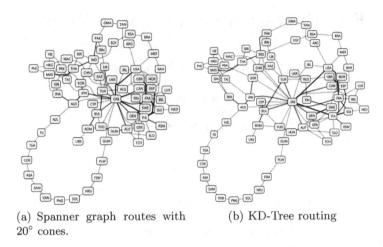

(a) Spanner graph routes with 20° cones.

(b) KD-Tree routing

Fig. 1. The "Olympic Torch Relay" graph from the GD'08 competition

rather than individual nodes. The second approach generates a sparse visibility-graph *spanner*. The two techniques are complementary, that is they can be used together to obtain even faster routing.

2 Related Work

Dobkin et al. [4] introduced visibility-graph methods for shortest-path edge routing into graph-drawing applications. They also considered the problem of fitting splines to the piecewise-linear path to obtain smooth curves.

Freivalds [8] gives a novel approach which treats edge routing as a problem of finding a low-cost path across a continuous cost function defined over the drawing area. A grid simplification is used so that the cost of routing one edge is $O(L^2 \log L)$ where L is the length of the path in grid units. The method is slow but is noteworthy in that adding additional routing criteria, such as perpendicular crossings between edges and slight offsets between collinear edges, is very easy.

Wybrow et al. [13] explored an efficient incremental implementation of a tangent-visibility graph for interactive graph manipulation or editing, for example, adding or removing a single node. However, their method is still $O(n^2 \log n)$ running time for n nodes in the static case. Faster algorithms for static construction of a visibility graph exist, but they are intricate and the asymptotic complexity improvement is not clear cut. For example, Ghosh and Mount [10] give an $O(E + V \log V)$ time algorithm for constructing a visibility graph with E edges over a set of obstacles with V vertices. Note that the usual tangent-visibility graph construction is not sensitive to the number of vertices V but rather to the number of obstacles n, and that the visibility graph may contain $O(n^2)$ edges.

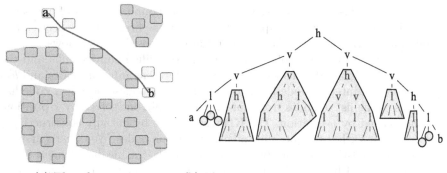

(a) The edge routing

(b) The corresponding KD-tree with compound obstacles enclosed.

Fig. 2. An edge between two nodes a and b, routed using a tangent-visibility graph over a simplified set of obstacles and the corresponding KD-tree. The labels indicate: 'h'-a horizontal split internal node; 'v'-a vertical split internal node; and 'l'-a KD-tree leaf-node.

3   A Spatial-Decomposition Routing Scheme

The most expensive part of the routing schemes described above is the $O(n^2 \log n)$ construction of the tangent-visibility graph over n nodes with convex boundaries. Therefore, the first new idea we explore in this paper is a scheme for routing over simpler visibility graphs using a spatial partitioning scheme. The intuition is to replace groups of nodes (especially those that are far from the end nodes of the edge being routed) with their convex hulls, thus reducing the number of obstacles to consider in construction of the visibility graph. For example, see Figure 2.

To achieve this we need to obtain a recursive spatial partitioning of the nodes such that the convex hulls of the nodes in each partition are not overlapping with their siblings in the partition hierarchy. To be precise, recursive application of a spatial partitioning to nodes positioned in the plane gives a tree structure where each tree node at level k in the tree has children on level $k + 1$. We use $desc(T)$ to denote the set of all leaf nodes (the original nodes in our graph) that are contained in a particular tree node T. We require that for any tree node U at level k in the tree, the convex hull of $desc(U)$ must not overlap with the convex hull of $desc(V)$ for any other tree node V also at level k, i.e. a sibling of U.

In order to achieve a reasonable asymptotic complexity, we also require that the tree be balanced. Obtaining such a tree for a given arrangement of nodes may be difficult or impossible. However, if we are willing to allow a little adjustment of node positions then we can enforce separation of siblings in a balanced KD-tree partitioning [2].

3.1   KD-Tree Partitioning

Our spatial-decomposition routing scheme begins with a starting configuration of nodes obtained with any layout algorithm, the examples in the paper were arranged using a fast-force directed approach. The bounding boxes of nodes can be initially overlapping as overlaps are removed by the first step, see Section 3.2. We build a KD-tree structure for these initial node positions as follows.

Fig. 2 shows an example of routing around simplified convex hulls and the KD-tree used to generate this routing. The KD-tree has *internal* nodes and *leaf* nodes, where an internal node has two child KD-tree nodes and a leaf node has two copies of the list of nodes from our original graph, one sorted by x-position, the other sorted by y-position. The KD-tree is built by initially constructing a single leaf node with lists containing all original graph nodes. We then recursively split the leaves either horizontally or vertically across the median element in the appropriate sorted list, and insert a new internal node as parent of these new leaves in the emerging hierarchy. We follow Lauther [11] in choosing to do a horizontal split if the bounding box of the elements in the leaf node is wider than tall, and vice versa otherwise, in order to keep the aspect ratio of leaf bounding boxes roughly square. We continue splitting until leaves are all smaller than some arbitrary bucket-size B. Initially sorting the n graph nodes by x- and y-position takes $O(n \log n)$ time. The sortedness of lists for new leaf nodes can be maintained by copying them in order from their parents. The tree is balanced since we always split across the median element, so $O(\log n)$ splits are performed. Thus, construction of the KD-tree requires $O(n \log n)$ time.

3.2 Removing Overlaps

The routing scheme that follows requires that nodes in the KD-tree do not overlap their siblings. We first remove overlaps between the B children of each leaf node in the KD-tree, i.e. the original graph nodes. A number of methods for effectively resolving overlaps between rectangular bounding boxes exist. We use the quadratic-programming based method of [5] since we find that it leads to relatively little displacement of nodes from their starting positions.

Next we must remove overlaps between the bounding boxes of the children of internal nodes. Each internal node i has two children as the result of a split. If the split was horizontal then we resolve overlap horizontally. That is, if the amount of horizontal overlap $o_h = rightSide(leftChild(i)) - leftSide(rightChild(i)) > 0$, then we translate $leftChild(i)$ by $-o_h/2$ and $rightChild(i)$ by $o_h/2$. We resolve overlap in the same manner vertically if i was constructed with a vertical split. All internal nodes are processed in this way, proceeding bottom-up.

Since we move each (graph) node up to $\log n$ times, the running time of this overlap removal step is $O(n \log n)$.

3.3 Computing Convex Hulls

The next step is to compute convex hulls around the descendents of each internal node. Again, this is computed bottom up. It is possible to compute the convex hulls of all internal nodes in the KD-Tree in $O(n \log n)$ total time using the linear time hull merging method of Preparata and Hong [12]. However, we use a naïve application of Graham Scan to calculate internal node hulls of the points in child hulls in $O(n \log^2 n)$ total time since the overall complexity of edge routing is dominated by the computation of visibility graphs anyway. In the sequel, $hull(i)$ refers to the precomputed convex hull of internal node i.

3.4 Simplified Visibility Graphs

Using the KD-tree of non-intersecting convex hulls described above we are able to construct a simplified visibility graph for all edges between a particular pair of leaves. Procedure *leaf-obstacles* returns a list of obstacles for any two leaves u and v in the KD-Tree T.

leaf-obstacles(u, v, T)
    $U \leftarrow$ the set of nodes of the shortest path between u and v in the KD-tree
    $w \leftarrow$ the lowest common ancestor of u and v
    $H \leftarrow \{hull(sibling(i))|i \in U \setminus \{u, v, w\}\}$
return $H \cup \{hull(c)|c \in children(u) \cup children(v)\}$

Where $sibling(i)$ returns the sibling of internal node i and $children(u)$ returns the original graph nodes that are children of leaf node u.

Lemma 1. *Procedure* leaf-obstacles *returns* $O(\log n)$ *obstacle hulls.*

Proof. If the maximum bucket size $B = 1$ then the height of the balanced KD-tree T is $\log n$ in which case the hulls returned by leaf-obstacles *are just the siblings of the ancestors of u and v up to the lowest common ancestor. The worst case is that the lowest common ancestor of u and v is the root of T, resulting in $2 \log n$ obstacles. In practice we use $B \approx 10$ which results in up to $2(\log n - \log B + B)$, i.e. also $O(\log n)$ for $B << n$.* $\square$

3.5 Routing Edges

First we group edges by the (unordered) pair of leaves in KD-Tree T of their end nodes. For each group of edges between KD-Tree leaves u and v we generate the visibility graph over *leaf-obstacles*(u, v, T). This takes time $O(\log^2 n \log \log n)$.

4 A Sparse Visibility-Graph Spanner

An alternative approach for edge routing in a large graph that we explore is the one suggested by [3]. This approach uses so called Yao graphs which are built by using fans of cones. A fan of cones is constructed at each vertex of an obstacle and only one edge of the visibility graph is chosen per cone. The resulting spanner graph contains only $O(\frac{\pi n}{\alpha})$ edges where n is the number of vertices of the graph and α is the cone angle. In spite of the graph sparseness for every $\epsilon > 0$ one can choose angle α such that for each shortest path in the full visibility graph the length of the corresponding shortest path in the spanner is at most $(1 + \epsilon)$ of the length of the former [3]. We construct the spanner graph by a direct method rather than by following the suggestion of [3] to build the conic Voronoi diagram first. We have not found the details of this algorithm for building such a spanner graph in [3] or in literature.

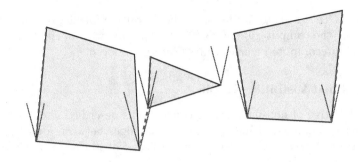

Fig. 3. Bold-dashed segments depict the edges created by one sweep

The technique presented here is a sweepline algorithm. One sweep finds edges inside cones constructed with the inner angle α and the bisector pointing to the direction of the sweep. By performing $\frac{\pi}{\alpha}$ sweeps we cover all possible edge directions from 0 to π. The input for the algorithm is the set P of convex mutually disjoint polygonal obstacles, the angle α and a vector representing the cone bisector direction (sweep direction). We say that a vertex u is visible from vertex v if $u \neq v$ and the line segment uv does not intersect the interior of a polygon in P. This way the neighbor vertices of an obstacle are visible to each other, and a side of an obstacle can be taken as an edge of the visibility graph. For the sake of simplicity the following explanation assumes further that the bisector coincides with the vector $(0, 1)$ and therefore the sweepline is horizontal and processing is from bottom to top. For points $a = (a_x, a_y)$ and $b = (b_x, b_y)$ we say the *cone distance* between them is $|b_y - a_y|$. Let V be the set of the vertices of P. For $v \in V$ we denote by C_v the cone with the apex at v, bisector $(0, 1)$, and angle α and by Vis_v the set of vertices from $V \cap C_v$ which are *visible* from v. For each $v \in V$ with $Vis_v \neq \emptyset$ the algorithm finds $u \in Vis_v$ which is closest to v in the cone distance, and adds the edge (v, u) to the spanner graph, see Fig. 3.

The algorithm works by processing events as the sweepline moves up. There are the following types of events (see Fig. 4(a)):

lowest vertex at the leftmost lowest vertex of an obstacle;

left vertex at a vertex that can be reached by a clockwise walk on obstacle edges starting from the vertex of a *lowest vertex* event and stopping at the rightmost highest vertex;

right vertex at a vertex that can be reached by a counterclockwise walk starting at the vertex of a *lowest vertex* event and stopping at the vertex before the rightmost highest vertex;

left intersection at the lowest intersection of a left cone side and an obstacle;

right intersection at the lowest intersection of a right cone side and an obstacle;

cone closure see below.

Events are kept in a priority queue Q with those sited at the lowest y-coordinate taking highest priority. For each cone participating in a sweep we keep pointers to its left and right sides. A cone side can be a default cone side, i.e. a ray starting at the cone apex at angle $\pm\frac{\alpha}{2}$ to the y-axis, or it could be a ray along an

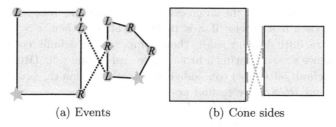

(a) Events   (b) Cone sides

Fig. 4. (a) Stars are *lowest vertex* events; circles labeled L and R are *left* and *right vertex* events; diamonds labeled L and R are *left* and *right intersection* events (b) Dotted lines show the default cone sides, Dashed lines - broken cone sides that are created at vertex events, solid lines - broken cone sides that are created at intersection events

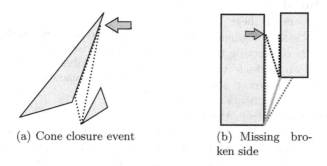

(a) Cone closure event   (b) Missing bro-
                                    ken side

Fig. 5. (a) The arrow points to a *cone closure* event. After such an event the cone is discarded. (b) The lower cone is discarded after discovering the vertex that completes the grey edge. The cone's left broken side is removed from LCS and the intersection marked by the arrow is not detected.

obstacle side if the cone is partially obscured by the obstacle as demonstrated by Fig. 4(b). We call such a cone side a *broken side*. Now we can define a *cone closure* event as an event happening when a broken side intersects a default cone side of the same cone, see Fig. 5(a). For a broken side we keep a pointer to its default cone side.

4.1   Balanced Trees of Active Cone Sides and Obstacle Segments

During the sweep we maintain a set of active cones. An active cone with its apex at a vertex is constructed at the vertex event. It is discarded when completely obstructed by an obstacle or when a visible vertex is discovered inside of the cone. We keep left cone sides of the active cones in a balanced tree LCS, and their right cone sides in a balanced tree RCS. The processing order guarantees that no two default active left (right) sides intersect. We also know that the cone side that we search for, insert into or remove from the tree must intersect the sweepline. This allows us to define the following order between cone sides a and b where x is the intersection of a with the sweepline. If x is to the left of b then $a < b$, else if x is to the right of b then $a > b$. Otherwise, if a and b are both broken sides, then compare the default cone sides they point to, else we are comparing a broken side and a cone side; (Rule A) in LCS (RCS) the default left(right) cone side is greater(less) than the broken cone side.

Trees LCS and RCS serve to find active cones "seeing" a vertex. Another function of the trees is calculation of intersection events. However, as shown at Fig. 5(b), a broken side containing an intersection event site can be removed before the intersection is found. To work around this we maintain two additional balanced trees called LS and RS. The members of these trees are called active obstacle segments; they are line segments connecting two adjacent vertices of an obstacle which are intersected by the current sweepline. Members of LS are segments traversed on the clockwise walk from the lowest

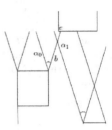

Fig. 6. FindConesSeeingEvent: Cone side b is the first in RCS not to the left of e. Side a first takes the value of a_0 then a_1. The two cones marked by an arc-segment "see" e.

vertex to the top of the obstacle and the remaining obstacle segments are members of RS. An active segment is added to the tree when its low vertex is processed and removed when its top is processed. Elements of LS (RS) are called left (right) active segments. The order of the segments in LS and RS is defined by the x-coordinates of the intersections of the segments and the sweepline. We call a segment *almost horizontal* if the absolute value of the difference between the y-coordinates of its start and end point is less than some small positive ϵ set in advance. Almost horizontal segments are not included in LS and RS since their intersections with the sweepline are not well defined, since we assume that the sweepline is horizontal. The order is well defined since the obstacles are disjoint.

4.2   Algorithm Description

The main loop of the algorithm is described below.

```
Sweep (P, α, bisector)
    initialize queue Q by all lowest vertex events
    while Q is not empty
        e ← pop event from Q
        ProcessEvent(e)
```

Routine **ProcessEvent** proceeds according to the event type. If the event is a *cone closure* then we discard the cone. However, it can happen that the cone of a *cone closure* event has been discarded earlier when a visible vertex was found inside the cone. To handle this we keep a Boolean flag associated with a cone and set it to true when the cone is removed. In the case of a *cone closure* event if flag is not set we remove the cone; that is, its left and right sides are removed from LCS and RCS respectively.

We describe in detail only the event handlers **LeftVertexEvent** and **LeftIntersectionEvent** since other events are symmetric, with the exception of *lowest vertex*. We explain this exception below.

LeftVertexEvent(e)
    move the sweepline to the site of e
    FindConesSeeingEvent(e)
    CloseConesByHorizontalSegment(e)
    remove from LS the segment incoming into e clockwise
    AddConeAtLeftVertex(e)

The procedure **FindConesSeeingEvent** finds all cones that "see" the site of e, creates the corresponding edges and discards the cones, see Fig. 6.

FindConesSeeingEvent(e)
    $b \leftarrow$ the first right cone side in RCS which is not to left of e
    if b exists
        $a \leftarrow$ the left side of the cone of b
        while a is defined and a is not to the right of e
            create the edge from the apex of the cone of a to the vertex and remove the cone
            $a \leftarrow$ the successor of a in LCS

Procedure **CloseConesByHorizontalSegment** handles the case when the obstacle side going clockwise and ending at the vertex of event v is almost horizontal. It finds all cones obstructed by the segment. Since by this time we have removed all cones "seeing" v or "seeing" the start of the segment, every cone with a side intersecting the segment is completely obstructed by it. Tree RCS, for example, can be used to find all such cones in an efficient manner.

In **AddConeAtLeftVertex** we try to create a cone, enqueue events, and add a segment to LS. Let v be the vertex of e, and u is the next one on the obstacle in the clockwise order. We enqueue a left vertex event for u if it is not below v. The cone is created at v only when it is not completely obscured by the obstacle. The left side of the cone is a default side: for this side we look for the intersection with the last segment of RS to the left of v. If the right cone side is a default cone side we look for the intersection of it with the first segment of LS to the right of v. If the right side of the cone is an obstacle side we check for its intersection with the last segment of RCS which is to left of v. If the cone is created we add new cone left (right) side to LCS (RCS). If the segment $[v, u]$ is not almost horizontal and points to the left of the default right cone side starting from v then the segment is added to LS as a left active segment. An obstacle segment which does not point to the left of the default right cone side is not inserted into LS since no default right cone with the apex different from v can intersect it without first intersecting the obstacle at some other segment.

At a lowest vertex event we do almost all the work of a left vertex event and a right vertex event. However, since the lowest vertex is the first one examined on the obstacle we do not try to close cones by the horizontal obstacle segments adjacent to the vertex. When processing a right vertex event we enqueue the next right vertex event only in the case when the segment from the event vertex to the next vertex going counterclockwise on the obstacle is not almost horizontal; this way we avoid processing a top vertex of an obstacle twice.

Let us describe the way **LeftIntersectionEvent** works. This procedure deals
with the intersection of a default left cone side and a right obstacle segment.

```
LeftIntersectionEvent(e)
    c ← the cone side of e
    x ← the intersection point of e
    s ← the obstacle segment of e
    u ← the top point of s
    if the cone of c is not removed
        if segment [x, u] is almost horizontal
            remove the cone of c
            move the sweepline to the event site
        else
            RemoveFromTree(c, LCS)
            move the sweepline to the event site
            t ← new broken side [x, u]
            replace c by t in the cone and insert t into LCS
            m ← the successor of t in LCS
            if m exists and intersects t
                enqueue the new left intersection event
            if t intersects the cone right side and the intersection point is below u
                enqueue a new cone closure event at the intersection point
```

It can happen that the sweepline passes through a left intersection event site
before calling **LeftIntersectionEvent** for this event. In this case **Remove-
FromTree** does not fail because of Rule A of the the order of LCS.

4.3   Performance of the Sweep

Let n be the number of vertices of obstacles from P. The number of events is
$O(n)$ since there are n vertices producing not more than n cones and each cone
creates at most three events: two intersection events and one closure event. The
trees never have more than n elements each. Each search on the trees takes
$O(\log n)$ steps. Operations "successor" or "predecessor" on the trees also take
$O(\log n)$ steps. In **FindConesSeeingEvent** and in **CloseConesByHorizontalSegment** we walk the tree by moving to the cone side successor until some
condition holds. We can potentially make $O(n \log n)$ steps per call. However, for
each processed cone side we remove the corresponding cone, so the routines cannot make more then $O(n \log n)$ steps during the whole algorithm run. Therefore
the overall number of steps of the algorithm is $O(n \log n)$.

5   Spline Refinement

At the final stage of our routing algorithm we "beautify" the spline. The detailed
discussion of this stage is beyond the scope and space limitations of this paper, but
on a very high level we do the following steps; *shortcutting, relaxation* and *fitting*.
In shortcutting we try to skip each internal vertex of the shortest path by removing
it and checking that the path still does not intersect the interior of an obstacle.
Intersections are checked efficiently using a binary space partitioning. In relaxation
we modify the path in such a way that it does not touch the obstacles anymore. In
fitting we inscribe cubic Bezier segments into the corners of the shortest path. We
have not carefully proven asymptotic complexity of these steps but in practice we
find only a fraction of the full routing time is spent in refinement.

6   Experimental Results

We tested routing over various combinations of spanner visibility graphs and KD-tree partitioning for several different graphs of very different sizes. In summary, we find that the two methods proposed in this paper are complementary or can be used in isolation to achieve significant speed-up. For a large graph with 1138 nodes and 1458 edges shortest-path edge routing over the standard tangent-visibility graph took around 95 seconds. Routing over a spanner visibility graph with $10°$ cones reduced this time to 43 seconds, including time spent in spline refinement. With $45°$ cones, this was further reduced to 34 seconds. Increasing cone size was found to increase the longest edge length - by up to 7% for $45°$ cones, however the short-cutting step in our spline refinement phase was very effective at keeping average edge lengths relatively short. At a cursory glance the quality of the spanner-visibility graph routing together with refinement is close to the optimal shortest path routing. Routes that are slightly longer than necessary (for example following the side of an obstacle when a more direct route is possible) are only noticeable with careful inspection, e.g. see Fig. 7.

Adding the KD-tree routing scheme was found to add a further, very significant, speed-up. Using a $45°$ cone spanner as well as KD-tree, routing the 1458 edges of our largest graph took only 5 seconds (compared to 95 seconds optimal routing). The extra "spreading-out" of nodes due to the spatial partitioning scheme, and the resultant increase in edge length (around 20% on average), was noticeable (e.g. see Fig. 1), but less so for the very large graph.

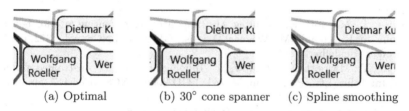

(a) Optimal   (b) 30° cone spanner   (c) Spline smoothing

Fig. 7. Detail from routing over the GD'08 "Companies" contest graph. (a) shows the optimal shortest path routing (b) an edge that follows the side of a shape rather than taking the optimal shortest path when routed using a 30° cone spanner (c) spline smoothing makes this path seem less bad. Even so, such non-optimal routes are relatively rare thanks to short-cutting (see Sec. 5).

7   Conclusion and Further Work

This paper represents the first attempt of which we are aware of using a spanner visibility graph scheme in routing of graph drawings. We achieve very significant speed-up with only marginal degradation in route quality so in future we intend to use it by default with a largish cone-size of $30°$ for all routing. The only disadvantage of the spanner visibility graph scheme is that it is quite complicated to implement. However, in this paper we have given more implementation details than we have found in the literature.

The KD-tree routing scheme is novel as far as we are aware. This gave us very significant speed improvement and was found to be particularly fast when used in combination with the spanner visibility graph scheme. The only disadvantage is that additional adjustment of nodes is required which may make it impractical (for example) in interactive scenarios where too much layout adjustment would spoil the user's mental map.

We were also pleased with the results of our spline refinement strategy when applied to spanner visibility graph routing. In the future we intend to do further analysis and improvement of algorithmic complexity of this step which currently could be high in the worst case, especially our short-cutting strategy.

References

1. Battista, G.D., Eades, P., Tamassia, R., Tollis, I.G.: Graph Drawing: Algorithms for the Visualization of Graphs. Prentice Hall, Englewood Cliffs (1999)
2. Bentley, J.L.: Multidimensional divide and conquer. Communications of the ACM 23(4), 214–229 (1980)
3. Clarkson, K.L.: Approximation algorithms for shortest path motion planning. In: STOC 1987: Nineteenth, New York (May 1987)
4. Dobkin, D.P., Gansner, E.R., Koutsofios, E., North, S.C.: Implementing a general-purpose edge router. In: DiBattista, G. (ed.) GD 1997. LNCS, vol. 1353, pp. 262–271. Springer, Heidelberg (1997)
5. Dwyer, T., Marriott, K., Stuckey, P.J.: Fast node overlap removal. In: Healy, P., Nikolov, N.S. (eds.) GD 2005. LNCS, vol. 3843, pp. 153–164. Springer, Heidelberg (2006)
6. Dwyer, T., Marriott, K., Wybrow, M.: Integrating edge routing into force-directed layout. In: Kaufmann, M., Wagner, D. (eds.) GD 2006. LNCS, vol. 4372, pp. 8–19. Springer, Heidelberg (2007)
7. Dwyer, T., Marriott, K., Wybrow, M.: Topology preserving constrained graph layout. In: Tollis, I.G., Patrignani, M. (eds.) GD 2008. LNCS, vol. 5417, pp. 230–241. Springer, Heidelberg (2009),
   http://www.csse.monash.edu.au/~tdwyer/topology.pdf
8. Freivalds, K.: Curved edge routing. In: Freivalds, R. (ed.) FCT 2001. LNCS, vol. 2138, pp. 126–137. Springer, Heidelberg (2001)
9. Gansner, E.R., North, S.C.: Improved force-directed layouts. In: Whitesides, S.H. (ed.) GD 1998. LNCS, vol. 1547, pp. 364–373. Springer, Heidelberg (1999)
10. Ghosh, S.K., Mount, D.M.: An output-sensitive algorithm for computing visibility. SIAM Journal on Computing 20(5), 888–910 (1991)
11. Lauther, U.: Multipole-based force approximation revisited - a simple but fast implementation using a dynamized enclosing-circle-enhanced k-d-tree. In: Kaufmann, M., Wagner, D. (eds.) GD 2006. LNCS, vol. 4372, pp. 20–29. Springer, Heidelberg (2007)
12. Preparata, F.P., Hong, S.J.: Convex hulls of finite sets of points in two and three dimensions. Communications of the ACM 20(2), 87–92 (1977)
13. Wybrow, M., Marriott, K., Stuckey, P.J.: Incremental connector routing. In: Healy, P., Nikolov, N.S. (eds.) GD 2005. LNCS, vol. 3843, pp. 446–457. Springer, Heidelberg (2006)

Leftist Canonical Ordering

Melanie Badent[1], Michael Baur[2], Ulrik Brandes[1], and Sabine Cornelsen[1]

[1] Department of Computer & Information Science, University of Konstanz
{Melanie.Badent,Ulrik.Brandes,Sabine.Cornelsen}@uni-konstanz.de
[2] Department of Computer Science, Universität Karlsruhe (TH)
baur@informatik.uni-karlsruhe.de

Abstract. Canonical ordering is an important tool in planar graph drawing and other applications. Although a linear-time algorithm to determine canonical orderings has been known for a while, it is rather complicated to understand and implement, and the output is not uniquely determined. We present a new approach that is simpler and more intuitive, and that computes a newly defined leftist canonical ordering of a triconnected graph which is a uniquely determined leftmost canonical ordering.

1 Introduction

Canonical vertex orderings were introduced by de Fraysseix, Pach, and Pollack [13,14] and are the backbone of several algorithms for planar graphs, including graph drawing algorithms [2,3,4,8,9,10,16,17,18,19,20,22,23,28,27,29,30], graph encoding [1,11,26], construction of realizers, spanners, or orderly spanning trees [5,6,7,15,31,32,33], and more [12,25,34].

Kant [28] generalized canonical orderings to triconnected graphs. While several implementations of the linear-time algorithm of Kant are available, this algorithm is neither trivial to code, nor is its correctness easily understood. Based on a simple and intuitive criterion, we present a new algorithm that might further broaden the scope of adoption and ease teaching.

Since a triconnected graph can have many canonical orderings, we introduce the leftist (and rightist) canonical ordering that is uniquely determined. The leftist canonical ordering is in particular a leftmost canonical ordering.

The main advantage of our algorithm compared to the algorithm in [28] is that we do not use the dual graph nor any face labels. Further, we compute the unique leftist canonical ordering from scratch, i. e., without any reordering, and we compute it from the low numbers to the high numbers contrary to the previous algorithm that builds the canonical ordering from the end by shelling off paths from the outer face. A similar approach for biconnected canonical orderings can be found in [24]. We also give a detailed pseudocode such that it can be easily implemented. Finally, our proof of correctness includes a new proof of the existence of a canonical ordering for triconnected graphs.

The paper is organized as follows. Canonical orderings are defined in Sect. 2. The new algorithm and its linear-time implementation are described in Sects. 3 and 4, respectively.

D. Eppstein and E.R. Gansner (Eds.): GD 2009, LNCS 5849, pp. 159–170, 2010.

2   Preliminaries

Throughout this paper, let $G = (V, E)$ be a simple undirected graph with n vertices, $n \geq 3$, and m edges. We assume that G is planar and triconnected, hence it has a unique planar embedding up to the choice of the outer face. For a subset $V' \subseteq V$ we denote by $G[V']$ the subgraph of G induced by V'. By $\deg_G(v)$ we denote the number of edges of G that contain v. A path is a sequence $P = \langle v_0, \ldots, v_\ell \rangle$ of distinct adjacent vertices, i.e., $\{v_i, v_{i+1}\} \in E$. We also denote the set $\{v_0, \ldots, v_\ell\}$ by P.

Canonical orderings were introduced originally for triangulated graphs by de Fraysseix et al. [13,14]. The following rephrases Kant's generalization to triconnected graphs [28].

Definition 1 (canonical ordering). *Let $\Pi = (P_0, \ldots, P_r)$ be a partition of V into paths and let $P_0 = \langle v_1, v_2 \rangle$, $P_r = \langle v_n \rangle$ such that v_2, v_1, v_n is a path on the outer face of G in clockwise direction. For $k = 0, \ldots, r$ let $G_k = G[V_k] = (V_k, E_k)$ be the subgraph induced by $V_k = P_0 \cup \cdots \cup P_k$, let C_k be the outer face of G_k. Partition Π is a* canonical ordering *of (G, v_1) if for each $k = 1, \ldots, r - 1$:*

1. *C_k is a simple cycle.*
2. *Each vertex z_i in P_k has a neighbor in $V \setminus V_k$.*
3. *$|P_k| = 1$ or $\deg_{G_k}(z_i) = 2$ for each vertex z_i in P_k.*

P_k is called a singleton *if $|P_k| = 1$ and a* chain *otherwise.*

A canonical ordering Π is refined to a *canonical vertex ordering* $v_1, \ldots, v_n$ by ordering the vertices in each P_k, $k > 0$, according to their clockwise appearance on C_k (see Figs. 1(a)-1(c)).

The following observations help build an intuitive understanding of canonical orderings. Note that Propositions 4 and 5 of Lemma 1 are part of Kant's original definition.

Lemma 1. *For $k = 1, \ldots, r - 1$:*

1. *P_k has no chord.*
2. *For each vertex v in P_k there is a v-v_n-path $v = v_{k_0}, \ldots, v_{k_\ell} = v_n$ where each v_{k_i} is in P_{k_i} and $k_i < k_j$ for $0 \leq i < j \leq \ell$. Especially:*
   (a) $G[V \setminus V_k]$ is connected.
   (b) If $\deg_{G_k}(v) = 2$, then v is in C_k.
   (c) P_k is on C_k.
3. *(a) A singleton P_{k+1} and a path of C_k bound some faces or*
   (b) a chain P_{k+1} and a path of C_k bound one face.
4. *G_k is biconnected.*
5. *If v, w is a separation pair of G_k, then both are on C_k.*

Proof. The properties are directly implied by the fact that G is triconnected and by the definition of a canonical ordering. □

Remark 1. Two incident faces of a triconnected planar graph share one vertex or one edge. Especially, no face has a chord.

2.1   Leftmost Canonical Ordering

Kant [28] introduced a leftmost and rightmost canonical ordering of G. Let $P_0, \ldots, P_k$ be a sequence of paths that can be extended to a canonical ordering of G. A path P of G is a *feasible candidate* for the step $k+1$ if also $P_0, \ldots, P_k, P$ can be extended to a canonical ordering. Let $v_1 = c_1, c_2, \ldots, c_q = v_2$ be the vertices from left to right on C_k. Let c_ℓ be the neighbor of P on C_k such that ℓ is as small as possible. We call c_ℓ the *left neighbor* of P.

Definition 2 (leftmost canonical ordering). *A canonical ordering $P_0, \ldots, P_r$ is called* leftmost (rightmost) *if for $k = 0, \ldots, r-1$ the following is true. Let c_ℓ be the left neighbor of P_{k+1} and let $P_{k'}, k+1 \leq k' \leq r$, be a feasible candidate for the step $k+1$ with left neighbor $c_{\ell'}$. Then either (1) $\ell \leq \ell' (\ell \geq \ell')$ or (2) there is an edge between P_{k+1} and $P_{k'}$ (see Fig. 1(b)).*

Note that once a canonical ordering is known a simple linear-time algorithm can be used to rearrange its paths so that it becomes leftmost [28]. Also note that Kant did not use Condition 2 of a leftmost canonical ordering in his definition, however, he used it in his reordering algorithm. While leftmost canonical orderings are particularly useful for many applications, we stress that the rearrangement is applicable to any canonical ordering and that a leftmost canonical ordering is only unique with respect to a given partiton.

2.2   Leftist Canonical Ordering

In the leftist canonical ordering we add in each step the leftmost possible path where the choice is not only within an already given partition.

Definition 3 (leftist canonical ordering). *A canonical ordering $P_0, \ldots, P_r$ is called* leftist (rightist) *if for $k = 0, \ldots, r-1$ the following is true. Let c_ℓ be the left neighbor of P_{k+1} and let P be a feasible candidate for the step $k+1$ with left neighbor $c_{\ell'}$. Then $\ell \leq \ell' (\ell \geq \ell')$ (see Figs. 1(c) and 1(a)).*

Note that a feasible candidate for the step $k+1$ needs not to be a feasible candidate for the step $k+2$ anymore. Also note that the leftist canonical ordering is unique irrespective of a given partition and it is a leftmost canonical ordering. A leftist canonical ordering can also be found by choosing always the rightmost face or singleton in the algorithm of Kant [28]. A similar concept related to Schnyder realizers without clockwise cycles was defined for triangulated graphs in [6].

3   New Algorithm

Starting from $P_0 = \langle v_1, v_2 \rangle$, we build the canonical ordering by adding $P_1, \ldots, P_r$ in this order. In step $k+1$, the "belt" around G_k, i. e., the sequence of vertices not in G_k that lie on faces incident to G_k is considered. Then, a candidate not causing any "self-intersection" within the belt is chosen. Before we give the details, we start with a recursive definition of which paths will be considered in the step $k+1$.

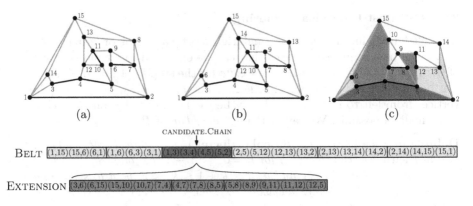

Fig. 1. Different canonical orderings (black paths are chains). (a) Rightist canonical ordering. (b) Leftmost canonical ordering respecting the ordering in (a). (c) Leftist canonical ordering and its construction. The light and dark grey faces are the belt of G_0. The next candidate in Algorithm 3 is $P_1 = \langle 3, 4, 5 \rangle$. Algorithm 5 substitutes the dark grey face by the middle grey faces, i. e., by the EXTENSION found by Algorithm 4.

Definition 4 (cut faces and locally feasible candidates). $P_0 = \langle v_1, v_2 \rangle$ *is a locally feasible candidate. Let $P_0, \ldots, P_k$ be a sequence of locally feasible candidates and V_k, G_k, and C_k as in Definition 1. A* cut face *f of G_k is an inner face of G that is incident to some vertex on C_k but is not a face of G_k. Let P_f be the clockwise sequence of vertices incident to f that are not in V_k. If f is incident to an edge on C_k, then f is called a* candidate face *and P_f is called a* candidate *for the step $k + 1$. A candidate face f and the candidate P_f are* locally feasible *for the step $k + 1$ if*

1. *v_n is not in P_f or $P_0, \ldots, P_k, P_f$ is a partition of V,*
2. *$G[V \setminus (V_k \cup P_f)]$ is connected, and*
3. *P_f is a singleton or the degree of each vertex of P_f in $G[V_k \cup P_f]$ is two.*

In the remainder of this section, we will see that the locally feasible candidates are exactly the feasible candidates. We start with the following lemma which is a direct consequence of Definitions 1 and 4 and the triconnectivity of G.

Lemma 2

1. *A canonical ordering is a sequence of locally feasible candidates.*
2. *If a sequence of locally feasible candidates partitions the whole vertex set of a triconnected graph, it is a canonical ordering.*

In what follows, we consider the vertices on C_k to be from left to right between v_1 and v_2. Accordingly, we also consider the cut faces from left to right: A *cut edge* of G_k is an edge of G that is incident to one vertex in V_k and one vertex in $V \setminus V_k$. Let f and f' be two cut faces. Let c and c', respectively, be the leftmost vertices on C_k that are incident to f and f', respectively. We say that f is to the left of f' if c is to the left of c' on C_k or if $c = c'$, then the cut edges of f are to the left of the cut edges of f' in the incidence list of c.

Algorithm 1. Leftist Canonical Ordering

---

begin

> Let $v_2, v_1, v_3, \ldots, v_\ell$ be the bound of the inner face incident to $\{v_1, v_2\}$.
> $P_0 \leftarrow \langle v_1, v_2 \rangle,\ P_1 \leftarrow \langle v_3, \ldots, v_\ell \rangle,\ k \leftarrow 1$
> **while** $|V_k| < n - 1$ **do**
> > Let f be the leftmost locally feasible candidate face
> > $P_{k+1} \leftarrow P_f$
> > $k \leftarrow k + 1$
>
> $P_{k+1} \leftarrow \langle v_n \rangle$

end

Corollary 1. *If Algorithm 1 terminates, it computes the leftist canonical ordering of a triconnected planar graph.*

Before we prove that in each step there exists a locally feasible candidate face, we describe locally feasible candidates in terms of "self-intersection" of the belt. Let $P_0, \ldots, P_k$ be a sequence of locally feasible candidates. The *belt* of G_k is the sequence of vertices not in G_k that are incident to the cut faces of G_k from left to right. I.e., let $f_1, \ldots, f_s$ be the cut faces of G_k ordered from left to right. Let P_{f_0} be the vertices in $V \setminus V_k$ that are incident to the outer face in counterclockwise order. Then, the concatenation of $P_{f_1}, \ldots, P_{f_s}$ and P_{f_0} is the belt of G_k. Consider Fig. 1(a). Then, $P_2 = \langle 6, 7 \rangle$, $P_3 = \langle 8 \rangle$, and the belt of G_3 is $15, 14|14|14, 15, 13, 12|12, 10|10, 11, 9|9|9, 11, 13|13, 15|15$.

Definition 5 (forbidden, singular, stopper). *A vertex v of the belt of G_k is*

- forbidden *if v does not occur consecutively in the belt of G_k,*
- singular *if v occurs more than twice in the belt of G_k and its occurrence is consecutive, and*
- *a stopper if it is forbidden or singular.*

In the above example, 15, 13, and 11 are forbidden and 14 and 9 are singular vertices. Note that v_n is always the first and last vertex of the belt. Hence, it remains forbidden until the end. It will turn out that the locally feasible candidates are those that do not contain a stopper or that are singular singletons.

Lemma 3. *Let $P_0, \ldots, P_k$ be a sequence of locally feasible candidates. Let f be a candidate face for the step $k + 1$ and let $P = P_f$.*

1. *If a vertex v of P is adjacent to more than two vertices in $V_k \cup P$, then v occurs more than twice in the belt.*
2. *If $G[V \setminus (V_k \cup P)]$ is not connected, then P contains a forbidden vertex.*
3. *If a vertex v of P is singular, then v is a locally feasible singleton.*
4. *If P contains a forbidden vertex v, then $G[V \setminus (V_k \cup P)]$ is not connected or P contains another vertex with more than two neighbors in $V_k \cup P$.*

Proof. 1. Let e be an edge incident to v and a vertex in $V_k \cup P$ that is not incident to f. By Remark 1, edge e is a cut edge and hence incident to two cut faces. Thus, v is incident to at least three cut faces.

2. Let W be the set of vertices in a connected component of the graph induced by $V \setminus (V_k \cup P)$ and not containing v_n. Since $V \setminus V_k$ was connected, W is adjacent to P and there is a path from P to v_n not intersecting W. By the triconnectivity of G, there is an edge between W and the part of C_k not contained in f. Further, there is at least a third vertex on $C_k \cup P$ adjacent to W. Let w be the rightmost vertex on $C_k \cup P$ that is adjacent to W and let v be the leftmost such vertex. Assume that w is on C_k. Then v is on P. Consider the face f' containing v and w. Then, the belt contains some vertices of W between the occurrences of v for the belt faces f and f' (see Fig. 2(a)).

3. If v is singular, then it is a candidate. By Proposition 2, $G[V \setminus (V_k \cup \{v\})]$ is connected.

4. Since v is forbidden, there is a cut face f' containing v and a cut face h between f and f' such that P_h contains a vertex $w \neq v$. If w is not incident to f, then w and v_n are in two connected components of $G[V \setminus (V_k \cup P)]$ (see Fig. 2(b)). So assume now that for all faces h' between f and f' the path $P_{h'}$ contains only vertices incident to f. Among these faces let h be the face that is next to f. By Remark 1, P_h consists of one vertex $w \neq v$ and w is singular (see Fig. 2(c)). $\qquad\square$

Corollary 2. *1. A candidate that is a chain is locally feasible if and only if it does not contain any stopper.*
2. A vertex of the belt is a locally feasible singleton if and only if it is singular.

For example, the locally feasible candidates for the step $k + 1 = 4$ in Fig. 1(a) are $\langle 14 \rangle$, $\langle 12, 10 \rangle$, and $\langle 9 \rangle$.

Theorem 1. *Algorithm 1 computes the leftist canonical ordering of a triconnected planar graph.*

Proof. By Lemma 1, it remains to show that in each step of the algorithm there is a locally feasible candidate. By Corollary 2.2, if there are any singular vertices, we have a locally feasible candidate. So, assume now we do not have any singular vertices. By Corollary 2, we have to show that there is a candidate that does not contain any forbidden vertex.

Let f be a candidate face and let $P = P_f$. Assume that P contains a forbidden vertex v. Let f' be a cut face containing v such that the belt contains a vertex other than v between the occurrence of v in P_f and the occurrence of v in $P_{f'}$. Let $f, h_1, \ldots, h_\alpha, f'$ be the sequence of cut faces between f and f'. We show by induction on the number of forbidden vertices in $P_{h_1}, \ldots, P_{h_\alpha}$ that there is a locally feasible candidate among $P_{h_1}, \ldots, P_{h_\alpha}$.

By the choice of f' and by triconnectivity of G, there is at least one $i = 1, \ldots, \alpha$ such that P_{h_i} is a candidate that does not contain v. If v is the only forbidden vertex in $P_{h_1}, \ldots, P_{h_\alpha}$, then P_{h_i} is locally feasible.

If P_{h_i} contains a forbidden vertex w (recall that by our assumption there are no singular vertices), there is a cut face $h \neq h_i$ among $f, h_1, \ldots, h_\alpha, f'$ incident to w such that the belt contains a vertex other than w between the occurrence

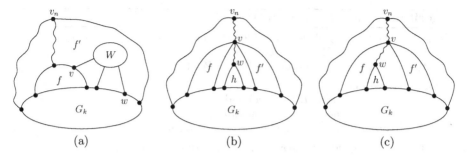

Fig. 2. Illustration of the proof of Lemma 3. (a) W is a connected component of $G[V \setminus (V_k \cup P_f)]$ not containing v_n. Faces f and f' are not consecutive in the belt of G_k. Thus, f contains a forbidden vertex v. (b, c) If v is forbidden, then (b) $G[V \setminus (V_k \cup P_f)]$ is not connected or (c) there is a singular vertex w.

of w in P_{h_i} and in P_h. The cut faces between h and h_i do not contain v. Hence, by the induction hypothesis, one of them is a locally feasible candidate. $\square$

4   Linear-Time Implementation

We will maintain a list BELT that represents the cut faces from left to right. For a simpler implementation, BELT contains lists of edges rather than one list of vertices and each cut face f is represented by a *belt item* which is a pair consisting of

- a list CHAIN of f's incident edges not in G_k in clockwise order and
- the rightmost stopper of P_f (if any).

We traverse the list BELT using a pointer CANDIDATE.

To decide whether a vertex is a stopper, we maintain two counters. Let CUTFACES(v) be the number of cut faces and CUTEDGES(v) the number of cut edges to which v is incident. In order to make the following lemma true also for v_n, we will count the outer face twice in CUTFACES(v_n).

Lemma 4. *A vertex v in the belt of G_k is*

- *forbidden if and only if CUTFACES(v) > CUTEDGES(v) + 1 and*
- *singular if and only if $2 <$ CUTFACES(v) = CUTEDGES(v) + 1.*

Proof. A vertex occurs once for each cut face it is incident to in the belt. Two occurrences of a vertex v in the belt are consecutive if and only if the corresponding cut faces share a cut edge incident to v. So, all occurrences of v in the belt are consecutive if and only if v is only incident to one more cut face than to cut edges. $\square$

The algorithm `canonicalOrdering` (see Algorithm 2) now works as follows. We start with a copy of G in which each undirected edge $\{v, w\}$ is replaced by the

---

Algorithm 2. Leftist Canonical Ordering

Input   : $G = (V, E)$ planar embedded triconnected undirected graph
                 $v_1 \in V$ on the outer face
Output : leftist canonical ordering $P_0, \ldots, P_k$ of (G, v_1)

canonicalOrdering

> replace each $\{v, w\} \in E$ by (v, w) and (w, v)
> $v_n \leftarrow$ clockwise neighbor of v_1 on outer face
> $v_2 \leftarrow$ counterclockwise neighbor of v_1 on outer face
>
> **for** $v \in V$ **do** CUTFACES$(v) \leftarrow 0$;   CUTEDGES$(v) \leftarrow 0$
> CUTFACES$(v_n) \leftarrow 1$
>
> mark (v_1, v_2) and (v_2, v_1)
> BELT $\leftarrow \langle (\langle (v_2, v_1), (v_1, v_2), (v_2, v_1) \rangle, \text{NIL}) \rangle$
>
> $k \leftarrow -1$
> CANDIDATE $\leftarrow$ first item in BELT
> **while** BELT $\neq \emptyset$ **do**
> > $k \leftarrow k + 1$
> > $P_k \leftarrow$ `leftmostFeasibleCandidate`
> > `updateBelt`

end

---

two directed edges (v, w) and (w, v). In the beginning, the belt is initialized by $(\langle (v_2, v_1), (v_1, v_2), (v_2, v_1) \rangle, \text{NIL})$. Thus, `leftmostFeasibleCandidate` (see Algorithm 3) chooses $P_0 = \langle v_1, v_2 \rangle$ as the first path.

In general, each iteration in Algorithm 2 consists of three steps: (1) We choose the new leftmost locally feasible candidate P_k, (2) we find the new cut faces incident to P_k, and (3) we replace P_k by its incident cut faces in the belt and update its neighbors (see Fig. 1(c)). In detail:

`leftmostFeasibleCandidate`. We traverse BELT from the current cut face CANDIDATE to the right doing the following: If CANDIDATE is a candidate face, traverse CANDIDATE.CHAIN from right to left until a stopper is found. If so, store it. If CANDIDATE.CHAIN contains no stopper or it is a singular singleton, it is the next locally feasible candidate. Otherwise, go to the next face. See Algorithm 3.

`beltExtension`. To find the new cut faces, we traverse CANDIDATE.CHAIN from left to right. The outer repeat-loop iterates over all vertices incident to two edges of CANDIDATE.CHAIN. Each iteration finds the new cut faces incident to such a vertex and increments the counter CUTEDGES. In the inner repeat-loop, we traverse all new edges of a new cut face and store them in the list CHAIN. Here the counter CUTFACES is incremented. Each list CHAIN is finally appended to the list EXTENSION that stores all new belt items incident to CANDIDATE.CHAIN. See Algorithm 4.

`updateBelt`. We replace CANDIDATE (and all its copies if it was a singleton) by the new cut faces found by `beltExtension`. The last edge of the predecessor and the first edge of the successor of CANDIDATE are removed since they are now contained in G_k. If the predecessor of CANDIDATE was not a candidate face before or it lost its stopper, then we go one step to the left in BELT and set CANDIDATE to its predecessor. See Algorithm 5.

Algorithm 3. Skip infeasible candidates

---

list leftmostFeasibleCandidate
$\quad$ FOUND ← **false**
$\quad$ **repeat**
$\quad\quad$ let $\langle (v_0, v_1), (v_1, v_2), \ldots, (v_\ell, v_{\ell+1}) \rangle :=$ CANDIDATE.CHAIN
$\quad\quad$ **if** $v_0 \neq v_{\ell+1}$ **then**
$\quad\quad\quad$ $j \leftarrow \ell$
$\quad\quad\quad$ **while** $j > 0$ **and not** $(forbidden(v_j)$ **or** $singular(v_j))$ **do** $j \leftarrow j - 1$
$\quad\quad\quad$ **if** $j > 0$ **then** CANDIDATE.STOPPER ← v_j
$\quad\quad\quad$ **if** $j = 0$ **or** $\qquad\qquad\qquad\qquad\qquad\qquad$ **then**
$\quad\quad\quad\quad$ $(singular($CANDIDATE.STOPPER$)$ **and** $\ell = 1)$
$\quad\quad\quad\quad$ FOUND ← **true**
$\quad\quad\quad\quad$ **for** $(v, w) \in$ CANDIDATE.CHAIN **do** mark (w, v)
$\quad\quad$ **if not** FOUND **then**
$\quad\quad\quad$ CANDIDATE ← $successor($CANDIDATE$)$
$\quad\quad\quad$ **if** CANDIDATE = NIL **then HALT: illegal input graph**
$\quad$ **until** FOUND
$\quad$ **return** $\langle v_1, \ldots, v_\ell \rangle$
end

Theorem 2. *Algorithm 2 computes the leftist canonical ordering of a triconnected planar graph in linear time.*

Proof. **Linear running time:** In the algorithm `beltExtension` each edge is touched at most twice. In the algorithm `leftmostFeasibleCandidate` each candidate is scanned from right to left until the first stopper occurs. All the scanned edges will have been deleted from the list when the candidate will be scanned the next time. In total only $2m$ edges will be added to BELT.

Correctness: While scanning BELT from left to right, we always choose the leftmost locally feasible candidate: Assume that at step $k + 1$ we scan a face f and there are no locally feasible candidates to the left of f. The face f is omitted because it is not a candidate or it contains a stopper. None of the two properties changes if no direct neighbor in BELT had been added to G_k. Hence, as long as f is not locally feasible, no face to the left of f has to be considered. Further, the number of incident cut faces or cut edges of a vertex never decreases. We show that a candidate can only become locally feasible after his rightmost stopper has become singular.

Let v be the rightmost stopper of P_f and assume v is forbidden. Let f_L be the leftmost and f_R be the rightmost cut face containing v. We can conclude by the proof of Theorem 1 that all occurrences of v between f_L and

---

Algorithm 4. Construct list of new belt items incident to P_k

---

list beltExtension(list $\langle e_0, \ldots, e_\ell \rangle$)
    EXTENSION $\leftarrow \emptyset$
    for $j \leftarrow 1, \ldots, \ell$ do /* check for new cut faces incident to source */
        $v_{start} \leftarrow source(e_j)$
        $v_{end} \leftarrow target(e_j)$
        FIRST $\leftarrow e_j$
        repeat
            FIRST $= (v, w) \leftarrow$ clockwise next in $N^+(v_{start})$ after FIRST
            CUTEDGES$(w) \leftarrow$ CUTEDGES$(w) + 1$
            if FIRST *not marked* then /* new cut face */
                CHAIN $\leftarrow \emptyset$
                $e \leftarrow (v, w)$
                repeat /* traverse clockwise */
                    mark e
                    append CHAIN $\leftarrow e$
                    CUTFACES$(w) \leftarrow$ CUTFACES$(w) + 1$
                    $e = (v, w) \leftarrow$ counterclockwise next in $N^+(w)$ after (w, v)
                until $w \in \{v_{start}, v_{end}\}$
                mark e
                append CHAIN $\leftarrow e$
                append EXTENSION $\leftarrow$ (CHAIN, NIL)
        until $w = v_{end}$
    return EXTENSION
end

Algorithm 5. Replace feasible candidate with incident faces

---

updateBelt
    if *singular(*CANDIDATE.STOPPER*)* then
        remove neighboring items with same singleton from BELT

    PRED $\leftarrow$ *predecessor*(CANDIDATE)
    SUCC $\leftarrow$ *successor*(CANDIDATE)
    if SUCC $\neq \emptyset$ then remove first edge from SUCC.CHAIN

    EXTENSION $\leftarrow$ **BeltExtension**(CANDIDATE.CHAIN)
    replace CANDIDATE by EXTENSION
    if EXTENSION $\neq \emptyset$ then
        CANDIDATE $\leftarrow$ first item of EXTENSION
    else
        CANDIDATE $\leftarrow$ SUCC

    if PRED $\neq \emptyset$ then
        remove last edge (v, w) from PRED.CHAIN
        if $v =$ PRED.STOPPER or $w = source(first\ edge\ of$ PRED.CHAIN) then
            PRED.STOPPER $\leftarrow$ NIL
            CANDIDATE $\leftarrow$ PRED
end

f are consecutive and that Algorithm 2 finds the locally feasible candidates between f and f_R in the belt until the belt contains only v between f and f_R. It follows that all occurrences of v in the belt are consecutive. Let now v be singular. Then, the only two incident cut faces $f = f_L$ and f_R of v would share a cut edge $\{v, w\}$ that would not have been a cut edge before. Hence, w would have been a stopper of f to the right of v. $\square$

Note that the algorithm for computing the leftist canonical ordering can also be used to compute the rightist canonical ordering. In that case, we store for each cut face the leftmost stopper and we scan the belt from right to left.

References

1. Barbay, J., Aleardi, L.C., He, M., Munro, I.: Succinct Representation of Labeled Graphs. In: Tokuyama, T. (ed.) ISAAC 2007. LNCS, vol. 4835, pp. 316–328. Springer, Heidelberg (2007)
2. Barequet, G., Goodrich, M.T., Riley, C.: Drawing Planar Graphs with Large Vertices and Thick Edges. J. of Graph Algorithms and Applications 8(1), 3–20 (2004)
3. Biedl, T.C.: Drawing Planar Partitions I: LL-Drawings and LH-Drawings. In: Proc. 14th Symp. on Computational Geometry, pp. 287–296. ACM Press, New York (1998)
4. Biedl, T.C., Kaufmann, M.: Area-Efficient Static and Incremental Graph Drawings. In: Burkard, R.E., Woeginger, G.J. (eds.) ESA 1997. LNCS, vol. 1284, pp. 37–52. Springer, Heidelberg (1997)
5. Bose, P., Gudmundsson, J., Smid, M.: Constructing Plane Spanners of Bounded Degree and Low Weight. Algorithmica 42(3-4), 249–264 (2005)
6. Brehm, E.: 3-Orientations and Schnyder 3-Tree-Decompositions. Master's thesis, FU Berlin (2000)
7. Chiang, Y.-T., Lin, C.-C., Lu, H.-I.: Orderly Spanning Trees with Applications to Graph Encoding and Graph Drawing. In: Proc. 12th ACM–SIAM Symp. on Discrete Algorithms (SODA 2001), pp. 506–515 (2001)
8. Chrobak, M., Kant, G.: Convex Grid Drawings of 3-Connected Planar Graphs. Int. J. of Computational Geometry and Applications 7(3), 211–223 (1997)
9. Chrobak, M., Nakano, S.-I.: Minimum-Width Grid Drawings of Plane Graphs. Computational Geometry 11(1), 29–54 (1998)
10. Chrobak, M., Payne, T.H.: A Linear-Time Algorithm for Drawing a Planar Graph on a Grid. Information Processing Letters 54(4), 241–246 (1995)
11. Chuang, R.C.-N., Garg, A., He, X., Kao, M.-Y., Lu, H.-I.: Compact Encodings of Planar Graphs via Canonical Orderings and Multiple Parentheses. In: Larsen, K.G., Skyum, S., Winskel, G. (eds.) ICALP 1998. LNCS, vol. 1443, pp. 118–129. Springer, Heidelberg (1998)
12. de Fraysseix, H., Mendez, P.O.: Regular Orientations, Arboricity, and Augmentation. In: Tamassia, R., Tollis, I.G. (eds.) GD 1994. LNCS, vol. 894, pp. 111–118. Springer, Heidelberg (1995)
13. de Fraysseix, H., Pach, J., Pollack, R.: Small Sets Supporting Fáry Embeddings of Planar Graphs. In: Proc. 20th ACM Symp. on the Theory of Computing (STOC 1988), pp. 426–433. ACM Press, New York (1988)
14. de Fraysseix, H., Pach, J., Pollack, R.: How to Draw a Planar Graph on a Grid. Combinatorica 10(1), 41–51 (1990)

15. Di Battista, G., Tamassia, R., Vismara, L.: Output-Sensitive Reporting of Disjoint Paths. Algorithmica 23(4), 302–340 (1999)
16. Di Giacomo, E., Didimo, W., Liotta, G.: Radial Drawings of Graphs: Geometric Constraints and Trade-offs. J. of Discrete Algorithms 6(1), 109–124 (2008)
17. Di Giacomo, E., Didimo, W., Liotta, G., Wismath, S.K.: Curve-Constrained Drawings of Planar Graphs. Computational Geometry 30(2), 1–23 (2005)
18. Dujmović, V., Suderman, M., Wood, D.R.: Really Straight Graph Drawings. In: GD 2004 [21], pp. 122–132
19. Erten, C., Kobourov, S.G.: Simultaneous Embedding of Planar Graphs with Few Bends. In: GD 2004 [21], pp. 195–205
20. Fößmeier, U., Kant, G., Kaufmann, M.: 2-Visibility Drawings of Planar Graphs. In: North, S.C. (ed.) GD 1996. LNCS, vol. 1190, pp. 155–168. Springer, Heidelberg (1997)
21. Pach, J. (ed.): GD 2004. LNCS, vol. 3383. Springer, Heidelberg (2005)
22. Goodrich, M.T., Wagner, C.G.: A Framework for Drawing Planar Graphs with Curves and Polylines. J. of Algorithms 37(2), 399–421 (2000)
23. Gutwenger, C., Mutzel, P.: Planar Polyline Drawings with Good Angular Resolution. In: Whitesides, S.H. (ed.) GD 1998. LNCS, vol. 1547, pp. 167–182. Springer, Heidelberg (1999)
24. Harel, D., Sardas, M.: An Algorithm for Straight-Line Drawing of Planar Graphs. Algorithmica 20, 119–135 (1998)
25. He, X.: On Floor-Plan of Plane Graphs. SIAM J. on Computing 28(6), 2150–2167 (1999)
26. He, X., Kao, M.-Y., Lu, H.-I.: Linear-Time Succinct Encodings of Planar Graphs via Canonical Orderings. SIAM J. on Discrete Mathematics 12(3), 317–325 (1999)
27. Kant, G.: Drawing Planar Graphs using the lmc-Ordering. Technical Report RUU-CS-92-33, Dep. of Information and Computing Sciences, Utrecht University (1992)
28. Kant, G.: Drawing Planar Graphs Using the Canonical Ordering. Algorithmica 16(4), 4–32 (1996)
29. Kant, G.: A More Compact Visibility Representation. Int. J. of Computational Geometry and Applications 7(3), 197–210 (1997)
30. Kant, G., He, X.: Regular Edge Labeling of 4-Connected Plane Graphs and its Applications in Graph Drawing Problems. Theoretical Computer Science 172(1-2), 175–193 (1997)
31. Miura, K., Azuma, M., Nishizeki, T.: Canonical Decomposition, Realizer, Schnyder Labeling and Orderly Spanning Trees of Plane Graphs. Int. J. of Foundations of Computer Science 16(1), 117–141 (2005)
32. Nakano, S.-I.: Planar Drawings of Plane Graphs. IEICE Transactions on Information and Systems E83-D(3), 384–391 (2000)
33. Schnyder, W.: Embedding Planar Graphs on the Grid. In: Proc. 1st ACM–SIAM Symp. on Discrete Algorithms (SODA 1990), pp. 138–148 (1990)
34. Wada, K., Chen, W.: Linear Algorithms for a k-Partition Problem of Planar Graphs. In: Hromkovič, J., Sýkora, O. (eds.) WG 1998. LNCS, vol. 1517, pp. 324–336. Springer, Heidelberg (1998)

Succinct Greedy Drawings Do Not Always Exist*

Patrizio Angelini, Giuseppe Di Battista, and Fabrizio Frati

Dipartimento di Informatica e Automazione – Roma Tre University, Italy
{angelini,gdb,frati}@dia.uniroma3.it

Abstract. A greedy drawing is a graph drawing containing a distance-decreasing path for every pair of nodes. A path $(v_0, v_1, \dots, v_m)$ is distance-decreasing if $d(v_i, v_m) < d(v_{i-1}, v_m)$, for $i = 1, \dots, m$. Greedy drawings easily support geographic greedy routing. Hence, a natural and practical problem is the one of constructing greedy drawings in the plane using few bits for representing vertex Cartesian coordinates and using the Euclidean distance as a metric. We show that there exist greedy-drawable graphs that do not admit any greedy drawing in which the Cartesian coordinates have less than a polynomial number of bits.

1 Introduction

In *geographic routing* nodes forward packets based on their geographic locations. A very simple geographic routing protocol is *greedy routing*, in which each node knows its location, the location of its neighbors, and the location of the packet's destination. Based on this information, a node forwards the packet to a neighbor that is *closer than itself* to the destination's geographic location.

Unfortunately, greedy routing has two weaknesses. First, GPS devices, typically used to determine coordinates, are expensive and increase the energy consumption of the nodes. Second, a bad interaction between the network topology and the node locations can lead to situations in which the communication fails because a *void* has been reached, i.e., a packet has reached a node whose neighbors are all farther from the destination than the node itself.

A brilliant solution to the greedy routing weaknesses has been proposed by Rao *et al.*, who in [13] proposed a protocol in which nodes are assigned *virtual coordinates* and the standard greedy routing algorithm is applied relying on such virtual locations rather than on the geographic coordinates. Clearly, virtual coordinates need not to reflect the nodes actual positions and, hence, they can be suitably chosen to guarantee that the greedy routing algorithm succeeds in delivering packets.

After the publication of [13], intense research efforts have been devoted to determine: (i) Which network topologies admit a virtual coordinates assignment such that greedy routing is guaranteed to work. (ii) Which distance metrics, which systems of coordinates, and how many dimensions are suitable for virtual coordinates. (iii) How many bits are needed to represent the vertex coordinates.

From a graph-theoretical point of view, Problem (i) can be stated as follows: Which are the graphs that admit a *greedy drawing*, i.e., a drawing such that, for every two nodes

* This work is partially supported by the Italian Ministry of Research, Grant number RBIP06BZW8, FIRB project "Advanced tracking system in intermodal freight transportation".

D. Eppstein and E.R. Gansner (Eds.): GD 2009, LNCS 5849, pp. 171–182, 2010.

u and v, there exists a *distance-decreasing path* from u to v? A path $(v_0, v_1, \ldots, v_m)$ is distance-decreasing if $d(v_i, v_m) < d(v_{i-1}, v_m)$, for $i = 1, \ldots, m$. This formulation of the problem gives a clear perception of how greedy routing can be seen as a "bridge" problem between the theory of routing and Graph Drawing, thus explaining why it attracted attention in both areas.

Concerning drawings in the plane adopting the Euclidean distance, Papadimitriou and Ratajczak [11] showed that $K_{k,5k+1}$ has no greedy drawing, for $k \geq 1$. Further, they observed that, if a graph G has a greedy drawing, then any graph containing G as a spanning subgraph has a greedy drawing. Dhandapani [2] showed, with an existential proof based on an application of the Knaster-Kuratowski-Mazurkievicz Theorem [8] to the Schnyder's methodology [14], that every *triangulation* admits a greedy drawing. Algorithms for constructing greedy drawings of triangulations and triconnected planar graphs have been proposed in [1,9]. In [9] it is also proved that there exist trees not admitting any greedy drawing.

Concerning Problem (ii), it has been shown that virtual coordinates guarantee greedy routing to work for every tree, and hence for every connected topology, when they can be chosen in the hyperbolic plane [7].

Unfortunately, the above mentioned algorithms construct greedy drawings that are not *succinct*, i.e., in the worst case they require $\Omega(n \log n)$ bits for representing the vertex coordinates (Problem (iii)). This makes them unsuitable for the motivating application of greedy routing. For solving this drawback, Eppstein and Goodrich [5] proposed an elegant algorithm for greedy routing in the hyperbolic plane representing vertex coordinates with $O(\log n)$ bits. However, the perhaps most natural question of whether greedy drawings can be constructed in the plane using $O(\log n)$ bits for representing vertex Cartesian coordinates and using the Euclidean distance as a metric was, up to now, open. This paper gives a negative answer to the above question.

Theorem 1. *For infinitely many n, there exists a $(3n + 3)$-node greedy-drawable tree that requires $\Omega(b^n)$ area in any greedy drawing in the plane using the Euclidean distance as a metric, under any finite resolution rule, for some constant $b > 1$.*

Observe the equivalence between stating the theorem in terms of area requirement of the drawing and in terms of number of bits required for the vertex Cartesian coordinates. Theorem 1 is one of the few results (e.g., [4]) showing that certain families of graph drawings require exponential area. Notice that greedy drawings are a kind of *proximity drawings* [3], a class of graph drawings, including Euclidean Minimum Spanning Trees [10,6], for which very little is known about the area requirement [12].

The paper is organized as follows. In Sect. 2 we introduce some definitions and preliminaries; in Sect. 3 we prove that there exists a tree T_n requiring exponential area in any greedy drawing; in Sect. 4 we show an algorithm for constructing a greedy drawing of T_n; finally, in Sect. 5 we conclude and present some open problems.

2   Definitions and Preliminaries

A *tree* is a connected acyclic graph. The *degree of a node* is the number of edges incident to it. A *leaf* is a node with degree 1. A *leaf edge* is an edge incident to a leaf. A *path* is a tree in which every node other than the leaves has degree 2. A *caterpillar* is a tree in which the removal of all the leaves and of all the leaf edges yields a path, called *spine* of the caterpillar, whose nodes and edges are called *spine nodes* and *spine edges*, respectively.

A *drawing* of a graph is a mapping of each node to a distinct point of the plane and of each edge to a Jordan curve between its endpoints. A *planar drawing* is such that no two edges intersect except, possibly, at common endpoints. A *straight-line drawing* is such that all the edges are straight-line segments. A planar drawing determines a circular ordering of the edges incident to each node. Two drawings of the same graph are *equivalent* if they determine the same circular ordering around each node. An *embedding* is an equivalence class of planar drawings.

The *area* of a straight-line drawing is the area of its convex hull. The concept of area of a drawing only makes sense for a fixed *resolution rule*, i.e., a rule that does not allow, e.g., vertices to be arbitrarily close (*vertex resolution rule*) or edges to be arbitrarily short (*edge resolution rule*). In fact, without any of such rules, one could construct arbitrarily small drawings with arbitrarily small area. In the following, we derive a lower bound valid under any of such rules. Namely, we prove that, in any greedy drawing of an n-node tree T_n, the ratio between the length of the longest edge and the length of the shortest edge is exponential in n, which implies that such a drawing requires exponential area when any resolution rule has been fixed.

We now state some basic properties of the greedy drawings of trees.

The *cell* of a node v in a drawing is the set of all the points in the plane that are closer to v than to any of its neighbors.

Lemma 1. *(Papadimitriou and Ratajczak [11]) A drawing is greedy if and only if the cell of each node v contains no node other than v.*

We remark that the cell of a leaf node v with parent u is the half-plane containing v and delimited by the axis of segment $\overline{uv}$, where the *axis* of a segment is the line perpendicular to the segment through its median point.

Lemma 2. *Given a greedy drawing Γ of a tree T, any subtree of T is represented in Γ by a greedy drawing.*

Proof: Suppose, for a contradiction, that a subtree T' of T exists not represented in Γ by a greedy drawing. Then, there exist two nodes u and v such that the only path from u to v in T' is not distance-decreasing. However, such a path is also the only path from u to v in T, a contradiction. □

Lemma 3. *Given a greedy drawing Γ of a tree T and given any edge (u, v) of T, the subtree T' of T that contains u and that is obtained by removing edge (u, v) from T completely lies in Γ in the half-plane containing u and delimited by the axis of segment $\overline{uv}$.*

Proof: Suppose, for a contradiction, that there exists a node w of T' that lies in Γ in the half-plane containing v and delimited by the axis of $\overline{uv}$. Then, $d(v, w) < d(u, w)$. The only path from v to w in T passes through u, hence it is not distance-decreasing, a contradiction. □

Lemma 4. *Any straight-line greedy drawing of a tree is planar.*

Proof: Suppose, for a contradiction, that there exists a tree T admitting a non-planar straight-line greedy drawing Γ. Let $e_1 = (u, v)$ and $e_2 = (w, z)$ be two edges that cross in Γ. Edges e_1 and e_2 are not adjacent, otherwise they would overlap and Γ would not be greedy. Then, there exists an edge $e_3 \neq e_1, e_2$ in the only path connecting u to w. Lemma 3 implies that e_1 and e_2 lie in distinct half-planes delimited by the axis of the segment representing e_3, hence they do not cross, a contradiction. $\square$

Corollary 1. *Consider a greedy drawing Γ of a tree T. For each edge, remove its drawing from Γ and substitute it with a straight-line segment connecting its endpoints. The resulting drawing is a straight-line planar greedy drawing of T.*

Because of Lemma 4 and of Corollary 1, in order to prove Theorem 1, we can restrict the attention to planar straight-line greedy drawings. In the following, all considered drawings will be planar and straight-line.

Lemma 5. *In any greedy drawing of a tree T, the angle between two adjacent segments is strictly greater than $60°$.*

Proof: Consider any greedy drawing of T in which the angle between two adjacent segments $\overline{w_1 w_2}$ and $\overline{w_2 w_3}$ is no more than $60°$. Then, $|\overline{w_1 w_3}| \leq |\overline{w_1 w_2}|$ or $|\overline{w_1 w_3}| \leq |\overline{w_2 w_3}|$, say $|\overline{w_1 w_3}| \leq |\overline{w_2 w_3}|$. Since $d(w_1, w_3) \leq d(w_2, w_3)$, the unique path (w_1, w_2, w_3) from w_1 to w_3 in T is not distance-decreasing. $\square$

In the following we define a family of trees with $3n + 3$ nodes, for every $n \geq 2$, that will be exploited in order to prove Theorem 1. Refer to Fig. 1.

Definition 1. *Let T_n be a caterpillar with spine $(v_1, v_2, \ldots, v_n)$ such that v_1 has degree 5 and v_i has degree 4, for each $i = 2, 3, \ldots, n$. Let $a_1, b_1, c_1,$ and d_1 be the leaves of T_n adjacent to v_1, let a_i and b_i be the leaves of T_n adjacent to v_i, for $i = 2, 3, \ldots, n-1$, and let $a_n, b_n,$ and c_n be the leaves of T_n adjacent to v_n.*

Distinct embeddings of T_n differ for the order of the edges incident to the spine nodes. More precisely, the clockwise order of the edges incident to each node v_i is one of the following: 1) (v_{i-1}, v_i), then a leaf edge, then (v_i, v_{i+1}), then a leaf edge: v_i is a *central node* (node v_n in Fig. 1.b); 2) (v_{i-1}, v_i), then two leaf edges, then (v_i, v_{i+1}): v_i is a *bottom node* (node v_2 in Fig. 1.b); or 3) (v_{i-1}, v_i), then (v_i, v_{i+1}), then two leaf edges: v_i is a *top node* (node v_3 in Fig. 1.b). Node v_1 is considered as a central node.

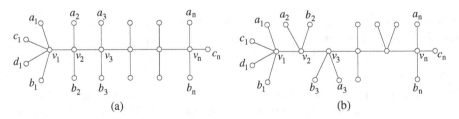

(a)   (b)

Fig. 1. Two embeddings of caterpillar T_n. In (a) all the spine nodes are central nodes. In (b) node v_2 is a bottom node and node v_3 is a top node.

3   The Lower Bound

In this section we prove that any greedy drawing of T_n requires exponential area.

The proof is based on the following intuitions: (i) For any central node v_i there exists a "small" convex region containing all the spine nodes v_j, with $j > i$, and their adjacent leaves (Lemma 6). (ii) Almost all the spine nodes are central nodes (Lemma 8). (iii) The slopes of edges (v_i, a_i), (v_i, v_{i+1}), and (v_i, b_i) incident to a central node v_i are in a certain range, which is more restricted for the edges incident to v_{i+1} than for those incident to v_i (Lemma 6). (iv) If the angle between (v_i, a_i) and (v_i, b_i) is too small, then v_j, a_j, and b_j, with $j \geq i + 2$, can not be drawn (Lemma 10). (v) If both the angles between (v_i, a_i) and (v_i, b_i), and between (v_{i+1}, a_{i+1}) and (v_{i+1}, b_{i+1}) are large enough, then the ratio between the length of the edges incident to v_i and the length of the edges incident to v_{i+1} is constant (Lemma 9).

First, we discuss some properties of the slopes of the edges in the drawing. Second, we argue about the exponential decrease of the edge lengths.

3.1   Slopes

Consider any drawing of v_1 and of its adjacent leaves; rename such leaves so that the counter-clockwise order of the vertices around v_1 is a_1, c_1, d_1, b_1, and v_2.

In the following, when we refer to an angle $\widehat{v_1 v_2 v_3}$, we mean the angle that brings the half-line from v_2 through v_1 to coincide with the half-line from v_2 through v_3 by a counter-clockwise rotation.

Property 1. $\widehat{b_1 v_1 a_1} < 180°$.

Proof: By Lemma 5, $\widehat{a_1 v_1 c_1} > 60°$, $\widehat{c_1 v_1 d_1} > 60°$, and $\widehat{d_1 v_1 b_1} > 60°$.   □

Now we argue that, for any central node v_i, there exists a "small" convex region that contains all the spine nodes v_j, with $j > i$, and their adjacent leaves.

Let v_i be a central node and suppose that $\widehat{b_i v_i a_i} < 180°$. Denote by R_i the convex region delimited by $\overline{v_i a_i}$, by $\overline{v_i b_i}$, and by the axes of such segments (see Fig. 2.b). Denote by p_i the intersection between the axes of $\overline{v_i a_i}$ and $\overline{v_i b_i}$, and by h_i^a (h_i^b) the midpoint of $\overline{v_i a_i}$ (resp. $\overline{v_i b_i}$).

Assume that $x(a_i) = x(b_i)$, $x(v_i) < x(a_i)$, and $y(a_i) > y(b_i)$. Such a setting can be achieved without loss of generality up to a rotation/mirroring of the drawing and a renaming of the leaves. In the following, whenever a central node v_i is considered, the drawing is rotated/mirrored and the leaves adjacent to v_i are renamed so that $x(a_i) = x(b_i)$, $x(v_i) < x(a_i)$, and $y(a_i) > y(b_i)$.

Let $slope(u, v)$ be the angle bringing the half-line from u directed downward to coincide with the half-line from u through v by a counter-clockwise rotation (see Fig. 2.a). Further, let $slope_\perp(u, v)$ be equal to $slope(u, v) - 90°$. We observe the following:

Property 2. $slope(v_i, b_i) < slope_\perp(b_i, p_i) < slope_\perp(p_i, a_i) < slope(v_i, a_i)$.

Proof: Inequality $slope(v_i, b_i) < slope_\perp(b_i, p_i)$ (and analogously $slope_\perp(p_i, a_i) < slope(v_i, a_i)$) holds since $slope(h_i^b, p_i) < slope(b_i, p_i)$. Inequality $slope_\perp(b_i, p_i) < slope_\perp(p_i, a_i)$ holds by assumption.   □

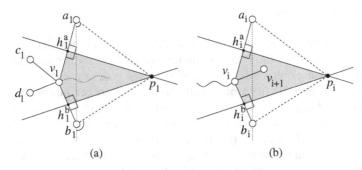

(a)   (b)

Fig. 2. (a) Region R_1 contains the drawing of $T_n \setminus \{a_1, b_1, c_1, d_1, v_1\}$. The slopes of $\overline{a_1 p_1}$ and $\overline{b_1 p_1}$ are shown. (b) Region R_i contains the drawing of path $(v_{i+1}, v_{i+2}, \ldots, v_n)$ and of its adjacent leaves.

Lemma 6. *Suppose that v_i is a central node. Then, the following properties hold: (i) $\widehat{b_i v_i a_i} < 180°$; (ii) the drawing of path $(v_{i+1}, v_{i+2}, \ldots, v_n)$ and of its adjacent leaves lies in R_i; and (iii) any edge (v_j, x), where $x \in \{a_j, b_j, v_{j+1}\}$ with $j > i$, is such that $slope_\perp(b_i, p_i) < slope_\perp(v_j, x) < slope_\perp(p_i, a_i)$. See Fig. 3.a.*

Proof: When $i = 1$, Property 1 ensures property (i). Further, Lemma 1 ensures property (ii), that is, the drawing of $T_n \setminus \{a_1, b_1, c_1, d_1, v_1\}$ lies in R_1 (see Fig. 2.a). In order to prove property (iii), suppose, for a contradiction, that an edge (v_j, x) exists, where $x \in \{a_j, b_j, v_{j+1}\}$ with $j > 1$, such that $slope_\perp(b_1, p_1) < slope_\perp(v_j, x) < slope_\perp(p_1, a_1)$ does not hold. Then, it is easy to see that the half-plane delimited by the axis of $\overline{v_j x}$ and containing x also contains at least one out of a_1, v_1, and b_1, thus providing a contradiction to the greediness of the drawing, by Lemma 3.

By induction, suppose that properties (i), (ii), and (iii) of the lemma hold for some i. Let k be the smallest index greater than i such that v_k is a central node. Then, by property (iii) of the inductive hypothesis and by Property 2, $slope(v_i, b_i) < slope_\perp(b_i, p_i) < slope(v_k, b_k) < slope(v_k, a_k) < slope_\perp(p_i, a_i) < slope(v_i, a_i)$ holds, which implies $\widehat{b_k v_k a_k} < \widehat{b_i v_i a_i} < 180°$, and property (i) of the lemma follows for k.

By Lemma 4, the drawing is planar; by Lemma 1, the cells of a_k and b_k do not contain any node other than a_k and b_k, respectively. Hence, if a node u is in R_k, then no node of any subtree of T_n containing u and not containing v_k lies outside R_k. Thus, v_{k-1} does not lie in R_k (since a subtree of T_n exists containing v_{k-1}, v_i, and not containing v_k); since v_k is a central node, then v_{k+1} lies on the opposite side of v_{k-1} with respect to the path composed of edges (v_k, a_k) and (v_k, b_k). Hence, v_{k+1} (and path $(v_{k+1}, v_{k+2}, \ldots, v_n)$ together with its adjacent leaves) lies inside R_k, and property (ii) of the lemma follows for k.

Property (iii) can be proved analogously as in the base case, by implicitly exploiting that properties (i) and (ii) hold for k. Namely, if $slope_\perp(b_k, p_k) < slope_\perp(v_j, x) < slope_\perp(p_k, a_k)$ does not hold, for some edge (v_j, x) with $j > k$, then the half-plane delimited by the axis of $\overline{v_j x}$ and containing x also contains at least one out of a_k, v_k, and b_k, thus implying that the drawing is not greedy, by Lemma 3. □

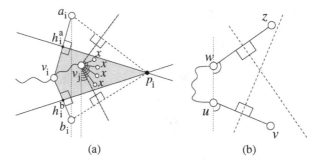

Fig. 3. (a) Possible slopes for an edge (v_j, x). (b) Illustration for the proof of Lemma 7.

Now we give a general property of a greedy drawing of a tree. Consider two edges (u, v) and (w, z) such that the path from u to w does not contain v and z. Suppose that v and z lie in the same half-plane delimited by the line through u and w. Suppose, without loss of generality up to a rotation/mirroring of the drawing, that $x(u) = x(w)$, $y(u) < y(w)$, and $0° < slope(u, v), slope(w, z) < 180°$. See Fig. 3.b.

Lemma 7. $slope(u, v) < slope(w, z)$.

Proof: Suppose, for a contradiction, that $slope(u, v) \geq slope(w, z)$. Then, either v lies in the half-plane delimited by the axis of $\overline{wz}$ and containing z, or z lies in the half-plane delimited by the axis of $\overline{uv}$ and containing v. Hence, by Lemma 2, the drawing is not greedy. □

3.2 Exponential Decreasing Edge Lengths

Now we are ready to go in the mainstream of the proof that any greedy drawing of T_n requires exponential area. Such a proof is in fact based on the following three lemmata. The first one states that a linear number of spine nodes are central nodes, in any greedy drawing of T_n.

Lemma 8. *Suppose that v_i is a central node, for some $i \leq n-3$. Then, v_{i+1} is a central node.*

Proof: Refer to Fig. 4. Suppose, for a contradiction, that v_{i+1} is not a central node. Suppose that v_{i+1} is a top node, the case in which it is a bottom node being analogous. Rename the leaves adjacent to v_{i+1} in such a way that the counter-clockwise order of the neighbors of v_{i+1} is v_i, b_{i+1}, a_{i+1}, and v_{i+2}. By property (i) of Lemma 6, $\widehat{b_i v_i a_i} < 180°$. By property (iii) of Lemma 6, by Property 2, and by the assumption that v_{i+1} is a top node, $slope(v_i, b_i) < slope(v_{i+1}, b_{i+1}) < slope(v_{i+1}, a_{i+1}) < slope(v_{i+1}, v_{i+2}) < slope(v_i, a_i)$. By Lemma 5, $\widehat{b_{i+1} v_{i+1} a_{i+1}} > 60°$. It follows that $\widehat{a_{i+1} v_{i+1} v_{i+2}} < 120°$.

Suppose that v_{i+2} is a central node (a top node; a bottom node). Rename the leaves adjacent to v_{i+2} in such a way that the counter-clockwise order of the neighbors of v_{i+2} is v_{i+1}, b_{i+2}, v_{i+3}, and a_{i+2} (resp. v_{i+1}, b_{i+2}, a_{i+2}, and v_{i+3}; v_{i+1}, v_{i+3}, b_{i+2}, and

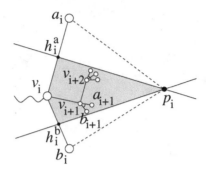

Fig. 4. Illustration for the proof of Lemma 8

a_{i+2}). Notice that node v_{i+3} exists since $i \leq n - 3$. By Lemma 7, $slope(v_{i+2}, b_{i+2}) > slope(v_{i+1}, a_{i+1})$ (resp. $slope(v_{i+2}, b_{i+2}) > slope(v_{i+1}, a_{i+1})$; $slope(v_{i+2}, v_{i+3}) > slope(v_{i+1}, a_{i+1})$). By property (iii) of Lemma 6, $slope(v_{i+2}, a_{i+2}) < slope(v_i, a_i)$ (resp. $slope(v_{i+2}, v_{i+3}) < slope(v_i, a_i)$; $slope(v_{i+2}, a_{i+2}) < slope(v_i, a_i)$). It follows that $\widehat{b_{i+2}v_{i+2}a_{i+2}} < 120°$ (resp. $\widehat{b_{i+2}v_{i+2}v_{i+3}} < 120°$; $\widehat{v_{i+3}v_{i+2}a_{i+2}} < 120°$), hence at least one of $\widehat{b_{i+2}v_{i+2}v_{i+3}}$ and $\widehat{v_{i+3}v_{i+2}a_{i+2}}$ (resp. of $\widehat{b_{i+2}v_{i+2}a_{i+2}}$ and $\widehat{a_{i+2}v_{i+2}v_{i+3}}$; of $\widehat{v_{i+3}v_{i+2}b_{i+2}}$ and $\widehat{b_{i+2}v_{i+2}a_{i+2}}$) is less than 60°. By Lemma 5, the drawing is not greedy.   □

The next lemma shows that, if the angles $\widehat{b_i v_i a_i}$ incident to each central node v_i are large enough, then the sum of the lengths of $\overline{v_i a_i}$ and $\overline{v_i b_i}$ decreases exponentially in the number of considered central nodes.

Lemma 9. *Let v_i be a central node, with $i \leq n - 3$. Suppose that both the angles $\widehat{b_i v_i a_i}$ and $\widehat{b_{i+1} v_{i+1} a_{i+1}}$ are greater than $150°$. Then, the following inequality holds: $|\overline{v_{i+1} a_{i+1}}| + |\overline{v_{i+1} b_{i+1}}| \leq (|\overline{v_i a_i}| + |\overline{v_i b_i}|)/\sqrt{3}$.*

Proof: Refer to Fig. 5.a. By Lemma 8, v_{i+1} is a central node. Denote by $l(v_{i+1})$ the vertical line through v_{i+1} and denote by $l(h_i^a)$ and $l(h_i^b)$ the horizontal lines through h_i^a and h_i^b, respectively.

By property (iii) of Lemma 6, we have that $slope_\perp(b_i, p_i) < slope(v_{i+1}, b_{i+1}) < slope(v_{i+1}, a_{i+1}) < slope_\perp(p_i, a_i)$. Hence, by Property 2, we have $slope(v_i, b_i) < slope(v_{i+1}, b_{i+1}) < slope(v_{i+1}, a_{i+1}) < slope(v_i, a_i)$. It follows that both a_{i+1} and b_{i+1} lie in the half-plane delimited by $l(v_{i+1})$ and not containing v_i. Denote by d_{i+1}^a (d_{i+1}^b) the intersection point between $l(v_{i+1})$ and $l(h_i^a)$ (resp. and $l(h_i^b)$). Observe that $|\overline{d_{i+1}^b d_{i+1}^a}| < (|\overline{v_i b_i}| + |\overline{v_i a_i}|)/2$. Denote by f_{i+1}^a (by f_{i+1}^b) the intersection point between $l(h_i^a)$ and the line through v_{i+1} and a_{i+1} (resp. between $l(h_i^b)$ and the line through v_{i+1} and b_{i+1}). Clearly, $|\overline{v_{i+1} a_{i+1}}| < |\overline{v_{i+1} f_{i+1}^a}|$ and $|\overline{v_{i+1} b_{i+1}}| < |\overline{v_{i+1} f_{i+1}^b}|$. Angles $\widehat{d_{i+1}^b v_{i+1} f_{i+1}^b}$ and $\widehat{f_{i+1}^a v_{i+1} d_{i+1}^a}$ are each less than 30°, namely such angles sum up to an angle which is 180° minus $\widehat{f_{i+1}^b v_{i+1} f_{i+1}^a}$, which by hypothesis is greater than 150°. Hence, $|\overline{v_{i+1} a_{i+1}}| < |\overline{v_{i+1} f_{i+1}^a}| < |\overline{v_{i+1} d_{i+1}^a}|/\cos(30)$ and $|\overline{v_{i+1} b_{i+1}}| < |\overline{v_{i+1} f_{i+1}^b}| < |\overline{v_{i+1} d_{i+1}^b}|/\cos(30)$.

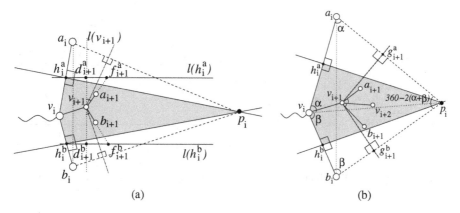

Fig. 5. Illustrations for the proofs of Lemma 9 (a) and Lemma 10 (b)

It follows that $|\overline{v_{i+1}a_{i+1}}| + |\overline{v_{i+1}b_{i+1}}| < (|\overline{v_{i+1}d_{i+1}^a}| + |\overline{v_{i+1}d_{i+1}^b}|)/\cos(30) < 2(|\overline{v_ib_i}| + |\overline{v_ia_i}|)/2\sqrt{3}$, thus proving the lemma. □

The next lemma shows that having large angles incident to central nodes is unavoidable for almost all central nodes.

Lemma 10. *No central node v_i, with $i \leq n-3$, is incident to an angle $\widehat{b_iv_ia_i}$ that is less than or equal to $150°$.*

Proof: Refer to Fig. 5.b. Suppose, for a contradiction, that there exists a central node v_i, with $i \leq n-3$, that is incident to an angle $\widehat{b_iv_ia_i} \leq 150°$. Denote by α and β the angles $\widehat{p_iv_ia_i}$ and $\widehat{b_iv_ip_i}$, respectively. Since triangles (v_i, p_i, h_i^a) and (a_i, p_i, h_i^a) are congruent, $\widehat{v_ia_ip_i} = \alpha$. Analogously, $\widehat{v_ib_ip_i} = \beta$. Summing up the angles of quadrilateral (v_i, a_i, p_i, b_i), we get $\widehat{a_ip_ib_i} = 360° - 2(\alpha + \beta)$.

By Lemma 8, v_{i+1} is a central node. Consider the line through v_{i+1} orthogonal to $\overline{a_ip_i}$ and denote by g_{i+1}^a the intersection point between such a line and $\overline{a_ip_i}$. Further, consider the line through v_{i+1} orthogonal to $\overline{b_ip_i}$ and denote by g_{i+1}^b the intersection point between such a line and $\overline{b_ip_i}$. By property (iii) of Lemma 6, $slope_\perp(b_i, p_i) < slope(v_{i+1}, b_{i+1}) < slope(v_{i+1}, a_{i+1}) < slope_\perp(p_i, a_i)$. Hence, $\widehat{b_{i+1}v_{i+1}a_{i+1}} < \widehat{g_{i+1}^bv_{i+1}g_{i+1}^a}$. Further, $\widehat{g_{i+1}^bv_{i+1}g_{i+1}^a} = 2\alpha + 2\beta - 180°$, as can be derived by considering quadrilateral $(g_{i+1}^b, v_{i+1}, g_{i+1}^a, p_i)$. Since, by hypothesis, $\alpha + \beta \leq 150°$, we have $\widehat{b_{i+1}v_{i+1}a_{i+1}} < \widehat{g_{i+1}^bv_{i+1}g_{i+1}^a} = 2\alpha + 2\beta - 180° \leq 120°$. However, since v_{i+1} is a central node, edge $(v_{i+1}v_{i+2})$, that exists since $i \leq n-3$, cuts angle $\widehat{b_{i+1}v_{i+1}a_{i+1}}$. It follows that at least one of angles $\widehat{b_{i+1}v_{i+1}v_{i+2}}$ and $\widehat{v_{i+2}v_{i+1}a_{i+1}}$ is less than $60°$. By Lemma 5, the drawing is not greedy. □

The previous lemmata immediately imply an exponential lower bound between the ratio of the lengths of the longest and the shortest edge of the drawing. Namely, node v_1 is a central node. By Lemma 8, v_i is a central node, for $i = 2, \ldots, n-3$. By Lemma 10, angle $\widehat{b_iv_ia_i} > 150°$, for each $i \leq n-3$. Hence, by Lemma 9, $|\overline{v_{i+1}a_{i+1}}| + |\overline{v_{i+1}b_{i+1}}| \leq$

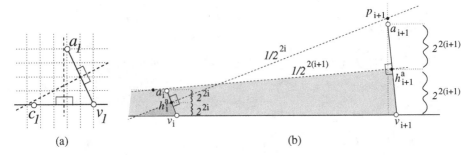

Fig. 6. Illustrations for the algorithm to construct a greedy drawing of T_n. (a) Base case. (b) Inductive case.

$(|\overline{v_i a_i}| + |\overline{v_i b_i}|)/\sqrt{3}$, for each $i \leq n - 4$; it follows that $|\overline{v_{n-3} a_{n-3}}| + |\overline{v_{n-3} b_{n-3}}| \leq (|\overline{v_1 a_1}| + |\overline{v_1 b_1}|)/(\sqrt{3})^{n-4}$. Since one out of $\overline{v_1 a_1}$ and $\overline{v_1 b_1}$, say $\overline{v_1 a_1}$, has length at least half of $|\overline{v_1 a_1}| + |\overline{v_1 b_1}|$, and since one out of $\overline{v_{n-3} a_{n-3}}$ and $v_{n-3} b_{n-3}$, say $\overline{v_{n-3} a_{n-3}}$, has length at most half of $|\overline{v_{n-3} a_{n-3}}| + |\overline{v_{n-3} b_{n-3}}|$, then $|\overline{v_1 a_1}|/|\overline{v_{n-3} a_{n-3}}| \geq \frac{1}{9}(\sqrt{3})^n$, thus implying the claimed lower bound.

4   Drawability of T_n

In Sect. 3 we have shown that any greedy drawing of T_n requires exponential area. Since in [11,9] it has been shown that there exist trees that do not admit any greedy drawing, one might ask whether the lower bound refers to a greedy-drawable tree or not. Of course, if T_n were not drawable, then the lower bound would not make sense. In this section we show that T_n admits a greedy drawing by providing a drawing algorithm, using a supporting exponential-size grid.

Since the algorithm draws the spine nodes in the order they appear on the spine with the degree-5 node as the last node, we revert the indices of the nodes with respect to Sects. 2 and 3, that is, node v_i of T_n is now node v_{n-i+1}.

The algorithm constructs a drawing of T_n in which all the spine nodes v_i are central nodes lying on the horizontal line $y = 0$. Since each leaf node a_i is drawn above line $y = 0$ and b_i is placed on the symmetrical point of a_i with respect to such a line, we only describe, for each $i = 1, \ldots, n$, how to draw v_i and a_i.

In order to deal with drawings that lie on a grid, in this section we denote by Δ_y/Δ_x the *slope* of a line (of a segment), meaning that whenever there is a horizontal distance Δ_x between two nodes of such a line (of such a segment), then their vertical distance is Δ_y.

The drawing is constructed by means of an inductive algorithm. In the base case, place v_1 at $(0, 0)$, h_1^a at $(-1, 2)$, a_1 at $(-2, 4)$, and c_1 at $(-9/2, 0)$ (see Fig. 6.a). At step i of the algorithm suppose, by inductive hypothesis, that: (i) The drawing of path $(v_1, v_2, \ldots, v_i)$ with its leaf nodes $a_1, a_2, \ldots, a_i$ is greedy, and (ii) $y(v_i) = 0$, $y(h_i^a) = 2^{2i}$, $y(a_i) = 2^{2i+1}$, and $x(v_i) - x(h_i^a) = x(h_i^a) - x(a_i) = 1$.

From the above inductive hypothesis it follows that the slope of segment $\overline{v_i a_i}$ is $-2^{2i}/1$ and the slope of its axis is $1/2^{2i}$. We show step $i + 1$ of the algorithm.

Place v_{i+1} at point $(x(v_i) + 2^{4i+3} - 2, 0)$, h^a_{i+1} at point $(x(v_i) + 2^{4i+3} - 3, 2^{2i+2})$, and a_{i+1} at point $(x(v_i) + 2^{4i+3} - 4, 2^{2i+3})$ (see Fig. 6.b). Such placements guarantee that part (ii) of the hypothesis is verified. The slope of segment $\overline{v_{i+1}a_{i+1}}$ is $-2^{2(i+1)}/1$. Hence, the slope of its axis is $1/2^{2(i+1)}$. Such an axis passes through point $q_i \equiv (x(v_i) - 3, 2^{2i+1})$. Since $0 < 1/2^{2(i+1)} < 1/2^{2i}$, it follows that path $(v_1, v_2, \ldots, v_i)$, together with nodes $a_1, a_2, \ldots, a_i$, lies below the axis of $\overline{v_{i+1}a_{i+1}}$. Finally, the axis of $\overline{v_i a_i}$ passes through point $p_{i+1} \equiv (x(v_i) + 2^{4i+3} - 4, 2^{2i} + 2^{2i+3} - 3/2^{2i})$. Thus, $y(p_{i+1}) > y(a_{i+1})$, since $2^{2i} + 2^{2i+3} - 3/2^{2i} > 2^{2i+3}$ as long as $2^{4i} > 3$, which holds for each $i \geq 1$. This implies that part (i) of the hypothesis is verified.

When the algorithm has drawn v_n and a_n (and symmetrically b_n), c_n and d_n still have to be drawn. However, this can be easily done by assigning to segments $\overline{v_n c_n}$ and $\overline{v_n d_n}$ the same length as segment $\overline{v_n a_n}$ and by placing them so that the angle $\widehat{b_n v_n a_n}$, which is strictly greater than $180°$, is split into three angles strictly greater than $60°$.

We remark that c_n and d_n are not placed at points with rational coordinates. However, they still obey to any resolution rule, namely their distance from any node or edge of the drawing is exponential with respect to the grid unit. Placing such nodes at grid points is possible after a scaling of the whole drawing and some non-trivial calculations. However, we preferred not to deal with such an issue since we just needed to prove that a greedy drawing of T_n exists.

5 Conclusions

In this paper we have shown that constructing succinct greedy drawings in the plane, when the Euclidean distance is adopted as a metric, may be unfeasible even for simple classes of trees. In fact, we proved that there exist caterpillars requiring exponential area in any greedy drawing, under any finite resolution rule. The proof uses a mixed geometric-topological technique that allows us to analyze the combinatorial space of the possible embeddings and to identify invariants of the slopes of the edges in any greedy drawing of such caterpillars.

Many problems remain open in this area. By the results of Leighton and Moitra [9], every triconnected planar graph admits a greedy drawing.

Problem 1. Which are the area requirements of greedy drawings of triconnected planar graphs?

While every triconnected planar graph admits a greedy drawing, not all biconnected planar graphs and not all trees admit a greedy drawing. For example, in [9] it is shown that a complete binary tree with 31 nodes does not admit any greedy drawing. Hence, the following problem is worth studying:

Problem 2. Characterize the class of trees (resp. of biconnected planar graphs) that admit a greedy drawing.

In this paper we argued about the relationship among greedy drawings, planarity, and straight-line drawability. We have shown, in Lemma 4, that every straight-line greedy drawing of a tree is planar. It would be interesting to understand whether trees are the only class of planar graphs with such a property.

Problem 3. Characterize the class of planar graphs such that every straight-line greedy drawing is planar.

References

1. Angelini, P., Frati, F., Grilli, L.: An algorithm to construct greedy drawings of triangulations. In: Tollis, I.G., Patrignani, M. (eds.) GD 2008. LNCS, vol. 5417, pp. 26–37. Springer, Heidelberg (2009)
2. Dhandapani, R.: Greedy drawings of triangulations. In: Huang, S.T. (ed.) SODA 2008, pp. 102–111 (2008)
3. Di Battista, G., Lenhart, W., Liotta, G.: Proximity drawability: a survey. In: Tamassia, R., Tollis, I.G. (eds.) GD 1994. LNCS, vol. 894, pp. 328–339. Springer, Heidelberg (1995)
4. Di Battista, G., Tamassia, R., Tollis, I.G.: Area requirement and symmetry display of planar upward drawings. Discrete & Computational Geometry 7, 381–401 (1992)
5. Eppstein, D., Goodrich, M.T.: Succinct greedy graph drawing in the hyperbolic plane. In: Tollis, I.G., Patrignani, M. (eds.) GD 2008. LNCS, vol. 5417, pp. 14–25. Springer, Heidelberg (2009)
6. Kaufmann, M.: Polynomial area bounds for MST embeddings of trees. In: Hong, S.-H., Nishizeki, T., Quan, W. (eds.) GD 2007. LNCS, vol. 4875, pp. 88–100. Springer, Heidelberg (2008)
7. Kleinberg, R.: Geographic routing using hyperbolic space. In: INFOCOM 2007, pp. 1902–1909 (2007)
8. Knaster, B., Kuratowski, C., Mazurkiewicz, C.: Ein beweis des fixpunktsatzes fur n dimensionale simplexe. Fundamenta Mathematicae 14, 132–137 (1929)
9. Leighton, T., Moitra, A.: Some results on greedy embeddings in metric spaces. In: FOCS 2008, pp. 337–346 (2008)
10. Monma, C.L., Suri, S.: Transitions in geometric minimum spanning trees. Discrete & Computational Geometry 8, 265–293 (1992)
11. Papadimitriou, C.H., Ratajczak, D.: On a conjecture related to geometric routing. Theoretical Computer Science 344(1), 3–14 (2005)
12. Penna, P., Vocca, P.: Proximity drawings in polynomial area and volume. Computational Geometry 29(2), 91–116 (2004)
13. Rao, A., Papadimitriou, C.H., Shenker, S., Stoica, I.: Geographic routing without location information. In: Johnson, D.B., Joseph, A.D., Vaidya, N.H. (eds.) MOBICOM 2003, pp. 96–108 (2003)
14. Schnyder, W.: Embedding planar graphs on the grid. In: SODA 1990, pp. 138–148 (1990)

Geometric Simultaneous Embeddings of a Graph and a Matching

Sergio Cabello[1], Marc van Kreveld[2], Giuseppe Liotta[3], Henk Meijer[4],
Bettina Speckmann[5], and Kevin Verbeek[5]

[1] Faculty of Mathematics and Physics, University of Ljubljana, Slovenia
sergio.cabello@fmf.uni-lj.si
[2] Dep. of Computer Science, Utrecht University, The Netherlands
marc@cs.uu.nl
[3] Dip. di Ing. Elettronica e dell'Informazione, Università degli Studi di Perugia, Italy
liotta@diei.unipg.it
[4] Roosevelt Academy, Middelburg, The Netherlands
h.meijer@roac.nl
[5] Dep. of Mathematics and Computer Science, TU Eindhoven, The Netherlands
speckman@win.tue.nl, k.a.b.verbeek@tue.nl

Abstract. The geometric simultaneous embedding problem asks whether two planar graphs on the same set of vertices in the plane can be drawn using straight lines, such that each graph is plane. Geometric simultaneous embedding is a current topic in graph drawing and positive and negative results are known for various classes of graphs. So far only connected graphs have been considered. In this paper we present the first results for the setting where one of the graphs is a matching.

In particular, we show that there exists a planar graph and a matching which do not admit a geometric simultaneous embedding. This generalizes the same result for a planar graph and a path. On the positive side, we describe algorithms that compute a geometric simultaneous embedding of a matching and a wheel, outerpath, or tree. Our proof for a matching and a tree sheds new light on a major open question: do a tree and a path always admit a geometric simultaneous embedding? Our drawing algorithms minimize the number of orientations used to draw the edges of the matching. Specifically, when embedding a matching and a tree, we can draw all matching edges horizontally. When embedding a matching and a wheel or an outerpath, we use only two orientations.

1   Introduction

The computation of node-link diagrams of two sets of relations on the same set of data is a recent and already well-established research direction in network visualization. The interest in this problem is partly due to its theoretical relevance and partly motivated by its importance in many application areas, such as software engineering, data bases, and social networks. There are various application scenarios where a visual analysis of dynamic and evolving graphs defined on the same set of vertices is useful, see [3,4] for detailed descriptions.

D. Eppstein and E.R. Gansner (Eds.): GD 2009, LNCS 5849, pp. 183–194, 2010.

Formally, the problem can be stated as follows: Let G_1 and G_2 be two graphs that share their vertex set, but which have different sets of edges. We would like to compute two readable drawings of G_1 and G_2 such that the locations of the vertices are the same in both visualizations. Cognitive experiments [9] prove that the readability of a drawing is negatively affected by the number of edge crossings and by the number of bends along the edges. Hence, if G_1 and G_2 are both planar, we want to compute plane drawings of the two graphs where the vertices have the same locations and edges are straight-line segments. Note that we allow edges from different graphs to cross.

In a seminal paper, Brass *et al.* define a *geometric simultaneous embedding* of two planar graphs sharing their vertex set as two crossing-free straight-line drawings that share the locations of their vertices [1]. Geometric simultaneous embedding is a current topic in graph drawing and positive and negative results are known for various classes of graphs. A comprehensive list can be found in Table 1 of a recent paper by Frati, Kaufmann, and Kobourov [7]. Specifically, Brass *et al.* [1] show that two paths, two cycles, and two caterpillars always admit a geometric simultaneous embedding. (A caterpillar is a tree such that the graph obtained by deleting the leaves is a path.) The authors also prove that three paths may not admit a geometric simultaneous embedding. Erten and Kobourov [5] prove that a planar graph and a path may not admit a geometric simultaneous embedding. Frati, Kaufmann, and Kobourov [7] extend this negative result to the case where the path and the planar graph do not share any edges. Geyer, Kaufmann, and Vrt'o [8] show that two trees may not have a geometric simultaneous embedding. A major open question in this area is the following: do a tree and a path always admit a geometric simultaneous embedding? Finally, Estrella-Balderrama *et al.* [6] prove that determining whether two planar graphs admit a geometric simultaneous embedding is NP-hard.

So far, only connected graphs have been considered and in particular, there are no results for one of the simplest classes of graphs, namely *matchings*. A matching is an independent set of edges. Clearly a geometric simultaneous embedding of two matchings always exists, since the union of two matchings is a collection of cycles and hence planar. But already the union of the edges of a path and a matching does not have to be planar: see Fig. 1 (left), which shows a path and a matching which form a subdivision of $K_{3,3}$.

Results. We study geometric simultaneous embeddings of a matching with various standard classes of graphs. In Section 2 we show that there exists a planar

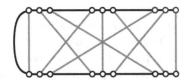

Fig. 1. Left: The union of a path (black) and a matching (gray), can be non-planar. Right: Two orientations of the matching edges (gray) are forced.

graph and a matching which do not admit a geometric simultaneous embedding. This generalizes the same result for a planar graph and a path [5].

On the positive side, we describe algorithms that compute a geometric simultaneous embedding of a matching and a wheel, outerpath, or tree. Specifically, in Section 3 we sketch a construction that computes a geometric simultaneous embedding of a wheel and a cycle, which immediately implies an embedding for a wheel and a matching. In Section 4 and 5 we describe algorithms to embed a matching together with two specific types of outerplanar graphs, namely *outerzigzags* and *outerpaths*. An outerzigzag is also known as a triangle strip. Its weak dual is a path and each of its vertices has degree at most 4. An outerpath is simply an outerplanar graph whose weak dual is a path. Our result for outerpaths of course subsumes the result for outerzigzags, but we nevertheless first present the construction for outerzigzags, to introduce our techniques on a conceptually simpler class of graphs. The algorithms for the wheel, the outerzigzag, and the outerpath, preserve the "natural" embedding of these graphs. That is, the center of the wheel is not incident to the outer face, and the embedding of outerplanar graphs is outerplanar. Note here, that an outerplanar graph and a path may not have a geometric simultaneous embedding if the circular ordering of the edges around the vertices of the outerplanar graph is fixed a-priori [7].

In Section 6 we present an algorithm that computes a geometric simultaneous embedding of a tree and a matching. This algorithm is inspired by and closely related to an algorithm by Di Giacomo *et al.* [2]. Since a path can be viewed as two matchings, our proof sheds some new light on the embeddabilty question for a tree and a path.

All our drawing algorithms minimize the number of orientations used to draw the edges of the matching. This may simplify the visual inspection of the data and of their relationships in practice. Consider the simple example in Fig. 1 (right). It immediately shows that a geometric simultaneous embedding of an outerpath or wheel with a matching requires the matching edges to have at least two orientations. Our constructions match this bound. When embedding a matching and a tree, we can even draw all matching edges horizontally.

2   Planar Graph and Matching

Theorem 1. *There exists a planar graph and a matching that do not admit a geometric simultaneous embedding.*

Consider the planar graph (black) and the matching (gray) depicted in Fig. 2. One can argue that either the subgraph induced by the vertices marked with boxes or the subgraph induced by the vertices marked with circles will always incur at least one crossing. The details can be found in the full paper.

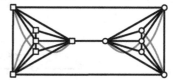

Fig. 2. A planar graph (black) and a matching (gray) that do not admit a geometric simultaneous embedding

3 Wheel and Matching

We can in fact compute a geometric simultaneous embedding of a wheel and a cycle, which immediately implies the result for a wheel and a matching. The construction is sketched in Fig. 3. The center of the wheel is marked with a box, the rim is drawn in black, and the cycle is drawn in gray. When we use this construction for a matching, then all but one matching edges are drawn vertically, the remaining edge (the one edge that is necessarily shared with a spoke of the wheel) is drawn horizontally. Details can be found in the full paper.

Theorem 2. *A wheel and a cycle always admit a geometric simultaneous embedding.*

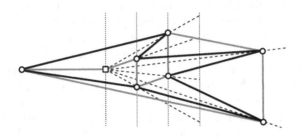

Fig. 3. A geometric simultaneous embedding of a wheel and a cycle

4 Outerzigzag and Matching

Recall that an outerzigzag is a triangle strip: it is a triangulated outerplanar graph, whose weak dual is a path and whose vertices have degree at most 4. More precisely, there are exactly two vertices of degree 2, two vertices of degree 3, and all other vertices have degree 4. Let $G_1 = (V, E_1)$ be an outerzigzag and let $G_2 = (V, E_2)$ be a matching. We first place the vertices of V in such a way, that their placement induces a regular plane drawing of G_1. We then move some of the vertices vertically to planarize G_2, while keeping the drawing of G_1 planar.

Specifically, we place the vertices of V on a grid of size $2n \times 4n$ at positions $(0,0)$, $(2,1)$, $(4,0)$, $(6,1)$, $(8,0)$, etc. One of the degree-2 vertices of G_1 is drawn at $(0,0)$, the remainder is drawn in such a way, that the edges of G_1 always

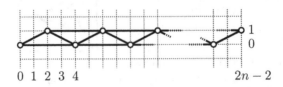

Fig. 4. Drawing an outerzigzag G_1 on a grid

connect two consecutive points on the line $y = 0$, or two consecutive points on the line $y = 1$, or two points at distance $\sqrt{5}$, see Fig. 4.

We classify the edges of the matching G_2 based on the placement of their vertices on the grid:

BB-edges connect any two vertices on the line $y = 0$.
TT-edges connect any two vertices on the line $y = 1$.
BT-edges connect two vertices $(i, 0)$ and $(j, 1)$ with $i < j$.
TB-edges connect two vertices $(i, 1)$ and $(j, 0)$ with $i < j$.

We then move half of the vertices of V vertically according to three simple rules:

1. Only the right vertex of a matching edge moves, the left vertex is fixed.
2. The right vertex of every BT-edge and every TT-edge is moved up until the edge has slope $+1$.
3. The right vertex of every TB-edge and every BB-edge is moved down until the edge has slope -1.

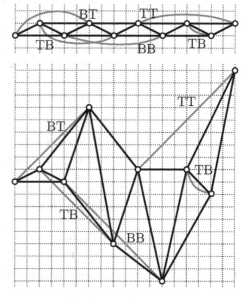

See Fig. 5 for an example. It is easy to see that the displacements preserve the planarity of the embedding of G_1, because vertices only move vertically. The displacements also make G_2 planar: all edges of E_2 with slope $+1$ are on parallel diagonal lines, so they cannot intersect. Symmetrically, all edges of E_2 with slope -1 cannot intersect. Finally, an edge with slope -1 and an edge with slope $+1$ from E_2 cannot intersect because their only y-overlap is between 0 and 1, and here they are sufficiently separated to prevent intersections. Clearly this construction uses only two orientations for the edges of the matching.

Theorem 3. *An outerzigzag and a matching always admit a geometric simultaneous embedding.*

Fig. 5. A geometric simultaneous embedding of an outerzigzag (black) and a matching (gray)

5   Outerpath and Matching

We now extend the approach for outerzigzags to outerpaths. First, we assume that the outerpath is triangulated. Since a triangulated outerpath has two vertices of degree 2, we can make them the ends of an initial placement. We place all vertices on two horizontal lines $y = 0$ and $y = 1$ in such a way, that we obtain

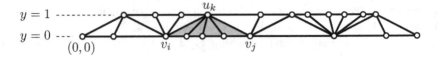

Fig. 6. Outerpath with one fan indicated in gray

a plane drawing of the outerpath G_1 (see Fig. 6). We say that vertices which lie on the line $y = 1$ are on the *top chain*, correspondingly, vertices which lie on the line $y = 0$ are on the *bottom chain*. The leftmost vertex is placed at $(0,0)$. This initial placement again induces a classification of the edges of the matching into TT-edges, TB-edges, BT-edges, and BB-edges. In the final drawing these edges have slopes of -1 or $+1$ as before. However, the embedding algorithm for outerpaths needs to move vertices not only vertically, but also horizontally and hence the x-order of the vertices in the initial placement is not preserved.

We view an outerpath as a sequence of *maximal fans* that are alternatingly directed upwards and downwards. A maximal fan shares its first and last edge with a neighboring fan. A downward fan is indicated in gray in Fig. 6. We denote by d the maximum degree of any vertex in the outerpath G_1.

Our algorithm works as follows: we treat one fan after the other, moving from left to right. When we treat a fan, we place its vertices at new locations to planarize G_2, while keeping the drawing of G_1 planar. In the following we explain the placement algorithm for a downward fan F, upward fans a treated similarly. We denote the single *apex* vertex of F by u_k and its sequence of *finger* vertices by $v_i, \ldots, v_j$ (see Fig. 6). Note that $i < j$; if $i = j$ then the outerpath was not triangulated or F was not maximal. We place all vertices of F, with the exception of v_i and v_j. Vertex v_i has already been placed, since it is the apex of the preceding fan (or it is the leftmost vertex which remains fixed). We do not place v_j since it is the apex of the following fan and will be placed when that fan is treated. We distinguish three cases, depending on the matching partner of the apex u_k. Case (1): the matching partner of u_k has already been placed, Case (2): the matching partner of u_k has not been placed yet and it is not among $v_{i+1}, \ldots, v_{j-1}$, and Case (3): the matching partner of u_k is among $v_{i+1}, \ldots, v_{j-1}$.

Case (1). Apex u_k has a matching partner that has already been placed. Hence the matching partner lies either on the top chain and has an index smaller than k, or it lies on the bottom chain and has an index smaller than or equal to i. Let X denote the total width (x-extent) of the construction so far. We place u_k at x-coordinate $2X + 1$ and then move u_k upwards until it lies on the line with slope $+1$ through its matching partner (see Fig. 7 (left)).

Next we place $v_{i+1}, \ldots, v_{j-1}$ at positions $(2X, 0)$, $(2X + 1/d, 0)$, ..., $(2X + (j - i - 2)/d, 0)$. Consider the $j - i - 1$ lines through u_k and each of $v_{i+1}, \ldots, v_{j-1}$. If we ensure that the final placements of $v_{i+1}, \ldots, v_{j-1}$ lie on these lines, then we will never invert any triangle of the fan. We now move those vertices of $v_{i+1}, \ldots, v_{j-1}$ that are right vertices of matching edges down on their lines until they reach the proper position, determined by the slope -1 lines through their matching partners. Those vertices of $v_{i+1}, \ldots, v_{j-1}$ that are left vertices of matching edges

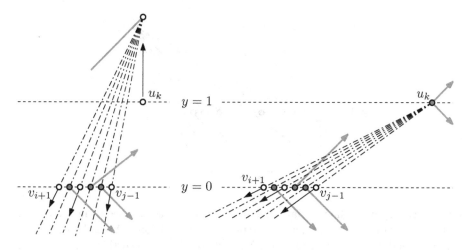

Fig. 7. Left: Case 1, u_k is a right vertex of a matching edge. Right: Case 2, u_k is a left vertex of a matching edge.

stay where they are; they define slope -1 or $+1$ lines on which their matching partners will be placed eventually. By construction, none of these lines intersect to the right of the vertices that defined them, also not with lines defined by vertices treated earlier (see Fig. 7 (left)).

See Fig. 8 for a global sketch of Case 1. Note that v_{i+1} stays to the right of all vertices placed before. This is true because the line defined by u_k and v_{i+1} has slope > 1, and the separation between v_{i+1} and the previously placed vertices is at least X. The value of X also bounds the y-extent for the previously placed vertices to the range $[-X, +X]$, since the edges of the matching have slopes -1 and $+1$. Further note that triangle $\triangle u_k v_{i+1} v_i$ is not inverted, regardless of where v_i is placed in the initial part and whether v_{i+1} is moved on its line. Finally, note that v_j can be placed anywhere on the line $y = 0$ lower, as long as its x-coordinate is at least that of u_k: the triangle $\triangle u_k v_j v_{j-1}$ will not be inverted.

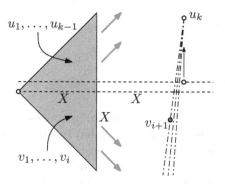

Fig. 8. Global situation of Case 1, previously placed vertices lie inside the gray triangle

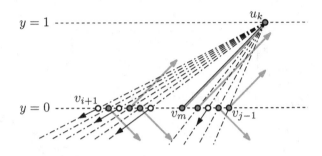

Fig. 9. Case 3

Case (2). Apex u_k has a matching partner that has not been placed yet and which is not among $v_{i+1}, \ldots, v_{j-1}$. We place u_k at position $(3X+2, 1)$, where X is again defined as the total width so far. Next we place the vertices $v_{i+1}, \ldots, v_{j-1}$ at positions $(3X, 0)$, $(3X + 1/d, 0)$, ..., $(3X + (j - i - 2)/d, 0)$. Consider the $j - i - 1$ lines through u_k and each of $v_{i+1}, \ldots, v_{j-1}$. We again move those vertices of $v_{i+1}, \ldots, v_{j-1}$ that are right vertices of matching edges down on their lines until they reach the proper position, determined by the slope -1 lines through their matching partners (see Fig. 7 (right)).

All lines on which the vertices move have slope at least $1/2$, implying that all vertices of $v_{i+1}, \ldots, v_{j-1}$ are placed to the right of all previously placed vertices, due to the x-separation of at least $2X$. Again we note that $\triangle u_k v_{i+1} v_i$ is not inverted, and that v_j may be placed anywhere to the right of u_k without the risk of inverting $\triangle u_k v_j v_{j-1}$.

Case (3). Apex u_k has a matching partner v_m that is among $v_{i+1}, \ldots, v_{j-1}$, see Fig. 9. We place u_k at position $(3X+2, 1)$ and v_m at position $(3X+1, 0)$, where X is again the total width so far. Note that the edge (u_k, v_m) is an edge of both G_1 and G_2. Next we place the vertices $v_{i+1}, \ldots, v_{m-1}$ at positions $(3X, 0)$, $(3X + 1/d, 0)$, ..., $(3X + (m - i - 2)/d, 0)$, and the vertices $v_{m+1}, \ldots, v_{j-1}$ at positions $(3X + 1 + 1/d, 0)$, $(3X + 1 + 2/d, 0)$, ..., $(3X + 1 + (j - m - 1)/d, 0)$. As before we now use the lines through u_k and each of $v_{i+1}, \ldots, v_{m-1}, v_{m+1}, \ldots, v_{j-1}$ to move vertices down if they are right vertices of matching edges.

Theorem 4. *An outerpath and a matching always admit a geometric simultaneous embedding.*

6   Tree and Matching

Our algorithm that computes a geometric simultaneous embedding for a tree and a matching is inspired by and closely related to an algorithm by Di Giacomo *et al.* [2], which computes a *matched drawing* of two trees. Matched drawings are a relaxation of geometric simultaneous embeddings. Specifically, two planar graphs G_1 and G_2 are *matched*, if they are defined on two vertex sets V_1 and V_2 of the same cardinality and if there is a one-to-one mapping between V_1

and V_2. A matched drawing of two matched graphs is a pair of planar straight-line drawings, such that matched vertices of G_1 and G_2 are assigned the same y-coordinate. A geometric simultaneous embedding of a tree and a matching is in essence a matched drawing of half of the vertices of the tree with the other half. And indeed, the algorithm by Di Giacomo *et al.* can be adapted in a straightforward manner to compute a geometric simultaneous embedding of a tree and a matching. However, the edges of the matching in the resulting drawing will in general not all have the same orientation. In the remainder of this section we show how to refine the construction from [2], to compute a simultaneous embedding where all matching edges are drawn horizontally.

We place the vertices one by one, always placing the two vertices of a matching edge consecutively at the same y-coordinate. We use y-coordinates $1, \ldots, n/2$, from the outside in. That is, at any point of the construction, there are two indices i and j with $1 \leq i \leq j \leq n/2$ such that the coordinates $1, \ldots, i-1$ and $j+1, \ldots, n/2$ have been used, and the coordinates $i, \ldots, j$ have not been used yet. At every even placement we decide if we should place the next vertex at the top or at the bottom, that is, at the highest or the lowest available y-coordinate.

Let T be the tree with some of its vertices already placed. The placed vertices partition the tree into connected components (subtrees); we call each component up to and including the placed vertices a *rope*. The placed vertices incident to a rope are called the *knots* of that rope. We maintain the following invariant: after every odd placement, every rope of T has one or two knots, but not more. After an even placement this invariant might be false for exactly one rope, which has three knots. There is a unique vertex, which we call the *splitter*, that lies on the three paths between the knots. We show below how to restore the invariant with the next odd placement by choosing the splitter as the next vertex to place.

Since we place vertices from the outside in, there are nine types of ropes which we encounter during the construction. They are the degree-1 ropes with one knot at the top or at the bottom, the degree-2 ropes with two knots at the top, or two at the bottom, or one at the top and one at the bottom, and the degree-3 ropes with zero, one, two, or three knots at the top and three, two, one, or zero knots at the bottom. We call these ropes T-rope, B-rope, TT-rope, BB-rope, TB-rope, BBB-rope, TBB-rope, TTB-rope, or TTT-rope.

Even placement. The invariant above implies that before an even placement there are only degree-1 and degree-2 ropes. Furthermore, there is exactly one edge (v, w) of the matching M that has one, but not both of its vertices placed. We assume that v has been placed and place w next, at the same y-coordinate as v. The exact placement depends on the type of rope w is part of, as well as the y-coordinate of v. Fig. 10 shows the cases for T-ropes, TB-ropes, and TT-ropes, B-ropes and BB-ropes are symmetric. Placing w can create at most two degree-2 ropes or one degree-3 rope, plus zero or more degree-1 ropes. New degree-1 ropes all have w as their knot. The new degree-2 ropes may have no internal vertices, in which case they are fully placed or *tight*, as they are a straight edge. Placing w creates a degree-3 rope, if w was part of a degree-2 rope but did not lie on the path between its two knots. In this case a new splitter s is identified

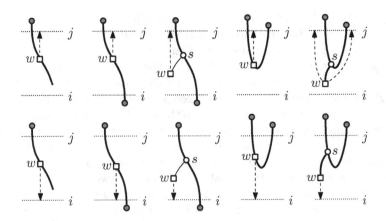

Fig. 10. Even placement for T-ropes, TB-ropes, and TT-ropes

(marked by a circle in Fig. 10), which is placed in the next odd placement. The top right case depicted in Fig. 10 shows two dashed arrows, indicating that there are two possible locations for w. Which of the two we have to use depends on the matching partner of the splitter s. We explain below how to make this decision.

Odd placement. Before an odd placement, all matching edges are either completely placed, or not placed at all. There are two cases: the previous even placement left us with a splitter, or not. If there is no splitter, then we place any unplaced vertex, whose placement does not create a splitter. Any vertex that is directly adjacent to an already placed vertex qualifies. If there is a splitter s, then we place it next. If s is part of a TTT-rope or a TTB-rope, then we place it at the lowest unused y-coordinate i, which creates two or three new TB-ropes and one or zero new BB-ropes. Symmetrically, if s is part of a TBB-rope or a BBB-rope, then we place it at the highest unused y-coordinate j.

There are two additional things to consider. Let u be the matching partner of the splitter s. By construction u has not been placed yet, but it will be placed in the next step, on the same y-coordinate as s.

(1) If s was part of a TTT-rope (or symmetrically, a BBB-rope), then placing s creates three new TB-ropes. If u is part of one of these TB-ropes, then we need to ensure that this particular TB-rope is one of the two "on the outside". The TTT-rope was created by placing a vertex w at y-coordinate j in the previous step (top right case in Fig. 10). Recall that we had two choices for the location of w. One of the two ensures that u is on the outside (see Fig. 11 (top)). Hence we look ahead and place w accordingly. Placing s might also have created one or more B-ropes. If u is part of one of these B-ropes, then we need to ensure again that this particular B-rope is on the outside. We can easily achieve that by ordering the degree 1-ropes with knot s accordingly.

(2) If s was part of a TTB-rope (or symmetrically, a TBB-rope), then placing s creates two TB-ropes and one BB-rope. If u is part of one of the TB-ropes, then we have to ensure again that this particular TB-rope is on the outside.

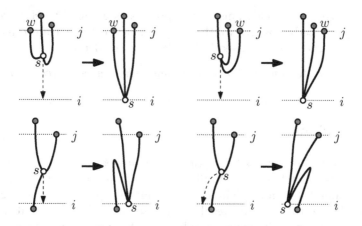

Fig. 11. Odd placement for a splitter and a TTT-rope or a TTB-rope

However, we have a choice of two possible locations for s, see Fig. 11 (bottom). One of the two ensures that u is on the outside, hence we place s accordingly. Again, placing s might also have created one or more B-ropes. In general we place these B-ropes between the two TB-ropes. But if u is part of a B-rope, then we need to place this particular rope on the outside.

Next we have to argue that there is actually space to draw the tree without crossings and with straight edges. For the matching this is obvious since its edges are horizontal and lie on different y-coordinates. We maintain the following invariant, which holds for every rope after every even placement. Let i and j be the lowest and highest unused y-coordinates. There exists a parallelogram between the horizontal lines i and j in which the whole rope can be drawn without crossings and with straight lines. The parallelograms have positive width and have an "alignment" that corresponds to the needs of the rope. In particular, the non-horizontal sides of the parallelogram have a slope s, such that any line with slope s and through a knot of the rope intersects the interior of the parallelogram. Hence every y-coordinate within the parallelogram can be reached by a straight line from its knots. Parallelograms of different ropes are disjoint.

It remains to show how to maintain this invariant as ropes are split. We find the new parallelograms of the sub-ropes inside the parallelogram of the parent. We might have to scale and shear parallelograms to make this work, creating

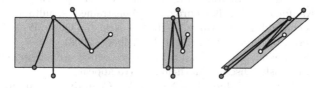

Fig. 12. A parallelogram and its rope can be scaled to become arbitrarily narrow and sheared to get different slopes

extremely narrow parallelograms in the process (see Fig. 12). However, they always keep a strictly positive width. The cases are many, but not difficult and very similar to the ones discussed in [2]. We omit the details.

Theorem 5. *A tree and a matching always admit a geometric simultaneous embedding.*

7   Conclusions

We presented the first results for geometric simultaneous embeddings where one of the graphs is a matching. Specifically, we showed that there exist planar graphs that do not admit a geometric simultaneous embedding with a matching. We do not know whether this negative result holds also under the additional constraint that the matching and the planar graph do not have any edges in common.

We also described algorithms that compute a geometric simultaneous embedding of a matching and a wheel, outerpath, or tree. Our drawing algorithms minimize the number of orientations used to draw the edges of the matching. The main remaining open question is: do an outerplanar graph and a matching always admit a geometric simultaneous embedding?

References

1. Braß, P., Cenek, E., Duncan, C., Efrat, A., Erten, C., Ismailescu, D., Kobourov, S., Lubiw, A., Mitchell, J.: On simultaneous planar graph embeddings. Computational Geometry: Theory & Applications 36(2), 117–130 (2007)
2. Di Giacomo, E., Didimo, W., van Kreveld, M., Liotta, G., Speckmann, B.: Matched drawings of planar graphs. In: Hong, S.-H., Nishizeki, T., Quan, W. (eds.) GD 2007. LNCS, vol. 4875, pp. 183–194. Springer, Heidelberg (2008)
3. Erten, C., Harding, P., Kobourov, S., Wampler, K., Yee, G.: Exploring the computing literature using temporal graph visualization. In: Proc. Conference on Visualization and Data Analysis, pp. 45–56 (2003)
4. Erten, C., Harding, P., Kobourov, S., Wampler, K., Yee, G.: GraphAEL: Graph animations with evolving layouts. In: Liotta, G. (ed.) GD 2003. LNCS, vol. 2912, pp. 98–110. Springer, Heidelberg (2004)
5. Erten, C., Kobourov, S.: Simultaneous embedding of planar graphs with few bends. Journal of Graph Algorithms and Applications 9(3), 347–364 (2005)
6. Estrella-Balderrama, A., Gassner, E., Jünger, M., Percan, M., Schaefer, M., Schulz, M.: Simultaneous geometric graph embeddings. In: Hong, S.-H., Nishizeki, T., Quan, W. (eds.) GD 2007. LNCS, vol. 4875, pp. 280–290. Springer, Heidelberg (2008)
7. Frati, F., Kaufmann, M., Kobourov, S.: Constrained simultaneous and near-simultaneous embeddings. In: Hong, S.-H., Nishizeki, T., Quan, W. (eds.) GD 2007. LNCS, vol. 4875, pp. 268–279. Springer, Heidelberg (2008)
8. Geyer, M., Kaufmann, M., Vrt'o, I.: Two trees which are self-intersecting when drawn simultaneously. Discrete Mathematics (to appear)
9. Ware, C., Purchase, H., Colpoys, L., McGill, M.: Cognitive measurements of graph aesthetics. Information Visualization 1(2), 103–110 (2002)

Algebraic Methods for Counting Euclidean Embeddings of Rigid Graphs

Ioannis Z. Emiris[1], Elias P. Tsigaridas[2], and Antonios E. Varvitsiotis[3]

[1] University of Athens, Athens, Greece
[2] INRIA Méditerranée, Sophia-Antipolis, France
[3] CWI, Amsterdam, The Netherlands

Abstract. The study of (minimally) rigid graphs is motivated by numerous applications, mostly in robotics and bioinformatics. A major open problem concerns the number of embeddings of such graphs, up to rigid motions, in Euclidean space. We capture embeddability by polynomial systems with suitable structure, so that their mixed volume, which bounds the number of common roots, to yield interesting upper bounds on the number of embeddings. We focus on $\mathbb{R}^2$ and $\mathbb{R}^3$, where Laman graphs and 1-skeleta of convex simplicial polyhedra, respectively, admit inductive Henneberg constructions. We establish the first general lower bound in $\mathbb{R}^3$ of about 2.52^n, where n denotes the number of vertices. Moreover, our implementation yields upper bounds for $n \leq 10$ in $\mathbb{R}^2$ and $\mathbb{R}^3$, which reduce the existing gaps, and tight bounds up to $n = 7$ in $\mathbb{R}^3$.

Keywords: Rigid graph, Euclidean embedding, Henneberg construction, polynomial system, root bound, cyclohexane caterpillar.

1 Introduction

Rigid graphs (or frameworks) constitute an old but still active area of research due to certain deep mathematical questions, as well as numerous applications, e.g. mechanism theory [8,9], and structural bioinformatics [4].

Given graph $G = (V, E)$ and a collection of edge lengths $l_{ij} \in \mathbb{R}_{>0}$, for $(i, j) \in E$, a *Euclidean embedding* in $\mathbb{R}^d, d \geq 1$ is a mapping of the vertices V to points in $\mathbb{R}^d$, such that l_{ij} equals the Euclidean distance between the images of the i-th and j-th vertices, for all edges $(i, j) \in E$. There is no requirement on whether the edges cross or not. G is *generically* rigid in $\mathbb{R}^d$ if, for generic edge lengths, it admits a finite number of embeddings in $\mathbb{R}^d$, modulo rigid motions. G is *minimally* rigid if it is no longer rigid once any edge is removed. In the sequel, generically minimally rigid graphs are referred to as rigid.

A graph is called Laman if $|E| = 2|V| - 3$ and, additionally, all of its induced subgraphs on $k < |V|$ vertices have $\leq 2k - 3$ edges. The Laman graphs are precisely the rigid graphs in $\mathbb{R}^2$; they also admit inductive constructions. In $\mathbb{R}^3$ there is no analogous combinatorial characterization of rigid graphs, but the 1-skeleta, or edge graphs, of (convex) simplicial polyhedra are rigid in $\mathbb{R}^3$, and admit inductive constructions.

D. Eppstein and E.R. Gansner (Eds.): GD 2009, LNCS 5849, pp. 195–200, 2010.

Fig. 1. The Desargues graph   **Fig. 2.** All simplicial polyhedra for $n = 6$

We deal with the problem of computing the maximum number of distinct planar and spatial Euclidean embeddings of rigid graphs, up to rigid motions, as a function of the number of vertices. To study upper bounds, we define a square polynomial system, expressing the edge length constraints, whose real solutions correspond precisely to the different embeddings. Here is a system expressing embeddability in $\mathbb{R}^3$, where (x_i, y_i, z_i) are the coordinates of the i-th vertex, and 3 vertices are fixed to discard translations and rotations:

$$\begin{cases} x_i = a_i, \ y_i = b_i, \ z_i = c_i, & i = 1, 2, 3, \quad a_i, b_i, c_i \in \mathbb{R}, \\ (x_i - x_j)^2 + (y_i - y_j)^2 + (z_i - z_j)^2 = l_{ij}^2, & (i, j) \in E - \{(1,2), (1,3), (2,3)\} \end{cases} \tag{1}$$

All nontrivial equations are quadratic; there are $2n - 4$ for Laman graphs, and $3n - 9$ for 1-skeleta of simplicial polyhedra, where n is the number of vertices. Bézout's bound on the number of complex roots equals the product of the degrees, hence 4^{n-2} and 8^{n-3}, respectively.

For the planar and spatial case, the best upper bounds are $\binom{2n-4}{n-2}$ and $\frac{2^{n-3}}{n-2}\binom{2n-6}{n-3}$ [2]. In applications, it is crucial to know the number of embeddings for small n. The main result in this direction was to show that the Desargues (or 3-prism) graph (Figure 1) admits 24 embeddings in $\mathbb{R}^2$ [2]. This led the same authors to lower bounds in $\mathbb{R}^2$: $24^{\lfloor (n-2)/4 \rfloor} \simeq 2.21^n$ by a "caterpillar" constructed by concatenating copies of the Desargues graph, and $2 \cdot 12^{\lfloor (n-3)/3 \rfloor} \simeq 2.29^n/6$ obtained by a Desargues "fan"[1].

Bernstein's bound on the number of roots of a polynomial system exploits the sparseness of the equations. It is bounded by Bézout's bound and typically much tighter. We have developed software to construct all rigid graphs up to isomorphism, for small n, and compute the Bernstein's bounds. Besides some straightforward upper bounds in Lemmas 1 and 6, our main contribution is twofold. We derive the first general lower bound in $\mathbb{R}^3$:

$$16^{\lfloor (n-3)/3 \rfloor} \simeq 2.52^n, \quad n \geq 9,$$

by designing a cyclohexane caterpillar. We also obtain improved upper and lower bounds for $n \leq 10$ in $\mathbb{R}^2$ and $\mathbb{R}^3$ (Tables 1 and 2). Moreover, we establish tight bounds for $n \leq 7$ in $\mathbb{R}^3$ by appropriately formulating the polynomial system. We apply Bernstein's Second theorem to show that the above polynomial system cannot yield tight bounds.

---

[1] We have corrected the exponent of the original statement.

The rest of the paper is structured as follows: Section 2 discusses the planar case, Section 3 presents our algebraic tools, Section 4 deals with $\mathbb{R}^3$, and we conclude with open questions.

Some results appeared in [5] in preliminary form.

2   Planar Embeddings of Laman Graphs

Laman graphs admit inductive constructions, starting with a triangle, and followed by a sequence of Henneberg-1 (or H_1) and Henneberg-2 steps (or H_2). Each step adds a new vertex: H_1 connects it to two existing vertices, H_2 connects it to 3 existing vertices having at least one edge among them, which is removed. We represent a Laman graph by its Henneberg sequence $\triangle s_4 \ldots, s_n$, where $s_i \in \{1, 2\}$. A Laman graph is called H_1 if and only if it can be constructed using only H_1; otherwise it is called H_2. Since two generic circles intersect in two real points, H_1 exactly doubles the maximum number of embeddings. It follows that a H_1 graph has 2^{n-2} embeddings.

One can easily verify that every $\triangle 2$ graph is isomorphic to a $\triangle 1$ graph and that every $\triangle 12$ graph is isomorphic to a $\triangle 11$ graph. Consequently, all Laman graphs with $n = 4, 5$ are H_1, with 4 and 8 embeddings, respectively. For $n = 6$, there are 3 possibilities: the graph is either H_1, $K_{3,3}$, or the Desargues graph. Since the $K_{3,3}$ graph has at most 16 embeddings [8,9] and the Desargues graph has 24 embeddings [2], the upper bound is 24.

Using our software (Section 3), we construct all Laman graphs for $n = 7, \ldots, 10$, compute the respective Bernstein's bounds, and bound the maximum number of Euclidean embeddings by 64, 128, 512 and 2048, respectively. Table 1 summarizes our results for $n \leq 10$. The lower bound for $n = 9$ follows from the Desargues fan. All other lower bounds follow from the fact that H_1 doubles the number of embeddings.

We now establish a "sparse" upper bound.

Lemma 1. *Let G be Laman, with $k \geq 4$ degree-2 vertices. Then, the number of its planar embeddings is at most $2^{k-4}4^{n-k}$.*

Proof. There are at least k removals of degree-2 vertices, since all created graphs are Laman. The final graph has a Bézout bound of 4^{n-k}.

3   An Algebraic Interlude

Given a polynomial f in n variables, its support is the set of exponents in $\mathbb{N}^n$ corresponding to nonzero terms (or monomials). The Newton polytope of f is the convex hull of its support and lies in $\mathbb{R}^n$. Consider polytopes $P_i \subset \mathbb{R}^n$ and $\lambda_i \in \mathbb{R}, \lambda_i \geq 0$, for $i = 1, \ldots, n$. Consider the Minkowski sum $\lambda_1 P_1 + \cdots + \lambda_n P_n \in \mathbb{R}^n$: its (Euclidean) volume is a homogeneous polynomial of degree n in the λ_i. The coefficient of $\lambda_1 \cdots \lambda_n$ is the *mixed volume* of $P_1, \ldots, P_n$. If $P_1 = \cdots = P_n$, then the mixed volume is $n!$ times the volume of P_1. We focus on $\mathbb{C}^* = \mathbb{C} - \{0\}$.

Theorem 2. [1] *Let $f_1 = \cdots = f_n = 0$ be a polynomial system in n variables with real coefficients, where the f_i have fixed supports. The number of isolated common solutions in $(\mathbb{C}^*)^n$ is bounded above by the mixed volume of (the Newton polytopes of) the f_i. This bound is tight for generic coefficients of the f_i's.*

Bernstein's Second theorem below was used in $\mathbb{R}^2$ [7]; we apply it to $\mathbb{R}^3$. Given $v \in \mathbb{R}^n - \{0\}$ and polynomial f_i, $\partial_v f_i$ is the polynomial obtained by keeping only the terms whose exponents minimize inner product with v; its Newton polytope is the face of the Newton polytope of f_i supported by v.

Theorem 3. [1] *If for all $v \in \mathbb{R}^n - \{0\}$ the face system $\partial_v f_1 = \ldots = \partial_v f_n = 0$ has no solutions in $(\mathbb{C}^*)^n$, then the mixed volume of the f_i exactly equals the number of solutions in $(\mathbb{C}^*)^n$, and all solutions are isolated. Otherwise, the mixed volume is a strict upper bound.*

In order to bound the number of embeddings of rigid graphs, we have developed specialized software that constructs all Laman graphs and all 1-skeleta of simplicial polyhedra with $n \leq 10$. Our computational platform is SAGE[2]. We construct all graphs using the Henneberg steps, which we implemented in Python, using SAGE's interpreter. We classify all graphs up to isomorphism using SAGE's interface with N.I.C.E., an open-source isomorphism check software, keeping, for each graph, the Henneberg sequence with the fewest H_1 steps.

For each graph we construct a system whose real solutions express all embeddings, by formulation (2). We bound the number of its (complex) solutions by mixed volume. For every Laman graph, to discard translations and rotations, we fix one edge to be of unit length, aligned with an axis, with one vertex at the origin. In $\mathbb{R}^3$, a third vertex is fixed in a coordinate plane. Depending on the choice of the fixed edge, we obtain different systems hence different mixed volumes, and we keep their minimum.

We used an Intel Core2, at 2.4GHz, with 2GB of RAM. We tested more that 20,000 graphs, computed the mixed volume of more than 40,000 systems, taking a total time of about 2 days. Tables 1 and 2 summarize our results.

4   Spatial Embeddings of 1-Skeleta of Simplicial Polyhedra

Let us examine 1-skeleta of (convex) simplicial polyhedra, which are rigid in $\mathbb{R}^3$ [6]. For such a graph (V, E), $|E| = 3|V| - 6$ and all induced subgraphs on $k < |V|$ vertices have $\leq 3k - 6$ edges. Consider any $k + 2$ vertices forming a cycle with $\geq k - 1$ diagonals, $k \geq 1$. The extended Henneberg-k step (or H_k), $k = 1, 2, 3$, corresponds to adding a vertex, connecting it to the $k + 2$ vertices, and removing $k - 1$ diagonals among them. A graph is the 1-skeleton of a simplicial polyhedron in $\mathbb{R}^3$ if and only if it has a construction starting with the 3-simplex, followed by any sequence of H_1, H_2, H_3 [3].

---

[2] http://www.sagemath.org/

Since 3 spheres intersect generically in two points, H_1 exactly doubles the maximum number of embeddings. In order to discard translations and rotations, we fix a (triangular) facet of the polytope; we choose, without loss of generality, the first 3 vertices and obtain system (1) of dimension $3n$. Let $v = (0, 0, 0, 0, 0, 0, 0, 0, 0, -1, \ldots, -1) \in \mathbb{R}^{3n}$, the face system is:

$$\begin{cases} x_i = a_i, y_i = b_i, z_i = c_i, & i = 1, 2, 3, \quad a_i, b_i, c_i \in \mathbb{R}, \\ (x_i - x_j)^2 + (y_i - y_j)^2 + (z_i - z_j)^2 = 0, & (i,j) \in E, \ i, j \notin \{1, 2, 3\}, \\ x_i^2 + y_i^2 + z_i^2 = 0, & (i,j) \in E : \ i \notin \{1, 2, 3\}, \ j \in \{1, 2, 3\}. \end{cases}$$

This system has $(a_1, b_1, c_1, \ldots, a_3, b_3, c_3, 1, 1, \gamma\sqrt{2}, \ldots, 1, 1, \gamma\sqrt{2}) \in (\mathbb{C}^*)^{3n}$ as a solution, where $\gamma = \pm\sqrt{-1}$. According to Theorem 3, the mixed volume is not a tight bound on the number of solutions in $(\mathbb{C}^*)^{3n}$. This was also observed, for $\mathbb{R}^2$, in [7]. To remove spurious solutions let $w_i = x_i^2 + y_i^2 + z_i^2$, for $i = 1, \ldots, n$. This yields an equivalent system, with lower mixed volume, which will be used in our computations:

$$\begin{cases} x_i = a_i, y_i = b_i, z_i = c_i, & i = 1, 2, 3, \\ w_i = x_i^2 + y_i^2 + z_i^2, & i = 1, \ldots, n, \\ w_i + w_j - 2x_i x_j - 2y_i y_j - 2z_i z_j = l_{ij}^2, & (i,j) \in E - \{(1,2), (1,3), (2,3)\}. \end{cases} \quad (2)$$

For $n = 4$, the only simplicial polytope is the 3-simplex, which has 2 embeddings. For $n = 5$, there is a unique 1-skeleton of a simplicial polyhedron [3], and is obtained from the 3-simplex by H_1, hence it has exactly 4 embeddings.

For $n = 6$, there are two non-isomorphic graphs G_1, G_2 (Figure 2) [3], yielding respective mixed volumes of 8 and 16. G_2 is the graph of the cyclohexane, which admits 16 different embeddings [4]. To see this, the cyclohexane is a 6-cycle with known lengths between vertices at distance 1 (adjacent) and 2. Alternatively, G_2 corresponds to a Stewart platform parallel robot with 16 configurations, where triangles define the platform and base, and 6 lengths link the triangles in a jigsaw shape. This proves:

Lemma 4. *The 1-skeleton of a simplicial polyhedron with $n = 6$ has at most 16 embeddings and this is tight.*

Let us glue copies of cyclohexanes sharing a triangle, each adding 3 vertices, thus obtaining the 1-skeleton of a simplicial polytope. Applying Lemma 4 we have:

Theorem 5. *There exist edge lengths for which the cyclohexane caterpillar construction has $16^{\lfloor (n-3)/3 \rfloor} \simeq 2.52^n$ embeddings, for $n \geq 9$.*

Table 2 summarizes our results for $n \leq 10$. The upper bounds for $n = 7, \ldots, 10$ are computed by our software. The lower bound for $n = 9$ is from Theorem 5. All other lower bounds are obtained by considering a H_1 construction.

We state without proof a result similar to Lemma 1.

Lemma 6. *Let G be the 1-skeleton of a simplicial polyhedron with $k \geq 9$ degree-3 vertices. The number of embeddings of G is bounded above by $2^{k-9}8^{n-k}$.*

Table 1. Bounds for Laman graphs

n =	3	4	5	6	7	8	9	10
lower	2	4	8	24	48	96	288	576
upper	2	4	8	24	64	128	512	2048

Table 2. Bounds for 1-skeleta of simplicial polyhedra

n =	4	5	6	7	8	9	10
lower	2	4	16	32	64	256	512
upper	2	4	16	32	160	640	2560

5   Further Work

The most important problem in rigidity theory is the combinatorial characterization of rigid graphs in $\mathbb{R}^3$. Since we rely on Henneberg constructions, it is crucial to determine the effect of each step on the number of embeddings: we conjecture that H_2 multiplies it by ≤ 4 and H_3 by ≤ 8, but these may not always be tight. Our conjecture has been verified for small n.

Acknowledgments. A.V. performed this work as part of his Master's thesis at University of Athens; he thanks G. Rote for discussions, and acknowledges partial support by project IST-006413-2 ACS: Algorithms for Complex Shapes. I.E. is partially supported by project PITN-GA-2008-214584 SAGA: Shapes, Geometry, and Algebra; part of this work was carried out while on leave at Ecole Normale de Paris, and Ecole Centrale de Paris. E.T. is partially supported by contract ANR-06-BLAN-0074 "Decotes".

References

1. Bernstein, D.N.: The number of roots of a system of equations. Fun. Anal. Pril. 9, 1–4 (1975)
2. Borcea, C., Streinu, I.: The number of embeddings of minimally rigid graphs. Discrete Comp. Geometry 31(2), 287–303 (2004)
3. Bowen, R., Fisk, S.: Generation of triangulations of the sphere. Math. of Computation 21(98), 250–252 (1967)
4. Emiris, I.Z., Mourrain, B.: Computer algebra methods for studying and computing molecular conformations. Algorithmica, Special Issue 25, 372–402 (1999)
5. Emiris, I.Z., Varvitsiotis, A.: Counting the number of embeddings of minimally rigid graphs. In: Proc. Europ. Workshop Comput. Geometry, Brussels (2009)
6. Gluck, H.: Almost all simply connected closed surfaces are rigid. Lect. Notes in Math. 438, 225–240 (1975)
7. Steffens, R., Theobald, T.: Mixed volume techniques for embeddings of Laman graphs. In: Proc. Europ. Workshop Comput. Geometry, Nancy, France, pp. 25–28 (2008); Final version accepted in Comp. Geom: Theory & Appl., Special Issue
8. Walter, D., Husty, M.: On a 9-bar linkage, its possible configurations and conditions for paradoxical mobility. In: IFToMM Congress, Besançon, France (2007)
9. Wunderlich, W.: Gefärlice Annahmen der Trilateration und bewegliche Fachwerke I. Zeitschrift für Angewandte Mathematik und Mechanik 57, 297–304 (1977)

Removing Independently Even Crossings

Michael J. Pelsmajer[1,*], Marcus Schaefer[2], and Daniel Štefankovič[3]

[1] DePaul University, Chicago, IL 60604, USA
mschaefer@cs.depaul.edu
[2] Illinois Institute of Technology, Chicago, IL 60616, USA
pelsmajer@iit.edu
[3] University of Rochester, Rochester, NY 14627, USA
stefanko@cs.rochester.edu

Abstract. We show that $cr(G) \leq \binom{2\,iocr(G)}{2}$ settling an open problem of Pach and Tóth [5,1]. Moreover, $iocr(G) = cr(G)$ if $iocr(G) \leq 2$.

1 Crossing Numbers

Pach and Tóth point out in "Which Crossing Number is It Anyway?" that there have been many different ideas on how to define a notion of crossing number including—using current terminology—the following (see [6,13]):

crossing number: $cr(G)$, the smallest number of crossings in a drawing of G,
pair crossing number:[1] $pcr(G)$, the smallest number of pairs of edges crossing in a drawing of G,
odd crossing number: $ocr(G)$, the smallest number of pairs of edges crossing oddly in a drawing of G.

We make the typical assumptions on drawings of a graph: there are only finitely many crossings, no more than two edges cross in a point, edges do not pass through vertices, and edges do not touch. (For a detailed discussion see [13].) What about adjacent edges though? Do we allow them to cross or not? Do we count their crossings? Tutte [17] wrote "We are taking the view that crossings of adjacent edges are trivial, and easily got rid of." While this is true for the standard crossing number, it is not at all obvious for other variants. Székely [13] later commented "We interpret this sentence as a philosophical view and not a mathematical claim."

In [5], Pach and Tóth suggest a systematic study of this issue (see also [1, Section 9.4]): they introduce two rules that can be applied to any notion of crossing number. "Rule +" restricts the drawings to drawings in which adjacent edges are not allowed to cross. "Rule −" allows crossings of adjacent edges, but does not count them towards the crossing number. Pairing ocr, pcr, and cr with any of these two rules gives a total of eight possible variants (since $cr_+ = cr$ as we

---

* Partially supported by NSA Grant H98230-08-1-0043.
[1] Recently, the book by Tao and Vu [15] on additive combinatorics defined the crossing number as pcr.

D. Eppstein and E.R. Gansner (Eds.): GD 2009, LNCS 5849, pp. 201–206, 2010.

mentioned above); one of them has its own name: iocr := ocr_-, the *independent odd crossing number*, introduced by Tutte. The figure below is based on a similar figure from [1].

Rule +	ocr_+	pcr_+	cr
	ocr	pcr	
Rule −	iocr $= \mathrm{ocr}_-$	pcr_-	cr_-

Not much is known about the relationships between these crossing number variants, apart from what immediately follows from the definitions: the values in the display increase monotonically as one moves from the left to the right and from the bottom to the top. Even $\mathrm{cr} = \mathrm{cr}_-$ is open. Pach and Tóth did show that $\mathrm{cr}(G) \leq \binom{2\,\mathrm{ocr}(G)}{2}$, and this implies that five of the variants, namely $\mathrm{ocr}_+, \mathrm{ocr}, \mathrm{pcr}_+, \mathrm{pcr}$, and cr cannot be arbitrarily far apart, but the result does not cover the "Rule −" variants. There are examples of graphs for which ocr and pcr differ [12,16]. Valtr [18] showed that $\mathrm{cr}(G) = O(\mathrm{pcr}(G)/\log\mathrm{pcr}(G))$, which was improved by Tóth [16] to $\mathrm{cr}(G) = O(\mathrm{pcr}(G)/\log^2\mathrm{pcr}(G))$.

In this paper, we show that all eight variants are within a square of each other:

Theorem 1. $\mathrm{cr}(G) \leq \binom{2\,\mathrm{iocr}(G)}{2}$.

This answers an open problem from [5, Problem 13]; also see [1, Problem 9.4.7]. Pach and Tóth asked whether there are functions f, g, h, so that $\mathrm{cr}(G) \leq f$ $(\mathrm{cr}_-(G))$, $\mathrm{pcr}(G) \leq g(\mathrm{pcr}_-(G))$, and $\mathrm{ocr}(G) \leq h(\mathrm{iocr}(G))$ for all graphs G. Theorem 1 implies that $f = g = h = \binom{2x}{2}$ will do, but this is probably not the optimal choice for f, g, and h, and quite possibly not for bounding cr in terms of iocr either.

Theorem 1 immediately implies that $\mathrm{iocr}(G) = \mathrm{cr}(G)$ if $\mathrm{iocr}(G) \leq 1$. We can improve this result:

Theorem 2. *If* $\mathrm{iocr}(G) \leq 2$, *then* $\mathrm{cr}(G) = \mathrm{iocr}(G)$.

The proof of Theorem 2 is too long to be included in this short note and will be contained in the journal version of the paper. It is based on an analysis of the "odd configurations" that can occur in a drawing. We performed such an analysis for ocr so that we could show that $\mathrm{ocr}(G) = \mathrm{cr}(G)$ if $\mathrm{ocr}(G) \leq 3$ [8]. The proof of Theorem 2 is much harder, since a bound on $\mathrm{iocr}(G)$ does not imply any a priori bound on the number of edges crossing some other edge oddly. Indeed, the following problem is open:

Is there a function f so that every graph G has a drawing with indepen-
dent odd crossing number $\mathrm{iocr}(G)$ *and at most* $f(\mathrm{iocr}(G))$ *crossings?*

For $\mathrm{ocr}(G)$ such a result can be established; the best upper bound f known in this case is exponential [11].

Theorem 2 generalizes the Hanani-Tutte theorem, which states that $\mathrm{iocr}(G) = 0$ implies that $\mathrm{cr}(G) = 0$ [2,17]. There are aspects of the Hanani-Tutte theorem which are still not well understood; for example, to what extent it relies on the underlying surface. Only recently was it extended to the projective plane [7], that is, it was shown that $\text{iocr}_{N_1}(G) = \text{cr}_{N_1}(G)$ if $\text{iocr}_{N_1}(G) = 0$. However, it is not clear how to extend this to the case that $\text{iocr}_{N_1}(G) \leq 1$ or how to prove the Hanani-Tutte theorem for surfaces beyond the projective plane. We do know that $\text{ocr}_S(G) = \text{cr}_S(G)$ if $\text{ocr}_S(G) \leq 2$ for arbitrary surfaces S [10].

The independent odd crossing number is implicit in Tutte's paper "Toward a Theory of Crossing Number" [17] which attempts to build an algebraic foundation for the study of the standard crossing number. From an algebraic point of view, ocr and iocr are more convenient parameters than the standard crossing number; for example, the paper by Pach and Tóth [6] shows that $\text{iocr} \leq k$ can be recast as a vector-space problem. Tutte's algebraic approach has been continued by Székely [13,14] and, along different lines, Norine [3] and van der Holst [19]. Theorem 1 justifies the approach of studying standard crossing number via independent odd crossing number, by showing that they are not too far apart; indeed, it is tempting to conjecture that $\text{cr}(G) = O(\text{iocr}(G))$. And in spite of the fact that determining the independent odd crossing number of a graph is **NP**-complete [9], we feel that due to its algebraic nature it offers an intriguing and underutilized alternative approach to algorithmic aspects of crossing number problems.

2   Removing Even More Crossings

An edge in a drawing of a graph is *odd* if it is part of an *odd pair*, which is a pair of edges that cross an odd number of times. Edges that are not odd are *even*, and they cross every edge an even number of times (possibly zero times). An edge in a drawing is *independently odd* if it is part of an *independently odd pair*, which is a pair of non-adjacent edges that cross an odd number of times. Edges that are not independently odd are *independently even*, or *iocr-0* for short; an iocr-0 edge crosses all non-adjacent edges evenly (possibly zero times), while it may cross adjacent edges arbitrarily. Throughout this paper graphs are simple, that is, they have no loops or multiple edges, unless we say otherwise.

Pach and Tóth showed that if E is the set of even edges in a drawing D of G, then G can be redrawn so that all edges in E are crossing-free. As a corollary, they obtained $\text{cr}(G) \leq \binom{2\,\text{ocr}(G)}{2}$ [6]. We strengthen the Pach-Tóth result to the case that E is the set of independently even edges. According to Pach and Sharir [4], this has been conjectured.

Lemma 1. *If D is a drawing of a graph G in the plane, then G has a redrawing in which the independently even edges of D are crossing-free, and every pair of edges crosses at most once.*

Proof (Theorem 1). Start with a drawing D of G that realizes $\text{iocr}(G)$, that is, $\text{iocr}(D) = \text{iocr}(G)$. If F is the set of independently odd edges in D, then $|F| \leq 2\,\text{iocr}(D)$. By Lemma 1, there is a drawing of G with at most $\binom{|F|}{2}$ crossings.   □

To prove Lemma 1, we adapt the following result (a different strengthening of the Pach-Tóth result) from even/odd edges to iocr-0/independently odd edges. The *rotation* of a vertex is the cyclic order in which edges leave the vertex in a drawing, read clockwise. The *rotation system* of a drawing is the collection of all vertex rotations.

Lemma 2 (Pelsmajer, Schaefer, and Štefankovič [8]). *If D is a drawing of G in the plane and F is the set of odd edges in D, then G has a redrawing with the same rotation system, in which $G - F$ is crossing-free and there are no new pairs of edges that cross an odd number of times.*

Splitting a vertex means creating two copies of the vertex with an edge between them so that any edge incident to the original vertex is incident to exactly one of the two copies. (According to this definition, it makes sense to talk about the edges of the original graph occurring in the graph after a vertex split, even though the incidences of edges will change.) With this operation we can now state our analogue of Lemma 2 for iocr-0/independently odd edges.

Lemma 3. *If D is a drawing of G in the plane, and F is the set of independently odd edges in D, then one can apply a sequence of vertex splits to obtain a graph G' with drawing D' and the set F' of independently odd edges in D', such that (1) there are no new independent odd pairs (and hence $F' \subseteq F$), and (2) every edge of $G' - F'$ that is not a cut-edge of $G' - F'$ is crossing-free in D'.*

An edge is a cut-edge if and only if it belongs to no cycles, so Property (2) can be restated as saying that the union of cycles in $G' - F'$ is crossing-free in D'. Also, if Y is the set of cut-edges of $G' - F'$, then $G' - (F' \cup Y)$ is crossing-free in D'.

Proof (Lemma 3). Fix a drawing D of $G = (V, E)$ and let F be the set of independently odd edges in D. We establish the theorem by induction. We need to modify G during the proof by splitting vertices, hence we will use induction over the *weight*

$$w(G) := \sum_{v \in V} d(v)^3$$

of G where $d(v)$ the degree of v in G. For two graphs of the same weight, we induct over the number of cycles that are not crossing-free.

Suppose that C is a crossing-free cycle, with a vertex u that is incident to more than one edge on the same side of C. We modify the graph by splitting u into u_1 (replacing u on C) and u_2 (attached to the edges on the side with more than one edge) and inserting an edge between u_1 and u_2. This operation results in a graph G' with smaller weight and it does not create new independently odd edges (since edges in the exterior of C cannot cross edges on the interior, as all edges along C are crossing-free). We can now apply induction to G' to obtain the result. Thus, we may assume that for every vertex u in a crossing-free cycle C, u is incident to at most one edge on the interior of C and at most one edge on the exterior of C. It follows that any two edges incident to a vertex u in a crossing-free cycle do not cross.

Suppose that C is a cycle made up of iocr-0 edges only, and C is not crossing-free. At each vertex u of C we can ensure that the two edges of C incident at u (say, e and f) cross evenly by modifying the rotation at u and redrawing G close to u. The rotation of the remaining edges at u can then be changed so that each of them crosses e and f evenly. After the redrawing, all the edges of C are even and we can apply Lemma 2 to remove all crossings with edges of C without changing the rotation system or adding new pairs of edges that cross oddly. Now C is crossing-free, and no new independently odd pairs have been added. Suppose that C' is a cycle that was crossing-free before the redrawing. If C and C' share a vertex u, then the rotation at u is not modified when making C crossing-free, so the drawing of C' near u is unchanged. C' remains crossing-free under the redrawing of Lemma 2, too. Thus we have decreased the number of cycles that are not crossing-free.

We can therefore assume that any cycle consisting of iocr-0 edges is crossing-free. Any other iocr-0 edge is a cut-edge in the graph restricted to iocr-0 edges. $\qquad\square$

With Lemma 3, we can now prove Lemma 1.

Proof (Lemma 1). Fix a drawing D of G and let F be the set of independently odd edges in D. Apply Lemma 3 to obtain a graph G' with drawing D', let F' be the set of independently odd edges in D', and let Y be the set of cut-edges in $G' - F'$. Since $F' \cup Y$ contains all crossings in D', $G' - (F' \cup Y)$ is crossing-free in D' and we can let S be the set of its faces. Within each face of S, the edges of Y contained in it can be redrawn one-by-one without creating any crossings, since no edge of Y can complete a path that cuts a face in two (because then it would be part of a cycle in $G' - F'$, which contradicts it being a cut-edge of $G' - F'$). This yields a crossing-free drawing of $G' - F'$, and each of its faces corresponds to a face of S, with boundary formed from the boundary of the face of S and the edges of Y in that face. Therefore, each edge of F' still has both endpoints incident to a face. Within each face, all such edges of F' can be drawn so that every pair of edges crosses at most once.

Since G' was obtained from G by a sequence of vertex splits, G can be obtained from G' by a sequence of edge contractions. The edges in $E(G') - E(G)$ are crossing-free, so applying that sequence of contractions to the current drawing of G' yields a drawing of G in which $G - F'$ is crossing-free and each pair of edges in F' crosses at most once. Since $F' \subseteq F$, this completes the proof. $\qquad\square$

References

1. Brass, P., Moser, W., Pach, J.: Research Problems in Discrete Geometry. Springer, New York (2005)
2. Chojnacki (Haim Hanani), C.: Über wesentlich unplättbare Kurven im dreidimensionalen Raume. Fundamenta Mathematicae 23, 135–142 (1934)

3. Norine, S.: Pfaffian graphs, T-joins and crossing numbers. Combinatorica 28(1), 89–98 (2008)
4. Pach, J., Sharir, M.: Combinatorial geometry and its algorithmic applications: The Alcalá lectures. Mathematical Surveys and Monographs, vol. 152. American Mathematical Society, Providence (2009)
5. Pach, J., Tóth, G.: Thirteen problems on crossing numbers. Geombinatorics 9(4), 194–207 (2000)
6. Pach, J., Tóth, G.: Which crossing number is it anyway? J. Combin. Theory Ser. B 80(2), 225–246 (2000)
7. Pelsmajer, M.J., Schaefer, M., Stasi, D.: Strong Hanani–Tutte on the projective plane. SIAM Journal on Discrete Mathematics 23(3), 1317–1323 (2009)
8. Pelsmajer, M.J., Schaefer, M., Štefankovič, D.: Removing even crossings. J. Combin. Theory Ser. B 97(4), 489–500 (2007)
9. Pelsmajer, M.J., Schaefer, M., Štefankovič, D.: Crossing number of graphs with rotation systems. In: Hong, S.-H., Nishizeki, T., Quan, W. (eds.) GD 2007. LNCS, vol. 4875, pp. 3–12. Springer, Heidelberg (2008)
10. Pelsmajer, M.J., Schaefer, M., Štefankovič, D.: Removing even crossings on surfaces. European Journal of Combinatorics 30(7), 1704–1717 (2009)
11. Pelsmajer, M.J., Schaefer, M., Štefankovič, D.: Crossing numbers and parameterized complexity. In: Hong, S.-H., Nishizeki, T., Quan, W. (eds.) GD 2007. LNCS, vol. 4875, pp. 31–36. Springer, Heidelberg (2008)
12. Pelsmajer, M.J., Schaefer, M., Štefankovič, D.: Odd crossing number and crossing number are not the same. Discrete Comput. Geom. 39(1), 442–454 (2008)
13. Székely, L.A.: A successful concept for measuring non-planarity of graphs: the crossing number. Discrete Math. 276(1-3), 331–352 (2004)
14. Székely, L.A.: An optimality criterion for the crossing number. Ars Math. Contemp. 1(1), 32–37 (2008)
15. Tao, T., Vu, V.: Additive combinatorics. Cambridge Studies in Advanced Mathematics, vol. 105. Cambridge University Press, Cambridge (2006)
16. Tóth, G.: Note on the pair-crossing number and the odd-crossing number. Discrete Comput. Geom. 39(4), 791–799 (2008)
17. Tutte, W.T.: Toward a theory of crossing numbers. J. Combinatorial Theory 8, 45–53 (1970)
18. Valtr, P.: On the pair-crossing number. In: Combinatorial and Computational Geometry. Math. Sci. Res. Inst. Publ., vol. 52, pp. 569–575. Cambridge University Press, Cambridge (2005)
19. van der Holst, H.: Algebraic characterizations of outerplanar and planar graphs. European J. Combin. 28(8), 2156–2166 (2007)

Manhattan-Geodesic Embedding of Planar Graphs

Bastian Katz[1], Marcus Krug[1], Ignaz Rutter[1], and Alexander Wolff[2]

[1] Faculty of Informatics, Universität Karlsruhe (TH), KIT, Germany
{katz,krug,rutter}@iti.uka.de
[2] Institut für Informatik, Universität Würzburg, Germany

Abstract. In this paper, we explore a new convention for drawing graphs, the *(Manhattan-) geodesic* drawing convention. It requires that edges are drawn as interior-disjoint *monotone* chains of axis-parallel line segments, that is, as geodesics with respect to the Manhattan metric. First, we show that geodesic embeddability on the grid is equivalent to 1-bend embeddability on the grid. For the latter question an efficient algorithm has been proposed. Second, we consider *geodesic point-set embeddability* where the task is to decide whether a given graph can be embedded on a given point set. We show that this problem is $\mathcal{NP}$-hard. In contrast, we efficiently solve *geodesic polygonization*—the special case where the graph is a cycle. Third, we consider geodesic point-set embeddability where the vertex–point correspondence is given. We show that on the grid, this problem is $\mathcal{NP}$-hard even for perfect matchings, but without the grid restriction, we solve the matching problem efficiently.

1 Introduction

In this paper we consider a new convention for drawing graphs. One of the most popular conventions is the *orthogonal* drawing convention, which requires edges to be drawn as interior-disjoint *rectilinear* chains, that is, chains of axis-parallel line segments. Restricting the number of edge directions potentially yields very clear drawings. We go a step further and insist that, additionally, edges are drawn as *monotone* chains. Such chains are called *Manhattan paths*. The idea behind monotonicity is that following the course of a monotone curve is potentially easier than following the course of a curve that is allowed to make detours. Manhattan paths are geodesics with respect to the Manhattan metric. Therefore we name our new convention the *(Manhattan-) geodesic* drawing convention.

In the Euclidean plane, geodesics are straight-line segments, and the classic result of König, Fáry, and Stein says that the class of graphs that have a straight-line drawing is exactly the class of planar graphs. Since there are efficient (linear-time) planarity-testing algorithms, we can decide efficiently whether a given graph has a Euclidean-geodesic drawing. We consider the same problem, which we call (MANHATTAN-) GEODESIC EMBEDDABILITY, with respect to the Manhattan distance. As an example take K_4, the complete graph on four vertices, which has a geodesic drawing in the Euclidean plane but not in the

D. Eppstein and E.R. Gansner (Eds.): GD 2009, LNCS 5849, pp. 207–218, 2010.

Manhattan plane. To avoid problems of drawing resolution, both questions are also interesting on the grid. The Euclidean case has been solved, for example, by Schnyder [14] who can draw any planar n-vertex graph on a grid of size $(n-2) \times (n-2)$, which is asymptotically optimal in the worst case.

Fixed point set. Next, we consider the setting where we are given not just a graph, but also a set of points (in the plane or on the grid) to which the vertices of the graph must be brought into correspondence. We call this problem GEODESIC POINT-SET EMBEDDABILITY. Kaufmann and Wiese [7] considered point-set embeddability (PSE) with respect to the *polyline drawing convention*. They showed that it is $\mathcal{NP}$-hard to decide whether a graph can be embedded on a point set with at most one bend per edge and that two bends suffice for any planar graph and any point set. Cabello [1] showed that it is $\mathcal{NP}$-hard to decide whether a planar graph has a straight-line embedding on a given point set.

A special case of both the straight-line and the orthogonal drawing convention has also been considered. Rappaport [12] showed that it is $\mathcal{NP}$-hard to decide whether a set P of n points has an *orthogonal polygonization*, that is, whether the n-cycle can be realized on P using horizontal or vertical edges only. O'Rourke [9] proved that if one forbids 180°-degree angles in the vertices, then there exists at most one simple rectilinear polygon with vertex set P. He also showed how to reconstruct the polygon from P in $O(n \log n)$ time. We refer to Demaine's survey [2] about problems related to polygonization.

PSE with the same drawing convention but with respect to a different graph class—perfect matchings—was considered by Rendl and Woeginger [13]. They showed that given a set of n points in the plane, one can decide in $O(n \log n)$ optimal time whether each point can be connected to exactly one other point with an axis-parallel line segment. They also showed that the problem becomes hard if one insists that the segments do not cross. Hurtado [5] gave a simple $O(n \log n)$-time algorithm for the same problem under the geodesic drawing convention. The idea is to alternatingly go up and down the occupied grid columns.

Fixed correspondence. We further restrict the placement of the vertices by making the bijection between vertices and points part of the input. We call the resulting problem LABELED GEODESIC PSE. A special case of this problem (where the graph is a perfect matching) has applications in VLSI layout. Insisting on geodesic connections makes sure that signals reach their destinations as fast as possible. For example, a popular, but more restrictive wiring technique in VLSI layout, *single-bend wiring*, uses special geodesic connections. Raghavan et al. [11] have shown that one can decide our perfect matching problem efficiently when insisting on at most one bend per edge.

For the same problem with given vertex–point correspondence but under the polyline drawing convention, Pach and Wagner [10] showed that it is possible to embed any planar graph on any set of points, but they also showed that some edges may require $\Omega(n)$ bends. Goaoc et al. [4] showed that it is $\mathcal{NP}$-hard to decide whether a given graph can be 1-bend embedded on a given set of points with given vertex–point correspondence.

Table 1. Overview over results in geodesic embeddability; hard is short for $\mathcal{NP}$-hard

	GEODESIC EMBEDDABILITY	GEODESIC POINT-SET EMBEDDABILITY		
		unrestricted	labeled (on grid)	labeled (off grid)
planar graph	$\mathcal{P}$ [Thm. 1]	hard [Thm. 2]	hard [Thm. 4]	open
matching	trivial	$\mathcal{P}$ [5]	hard [Thm. 4]	$\mathcal{P}$ [Thm. 5]
polygonization	trivial	$\mathcal{P}$ [Thm. 3]	open	open

Our Contribution. Drawing graphs with (Manhattan) geodesics opens up a large new field of research; we have done the following first steps.

- We show that GEODESIC EMBEDDABILITY on the grid is equivalent to deciding whether the given graph has a rectilinear *one*-bend drawing on the grid, see Section 2. Liu et al. [8] proposed an algorithm to decide the latter question efficiently. It is easy to see that a rectilinear one-bend drawing of an n-vertex graph fits on the $n \times n$ grid.
- We then prove that GEODESIC PSE is $\mathcal{NP}$-hard on (and off) the grid, reducing (in two steps) from HAMILTONIAN CYCLE, see Section 3. In contrast, we give a complete and easy-to-check characterization of all *yes*-instances of GEODESIC POLYGONIZATION, which is the special case of GEODESIC PSE where the input graph is restricted to a cycle.
- We show that LABELED GEODESIC MATCHING on the grid is $\mathcal{NP}$-hard by reduction from 3-PARTITION, see Section 4. This implies hardness of LABELED GEODESIC PSE on the grid. Our proof vitally exploits the space limitation of the grid. On the other hand, we show that LABELED GEODESIC MATCHING becomes easy if we loosen or drop this limitation.

We give a list of results and open questions in geodesic embeddability in Table 1. In the remainder of the paper, by a *grid geodesic* (or, even shorter, a *geodesic*) we mean a Manhattan-geodesic connecting two grid points on the grid. A *geodesic grid embedding* (or *geodesic embedding* for short) of a graph G is a drawing of G such that the vertices of G are mapped to grid points and the edges of G are mapped to interior-disjoint grid geodesics.

2   Geodesic Embeddability

In this section we ask whether a given planar graph has a geodesic embedding on the grid, that is, we allow the vertices to be mapped to arbitrary grid points. Clearly, this question makes only sense for graphs of maximum degree 4, but K_4, for instance, does *not* have a geodesic embedding on the grid.

In the following, we show that a graph has a geodesic embedding on the grid if and only if it has an orthogonal embedding on the grid with at most one bend per edge. Liu et al. [8] characterized planar graphs which are orthogonally 1-bend embeddable and proposed an efficient decision algorithm for this problem. Hence, we have the somewhat surprising result that we can efficiently recognize graphs that admit a geodesic embedding on the grid.

Theorem 1. *Let $G = (V, E)$ be a planar graph. Then G has a geodesic embedding on the grid if and only if G is 1-bend embeddable on the grid.*

Proof. The "if"-direction is trivially true, so we immediately turn to the "only if"-direction. Suppose that G has a geodesic embedding $\mathcal{E}$ on the grid. We turn $\mathcal{E}$ into an *orthogonal representation* as introduced by Tamassia [15]. Such a representation consists of lists, one for each face of the given embedding. The list for a face f has, for each edge e incident to f, an entry describing (a) the shape of e in terms of left ($-90°$) and right ($+90°$) turns, and (b) the angle that the edge makes with its successor in the cyclic order of the edges around f.

Since $\mathcal{E}$ is geodesic, the angles along each edge sum up to a value in $\{-90°, 0°, +90°\}$. From the representation of $\mathcal{E}$ we compute a new representation where we replace the shape entry of each edge by the corresponding sum. The result is a valid representation since for each face the sum of the inner angles remains the same and for each vertex the sum of the angles between consecutive incident edges also remains the same. Since the new representation is valid, Tamassia's flow network [15] yields the corresponding (1-bend) embedding of G.   □

3   Geodesic Point-Set Embeddability

In this section, we ask whether a given planar graph can be embedded on a given set of grid points. We assume that we are not given a bijection between vertices and points.

First, we show that this problem, GEODESIC PSE, is $\mathcal{NP}$-hard by reduction from the problem HAMILTONIAN CYCLE COMPLETION (HCC), which is $\mathcal{NP}$-hard [6]. Our proof also works in the case where the (Manhattan-) geodesics are not restricted to the grid. HCC is defined as follows. Given a non-Hamiltonian cubic graph G, decide whether G has two vertices u and v such that $G + uv$ (i) is planar, (ii) has a Hamiltonian cycle H, and (iii) has an embedding such that u and v are adjacent to at most two faces on the same side of H.

Theorem 2. GEODESIC PSE *is $\mathcal{NP}$-hard, even for subdivisions of cubic graphs.*

Proof. Our proof is by reduction from HCC. Given an instance $G = (V, E)$ of HCC, note that $n = |V|$ is even and let $k = \frac{n}{2} + 1$. Given three non-negative integers k_0, k_1, k_2, let $P_0 = \{(-j, 0) \mid j = 0, \ldots, k_0 - 1\}$, $P_1 = \{(j, nj) \mid j = 1, \ldots, k_1\}$, $P_2 = \{(j, -nj) \mid j = 1, \ldots, k_2\}$, and $P(k_0, k_1, k_2) = P_0 \cup P_1 \cup P_2$, see Fig. 1a. Note that the points in $P(k_0, k_1, k_2)$ are placed such that between any two consecutive non-empty rows of the integer grid there are $n - 1$ empty rows. We now construct a graph $G' = (V', E')$ by splitting every edge of G by a vertex of degree 2. This yields $|V'| = |V| + |E| = 2n - 1 + k$. In the following, we show that G' can be embedded on $P(2n - 1, k_1, k_2)$ for some k_1, k_2 with $k_1 + k_2 = k$ if and only if G is a *yes*-instance of HCC.

Assume G is a *yes*-instance of HCC. Then there is a pair $\{u, v\}$ of vertices such that $G + uv$ contains a Hamiltonian cycle and u and v are incident to two faces on either side of this cycle. Without loss of generality, we can assume that

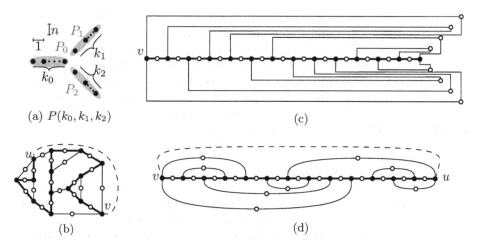

Fig. 1. Reduction of HCC to GEODESIC PSE

uv is incident to the outer face. An example of a plane graph G' is depicted in Fig. 1b; the splitting nodes are marked with circles, the original nodes of G with black disks. Maintaining the combinatorial embedding, we can embed the Hamiltonian path connecting u and v including its splitting nodes on a set of $2n - 1$ points on a horizontal line as in Fig. 1d. We embed the faces inside the cycle above the path and the faces outside the cycle below. Since each vertex of G' has degree at most 3, each vertex has at most one edge going up or down—except u and v, which both have exactly one edge going up and one going down. Set k_1 and k_2 to the numbers of edges inside and outside the cycle, respectively. Then we can map the splitting vertices of the remaining edges to the point sets P_1 and P_2, and route the edges as follows, see Fig. 1c. Each splitting node v that is mapped to a point in $P_1 \cup P_2$ has two neighbors, a left neighbor v^- and a right neighbor v^+ (according to their x-coordinates). We route the edge vv^- with one bend and the edge vv^+ with two bends. Note that the empty rows leave enough space for all horizontal edge segments.

Conversely, assume G' has a geodesic embedding on $P(2n - 1, k_1, k_2)$ with $k_1 + k_2 = k$. Then, the k vertices that are mapped to points in $P_1 \cup P_2$ are incident to at most $2k = n + 2$ edges. This is due to the fact that each such edge has its lexicographically larger endpoint in either P_1 or P_2, and we claim that no point in $P_1 \cup P_2$ can be adjacent to more than two lexicographically smaller points. To see the claim, note that for any point $v \in P_1$ the set of lexicographically smaller points is contained in the third quadrant with respect to v. Clearly, at most two geodesics can go from v to points in any fixed quadrant. For points in P_2, the argument is symmetric. Thus our claim holds.

Since G is cubic, G' has $3n$ edges. This leaves $3n - (n + 2) = 2n - 2$ edges incident to points in P_0 only. Since $|P_0| = 2n - 1$, P_0 induces a path π that alternates between vertices of degree 3 (original nodes) and degree 2 (splitting nodes). There are two possibilities: either both endpoints—call them s and t—

have degree 2 or both have degree 3. In the former case, π would contain $n - 1$ degree-3 vertices, and s and t would be adjacent to the only remaining degree-3 vertex (not in P_0). This would mean that G is Hamiltonian—contradiction.

Thus we may assume that s and t have degree 3. In this case, π witnesses a Hamiltonian path connecting s and t in G. This Hamiltonian path can be completed to a Hamiltonian cycle by an edge through the outer face of G. Since both u and v are incident to one edge pointing up and one edge pointing down from the path, they are incident to two faces on either side of the cycle in this embedding. This shows that G is indeed a *yes*-instance of HCC. $\qquad\square$

Now we turn to the case in which the instance consists of a simple cycle. We show that this problem, which we name GEODESIC POLYGONIZATION, can be solved efficiently. We start with a simple characterization of the *yes*-instances. To this end, we partition the grid points into two groups as follows. Let B be an axis-aligned rectangle. We say that a grid point p in B is *even* if its rectilinear distance to the lower left corner of B is even. Otherwise, we say that p is *odd*. We call a set of points *degenerate* if the set is contained in an axis-parallel line. It is clear that a degenerate point set does not have a polygonization. We now characterize all point sets that do have a polygonization. The proof, which is omitted here due to space limitations, is constructive, see our technical report [6]. It yields an efficient algorithm that computes a geodesic polygonization for any set of grid points with the given properties.

Theorem 3. *Let P be a non-degenerate set of points on the grid, let $\mathcal{B}(P)$ be the bounding box of P, and let h and w be the number of rows and columns spanned by $\mathcal{B}(P)$, respectively. Then P has a geodesic polygonization if and only if either (i) h or w is even or (ii) P does not contain all even points w.r.t. $\mathcal{B}(P)$.*

4   Labeled Geodesic Point-Set Embeddability

In Section 3, we showed that GEODESIC PSE is $\mathcal{NP}$-hard. In this section, we study the variant where the vertex–point correspondence is given. First, we settle the complexity of the problem on the grid.

Theorem 4. LABELED GEODESIC PSE *on the grid is $\mathcal{NP}$-hard, even if the given graph is a perfect matching.*

Proof. We reduce 3-PARTITION to LABELED GEODESIC MATCHING (LGM), which is a special case of LABELED GEODESIC PSE. An instance of 3-PARTITION consists of a multiset $A = \{a_1, \ldots, a_{3m}\}$ of $3m$ positive integers, each in the range $(B/4, B/2)$, where $B = (\sum A)/m$, and the question is whether there exists a partition of A into m subsets $A_1, \ldots, A_m$ of A, each of cardinality three, such that the sum of the numbers in each subset is B. Since 3-PARTITION is *strongly* $\mathcal{NP}$-hard [3], we may assume that B is bounded by a polynomial in m.

Based on an instance A of 3-PARTITION, we now construct an instance M of LGM consisting of pairs of grid points such that M is a *yes*-instance of LGM

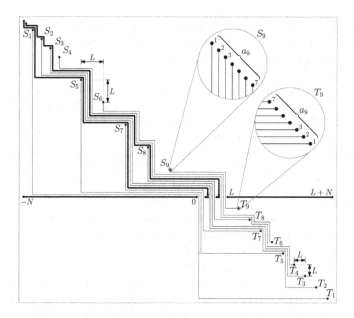

Fig. 2. Example of the reduction from 3-PARTITION to LGM using $A_1 = \{a_1, a_5, a_7\}$, $A_2 = \{a_2, a_3, a_8\}$, and $A_3 = \{a_4, a_6, a_9\}$ (not to scale)

if and only if A is a *yes*-instance of 3-PARTITION. Figure 2 shows an example instance M. The instance M consists of three types of point pairs.

The first type represents the numbers in A. We define $3m$ sets $S_1, \ldots, S_{3m}$ of grid points, all lying on the diagonal $\ell : y = -x$, in this order from left to right. For $1 \le i \le 3m$, the points in S_i occupy a_i consecutive grid points, and two consecutive sets are separated by a large gap of $L = Bm + m - 1$ grid points. The gap between the last point of S_{3m} and the origin is also L. The points in the sets $T_{3m}, \ldots, T_1$ lie on the line $\ell' : y = -x + L$, in this order from left to right. Again, points within a set are consecutive grid points, and between consecutive sets there are large gaps of L grid points. The matching is as follows. For $1 \le i \le 3m$ and for $1 \le j \le a_i$, the j-th point in S_i (counting from the left) matches the j-th point in T_i (counting from the *right*). The a_i point pairs in $S_i \cup T_i$ represent the number a_i.

The second type of point pairs forms a sort of "dot mask", the heart of our construction. These pairs lie on the x-axis. The geodesics between them are obviously line segments and pairwise disjoint. The leftmost segment goes from $-N$ to 0, where $N = 3mL + mB + 2(m - 1)$. The following $m - 1$ segments have unit length and leave gaps of width B. The rightmost segment goes from L to $L + N$.

The third type of point pairs gives rise to geodesics that resemble "fences" ensuring that all geodesics that represent a number from A go through the same gap in the mask. There are $m - 1$ such pairs. Their upper endpoints are consecutive grid points on the diagonal ℓ. They lie above the points in S_1, leaving a gap of $m-1$ grid points. The corresponding lower endpoints lie one unit above the, say, right endpoints of the unit-length segments on the x-axis. The matching is as follows: from left to right and for $1 \leq j \leq m-1$, the j-th upper endpoint matches the j-th lower endpoint.

It is easy to see that any geodesic embedding of M induces a partition of A: due to the fences, all edges corresponding to the same element of A must be routed through the same gap of the dot mask, each of the m gaps has width B, and each of the mB edges must go through some gap.

Conversely, given a partition, we construct a geodesic embedding of the matching. We start by drawing the dot mask whose layout only depends on the numbers B and m. Then, we analyze the first subset of the partition, A_1, and connect the points $S^1 = \bigcup_{a_j \in A_1} S_j$ to the corresponding points in $T^1 = \bigcup_{a_j \in A_1} T_j$, starting with the leftmost point in S^1 and the rightmost point in T^1. For each connection, we use the bottommost geodesic that goes above all geodesics we have drawn so far. Next, we draw the first (that is, leftmost) fence. Also in this case, we use the bottommost geodesic that goes above all geodesics we have drawn so far. We repeat these two steps, connecting the points corresponding to a subset of the partition and drawing a fence. Since we left enough horizontal and vertical space, this process does not get stuck. The fences direct the next B connections into the gaps, which have exactly the right width.

Since we assumed that B is polynomial in m, the numbers L and N, which determine the grid size needed by M, are also polynomial in m. Given an embedding, the partition can be constructed from it efficiently, and vice versa. Thus our reduction is polynomial.   □

Next, we show that LGM becomes easy if we loosen or drop the space limitation of the grid. We call an instance of LGM—a set M of n pairs of grid points (we call such pairs also *edges*)—*sparse* if the minimum distance between any two occupied columns and between any two occupied rows is at least $n+1$. In the remainder of this section, we give an efficient algorithm that solves sparse instances of LGM. Clearly, the algorithm can also be used for an instance that does not "live" on the grid, by underlaying the instance with a fine enough grid.

We say that an edge $e \in M$ is *downward* if its lexicographically larger endpoint e^+ lies below its lexicographically smaller endpoint e^-, otherwise e is *upward*. Clearly, M does not have a geodesic embedding if the bounding box of an edge *crosses* (that is, splits into two connected components) the bounding box of another edge. This can be tested easily, so from now on we assume that M is *non-crossing*, that is, no two bounding boxes of edges in M cross.

Let g and g' be any two geodesics. We say that g is *below* g' if there is a vertical line that intersects g below g'. We say that an edge $e \in M$ is *strictly below* an edge $e' \in M$ if, for any geodesic embedding γ of the two edges, $\gamma(e)$ is below $\gamma(e')$.

The *precedence graph* π_M is a directed graph whose vertex set is M and whose edges represent the strictly-below relationship. The precedence graph can be computed efficiently by a simple line sweep. It is clear that M does not have

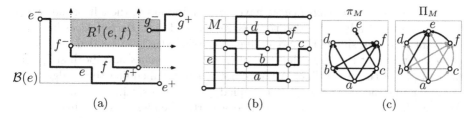

Fig. 3. (a) Critical region $R^\uparrow(e,f)$ with $g \in M^\uparrow(e,f)$, (b) a matching M, (c) the precedence graph π_M and the extended precedence graph Π_M of M

a geodesic embedding if π_M contains a cycle. Now we construct a supergraph of π_M whose acyclicity is *equivalent* to the realizability of M.

For any point $a = (x_a, y_a) \in \mathbb{R}^2$, let $Q_1(a) = \{(x,y) \in \mathbb{R}^2 \mid x_a \leq x, y_a \leq y\}$ be the *first quadrant w.r.t.* a and define the other three quadrants w.r.t. a accordingly, in counterclockwise order. Let e be a downward edge and let f be any other edge in M. For such a pair (e, f), we define the *upper critical region of e and f* as $R^\uparrow(e,f) = (Q_1(f^-) \cup Q_1(f^+)) \cap \mathcal{B}(e)$ (see Fig. 3a) and the *lower critical region of e and f* as $R^\downarrow(e,f) = (Q_3(f^-) \cup Q_3(f^+)) \cap \mathcal{B}(e)$. The critical regions for upward edges are defined by replacing Q_1 by Q_2 and Q_3 by Q_4. Let $M^\uparrow(e,f)$ and $M^\downarrow(e,f)$ be the sets of edges in M with at least one endpoint in $R^\uparrow(e,f)$ and $R^\downarrow(e,f)$, respectively.

Let $G = (M, E)$ be a directed graph with vertex set M. We say that an edge (e, f) of G *produces* the edge (e, g) if $g \in M^\uparrow(e,f)$ and the edge (g, f) if $g \in M^\downarrow(f, e)$. Now the *extended precedence graph* Π_M is the closure of π_M with respect to production.

Lemma 2. *If Π_M contains a cycle, M does not admit a geodesic embedding.*

Proof. We claim that an edge (e, f) in Π_M means that if M has some geodesic embedding γ, then $\gamma(e)$ is below $\gamma(f)$. Clearly, the claim holds for every edge in π_M. Now suppose edge (e, g) has been produced by the edge (e, f) and M has a geodesic embedding γ. Then we know that $g \in M^\uparrow(e, f)$. Assume that e is downward. By definition, at least one of the endpoints of g—call it q—lies in $Q_1(p) \cap \mathcal{B}(e)$, where p is one of the endpoints of f. (In particular, p lies in $\mathcal{B}(e)$, otherwise $Q_1(p) \cap \mathcal{B}(e)$ would be empty.) Due to the existence of edge (e, f), $\gamma(e)$ is below $\gamma(f)$ and thus below p. Since e is downward, $\gamma(e)$ must also be below q and hence below $\gamma(g)$. The case that e is upward and the case that (e, f) has produced an edge (g, f) can be argued symmetrically. Now induction yields the claim.   □

By Lemma 2, for M to have a geodesic embedding, it is necessary that Π_M is acyclic. We now show that this condition is also sufficient, by giving an algorithm that computes an embedding if Π_M is acyclic.

The algorithm sweeps a vertical line from left to right over the plane. Events occur only at the vertices of M. During the sweep, we partition the edges in M

into three groups. *Completed edges* have both endpoints to the left of the sweep-line. We have already embedded these edges as geodesics. *Partial edges* have one endpoint on either side of the sweep-line. A partial edge is embedded as a *partial geodesic* ending at the sweep-line. Finally, *untouched edges* have both endpoints to the right of the sweep-line. We have not started embedding these edges yet.

Let c and c' be two consecutive occupied grid columns, with c to the left of c'. Assume that we have already computed a partial geodesic embedding up to c. Let $u_1, \ldots, u_s$ be the set of upward partial edges which do not end at c, sorted from bottom to top (including the edges starting at c). We process the edges in this order. Assuming that we have already embedded $u_1, \ldots, u_{i-1}$, we proceed depending on whether u_i ends at c' or not.

If u_i ends at c', we embed u_i as the bottommost geodesic just above all edges $u_1, \ldots, u_{i-1}$, that is, there is no geodesic for u_i containing a point strictly below this geodesic. Hence, u_1 has a vertical segment only on the last unoccupied column to the left of c'. By induction, u_i has vertical segments only on the last i unoccupied columns.

If u_i does not end at c', let U_i denote the set of edges preceding u_i in Π_M. Then u_i must necessarily be embedded above all edges in U_i. We embed u_i as the bottommost geodesic above all edes $u_1, \ldots, u_{i-1}$ and above all endpoints of edges in U_i which are on c'. If there is no such restriction, we embed u_i as a straight-line segment.

We then proceed similarly with the downward edges. Let $d_1, \ldots, d_t$ be the set of partial downward edges that do not end at c, sorted from top to bottom, that is, sorted inversely to the upward edges. Let D_i denote the set of edges succeeding d_i in Π_M. We embed each edge d_i as the topmost geodesic below the geodesics $d_1, \ldots, d_{i-1}$ and below all endpoints of edges in D_i. As before, d_i has vertical segments only on the last i columns left of c'. A sample output of the algorithm is illustrated in Fig. 3b.

Since there are at most n edges by the definitions of the top- and bottommost geodesics, we need at most n unoccupied columns between c and c'. Since M is sparse, there are at least n unoccupied rows between two occupied rows on c', so we can embed the given edges between two occupied points on c'.

Theorem 5. *Let M be a sparse non-crossing matching with n edges on the grid. Then M has a geodesic embedding if and only if Π_M is acyclic. In $\mathcal{O}(n^3)$ time, we can compute a geodesic embedding of M or prove that no such embedding exists.*

Proof. The "only if" part has been proved in Lemma 2, so we immediately turn to the "if" part. We first compute Π_M. If Π_M contains a cycle, we reject. Otherwise, we use the above embedding algorithm to compute an embedding γ of M.

Concerning running time, it is clear that π_M can be computed by a simple plane sweep in $\mathcal{O}(n^2)$ time. For computing Π_M, we need $\mathcal{O}(n^2)$ iterations, one for each edge. An iteration takes linear time since all endpoints in the corresponding two critical regions can be reported in linear total time. The embedding algorithm runs in $\mathcal{O}(n^2)$ time.

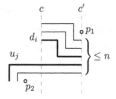

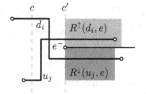

Fig. 4. Relative positions of downward/ upward partial edges

Fig. 5. Impossible crossing of upward/ downward partial edges

To show that γ is plane and geodesic, we maintain the following invariants during the execution of the algorithm.

1. All completed and partial edges are (partially) embedded as geodesics.
2. For every partial downward edge the partial embedding is not upward; vice versa for partial upward edges.
3. If the left endpoint of some downward edge e is above the left endpoint of an edge e' and the (partial) geodesic for e is below e' in the embedding, then Π_M contains a path from e to e'. A symmetric statement holds for upward edges.
4. The partial embedding respects all constraints corresponding to edges in Π_M.
5. No two (partial) geodesics intersect.

It is easy to see that invariants 1–4 are maintained by the algorithm. Invariant 1 yields that γ is geodesic. It remains to show that γ is plane (invariant 5).

Suppose that the algorithm introduces a crossing when going from grid column c to grid column c' and there is no crossing to the left of c. By definition of the top- and bottommost geodesic there is no intersection between two upward or two downward edges, respectively. That is, the algorithm can only introduce intersections between an upward and a downward edge. Let d_i be a downward edge and let u_j be an upward edge such that d_i and u_j intersect. Then d_i must be above u_j on c, otherwise there would be no crossing between the two edges. We now make a case distinction depending on whether or not there is an edge $e \in M$ with left endpoint e^- on c' such that d_i must be embedded below e and u_j must be embedded above e, that is, (d_i, e) and (e, u_j) are in Π_M.

First assume that there is no such edge e. Let V'_M be the points of V_M that lie to the left of or on c'. Let p_1 be the lowest point of V'_M such that d_i lies below p_1. Similarly, let p_2 be the highest point of V'_M such that u_j lies above p_2. Clearly, p_2 is below p_1 and by assumption there are at least n unoccupied rows between p_1 and p_2. By definition of the top- and bottommost geodesic for d_i and u_j the two edges do not cross, see Fig. 4.

Now assume that there is an edge e with (d_i, e) and (e, u_j) in Π_M. If the left endpoints d_i^- and u_j^- are in the same column, there is an edge (u_j, d_i) in π_M, which induces a cycle in Π_M, contradicting the assumption that Π_M is acyclic. Otherwise, the endpoints d_i^- and u_j^+ are in different columns.

Assume that d_i^- is to the left of u_j^-. Since d_i is above u_j, its right endpoint d_i^+ cannot be in the critical region $R^\downarrow(u_j, e)$ since this would imply that d_i must be below u_j, which violates invariant 4 at c. Hence, d_i^+ must be to the right of u_j^+. In this case, however, u_j^+ is in $R^\uparrow(d_i, e)$, that is, d_i must be below u_j, which again violates invariant 4 at c (see Fig. 5). The case that d_i^- is to the right of u_j^- is similar. $\qquad\qquad\Box$

Acknowledgments. We thank the anonymous referees for their detailed and helpful comments. AW thanks Ferran Hurtado for inviting him to a workshop in 2006, Manuel Abellanas for bringing LABELED GEODESIC PSE to his attention there, and Stefan Langerman and Pat Morin for discussions about the off-the-grid version of GEODESIC MATCHING.

References

1. Cabello, S.: Planar embeddability of the vertices of a graph using a fixed point set is NP-hard. J. Graph Algorithms Appl. 10(2), 353–363 (2006)
2. Demaine, E.: Simple polygonizations (2007), http://erikdemaine.org/polygonization/ (Accessed May 30, 2009)
3. Garey, M.R., Johnson, D.S.: Computers and Intractability. A Guide to the Theory of NP-Completeness. W.H. Freeman and Company, New York (1979)
4. Goaoc, X., Kratochvíl, J., Okamoto, Y., Shin, C.-S., Spillner, A., Wolff, A.: Untangling a planar graph. Discrete Comput. Geom (2009), http://dx.doi.org/10.1007/s00454-008-9130-6
5. Hurtado, F.: Personal communication (2006)
6. Katz, B., Krug, M., Rutter, I., Wolff, A.: Manhattan-geodesic point-set embeddability and polygonization. Technical Report 2009-17, Universität Karlsruhe (2009), http://digbib.ubka.uni-karlsruhe.de/volltexte/1000012949
7. Kaufmann, M., Wiese, R.: Embedding vertices at points: Few bends suffice for planar graphs. J. Graph Algorithms Appl. 6(1), 115–129 (2002)
8. Liu, Y., Marchioro, P., Petreschi, R., Simeone, B.: Theoretical results on at most 1-bend embeddability of graphs. Acta Math. Appl. Sinica (English Ser.) 8(2), 188–192 (1992)
9. O'Rourke, J.: Uniqueness of orthogonal connect-the-dots. In: Toussaint, G. (ed.) Computational Morphology, pp. 97–104. North-Holland, Amsterdam (1988)
10. Pach, J., Wenger, R.: Embedding planar graphs at fixed vertex locations. Graph. Combinator. 17(4), 717–728 (2001)
11. Raghavan, R., Cohoon, J., Sahni, S.: Single bend wiring. J. Algorithms 7(2), 232–257 (1986)
12. Rappaport, D.: On the complexity of computing orthogonal polygons from a set of points. Technical Report SOCS-86.9, McGill University, Montréal (1986)
13. Rendl, F., Woeginger, G.: Reconstructing sets of orthogonal line segments in the plane. Discrete Math 119(1-3), 167–174 (1993)
14. Schnyder, W.: Embedding planar graphs on the grid. In: Proc. 1st ACM-SIAM Symp. on Discrete Algorithms (SODA 1990), pp. 138–148 (1990)
15. Tamassia, R.: On embedding a graph in the grid with the minimum number of bends. SIAM J. Comput. 16(3), 421–444 (1987)

Orthogonal Connector Routing

Michael Wybrow[1], Kim Marriott[1], and Peter J. Stuckey[2]

[1] Clayton School of Information Technology,
Monash University, Clayton, Victoria 3800, Australia
{Michael.Wybrow,Kim.Marriott}@infotech.monash.edu.au
[2] National ICT Australia, Victoria Laboratory,
Department of Computer Science & Software Engineering,
University of Melbourne, Victoria 3010, Australia
pjs@csse.unimelb.edu.au

Abstract. Orthogonal connectors are used in a variety of common network diagrams. Most interactive diagram editors provide orthogonal connectors with some form of automatic connector routing. However, these tools use *ad-hoc* heuristics that can lead to strange routes and even routes that pass through other objects. We present an algorithm for computing optimal object-avoiding orthogonal connector routings where the route minimizes a monotonic function of the connector length and number of bends. The algorithm is efficient and can calculate connector routings fast enough to reroute connectors during interaction.

1 Introduction

Most interactive diagram editors provide some form of automatic connector routing between shapes whose position is fixed by the user. Usually the editor computes an initial automatic route when the connector is created and updates this each time the connector end-points (or attached shapes) are moved. Orthogonal connectors, which consist of a sequence of horizontal and vertical line segments, are a particularly common kind of connector, used in ER and UML diagrams among others. However, in current diagramming tools that we are aware of, automatic routing of orthogonal connectors uses *ad-hoc* heuristics that lead to aesthetically unpleasing routes and unpredictable behaviour.

For example, the graphic editors OmniGraffle Pro 5.1.1, and Dia 0.97, provide automatic orthogonal connector routing but these routes may overlap other objects in the diagram. Both Microsoft Visio 2007, and ConceptDraw Pro 5 provide object-avoiding orthogonal connector routing but in both applications connector routing does not use a predictable heuristic, such as minimizing distance or number of segments. Furthermore, the routes are mostly updated only after object movement has been completed, rather than as the action is happening. The Graph layout library yFiles[1] and demonstration editor yEd offers orthogonal edge routing but routing is not maintained throughout further editing.

Thus, we know of no interactive diagram authoring tool which ensures that the orthogonal connectors are optimally routed in any meaningful sense. On the

---

[1] http://www.yworks.com/products/yfiles/

D. Eppstein and E.R. Gansner (Eds.): GD 2009, LNCS 5849, pp. 219–231, 2010.
© Springer-Verlag Berlin Heidelberg 2010

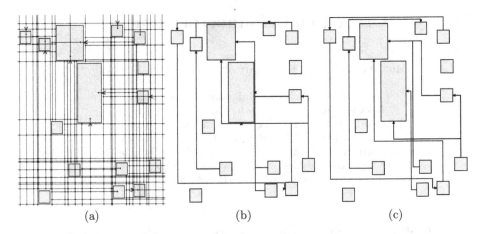

(a)   (b)   (c)

Fig. 1. Three stage routing: (a) the orthogonal visibility graph, (b) the optimal connector routes, (c) the final routes after centering and nudging. Arrows indicate routing direction for connectors.

other hand, automatic routing of poly-line connectors is better supported: two tools, Dunnart and Inkscape, provide real-time poly-line connector routing which is optimal in the sense that it minimizes edge bends and connector length. Both use the connector routing library `libavoid`[2] which has three steps in connector routing [1]. The first stage is to compute a visibility graph for the diagram which contains a node for each vertex of each object in the diagram and an edge between two nodes iff they are mutually visible. The second stage uses A* search to find the optimal route through the visibility graph for each connector. The third stage computes the visual representation of the connector. This three step approach is also used in the Spline-o-matic library[3] developed for GraphViz which supports poly-line and Bezier curve edge routing [2].

In this paper we describe how we have extended the connector routing library `libavoid` to support orthogonal connector routing. The main contribution is to show that a similar three step process to that used for poly-line connector routing can also be used for optimal orthogonal connector routing. We introduce the *orthogonal visibility graph* in which edges in the graph represent horizontal or vertical lines of visibility from the vertices and connector ports of each object (Section 3). Connector routes are found using an A* search through the orthogonal visibility graph (Section 4). The algorithm is guaranteed to find a route for each connector that is optimal in the sense that it minimizes bends and overall connector length. Finally, the actual visual route is computed (Section 5). This step orders and *nudges* apart the connectors in shared segments so as to ensure that unnecessary crossings are not introduced and that crossings occur at the start or end of the shared segment. It also tries to ensure that connectors pass down the middle of "alleys" in the diagram when this does not lead to additional cost. Figure 1 shows the three step process for an example layout. Our algorithms are surprisingly efficient and fast enough to reroute connectors even during

---

[2] http://adaptagrams.sourceforge.net/libavoid/
[3] http://www.graphviz.org/Misc/spline-o-matic/

direct manipulation of reasonably sized diagrams, thus giving instant feedback
to the diagram author (Section 6).

2   Problem Statement

For simplicity we model objects by their bounding rectangle and assume for
the purposes of complexity analysis that the number of connector points on
each object is a fixed constant. Also for simplicity, we assume that connectors
must start and end at distinct connection points. In practice, connectors are not
always connected to objects and may have end-points which are not connection
points. This can be handled by adding an extra node to the visibility graph for
this endpoint.

We are interested in finding a poly-line route of horizontal and vertical segments for each connector. We wish to find routes that are short and which
have few bends. While we also wish to reduce connector crossings we will delay
consideration of this until Section 8. We assume our penalty function $p(R)$ for
measuring the quality of a particular route R is a monotonic function f of the
length of the path, $||R||$, and the number of bends (or equivalently segments) in
R, $bends(R)$, i.e. $p(R) = f(||R||, bends(R))$. We require that the routes are *valid*:
they do not pass through objects and only contain right-angle bends.

We use the *Manhattan distance* $||(v_1, v_2)||_1 = |x_1 - x_2| + |y_1 - y_2|$ to measure
the shortest orthogonal route between points $v_1 = (x_1, y_1)$ and $v_2 = (x_2, y_2)$.
We make use of 4 cardinal directions: $\mathbb{N}, \mathbb{S}, \mathbb{E}, \mathbb{W}$. We assume the functions *right*,
left, and *reverse* defined by the mappings:

$$right = \{\mathbb{N} \mapsto \mathbb{E}, \ \mathbb{E} \mapsto \mathbb{S}, \ \mathbb{S} \mapsto \mathbb{W}, \mathbb{W} \mapsto \mathbb{N}\}$$
$$left = \{\mathbb{N} \mapsto \mathbb{W}, \mathbb{E} \mapsto \mathbb{N}, \ \mathbb{S} \mapsto \mathbb{E}, \ \mathbb{W} \mapsto \mathbb{S}\}$$
$$reverse = \{\mathbb{N} \mapsto \mathbb{S}, \ \mathbb{E} \mapsto \mathbb{W}, \mathbb{S} \mapsto \mathbb{N}, \ \mathbb{W} \mapsto \mathbb{E}\}$$

We define the directions of point $v_2 = (x_2, y_2)$ from $v_1 = (x_1, y_1)$ as:

$$dirns(v_1, v_2) = \{\mathbb{N} \mid y_2 > y_1\} \cup \{\mathbb{E} \mid x_2 > x_1\} \cup \{\mathbb{S} \mid y_2 < y_1\} \cup \{\mathbb{W} \mid x_2 < x_1\}$$

Note $dirns(v_1, v_2) = \{D\}$ means v_2 is on the line in direction D drawn from v_1.

3   Orthogonal Visibility Graph

The basis for our approach is the observation that when finding routes minimizing the penalty function we need only consider routes in the orthogonal visibility
graph. This is defined as follows.

Let I be the set of *interesting points* (x, y) in the diagram, i.e. the connector
points and corners of the bounding box of each object. Let X_I be the set of x
coordinates in I and Y_I the set of y coordinates in I. The *orthogonal visibility
graph* $VG = (V, E)$ is made up of nodes $V \subseteq X_I \times Y_I$ s.t. $(x, y) \in V$ iff there
exists y' s.t. $(x, y') \in I$ and there is no intervening object between (x, y) and
(x, y') and there exists x' s.t. $(x', y) \in I$ and there is no intervening object
between (x, y) and (x', y). There is an edge $e \in E$ between each point in V to its
nearest neighbour to the north, south, east and west iff there is no intervening
object in the original diagram.

An example orthogonal visibility graph is shown in Figure 1(a). It is quite different to the standard (non-orthogonal) visibility graph used for poly-line routing. In particular, the standard visibility graph has $O(n)$ nodes if there are n objects in the diagram while the orthogonal visibility graph has $O(n^2)$ nodes. Both have $O(n^2)$ edges.

Observation: *Let R be a valid orthogonal route for a connector c. Then there exists a valid orthogonal route R' using edges in the orthogonal visibility graph for c s.t. $p(R') \leq p(R)$.*

Proof. We simply take the route R and "shrink" each segment on the route onto a path in the visibility graph to give R'. By construction R' is no longer than R and has no additional bends. □

The orthogonal visibility graph can be constructed using the following algorithm. It has three steps:

1. Generate the *interesting horizontal segments*

$$H_I = \{ ((x, y), (x', y)) \mid (x, y), (x', y) \in I \text{ s.t. } x \leq x'$$
$$\text{and there is no intervening object between } (x, y) \text{ and } (x', y) \}.$$

2. Generate the *interesting vertical segments*

$$V_I = \{ ((x, y), (x, y')) \mid (x, y), (x, y') \in I \text{ s.t. } y \leq y'$$
$$\text{and there is no intervening object between } (x, y) \text{ and } (x, y') \}.$$

3. Compute the orthogonal visibility graph by intersecting all pairs of segments from H_I and V_I. We note that this could be done lazily, however for simplicity we construct the entire visibility graph in one step.

Theorem 1. *The orthogonal visibility graph can be constructed in $O(n^2)$ time for a diagram with n objects using the above algorithm.*

Proof. The interesting horizontal segments can be generated in $O(n \log n)$ time where n is the number of objects in the diagram by using a variant of the line-sweep algorithm from [3,4]. This uses a vertical sweep through the objects in the diagram, keeping a horizontal "scan line" list of open objects with each node having references to its closest left and right neighbors. Interesting, horizontal segments are generated, when an object is opened, closed, or a connection point is reached. Dually, the interesting vertical segments can generated in $O(n \log n)$ time by using a horizontal sweep. The last step takes $O(n^2)$ time since there are $O(n)$ interesting horizontal and vertical segments. □

4   Routing the Connector

We use an $A^\star$ algorithm which iteratively builds longer and longer partial paths that start from the *source* node s until the *destination* node d is reached. Partial paths are stored in a priority queue and at each step the partial path with lowest *cost* is taken from the queue and expanded. The expanded nodes are placed in the queue. The process stops when the path chosen for expansion is already at

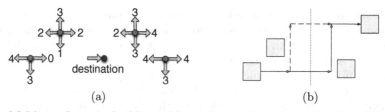

Fig. 2. (a) Minimal required additional bends for reaching the destination with correct direction from each point and direction. (b) The solid path is preferred to the dashed path since it is the "initially straighter" path. The dotted line shows the middle of the "alley" of possible paths for the middle segment of the connector.

d. The cost associated with each partial path is the cost of the partial path so far plus a lower bound on the remaining cost to the destination.

If we are only trying to minimize connector length, the only state we need to know about the partial path is the position of its end. However, if the number of bends is also part of the cost we also need to know the direction of the path. Thus, entries in the priority queue have form (v, D, l_v, b_v, p, c_v) where v is the node in the orthogonal visibility graph, D is the "direction of entry" to the node, l_v is the length of the partial path from s to v and b_v the number of bends in the partial path, p a pointer to the parent entry (so that the final path can be reconstructed), and c_v the cost of the partial path. There is at most one entry popped from the queue for each (v, D) pair. When an entry (v, D, l_v, b_v, p, c_v) is scheduled for addition to the priority queue, it is only added if no entry with the same (v, D) pair has been removed from the queue, i.e. is on the closed list. And only the entry with lowest cost for each (v, D) pair is kept on the priority queue. When we remove entry (v, D, l_v, b_v, p, c_v) from the priority queue we

1. add the neighbour (v', D) in the same direction with priority f$(l_v + ||(v, v')||_1 + ||(v', d)||_1, s_v + s_d)$;
2. conditionally add the neighbours $(v', right(D))$ and $(v', left(D))$ at right angles to the entry with priority f$(l_v + ||(v, v')||_1 + ||(v', d)||_1, s_v + 1 + s_d)$;

where s_d is the estimation of the remaining segments required for the route from (v', D') to (d, D_d). Since some edges are useful only when searching in a particular direction or for a specific point, we don't add the left or right neighbours if the extension of that visibility line neither passes right by an obstacle nor contains the target endpoint. The estimation of the remaining segments required is: $s_d =$

0. if $D' = D_d$ and $dirns(v', d) = \{D'\}$;
1. if $left(D_d) = D' \vee right(D_d) = D'$ and $D' \in dirns(v', d)$;
2. if $D' = D_d$ and $dirns(v', d) \neq \{D'\}$ but $D' \in dirns(v', d)$, or $D' = reverse(D_d)$ and $dirns(v', d) \neq \{D_d\}$;
3. if $left(D_d) = D' \vee right(D_d) = D'$ and $D' \notin dirns(v', d)$; and
4. if $D' = reverse(D_d)$ and $dirns(v', d) = \{D_d\}$, or $D' = D_d$ and $D' \notin dirns(v', d)$.

Figure 2(a) shows all the possible scenarios for determining the remaining minimal number of bends. We note that Miriyala *et al.* [5] use a similar cost.

Even taking into account number of bends, there are usually many alternate routes of the same cost from source to destination. To make the routing behaviour more predictable and faster we add a tie break for equal cost routes based on a time stamp of when the entry was added to the priority queue. This means that because the order in which neighbours is added is deterministic—straight, right, left—there is a slight preference for right turns and also that the latest path is extended in preference to earlier paths. See Figure 2(b).

The worst-case complexity of the A* algorithm is that of a priority queue based implementation of the shortest path algorithm over the orthogonal visibility graph. Thus:

Theorem 2. *The above algorithm will find an optimal valid route for a single connector through the orthogonal visibility graph in $O(n^2 \log n)$ time where the diagram has n objects.* □

5   Computing the Visual Representation

The third and last step in orthogonal connector routing is "nudging" of the connectors to compute their actual position in the drawing. The importance of this step is often overlooked, but feedback from users of Dunnart and Inkscape suggests that it has a significant impact on the perception of layout quality. It has two steps.

5.1   Ordering Shared Edges

The first aspect is determining the relative ordering of connectors in shared edges. A consequence of routing connectors along the orthogonal visibility graph is that multiple connectors will share edges of their paths. In order to make the connector route clearer we want to nudge these paths apart to make the distinct paths clear. It is important to do so in a manner which does not introduce unnecessary crossings or bends in segments.

We now explain our algorithm to generate a relative ordering of connectors in shared edges. Initially we construct the graph of shared edges, that is the subset of the edges in the visibility graph that have two or more connectors routed along that edge (plus their incident nodes). We process each connected component in the graph separately since each defines an independent subproblem in terms of the parts of connectors whose routes enter and exit this connected component of shared edges. Note that one connector may enter and exit the connected component multiple times in which case each sub-route is treated as a separate connector. Processing of each connected component has two steps.

We first try and assign a uniform *pseudo direction* for each of these connector sub-routes. This pseudo direction is independent of the actual direction of the connector—it is simply used for route adjustment. Choose an arbitrary connector sub-route A and fix its pseudo direction in an arbitrary direction. Now fix the pseudo direction of a connector sub-route that shares an edge with A to have the same direction as one of the shared edges. Follow the sub-route assigning the same pseudo-direction until there is a conflict in which case we mark the sub-route with a split point, reverse the pseudo-direction and continue following the sub-route. Continue this until all sub-routes segments have a pseudo direction. The whole process is $O(e)$ where e is the number of edges in all the sub-routes appearing in this tree.

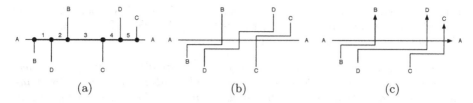

Fig. 3. (a) A set of orthogonal connectors which share edges, and (b) an ordering of shared edges to minimize crossings, (c) an order of shared edges that minimizes crossings and does not introduce additional segments

In practice, we have found that the pseudo-direction assignment for each connector sub-route is almost always *consistent*. We say a set of connector paths is *path consistent* if the pseudo-direction assignment is consistent for each connected component of the shared edge graph.

The next step is to determine for each shared edge a relative order, left to right along the pseudo direction, of each of the connectors that share that edge. We do this in an incremental fashion. Each edge starts with an empty sequence of connectors. We choose an, as yet unconsidered, connector sub-route and process each of its consistent sub-sections, one at a time (i.e. the sub-sections without split point. We insert this consistent sub-section in the ordering for each shared edge it makes use of. The key is that we will ensure that the necessary crossings of this sub-section with other connectors only occur at the end of the last (in the pseudo direction) shared edge between them, which is either at the end of the connector or at a split point.

Consider adding a connector c to a shared edge order O for edge e. We need to insert c in O in the appropriate place. There are three subsequences of connectors in $O = L +\!\!+ S +\!\!+ R$. Those that enter e (along the pseudo direction) from the left of c, L, those that enter in the same direction as c, S, and those that enter from the right, R. Now c already shares an edge e' with connectors in S, so we can project the order O' for this edge onto the connectors in $S \cup \{c\}$ to determine an order $S_L +\!\!+ [c] +\!\!+ S_R$.[4] The new order for edge s is hence $L +\!\!+ S_L +\!\!+ [c] +\!\!+ S_R +\!\!+ R$. This step is $O(e^2)$.

Example 1. Consider the tree of shared edges shown if Figure 3(a). The ordering of shared edges shown in Figure 3(b), has minimal connector crossings but adds two extra segments in the route for D. The algorithm proceeds as follows. We assign the pseudo direction left to right to connector A, and this propagates to the other edges as shown by the arrow heads in Figure 3(c). The tree is path consistent. We first add connector A as the unique route in each of edges 1–5. Next we add connector B. Since it enters from below it is ordered after A in edge 1 and 2, implicitly crossing A after edge 2. Similarly we add connector C. The resulting ordering is ([A,B],[A,B],[A],[A,C],[A,C]). Next we add connector D, it is added

---

[4] It may be that when starting from a split point that while c already shares an edge e' with connectors in S the ordering is not yet decided, in which case the ordering is determined by following the sub-route back across the split point and along the shared path to find the input ordering.

last on edges 2 and 3 but since it enters above C it is ordered between A and C in edge 4. The final resulting ordering is ([A,B],[A,B,D],[A,D],[A,D,C],[A,C]). The resulting diagram is shown in Figure 3(c).   □

Theorem 3. *If the shared edge graph is path consistent the above ordering algorithm produces segment orders with the minimal number of connector crossings, and all connector crossings are produced at the end of the last (in the pseudo direction) shared edge of the two connectors.*

Proof. (Sketch) Consider any pair of (sub-routes of) connectors A and B in a tree of shared edges. The algorithm ensures that the relative order of A and B is fixed in all their shared edges. By definition this order is defined by their left to right order on entry to the shared edge. Hence the two connectors can only cross at the exit of the shared edge, and only do so if that is necessary.   □

Theorem 4. *If the shared edge graph is planar then the above ordering algorithm produces a planar layout.*

Proof. (Sketch) The relative order of shared edged is always preserved from one endpoint of the connectors, thus a crossing will only be inserted if the relative order of the two endpoints is different, in which case the graph is not planar. □

5.2   Final Placement

The final step in the layout is to determine the exact coordinates of the orthogonal connector segments. This nudges connector routes a minimum distance apart to show the relative order of connectors with shared segments and also ensures that connectors pass down the middle of "alleys" in the diagrams when this does not lead to additional cost or additional edge crossings.

We collapse collinear segments in the connector routes into maximal horizontal and vertical segments. This means that segments in the path alternate horizontal and vertical alignment. We compute the horizontal and vertical position in separate passes. The horizontal pass works as follows and the vertical pass is symmetric.

1. Determine a desired horizontal position for all non-end segments in the connector. For the middle segment in an "S" or "Z" bend, this is the middle of the "alley" that the segment is in. For example, for the solid connector route shown in Figure 2(c), the dotted line shows the desired position for this segment. For the middle segment in an "⊐" or "⊏" bend, this is that of the vertex of the object that the segment bends around.
2. Generate a set of horizontal separation constraints to ensure that segments maintain their current relative horizontal ordering with each other and with the other objects in the diagram. In the case of shared segments the separation constraints impose the ordering determined previously. The constraints are designed to enforce non-overlap and also to stop segments passing through each other and so introducing additional connector crossings.
3. Project the desired values on to the separation constraints to find the horizontal position of the segments using the approximate projection algorithm *satisfy_VPSC* from [3,4].

The constraints and desired positions can be generated using a variant of the line-sweep algorithm from [3] in $O((n+s)\log(n+s))$ time where n is the number of diagram objects and s the total number of vertical connector segments. The approximate projection algorithm has $O((s+n)^2)$ worst-case complexity but in practice $O((s+n)\log(s+n))$ complexity [4].

6   Evaluation

We have implemented all algorithms in the open source libavoid library and call them from the Dunnart diagram editor.[5] The library is written in C++ and compiled with gcc 4.2.1 at -O3. To investigate the performance of the algorithms we have timed libavoid computing connector routes for a variety of diagrams.

We used diagrams with various sized grid arrangements of nodes, where each outside node is connected to the diagonally opposite node, and each node except those on the right and bottom edge is connected to the node directly down and to the right. We also used two larger random graphs and a protein topology diagram. Figure 4 shows the layout of three of these evaluation diagrams.

We measured the time to construct the orthogonal visibility graph, the time to find all connector routes using the A* algorithm, the time to centre routes in channels and perform nudging. The experiment was run on a MacBook Pro with a 2.53 GHz Intel Core 2 Duo processor and 2GB of memory. The results are shown in Table 1. We found for smaller diagrams with fewer than 100 nodes and edges the routing process for the entire diagram can be performed in a fraction of a second. Such smaller diagrams typify the sort of diagrams constructed in interactive authoring software. The size and construction time for the visibility graph in grid examples is notably smaller, as would be expected since the shapes have visibility just to their neighbours. The time required to centre routes is negligible, so should always be performed since it leads to more predictable routes. The Graph-compact example is notable for the smaller visibility graph and higher routing time, where both are due to the fact that many of the nodes in this graph are close together and obscure visibility or block the optimal routes of connectors.

Table 1. Average time taken to construct the orthogonal visibility graph, route all connectors, and compute final positions of all connectors for grids and random graphs

Diagram	Diagram size $|V|$	Diagram size $|E|$	VisGraph size $|V|$	VisGraph size $|E|$	Times (in msec.) to compute VisGraph	RouteConns	FinalPos	Total
Protein-1ABI	26	25	2,138	3,602	7	4	1	12
Grid-6x6	36	35	641	542	1	3	6	10
Grid-8x8	64	63	1,109	928	5	13	22	40
Grid-10x10	100	99	2,261	2,474	5	58	48	111
Grid-12x12	144	143	2,429	2,012	17	63	94	174
Graph-sparse	231	276	52,318	101,589	143	38	149	330
Graph-compact	305	413	51,187	98,801	144	261	605	1,010

---

[5] Dunnart is available for download from http://www.dunnart.org/

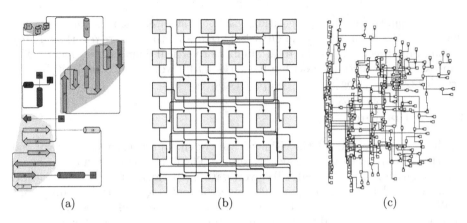

(a)   (b)   (c)

Fig. 4. Evaluation diagrams: (a) Protein-1ABI, (b) Grid-6x6 and (c) Graph-sparse

7   Related Work

Our work extends the previously mentioned research into three-step optimal routing of poly-line connectors to handle orthogonal connector routing. There has been some previous work on finding good orthogonal connector routings between fixed position shapes. The most closely related work in the graph drawing literature is that of Miriyala *et al.* [5] who also use an $A^\star$ algorithm for computing orthogonal connector paths. The main difference is that they search through the *rectangulation* of the diagram rather than the orthogonal visibility graph. The rectangulation is obtained by drawing vertical lines from the vertices of each shape. The search is then through these rectangles. While superficially similar, the rectangulation is actually quite different to the orthogonal visibility graph. The key difference is that the rectangulation does not directly model horizontal visibility. This means that their algorithm is heuristic and routes are not guaranteed to be optimal in any meaningful sense even if minimizing edge crossings is ignored. A disadvantage of our approach is that the rectangulation is $O(n)$ in size while the orthogonal visibility graph is $O(n^2)$ in size for n shapes.

Other related work includes algorithms for orthogonal graph layout. A standard technique is to solve a network flow problem in order to compute an orthogonal representation for the graph which minimizes the total number of connector bends [6]. A compaction step is then applied to the orthogonal representation to assign positions to the nodes which minimize the area of the drawing but do not introduce additional crossings or overlap. The orthogonal graph layout algorithm used in Tom Sawyer [7] uses a different approach. It first uses a heuristic to position nodes on a grid so that connected nodes are close and no nodes (treated as points) are on the same horizontal or vertical grid line. Next the edge routing is chosen using a heuristic. Nodes may be expanded in size in this step to ensure that edges require at most one bend. A compaction step is then applied.

The key difference between orthogonal graph layout and the problem we address is that in orthogonal graph layout the layout algorithm is responsible for positioning nodes so as to minimize bends, while in our context nodes, i.e. shapes, have fixed position and dimensions. We also mention incremental approaches to orthogonal graph layout which incrementally construct the layout as vertices (and all of their associated edges) are added one at a time [6,7]. Again nodes are allowed to move and the focus is on bend minimization.

Orthogonal connector routing has been extensively studied in computational geometry, in part because of its applications to circuit design. Lee *et al.* [8] provides an extensive survey of algorithms for orthogonal connector routing, while Lenguauer [9] provides an introduction to the algorithms used in circuit layout.

One of the earliest approaches in circuit design was so-called *maze running* in which objects were assumed to be laid out on a uniform grid and a shortest path algorithm was employed to find the shortest path in the grid [10]. The complexity is proportional to the size of the grid. In our context, the grid needs to be very fine because the user is free to place elements where they like and so the time complexity is prohibitively high. Our approach can be considered a modification to maze running in which we use a non-uniform grid whose mesh size is tailored to the geometry of the diagram.

Another approach used in circuit layout is to construct a *track graph* [11]. The construction is very similar to that of the orthogonal visibility graph. The difference is that not all intersections of the interesting horizontal and vertical segments are placed in the graph, thus the track graph is a sub-graph of the orthogonal visibility graph. The track graph is not suitable for our purposes because it is only intended for finding minimal length connector paths (the primary concern in circuit design).

The problem we are addressing is finding a *minimum-cost* path (MCP) where the cost is a function of the number of bend points and path length. Algorithms with $O(n \log^{3/2} n)$ complexity for routing a single connector where n is the number of objects (assuming they are rectangles) are known for this problem [8]. However, it is fair to say that these algorithms are quite complex and dependent on the kind of penalty function. Our approach has the advantage of simplicity and, since A* search is a generic technique, it can be extended to more complex penalty functions such as one, for instance, penalizing connector crossings (see Section 8). Furthermore, our approach is analogous to the poly-line connector routing approach already used in `libavoid` and so implementation effort is reduced. While the worst-case complexity of our approach is $O(n^2 \log n)$, in practice because of the good heuristic used in the A* search, we have found performance is perfectly acceptable.

Another significant contribution of our paper is our algorithm for "nudging" orthogonal connectors apart so as to improve the legibility of the layout. While Miriyala *et. al.* do consider nudging, the issues and approach are quite different. They do not consider the problem of how to avoid introducing unnecessary crossings when separating connectors with a shared path. Our algorithm for ordering connectors in shared paths so as to avoid introducing unnecessary crossings is related to algorithms for metro-line crossing [12,13]. The main difference is that we have the additional requirement that the ordering should not introduce unnecessary bends in the layout and so crossings are only allowed to occur when a connector enters or leaves the shared path, but not in the shared path itself.

8 Conclusion

Most diagram editors and graph construction tools provide some form of automatic orthogonal connector routing. However the routes are typically computed using *ad-hoc* techniques and are not usually updated during direct manipulation. We present algorithms for computing optimal object-avoiding orthogonal connector routings where optimality is w.r.t some monotonic function of the connector length and number of bends. Our approach is based on first computing an orthogonal visibility graph for the diagram, then an optimal route using an A* search, followed by computation of the precise connector path. The approach is surprisingly fast and allow us to recalculate optimal connector routings fast enough to reroute connectors interactively for use in authoring applications.

We plan to extend our work to incorporate a cost for edge crossings in the penalty function. An advantage of using the orthogonal visibility graph and the shared path ordering step is that it allows easy identification of edge crossings. Thus computing how many times a new connector route crosses the previously routed connectors can be done simply and with little additional overhead.

Acknowledgments. NICTA is funded by the Australian Government as represented by the Department of Broadband, Communications and the Digital Economy and the Australian Research Council. We acknowledge the support of the ARC through Discovery Project Grant DP0987168.

References

1. Wybrow, M., Marriott, K., Stuckey, P.J.: Incremental connector routing. In: Healy, P., Nikolov, N.S. (eds.) GD 2005. LNCS, vol. 3843, pp. 446–457. Springer, Heidelberg (2006)
2. Dobkin, D.P., Gansner, E.R., Koutsofios, E., North, S.C.: Implementing a general-purpose edge router. In: DiBattista, G. (ed.) GD 1997. LNCS, vol. 1353, pp. 262–271. Springer, Heidelberg (1997)
3. Dwyer, T., Marriott, K., Stuckey, P.: Fast node overlap removal. In: Healy, P., Nikolov, N.S. (eds.) GD 2005. LNCS, vol. 3843, pp. 153–164. Springer, Heidelberg (2006)
4. Dwyer, T., Marriott, K., Stuckey, P.: Fast node overlap removal—correction. In: Kaufmann, M., Wagner, D. (eds.) GD 2006. LNCS, vol. 4372, pp. 446–447. Springer, Heidelberg (2007)
5. Miriyala, K., Hornick, S.W., Tamassia, R.: An incremental approach to aesthetic graph layout. In: CASE 1993, pp. 297–308. IEEE Computer Society, Los Alamitos (1993)
6. Di Battista, G., Eades, P., Tamassia, R., Tollis, I.G.: Graph Drawing: Algorithms for the Visualization of Graphs. Prentice-Hall, Inc., Englewood Cliffs (1999)
7. Biedl, T.C., Madden, B., Tollis, I.G.: The three-phase method: A unified approach to orthogonal graph drawing. IJCGA 10(6), 553–580 (2000)
8. Lee, D., Yang, C., Wong, C.: Rectilinear paths among rectilinear obstacles. Discrete Applied Mathematics 70(3), 185–216 (1996)
9. Lengauer, T.: Combinatorial Algorithms for Integrated Circuit Layout. John Wiley & Sons, Inc., New York (1990)
10. Lee, C.Y.: An algorithm for path connections and its applications. IRE Transactions on Electronic Computers EC-10(2), 346–365 (1961)

11. Wu, Y.F., Widmayer, P., Schlag, M.D.F., Wong, C.K.: Rectilinear shortest paths and minimum spanning trees in the presence of rectilinear obstacles. IEEE Transactions on Computers 36(3), 321–331 (1987)
12. Argyriou, E., Bekos, M., Kaufmann, M., Symvonis, A.: Two polynomial time algorithms for the metro-line crossing minimization problem. In: Tollis, I.G., Patrignani, M. (eds.) GD 2008. LNCS, vol. 5417, pp. 336–347. Springer, Heidelberg (2009)
13. Bekos, M., Kaufmann, M., Potika, K., Symvonis, A.: Line crossing minimization on metro maps. In: Hong, S.-H., Nishizeki, T., Quan, W. (eds.) GD 2007. LNCS, vol. 4875, pp. 231–242. Springer, Heidelberg (2008)

On Rectilinear Drawing of Graphs

Peter Eades[1,*], Seok-Hee Hong[2,**], and Sheung-Hung Poon[3,***]

[1] School of Information Technologies, University of Sydney, Australia
{peter,shhong}@it.usyd.edu.au
[2] Department of Computer Science, National Tsing Hua University, Taiwan, R.O.C.
spoon@cs.nthu.edu.tw

Abstract. A *rectilinear drawing* is an orthogonal grid drawing without bends, possibly with edge crossings, without any overlapping between edges, between vertices, or between edges and vertices. Rectilinear drawings without edge crossings (*planar* rectilinear drawings) have been extensively investigated in graph drawing. Testing rectilinear planarity of a graph is NP-complete [10]. Restricted cases of the planar rectilinear drawing problem, sometimes called the "no-bend orthogonal drawing problem", have been well studied (see, for example, [13,14,15]).

In this paper, we study the problem of general *non-planar* rectilinear drawing; this problem has not received as much attention as the planar case. We consider a number of restricted classes of graphs and obtain a polynomial time algorithm, NP-hardness results, an FPT algorithm, and some bounds.

We define a structure called a "4-cycle block". We give a linear time algorithm to test whether a graph that consists of a single 4-cycle block has a rectilinear drawing, and draw it if such a drawing exists. We show that the problem is NP-hard for the graphs that consist of 4-cycle blocks connected by single edges, as well as the case where each vertex has degree 2 or 4. We present a linear time fixed-parameter tractable algorithm to test whether a degree-4 graph has a rectilinear drawing, where the parameter is the number of degree-3 and degree-4 vertices of the graph. We also present a lower bound on the area of rectilinear drawings, and a upper bound on the number of edges.

1 Introduction

A *rectilinear drawing* is an orthogonal grid drawing without bends, possibly with edge crossings, without any overlapping between edges and vertices. A graph is called a *rectilinear graph* if it admits a rectilinear drawing.

Rectilinear drawings without edge crossings (*planar* rectilinear drawings) have been extensively investigated in graph drawing. An undirected graph is *rectilinear planar* if it can be drawn in the plane such that every edge is a horizontal

* Supported in part by grants from National Science Council (NSC), National Taiwan University and Academia Sinica in Taiwan, R.O.C.
** Supported in part by grant 98R0036-03 from National Taiwan University in Taiwan, R.O.C.
*** Supported in part by grant NSC 97-2221-E-007-054-MY3 in Taiwan, R.O.C.

D. Eppstein and E.R. Gansner (Eds.): GD 2009, LNCS 5849, pp. 232–243, 2010.
© Springer-Verlag Berlin Heidelberg 2010

or vertical segment and no two edges cross. Garg and Tamassia [10] proved that testing rectilinear planarity of a graph is NP-complete. Restricted cases of the planar rectilinear drawing problem, sometimes called the "no-bend orthogonal drawing problem", have been well studied. Significant examples include linear-time algorithms to construct planar rectilinear drawings of *plane* graphs G of maximum degree three [13], subdivisions of planar triconnected cubic graphs [14], and series-parallel graphs of the maximum degree three [15].

Vijayan and Wigderson [16] considered the problem of rectilinear planar embedding with edge direction constraints. They gave a linear time testing algorithm and an $O(n^2)$ time embedding algorithm to construct such a drawing. Hoffman and Kriegel [11] improved the running time by presenting a linear time embedding algorithm. Bodlaender and Tel studied the connection between rectilinearity and angular resolution of planar graphs [5]. Recently, Eppstein [8] studied bendless orthogonal drawing problem in three dimensions, and showed that it is NP-complete to determine whether an arbitrary graph has such an embedding.

Many methods have been developed for constructing orthogonal drawings, aiming to minimize crossings as well as bends; see, for example, the original work of Batini et al. [2], or the "three phase method" of Biedl et al. [3]. However, non-planar rectilinear drawing has not been so well studied. Formann et al. [9] proved that given a graph G of maximum degree 4, it is NP-hard to decide whether G has a straight-line drawing with angular resolution $\frac{\pi}{2}$. In this paper, we investigate the problem of general non-planar rectilinear drawing.

Our work was also motivated by the recent development of *RAC (Right Angle Crossing)* drawing [6]. A RAC-drawing is a straight-line drawing of a graph, where all the crossings are at right angles. Research on RAC drawing arises from the controversial human experiments on the effects of crossing angles on performance of path tracing tasks. In 2006, Huang et al. [12] found that task response times decrease as the crossing angle increases, implying that drawings with large crossing angles are better for visualization. A rectilinear drawing can be regarded as an orthogonal-RAC drawing, that is, an orthogonal drawing with right angle crossings.

In this paper we present NP-completeness results and a linear time algorithm. The line between NP-completeness and a linear time algorithm is drawn with the concept of a "4-cycle block", defined in Section 2. We prove that one can test whether a graph that consists of a single 4-cycle block has a rectilinear drawing in linear time, and we can construct such a drawing in linear time. In contrast, we show that it is NP-complete to test whether a graph has a rectilinear drawing, even when it consists of a set of 4-cycle blocks connected by single edges. The NP-hardness remains even when the input graph consists of only degree-2 and degree-4 vertices.

Further, we present a linear time fixed-parameter tractable algorithm to test whether a degree-4 graph has a rectilinear drawing, where the parameter is the number of degree-3 and degree-4 vertices of the graph.

Note that the use of term "rectilinear drawing" is somewhat inconsistent in the literature. In 1941, Birkhoff [4] used the term in discussing density functions that can be approximated by sets of straight lines. In the Graph Theory literature the term has sometimes been used to mean the same as "straight-line drawing" (that is, without the requirement for orthogonality). In this paper we use the common meaning from the graph drawing literature, that is, as an orthogonal straight-line drawing.

This paper is organized as follows. In Section 2, we describe some basic properties of rectilinear drawings, including bounds on the area and density. Section 3 presents a linear time algorithm to test whether a graph with a "connected 4-cycle cover" has a rectilinear drawing. Section 4 presents hardness results on rectilinear drawings. In Section 5, we describe a fixed-parameter tractable algorithm to test whether a graph has a rectilinear drawing, where the parameter is the number of degree-3 and degree-4 vertices of the graph. Section 6 concludes with some open problems.

2   Some Basic Properties

In this Section we first describe some basic concepts and properties of rectilinear drawings, and then prove some simple bounds on their area and density.

2.1   Four-Cycle Covers and Blocks

Firstly we mention some simple consequences of the assumption that the drawing has no edge overlaps.

Lemma 1. *Suppose that G is a rectilinear graph. Then every vertex of G has degree at most 4 in G, no two 4-cycles of G share more than one edge, and no three 4-cycles share an edge.*

Motivated by Lemma 1, we say that a *4-cycle cover* C_4 of a graph $G = (V, E)$ is a set of 4-cycles that covers every edge, that is, every edge of G is in C_4. If G has a 4-cycle cover, then the *4-cycle incidence graph* is a graph G_4 with a vertex for each 4-cycle $c \in C_4$ and an edge (c, c') when the two 4-cycles c and c' share an edge. We say that C_4 is a *connected-4-cycle cover* if G_4 is connected.

Suppose that G is a graph of maximum degree 4 and that G' is a subgraph with a connected 4-cycle cover. If G' is maximal (that is, if edges or vertices are added then it no longer has a connected 4-cycle cover) then we say that G' is a *4-cycle block*. The concept of a 4-cycle block is critical in determining the line between polynomial time and NP-completeness.

A rectilinear drawing of a graph $G = (V, E)$ induces a partition $E = E_{hor} \cup E_{vert}$ of the edges into horizontal and vertical edges. Let $G_{hor} = (V, E_{hor})$ and $G_{vert} = (V, E_{vert})$. This partition has several useful properties.

Lemma 2. *Suppose that G is a rectilinear graph.*

(a) Both G_{hor} and G_{vert} are sets of disjoint paths.
(b) A path in G_{hor} meets a path in G_{vert} in at most one vertex.

(c) Every vertex of G is in exactly one path of G_{hor} and exactly one path of G_{vert}. (Note that we regard a single vertex as a trivial path).

Proof. Items (a) and (b) follow from the assumption that edges cannot overlap, and (c) is immediate.   ⊡

From Lemma 2 we can deduce some properties of cycles.

Lemma 3. *Suppose that G is a rectilinear graph. Then every cycle c in G contains at least two vertical edges and at least two horizontal edges. If c is a 4-cycle then the edges of c are alternately horizontal and vertical around c.*

Proof. Direct the cycle c clockwise. It is clear that every leftward horizontal edge has at least one corresponding rightward horizontal edge, and so c has at least two horizontal edges. Similarly, c has at least two vertical edges. If c has 4 edges then they must alternate.   ⊡

2.2 Density

From Lemma 1 we can deduce that if $G = (V, E)$ is a rectilinear graph with $n = |V|$ and $m = |E|$, then $m \leq 2n$. We can show a tighter bound as follows.

Lemma 4. *If $G = (V, E)$ is a rectilinear graph with $n = |V|$ and $m = |E|$, then $m \leq 2n - 2\sqrt{n}$. Further, for every n there is a rectilinear graph with n vertices and $2n - 2\lceil\sqrt{n}\rceil$ edges.*

Proof. Suppose that X_i is the set of vertices with x coordinate i; note that the induced subgraph on X_i is a set of paths and thus has at most $X_i - 1$ edges. Summing over all the sets X_i shows that the number of horizontal edges is at most $n - k$, where k is the number of such sets X_i, that is, the number of different x-coordinates of vertices. Similarly, if ℓ is the number of different y-coordinates of vertices, then the number of vertical edges is at most $n - \ell$. Thus

$$m \leq 2n - k - \ell. \tag{1}$$

Let x^* denote $\max_i |X_i|$. Now $\sum_i |X_i| = n$, and so $k \geq n/x^*$. Also, $\ell \geq x^*$, since no two vertices are at the same location; it follows that $k \geq n/\ell$. We can deduce from (1) that $m \leq 2n - (\ell + n/\ell)$. Minimizing the term $\ell + n/\ell$ over $1 \leq \ell \leq n$ gives the upper bound in the Lemma.

We can obtain the lower bound from grid graph of dimensions $\lceil\sqrt{n}\rceil \times \lceil\sqrt{n}\rceil$; this has at least n vertices and $2n - 2\lceil\sqrt{n}\rceil$ edges.   ⊡

2.3 Area

Area bounds for *planar* rectilinear drawings are have a long history (see, for example, [7]). It is straightforward to show that any rectilinear graph with n vertices has a rectilinear drawing on an $n \times n$ grid, and for planar graphs there is a corresponding lower bound. It could be tempting to suggest that allowing edge crossings allows smaller area rectilinear drawings. However, there is a graph on n vertices for which the minimum area grid drawing (planar or not) has area $\Omega(n^2)$; this is illustrated in Figure 1.

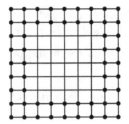

Fig. 1. The area of the rectilinear drawing is $\Omega(n^2)$

3   Graphs with a Connected-4-Cycle Cover

In this section we prove that rectilinearity can be tested in linear time is the graph has a connected-4-cycle cover, that is, it consists of a single 4-cycle block.

Theorem 1. *There is a linear time algorithm that tests whether a graph with a connected-4-cycle cover has a rectilinear drawing, and gives the drawing if it exists.*

The first step is check the necessary conditions of Lemma 1; if any of these conditions fails to hold, then we reject the input graph G. These conditions can be checked using a simple search in the set of vertices at distance at most 2 from each vertex; since the degree is bounded this takes linear time.

Next we partition the edge set E into E_{hor} and E_{vert}. This can be done with a depth-first traversal of the 4-cycle incidence graph G_4. At each step, we label the edges of a 4-cycle c as horizontal or vertical so that the labels alternate around c; we reject G if any inconsistency in labels is found. Since G_4 is connected, this labels every edge; also, the traversal takes linear time. Further note that the labeling is unique after the choice of labels on the first 4-cycle.

Next we check that the partition satisfies the conditions of Lemma 2. This is straightforward. Subsequently we assume that G_{hor} consists of paths $\{p_0, p_1, \ldots, p_{k-1}\}$ and G_{vert} consists of paths $\{q_0, q_1, \ldots, q_{\ell-1}\}$.

The next step is to assign a direction for each of the nontrivial paths in G_{hor}. Looking ahead, this direction induces a partial order on V; another partial order can be obtained from the direction of nontrivial paths in G_{vert}. By topologically sorting on each partial order, we can obtain $x-$ and y-coordinates for each vertex. However, we must be careful how the directions are assigned. Consider Figure 2. If the paths are directed as in Figure 2(a), then it is easy to see that the x-coordinates can be assigned by a topological sort. However, if a path happens to be in the wrong direction, as in Figure 2(b), then it is difficult to see how to assign x-coordinates. So we must choose a direction for each path.

We begin by assigning an arbitrary direction to each p_i; this gives an initial partial order on V. Let $\bar{p}_i$ denote the reverse of the path p_i. If there are 4 vertices a, b, c, d where $a, b \in p_i$, $c, d \in p_j$ such that $a, c \in q_s$ and $b, d \in q_t$ for some s, t, then p_i and p_j have a *compatibility* relationship, defined as follows. If $a < b$ and

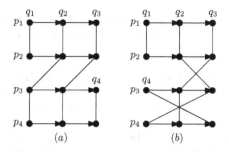

Fig. 2. The directed paths

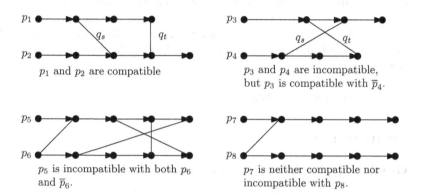

Fig. 3. Path direction examples

$c > d$ then we say that p_i is *incompatible* with p_j; if $a < b$ and $c < d$ then we say that p_i is *compatible* with p_j. In Figure 3, p_1 is compatible with p_2, and p_3 is incompatible with p_4; note, however, that p_3 is compatible with $\bar{p}_4$. The relationship of compatibility does not apply to the pair p_7, p_8, because these two paths share less than two paths in Q. Note that p_5 and p_6 are incompatible, further, p_5 and $\bar{p}_6$ are incompatible.

If there are two incompatible paths p_i and p_j, for which p_i and $\bar{p}_j$ are incompatible, then we reject G. Subsequently we assume that there are no such pairs of paths. In other words, we are assuming that for each pair p_i, p_j, if the compatibility relationship is defined then p_i is compatible with either p_j or $\bar{p}_j$.

Then we associate a boolean formula f with G as follows. For each path p_i we have a boolean variable x_i. If p_i is compatible with p_j then we insert the clause $(x_i = x_j)$ into f; if p_i is compatible with $\bar{p}_j$ then we insert the clause $(x_i \neq x_j)$ into f. Now $(x_i = x_j) = (x_i \vee \bar{x}_j) \wedge (\bar{x}_i \vee x_j)$, and $(x_i \neq x_j) = (x_i \vee x_j) \wedge (\bar{x}_i \vee \bar{x}_j)$, and thus f is a 2SAT formula. Satisfiability for this formula can be tested in linear time. If f is not satisfiable, then we reject G; otherwise, a satisfying set of values for the x_i gives a direction for each path in G_{hor} so that every pair of paths is compatible.

Similarly, we can obtain a direction for each of the paths $q_1, q_2, \ldots, q_\ell$ of G_{vert} such that each pair q_i, q_j is compatible (if there is no such direction, then we reject G).

Lemma 5. *Suppose that G satisfies the conditions of Lemmas 1, 2, and 3. Further suppose that each pair of p_i, p_j of paths in G_{hor} is compatible, and each pair q_i, q_j of paths in G_{vert} is compatible. Then G has a rectilinear representation.*

Proof. We define a partial order on the paths of G_{vert} as follows. Suppose that p_i is a path in G_{hor}. Recall that every vertex in p_i is in exactly one path of G_{vert}. Suppose that $e = (u, v)$ is a (directed) edge of p_i, where $u \in q_j$ and $v \in q_{j'}$ for two paths q_j and $q_{j'}$ of G_{vert}; then we define $q_j < q_{j'}$. Since pairs of paths in G_{hor} are compatible, this relation is a partial order. Thus using a topological sort we can assign an x-coordinate to each path q_j in G_{vert}, and thus to each vertex in G (recall that every vertex of G is in exactly one path of G_{vert}). Similarly, one can assign a y-coordinate to each vertex in G. The resulting drawing is rectilinear. ☐

This completes the proof of Theorem 1.

4   Hardness Results

4.1   Graphs of 4-Cycle Blocks Connected by Edges

The previous section gives a linear time algorithm for testing rectilinearity when the graph consists of a single 4-cycle block. In this section, we show that a slight relaxation of this condition leads to NP-completeness.

Theorem 2. *The decision problem whether a graph consisting of a set of 4-cycle blocks connected by single edges has a rectilinear drawing is NP-complete.*

Proof. First we show that the problem is in NP. Note that a rectilinear graph has a rectilinear drawing on an $n \times n$ grid. Thus we can guess a location for each vertex on an $n \times n$ grid, and then check to see whether it is a rectilinear drawing.

We reduce from the 3SAT problem. The input instance for the problem is a set $\{x_1, x_2, \ldots, x_n\}$ of n variables, and a collection $\{c_1, c_2, \ldots, c_m\}$ of m clauses, where each clause consists of exactly three literals. The 3SAT problem is to determine whether there exists a truth assignment to the variables so that each clause has at least one true literal. In the following, we will describe our polynomial-time reduction.

First we construct a skeleton which contains ports connecting to the variable towers and the clause gadgets. The main component of the skeleton consists of a series of 4-cycles connecting together to form an L-shaped backbone. See Figure 4 for the illustration of its construction.

The upward spikes are ports connecting to the variable towers, and the 4-cycles hanging on the right hand side of the skeleton are ports connecting to the clause gadgets. The variable tower for variable x_i is constructed as shown in

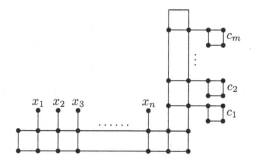

Fig. 4. Skeleton containing ports for connecting to variable towers and clause gadgets

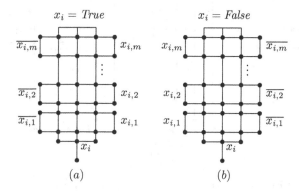

Fig. 5. Variable tower for x_i and the representation of its truth values

Figure 5(a), where we create a row of four consecutive 4-cycles R_j for each clause c_j and we set the two rightmost (resp. leftmost) vertices as connection ports for literal $x_{i,j}$ (resp. $\overline{x_{i,j}}$), where $x_{i,j}$ takes the same truth value as x_i. Moreover, we connect these m rows of consecutive 4-cycles by two adjacent columns of consecutive 4-cycles. See Figure 5(a).

This completes the construction of the variable tower for variable x_i.

We then connect this variable tower to the corresponding upward spike. The tower in Figure 5(a) represents the assignment of a *true* value to variable x_i, and its mirror image in Figure 5(b) represents the assignment of a *false* value to the variable x_i.

We then proceed to see how we construct the clause gadgets. Suppose that we want to construct the gadget for clause $c_j = x_i \vee \overline{x_k} \vee x_l$. First, let s_1^j, s_2^j be the two vertices on the 4-cycle σ_j at the connection port for c_j and adjacent to the connection edge between cycle σ_j and the skeleton. See Figure 6(a).

Note that the order of s_1^j and s_2^j are not essential since they are interchangeable. Now we connect vertex s_1^j to the upper ports of the two literals $x_{i,j}$ and $\overline{x_{k,j}}$, respectively, by a two-bend path. Moreover, at each of the two bending vertices of this connection path, we attach a 4-cycle. Then we perform the

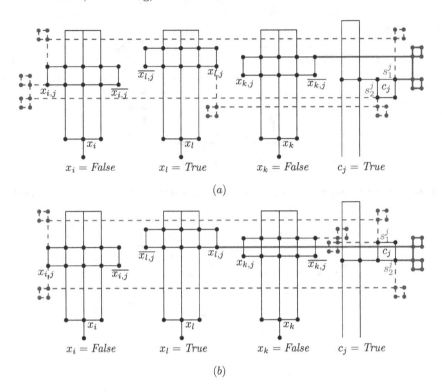

Fig. 6. Clause gadget for clause $c_j = x_i \vee \overline{x_k} \vee x_l$

similar construction by connecting vertex s_2^j to the lower ports of the two literals $x_{i,j}$ and $x_{l,j}$. This completes the construction of the gadget for clause c_j. We construct all clauses in the same fashion, and we finish the construction step. Clearly, our construction takes $O(mn)$ time. We let G be the graph we constructed. Below we proceed to show that the given 3SAT formula is satisfiable if and only if graph G has a rectilinear drawing.

We first consider the forward direction. Suppose the given 3SAT formula is satisfiable with some truth assignment. According to this truth assignment, we determine the embedding positions for the n variable towers according to the scenario in Figure 5. Then we will proceed to embed the clause gadgets one by one. Suppose that we are embedding the gadget for clause $c_j = x_i \vee \overline{x_k} \vee x_l$. As an example, assume that $\overline{x_k}$ is the true literal causing clause c_j to become true. As $\overline{x_k}$ is connected to s_1^j in our construction, we will rotate the 4-cycle σ_j at the connection port of clause c_j so that s_1^j is the rightmost vertex on cycle σ_j. See Figure 6(a). And we can connect the rightward port of literal $\overline{x_{k,j}}$ to the rightward port of s_1^j using the two-bend path between $\overline{x_k}$ and s_1^j. Other connection paths for this gadget can also be embedded accordingly. Figure 6(b) shows another example of embedding the gadget for clause c_j at the time that x_l is the true literal for c_j. Thus graph G has a rectilinear drawing.

We then consider the backward direction. Suppose G has a rectilinear drawing. Consider the embedded scenario of the gadget for clause $c_j = x_i \vee \overline{x_k} \vee x_l$. Without loss of generality, suppose that s_1^j is a rightmost vertex of 4-cycle σ_j. See Figure 6(a). As each connection path consists of two bends, the rightward port of s_1^j must connect to some rightward port of some variable tower. As such, if we set the truth values for the variables according to the embedding positions of their towers, all clauses will result in true values. Thus the given 3SAT formula is satisfiable. This finishes our NP-completeness proof.   □

4.2   Graphs with Degree-2 and Degree-4 Vertices

With our result on graphs with connected-4-cycle cover above, at the first glance, it is tempting to claim that the rectilinear drawability of a graph with only degree-2 and degree-4 vertices may be polynomial-time solvable. However, by slight modification (see below) of the NP-hardness proof in [9], we obtain its NP-completeness proof for this problem. The vertices with degree one or three can be eliminated by connecting pairs of them by short paths. Finally, only degree-2 and degree-4 vertices remain. We conclude in the following theorem.

Theorem 3. *Let G be a graph with only degree-2 and degree-4 vertices. The decision problem whether G has a rectilinear drawing is NP-complete.*

5   Fixed-Parameter Algorithm

In this section, we show that the rectilinear drawing problem is fixed-parameter tractable, where the parameter is the number of degree-3 and degree-4 vertices in the graph.

Theorem 4. *Let G_k be a degree-4 graph with k vertices of degree 3 or 4 for some constant k. The decision problem whether G_k has a rectilinear drawing can be answered in linear time, or more precisely, in $O(24^k \cdot k^{2k} \cdot n)$ time.*

Proof. Let K be the set of these k degree-3 or degree-4 vertices. The vertices in K can be distributed among k vertical columns. Considering these columns as boxes, we can have at most k^k different ways to put these vertices into the boxes. On the other hand, the vertices in K also can be distributed among k horizontal rows, and there are also at most k^k different ways to put these vertices into the corresponding rows. Hence, there are essentially at most k^{2k} ways to embed the vertices in K on the plane.

Now suppose that K_ϵ is one such embedding of K in the plane. Each vertex v has degree at most 4 and there are only 4 directions in which edges incident to v can have; thus there are at most $4! = 24$ possible choices of directions for the edges incident to v. Therefore there are at most 24^k ways to select the directions of all edges incident to the vertices of K_ϵ.

We can check the embeddability of the whole graph for each selection of edge directions at each vertex of K. Suppose that we have selected the direction for all edges incident to the vertices in K_ϵ. We proceed to embed all paths connecting between vertices in K_ϵ. For instance, suppose that we are going to embed the path $\eta_{u,v}$ between vertices u and v in K_ϵ, and we let e_u, e_v be the edges of path $\eta_{u,v}$ incident to u, v, respectively. We have known the directions for e_u and e_v, and we are going to decide the directions of other edges on path $\eta_{u,v}$. Note that routing of such paths can be done independently from each other, because planarity is not required. The feasibility of routing path $\eta_{u,v}$ from vertex u through e_u to vertex v through e_v depends on the length of $\eta_{u,v}$, and can be decided in time proportional to the length of $\eta_{u,v}$. Thus testing the feasibility of routing all paths in the graph G_k takes $O(n)$ time. In all, the decision problem whether G_k has a rectilinear drawing can be answered in $O(24^k \cdot k^{2k} \cdot n)$ time. $\square$

6 Conclusion

We consider the problem of general non-planar rectilinear drawing for restricted class of graphs and obtain a polynomial time algorithm and NP-hardness results. The boundary between NP-completeness and polynomial time is determined by the structure of 4-cycle blocks in the graph. We also present tight bound on the drawing area and an FPT algorithm for the parameter of the number of vertices of degree ≥ 3. It is feasible that there is an FPT algorithm where the exponential complexity depends on the number of 4-cycle blocks in the graph; this is left as an open problem.

There are two directions that need to be explored before commercial visualization software can make use of this work. Firstly, the algorithm in Section 3 is restricted graphs with connected 4-cycle covers. This is a relatively restrictive condition, and it would be wise to investigate whether there are polynomial-time algorithms for some wider classes. Secondly, many graphs do not have a rectilinear drawing; it would be interesting to investigate methods that find large subgraphs with rectilinear drawings (see, for example, [3]).

Finally, we note that Eppstein [8] studied bendless orthogonal drawing problem in three dimensions, and showed that it is NP-complete to determine whether an arbitrary graph has such an embedding. It would be interesting to extend the concept of 4-cycle block to three dimensions and determine the boundary between polynomial time algorithms and NP-completeness in three dimensions.

References

1. Aspvall, B., Plass, M., Tarjan, R.: A linear-time algorithm for testing the truth of certain quantified boolean formulas. Information Processing Letters 8(3), 121–123 (1979)
2. Batini, C., Talamo, M., Tamassia, R.: Computer aided layout of entity relationship diagrams. Journal of Systems and Software 4(2-3), 163–173 (1984)
3. Biedl, T., Madden, B., Tollis, I.: The three-phase method: a unified approach to orthogonal graph drawing. Int. J. Comput. Geometry Appl. 10(6), 553–580 (2000)
4. Birkhoff, G.: Rectilinear drawing, in Three public lectures on scientific subjects. Three public lectures on scientific subjects 28(1), 51–76 (1941)

5. Bodlaender, H.L., Tel, G.: A note on rectilinearity and angular resolution. Journal of Graph Algorithms and Applications 8, 89–94 (2004)

6. Didimo, W., Eades, P., Liotta, G.: Drawing graphs with right angle crossings. In: Dehne, F., et al. (eds.) WADS 2009. LNCS, vol. 5664, pp. 206–217. Springer, Heidelberg (2009)

7. Dolev, D., Leighton, F.T., Trickey, H.: Planar embedding of planar graphs. Advances in Computing Research 2, 147–161 (1984)

8. Eppstein, D.: The topology of bendless three-dimensional orthogonal graph drawing. In: Tollis, I.G., Patrignani, M. (eds.) GD 2008. LNCS, vol. 5417, pp. 78–89. Springer, Heidelberg (2008)

9. Formann, M., Hagerup, T., Haralambides, J., Kaufmann, M., Leighton, F.T., Symvonis, A., Welzl, E., Woeginger, G.J.: Drawing graphs in the plane with high resolution. In: Proceedings of FOCS, pp. 86–95. IEEE, Los Alamitos (1990); SIAM Journal on Computing 22(5), 1035–1052 (1993)

10. Garg, A., Tamassia, R.: On the computational complexity of upward and rectilinear planarity testing. SIAM Journal on Computing 31(2), 601–625 (2002)

11. Hoffman, F., Kriegel, K.: Embedding rectilinear graphs in linear time. Information Processing Letters 29(2), 75–79 (1988)

12. Huang, W., Hong, S., Eades, P.: Effects of crossing angles. In: Proceedings of PacificVis 2008, pp. 41–46. IEEE, Los Alamitos (2008)

13. Rahman, M. S., Naznin, M., Nishizeki, T.: Orthogonal drawings of plane graphs without bends. In: Mutzel, P., Jünger, M., Leipert, S. (eds.) GD 2001. LNCS, vol. 2265, pp. 392–406. Springer, Heidelberg (2002)

14. Rahman, M.S., Egi, N., Nishizeki, T.: No-bend orthogonal drawings of series-parallel graphs. In: Liotta, G. (ed.) GD 2003. LNCS, vol. 2912, pp. 387–392. Springer, Heidelberg (2004)

15. Rahman, M.S., Egi, N., Nishizeki, T.: No-bend orthogonal drawings of series-parallel graphs. In: Healy, P., Nikolov, N.S. (eds.) GD 2005. LNCS, vol. 3843, pp. 409–420. Springer, Heidelberg (2006)

16. Vijayan, G., Wigderson, A.: Rectilinear graphs and their embeddings. SIAM Journal on Computing 14(2), 355–372 (1985)

Semi-bipartite Graph Visualization for Gene Ontology Networks

Kai Xu[1], Rohan Williams[2], Seok-Hee Hong[3], Qing Liu[1], and Ji Zhang[4]

[1] CSIRO, Australia
[2] Australian National University, Australia
[3] School of Information Technologies, University of Sydney, Australia
[4] The University of Southern Queensland, Australia

Abstract. In this paper we propose three layout algorithms for *semi-bipartite graphs*—bipartite graphs with edges in one partition—that emerge from microarray experiment analysis. We also introduce a method that effectively reduces visual complexity by removing less informative nodes. The drawing quality and running time are evaluated with five real-world datasets, and the results show significant reduction in crossing number and total edge length. All the proposed methods are available in visualization package GEOMI [1], and are well received by domain users.

1 Introduction

Expression microarrays [2] have been widely used to measure *gene expression level*—the activity level of genes—in biological experiments. A typical microarray experiment involves comparing the gene expression levels of diseased (e.g. cancerous) and healthy tissue. The genes that show opposite expression levels—active in cancer tissue but not in normal, or vice versa—are known as *differentially expressed genes*. They are commonly used as candidates to study the genetic cause of the disease. The Gene Ontology [3] is a directed acyclic graph containing known gene functions (*terms*). A *parent term* describes an abstract function shared by its *child terms*, which represent more specific functions. All the edges point from the parent to its children. A gene is *annotated* by a term if it has that function, or in other words, the term is an *annotation* of the gene.

It is a rare case that the functions in the Gene Ontology provide a direct answer to the genetic cause of a disease. More complex analyses are generally required, such as how genes regulate each other, and the global function of a gene group. Many existing tools, such as the family of over-representation methods [4], only list terms that are statistically important. More recent tools, such as BiNGO [5], start to show the hierarchical structures among terms. However, genes are not shown and their performance or visual quality are not properly evaluated. In this paper, we propose novel methods that address these problems. Our main contributions are:

- Introduction of a new graph type: semi-bipartite graph;
- Proposing three layout algorithms for the semi-bipartite graphs;

D. Eppstein and E.R. Gansner (Eds.): GD 2009, LNCS 5849, pp. 244–255, 2010.
© Springer-Verlag Berlin Heidelberg 2010

- A visual complexity reduction technique for semi-bipartite graphs;
- Implementation and evaluation of the proposed methods.

2   Related Work

2.1   Gene Ontology Visualization

To date most work on Gene Ontology analysis has focused on statistical models designed to identify terms that occur at a higher proportion than random expectation [4]; there are several methods that aim to complement such analysis with visualization. In the work by Baehrecke et al. [6], a TreeMap [7] is used to display the part of the Gene Ontology hierarchy identified by the over-represented terms. The terms with multiple parents have to be duplicated under each parent to convert the Gene Ontology hierarchy into a tree. The GObar [8] uses the Graphviz package [9] to produce a layered drawing of the Gene Ontology hierarchy, but only with graphs less than 20 nodes. In SpindleViz [10] a variation of the Sugiyama method [11] is proposed to display the Gene Ontology hierarchy in three dimensions. Other hierarchy visualization methods can also be applied to the Gene Ontology, such as those surveyed by Katifori et al. [12].

All the work on Gene Ontology visualization illustrates its importance. However, the genes and how they are annotated are missing from these methods. It has been shown that such information is important to gene functional analysis [13]. This issue is attempted by Robinson et al. [14] by showing the number of annotations each term has. More details are provided in GO PaD [15], which visualizes both genes and Gene Ontology terms in a network with genes linked to their annotations. However, only a small number of selected terms, which are not experiment specific, are shown in the visualization. The BiNGO plug-in [5] for CytoScape [16] uses the gray-scale of a term color to overlay the over-representation information onto the Gene Ontology hierarchy. The genes and how they are annotated are only shown implicitly.

2.2   Layered Drawings and Sugiyama Method

Layered layout algorithms are natural choices for visualizing the Gene Ontology because of its hierarchical structure. Many existing algorithms are based on the 4-step framework first proposed by Sugiyama et al. [11]. Various algorithms have been proposed for each step; the details can be found in the book by Di Battista et al. [17] and Kaufmann and Wagner [18]. An important part of the Sugiyama method is the bipartite graph cross minimization problem, which is also particularly relevant to the layout algorithms we propose in this paper. Given a bipartite graph $G = (V, W, E)$ with two parallel straight-lines L_1 and L_2, a *two-layered* drawing consists of placing vertices in the vertex set V on L_1 and W on L_2 respectively. Each edge is a line segment joining one vertex in V and one in W. The embedding is fully determined by the vertex orderings of V and W. The *one-sided crossing minimization* problem has a fixed ordering of vertices in W on L_2, and the problem is shown to be NP-complete [19,20]. A number of heuristics, approximation, and exact algorithms have been proposed [11,19,21]. Eades and Wormald [19] proposed a *Median method*, which produces a 3-approximate solution. The *Barycenter method* by Sugiyama et al. [11] is an $O(\sqrt{n})$-approximation algorithm [19]. Currently, the best known approximation algorithm is given by Nagamochi [21] that delivers a drawing with a 1.4664 factor approximation.

3   Semi-bipartite Graph and Gene-Term Network

The data used in the Gene Ontology analysis can be treated as a bipartite graph, with the genes being one partition and the ontology being the other. However, the Gene Ontology partition is a directed acyclic graph itself. To accommodate this, we define a *semi-bipartite graph* as a graph $G = (V, W, E, F)$, where V and W are two sets of nodes, E is the set of edges between V and W, i.e., $E = \{(v_i, w_j) \mid v_i \in V, w_j \in W\}$, and F is the set of edges between the nodes in W, i.e., $F = \{(w_i, w_j) \mid w_i, w_j \in W\}$.

With this definition, we introduce the *Gene-Term Network* [13]. It contains two types of nodes: genes and Gene Ontology terms. The genes are those in the most differentially expressed list, and the terms are those in the *induced Gene Ontology hierarchy* (*induced hierarchy* for short) that includes terms that are annotated to the genes of interest and all their ancestors in the Gene Ontology together with induced edges. There are two types of edges: *annotation edges* that connect genes to their annotation terms and *term edges* that link the terms in the induced Gene Ontology hierarchy. The gene-term network is a semi-bipartite graph with one partition being a directed acyclic graph. Formally, given a set of genes $V = \{v_1, v_2, \ldots, v_n\}$ and the Gene Ontology hierarchy $GO = \{GO_T, GO_E\}$ (where GO_T and GO_E are the set of Gene Ontology terms and edges respectively), a gene-term network is a semi-bipartite graph: $G = (V, W, E, F)$ where

- $W = \{w \mid w \in GO_T, \exists v \in V, w \in t(v) \text{ or } w \in a(v)\}$, where $t(v)$ is the set of annotations of a gene v and $a(v)$ is the set of ancestor terms of $t(v)$.
- $E = \{(v, w) \mid v \in V, w \in t(v)\}$ is a set of *annotation edges* linking genes and their annotations;
- $F = GO_E \cap (W \times W)$ is a set of *term edges* induced from GO.
- The induced hierarchy $P = (W, F)$ is a directed acyclic graph.

4   Layout Algorithms for Semi-bipartite Graphs

4.1   Extended Bipartite Algorithms

This algorithm extends the barycenter method [11] and starts by drawing the induced hierarchy P using the Sugiyama method, treating it as the partition with fixed order. The gene nodes V are placed on a parallel layer and ordered according to the horizontal position of their annotation terms to minimize inter-partition edge crossings (we assume the drawing is top-down with horizontal layers and the same applies to other algorithms in this paper). The gene nodes are placed on the bottom layer because many of them annotate to the leaf terms of P; placing gene nodes above the hierarchy will introduce extra edge crossings. Please refer to [22] for algorithm details.

The running time of the algorithm is the sum of that of the Sugiyama method and barycenter computation. The former depends on the algorithms used for the various stages. In our implementation, we use the layering method by Gansner et al. [23] that minimizes the number of dummy nodes and requires polynomial time in the worst case. The crossing reduction algorithm is based on the median method [19] that runs in linear time ($O(|W| + |F|)$). Finally, the horizontal coordinate assignment uses a heuristic algorithm based on network simplex formulation [24] that requires polynomial time in the worst case. The barycenter can be computed in time linear to the size of gene node set and the number of annotation edges [19], i.e., $O(|V| + |E|)$. Therefore, the overall running time is $O(T(|W|, |F|)) + O(|V| + |E|)$, where $O(T(|W|, |F|))$ is the running time of the Sugiyama method for the induced hierarchy P.

4.2   Sub-hierarchy Barycenter Algorithm

This algorithm aims to reduce crossings caused by annotation edges and achieves this by adjusting the child term order in the induced hierarchy according to the gene node order. It is based on the following observations:

- Assume a term w has two child terms w_1 and w_2 with w_1 to the left of w_2 in the drawing. If most genes annotated to w_1 are to the right of w_2, changing the order of w_1 and w_2 is likely to reduce edge crossing. However, such a change will also affect all the descendants of w_1 and w_2. Therefore, the decision to change the order should be based on how genes are annotated to the sub-hierarchy rooted at w_1 and w_2.
- Following the previous example, assume w_1 and w_2 both have two child terms. The order change between w_1 and w_2 will affect the order among their child terms, but the reverse is not true. Therefore, the change of child order should be breadth-first.

Before describing the algorithm, we need to introduce *sub-hierarchy barycenter*, which is based on the position of all the gene nodes annotated to the terms in the sub-hierarchy rooted at a term (including the term itself). Formally, for a term w, its sub-hierarchy barycenter $b(w)$ is:

$$b(w) = average(x(v_i)), v_i \in V, \exists w_j \in sub(w), (v_i, w_j) \in F$$

where $average()$ computes the average value, $x(v_i)$ is the x-coordinate of v_i, and $sub(w)$ is the sub-hierarchy in the induced hierarchy rooted at w.

The *sub-hierarchy barycenter algorithm* starts with drawing the gene-term network with the extended bipartite algorithm. Then, it traverses the induced hierarchy bottom up to compute the sub-hierarchy barycenter for each term. After that, it traverses the induced hierarchy again breadth first to re-order the child terms accordingly. Finally, the gene nodes are re-ordered according to the new child term order. The whole process is repeated a fixed number of times. Because there is a total ordering of sub-hierarchy barycenter, there will be no ordering conflict when a term has multiple parents. Similarly, common child terms only need to be sorted once when they are shared by multiple parents. The details of the algorithm can be found in [22].

Besides the extended bipartite algorithm and Sugiyama method, the running time consists of three parts: that of the sub-hierarchy barycenter computation, the child term re-ordering, and the gene nodes ordering. During barycenter computation, each term w is visited $1+|parent(w)|$ times ($|parent(w)|$ is the number of parents of w): once for computing its own barycenter; $|parent(w)|$ times for all its parents. Because each term edge is visited exactly once when computing the barycenter of the parent term, the total number of term visits is:

$$\sum_{w \in W} (|parent(w)| + 1) = \sum_{w \in W} |parent(w)| + \sum_{w \in W} = |F| + |W|$$

During barycenter computation, each gene v is visited $|t(v)|$ times: once for each term it annotates to. Therefore the total number of gene visits is $\sum_{v \in V} t(v) = |E|$, and the running time for sub-hierarchy barycenter computation is $O(|E| + |F| + |W|)$. For child term re-ordering, each term is only ordered once, so the running time is expected to be $O(|W| \log |W|)$. Similarly, the running time of gene nodes re-ordering is $O(|V| \log |V|)$. Therefore, the overall running time of the sub-hierarchy barycenter algorithm is $O(T(|W|, |F|)) + O(|E| + |F|) + O(|W| \log |W|) + O(|V| \log |V|)$.

4.3   Partition Merge Algorithm

In the previous algorithm, the crossing reduction is achieved by changing the induced hierarchy embedding and gene node ordering separately. The *partition merge algorithm* makes global crossing reduction possible by merging two partitions. There are two variations of the algorithm. The first one places all gene nodes on the layer beneath the induced hierarchy, which is achieved by assigning the direction of all annotation edges pointing to gene nodes. No change is required for the layered layout algorithm if the layering step starts with the sinks (gene nodes), such as the Longest Path Layering [17], otherwise a constraint is required that all gene nodes must be placed on the bottom layer.

Relaxing the bottom layer constraint can potentially reduce the total edge length, which is another factor important to graph readability [25]. To avoid comparing all possible edge direction assignment permutations, we propose a *level barycenter* heuristic that assigns a gene to a layer that is the average of the layer of its annotation terms. When term layer equals level barycenter value, the edge direction is from term to gene so that gene with one annotation is placed under the term. The algorithm details can be found in [22]. Moving gene nodes up the hierarchy may introduce extra edge crossings, but it is counter balanced by shorter annotation edges that are less likely to intersect with other edges. Overall, the level barycenter algorithm performs very well in terms of edge crossings. Please refer to the evaluation results in Section 6 for the details.

The running time of the first variation of the Partition Merge algorithm is similar to that of the Extended Bipartite algorithm with the addition of the edge direction assignment step, which requires $O(|E|)$ time. Therefore, its total running time is $O(|E|) + O(T(|V + W|, |E + F|))$. For the second variation, the level barycenter computation requires $O(|V| + |E|)$ time and then the direction assignment takes $O(|E|)$ time, which leads to a total running time of $O(|V| + |E|) + O(T(|V + W|, |E + F|))$.

5   Term Reduction

During this study we found that as the data size increases, the size of the induced hierarchy increases at a much higher rate than that of the gene nodes. The reason is that each new gene usually annotates to several new terms, and each new term in turn introduces several new ancestors. In an effort to reduce the size of the induced hierarchy, we observe that many terms in the top part of the hierarchy are abstract and not informative for functional analysis, so they can be removed to reduce visual complexity. To identify such terms, we introduce the concept of the *subordinate term*, which are Gene Ontology terms that have no gene annotated to them and do not show new relationships between terms with gene annotation. Formally, we define the *indirect gene set* $g'(w)$ of a term w as the set of genes annotated to the term in its sub-hierarchy but not the term itself, i.e.,

$$g'(w) = \{v_i \mid v_i \in V, \exists\, w_j \in sub(w), (v, w_j) \in E, w_j \neq w\}$$

Similarly, the *direct gene set* of a term is the set of genes annotate to it, i.e.,

$$g(w) = \{v_i \mid v_i \in V, (v, w) \in E\}$$

Now, we can define a term w is *subordinate* if

1. $g(w) = \emptyset$, and
2. $\exists\, w_i \in sub(w), g'(w_i) = g'(w), w_i \neq w$

This means that a subordinate term does not have any genes annotated to it, and its indirect gene set is the same as one of its descendants'. The latter means the term is not the first common ancestor of two terms with direct gene annotated to them, i.e., is not of structural importance in function analysis.

However, there are many cases in Gene Ontology where a term can be both a parent and grandparent of the same term: term w_1 is a parent of term w_2 and w_3, and w_2 is also a parent of w_3. In this example, w_1 always has the same indirect gene set as w_2 if it has no other child term. According to the previous definition, w_1 is a subordinate term if no genes are annotated to it. However, these cases are considered biologically important under some conditions. To accommodate such cases, we define that a term is *semi-subordinate* if

1. $g(w) = \emptyset$, and
2. $\forall \, w_i \in child(w), g'(w_i) = g'(w)$.

The revised Condition 2 means that a term is semi-subordinate only if its indirect gene set is the same as all of its childrens.

The algorithm of identifying subordinate terms traverses the induced hierarchy bottom up. It starts with a queue that contains only the leaf terms. Once a term is checked, all its parents are appended to the queue, and a term is always checked after all its child terms. The indirect gene set of a term is the union of the direct gene set of all its children. A term is subordinate if its indirect gene set coincides with one of its childrens, and its direct gene set is empty. The process is repeated until the queue is empty. The algorithm details are available in [22]. The only change required for semi-subordinate terms is the check condition.

Similar to the sub-hierarchy barycenter algorithm, each term w is visited $1 + |parent(w)|$ times: once for computing its indirect gene set, $|parent(w)|$ times for all its parents. Each gene v is also visited $|t(v)|$ times: once for each term it annotates to. Note that the coincidence check can be done in constant time, because the indirect gene set of a child is always a subset of that of its parent and equal size is sufficient to show that they coincide. Therefore, the total running time is $O(|E| + |F| + |W|)$. The semi-subordinate algorithm has the same running time because the change it needs does not affect running time.

6   Evaluation

6.1   Dataset

The five sets of genes used in the evaluation were generated from the results of 91 microarray experiments with each measuring about 22000 mouse genes. Full details of the experiment and its findings can be found in [26]. Table 1 shows the size of the gene-ontology network for every dataset.

Table 1. Dataset size

Dataset	Total node $|V| + |W|$	Total edge $|E| + |F|$	Gene node $|V|$	Term node $|W|$	Annotation edge $|E|$	Term edge $|F|$
cgg5	49	65	4	45	7	58
cgg4	150	243	10	140	19	224
cgg3	221	394	13	208	38	356
cgg2	374	627	43	331	89	538
cgg1	447	864	25	422	81	783

6.2   Implementation

All algorithms and the term reduction method are implemented in GEOMI [1], which is a Java-based graph visualization system. Figure 1 is the user interface:

The layout is shown in the left panel with navigation functions including zoom in/out, pan, and rotate. In the right panel, users can choose among different layouts and two levels of term reduction. The blue nodes are genes, green nodes are terms, and the red node is the over-represented term. The same implementation of the Sugiyama method is used in all the algorithms, and the method used in each step is the same as described in Section 4.1. In the sub-hierarchy barycenter algorithm, the re-ordering of the child terms and gene nodes are repeated 20 times to find the embedding with minimal number of edge crossings.

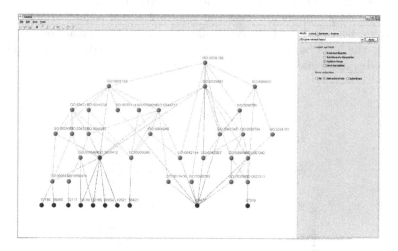

Fig. 1. The GEOMI plug-in with all four algorithms and term reduction (gene: blue, term: green, over-represented: red). User can choose the layout and term reduction from the panel on the right.

6.3   Layout Quality

We use edge crossings as one of the visual quality measurements, because it has been shown to be the most important layout aesthetics [25] for graph readability. Table 2 shows the number of edge crossings of the different layout methods. The

Table 2. Edge crossings of different layout methods

Dataset	Extended Bipartite	Sub-Hierarchy Barycenter	Partition Merge	Level Barycenter
cgg5	3	0	0	0
cgg4	162	135	119	126
cgg3	870	798	605	646
cgg2	2542	2028	1929	1918
cgg1	4164	3440	3883	3765

two variations of the partition merge algorithm are shown as Partition Merge (all the gene nodes on the bottom layer) and Level Barycenter respectively. The extended bipartite algorithm—the baseline method—is outperformed by the other algorithms in all cases. The sub-hierarchy barycenter, partition merge, and level barycenter algorithm have similar performance, providing about 20% less edge crossings than the extended bipartite algorithm. The sub-hierarchy barycenter performs better for the larger datasets (*cgg2* and *cgg1*) than the smaller ones (*cgg4* and *cgg3*), while the opposite is true for the partition merge and level barycenter algorithm. The latter performed well with regards to edge crossings given its main goal is to reduce total edge length.

The total edge lengths are shown in Figure 2. Term edges are not included in the tests, because level barycenter algorithm has little impact on them. It is clear that the level barycenter provides a considerable reduction in total edge length and is more effective as the graph size increases.

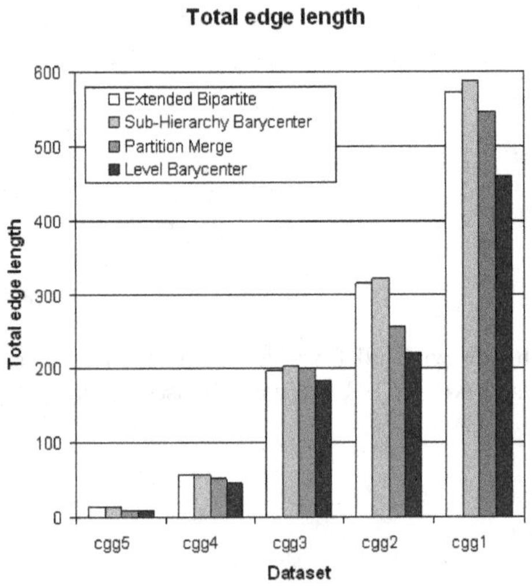

Total edge length

Fig. 2. The total annotation edge length

6.4 Term Reduction

The results of the term reduction are shown in Figure 3. Gene nodes are not included because they are not affected and the induced hierarchy accounts for the majority of the gene-term network. Term reduction is effective in reducing network size: the number of term nodes is reduced to at least half of what they were. Besides reduced graph size, other benefits of term reduction include less edge crossings and faster running time for the layout algorithms.

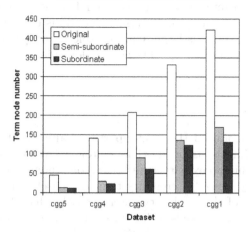

Fig. 3. The number of term nodes after term reduction

6.5   Running Time

Table 3 shows the running time of the four algorithms. All the tests were per-formed on a laptop equipped with an Intel Core 2 Duo 2.5 GHz CPU with 2 GB memory. The time is the average of 20 repetitions. In most cases all algo-rithms completed within one second, which means they scale well to graph with 447 nodes and 864 edges (*cgg1*). The relative speed among the algorithms is consistent with the running time analysis in Section 4.

Table 3. Running time of the algorithms in seconds

Dataset	Extended Bipartite	Sub-Hierarchy Barycenter	Partition Merge	Level Barycenter
cgg5	0.128	0.406	0.11	0.25
cgg4	0.126	0.625	0.109	0.328
cgg3	0.338	0.719	0.297	0.453
cgg2	0.378	0.969	0.375	0.609
cgg1	0.607	1.375	0.594	1.016

6.6   User Feedback

A pilot user study has been conducted to gain feedback from domain experts. We asked two domain experts: one bioinformatician with statistics background and one molecular biologist, who are the most likely users. The users were asked to compare the proposed methods (using GEOMI) against the way they currently perform function analysis. They found the extra information gene-term network provides very useful. The layout produced by the sub-hierarchy barycenter and partition merge algorithm have similar visual quality; both are better than that of the extended bipartite algorithm. Term reduction is effective and necessary, especially when the dataset is large. One possible improvement is to keep the term levels as they are in the original Gene Ontology hierarchy, which is not the case currently.

7 Conclusions

In this paper we introduced the semi-bipartite graph for the visual analysis of microarray experiments using the Gene Ontology. Among the proposed layout algorithms, sub-hierarchy barycenter, partition merge, and level barycenter produce considerably less edge crossings than the baseline method—extended bipartite. The result of the level barycenter algorithm has the least total edge length. All the algorithms take less than or close to one second to complete for graphs with size up to 447 nodes and 864 edges. The term reduction technique effectively reduces graph size by more than half. All the methods have been implemented in GEOMI and well received by the domain experts.

References

1. Ahmed, A., Dwyer, T., Forster, M., Fu, X., Ho, J., Hong, S.H., Koschützki, D., Murray, C., Nikolov, N.S., Taib, R., Tarassov, A., Xu, K.: GEOMI: Geometry for maximum insight. In: Healy, P., Nikolov, N.S. (eds.) GD 2005. LNCS, vol. 3843, pp. 468–479. Springer, Heidelberg (2006)
2. Schena, M., Shalon, D., Davis, R.W., Brown, P.O.: Quantitative monitoring of gene expression patterns with a complementary dna microarray. Science 270(5235), 467–470 (1995)
3. Ashburner, M., Ball, C.A., Blake, J.A., Botstein, D., Butler, H., Cherry, J.M., Davis, A.P., Dolinski, K., Dwight, S.S., Eppig, J.T., Harris, M.A., Hill, D.P., Issel-Tarver, L., Kasarskis, A., Lewis, S., Matese, J.C., Richardson, J.E., Ringwald, M., Rubin, G.M., Sherlock, G.: Gene ontology: tool for the unification of biology. Nature Genetics 25(1), 25–29 (2000)
4. Khatri, P., Draghici, S.: Ontological analysis of gene expression data: current tools, limitations, and open problems. Bioinformatics 21(18), 3587–3595 (2005)
5. Maere, S., Heymans, K., Kuiper, M.: Bingo: a cytoscape plugin to assess overrepresentation of gene ontology categories in biological networks. Bioinformatics 21(16), 3448–3449 (2005)
6. Baehrecke, E.H., Dang, N., Babaria, K., Shneiderman, B.: Visualization and analysis of microarray and gene ontology data with treemaps. BMC Bioinformatics 5, 84 (2004)
7. Shneiderman, B.: Tree visualization with treemaps: A 2d space-filling approach. ACM Transactions on Graphics 11(1), 92–99 (1992)
8. Lee, J.S.M., Katari, G., Sachidanandam, R.: GObar: A gene ontology based analysis and visualization tool for gene sets. BMC Bioinformatics 6, 189 (2005)
9. Gansner, E.R., North, S.C.: An open graph visualization system and its applications to software engineering. Software Practice and Experience 30(11), 1203–1233 (2000)

10. Joslyn, C.A., Mniszewski, S.M., Smith, S.A., Weber, P.M.: Spindleviz: A three dimensional, order theoretical visualization environment for the gene ontology. In: Proceedings of Joint BioLINK and 9th Bio-Ontologies Meeting (2006)
11. Sugiyama, K., Tagawa, S., Toda, M.: Methods for visual understanding of hierarchical system structures. IEEE Transactions on Systems, Man, and Cybernetics 11(2), 109–125 (1981)
12. Katifori, A., Halatsis, C., Lepouras, G., Vassilakis, C., Giannopoulou, E.: Ontology visualization methods—a survey. ACM Comput. Surv. 39(4), 10 (2007)
13. Xu, K., Huang, X.X., Cotsapas, C., Hong, S.H., McCaughan, G., Gorrell, M., Little, P., Williams, R.: Combined visualisation and analysis of gene ontology annotations using multivariate representations of annotations and bipartite networks. Technical Report 09/166, CSIRO (2009)
14. Robinson, P.N., Wollstein, A., Bohme, U., Beattie, B.: Ontologizing gene-expression microarray data: characterizing clusters with gene ontology. Bioinformatics 20(6), 979–981 (2004)
15. Alterovitz, G., Xiang, M., Mohan, M., Ramoni, M.F.: GO PaD: the gene ontology partition database. Nucleic Acids Research 35, D322–D327 (2007)
16. Shannon, P., Markiel, A., Ozier, O., Baliga, N.S., Wang, J.T., Ramage, D., Amin, N., Schwikowski, B., Ideker, T.: Cytoscape: A software environment for integrated models of biomolecular interaction networks. Genome Research 13, 2498–2504 (2003)
17. Battista, G.D., Eades, P., Tamassia, R., Tollis, I.G.: Graph Drawing: Algorithms for the Visualization of Graphs. Prentice Hall, Englewood Cliffs (1999)
18. Kaufmann, M., Wagner, D. (eds.): Drawing Graphs, Methods and Models. LNCS, vol. 2025. Springer, Heidelberg (2001)
19. Eades, P., Wormald, N.C.: Edge crossings in drawings of bipartite graphs. Algorithmica 11(4), 379–403 (1994)
20. Garey, M.R., Johnson, D.S.: Crossing number is NP-complete. SIAM J. Algebraic and Discrete Methods 4(3), 312–316 (1983)
21. Nagamochi, H.: An improved bound on the one-sided minimum crossing number in two-layered drawings. Discrete & Computational Geometry 33(4), 569–591 (2005)
22. Xu, K., Williams, R., Hong, S.H., Liu, Q., Zhang, J.: Semi-bipartite graph visualization for gene ontology networks. Technical Report EP091883, CSIRO (2009), http://www.it.usyd.edu.au/~visual/kaixu/papers/gene-ontology-layout.pdf
23. Gansner, E.R., Koutsofios, E., North, S.C., Vo, K.P.: A technique for drawing directed graphs. IEEE Transactions on Software Engineering 19(3), 214–230 (1993)
24. Chvatal, V.: Linear Programming. W.H. Freeman, New York (1983)
25. Purchase, H.C.: Which aesthetic has the greatest effect on human understanding? In: Proceedings of the 5th International Symposium on Graph Drawing, London, UK, pp. 248–261. Springer, Heidelberg (1997)
26. Cowley, M.J., Cotsapas, C.J., Williams, R.B.H., Chan, E.K.F., Pulvers, J.N., Liu, M.Y., Luo, O.J., Nott, D.J., Little, P.F.R.: Intra- and inter-individual genetic differences in gene expression. Nature Proceedings (2008), http://hdl.handle.net/10101/npre.2008.1799.1

On Open Problems in Biological Network Visualization

Mario Albrecht[1], Andreas Kerren[2], Karsten Klein[3], Oliver Kohlbacher[4],
Petra Mutzel[3], Wolfgang Paul[3,*], Falk Schreiber[5], and Michael Wybrow[6]

[1] Max Planck Institute for Informatics, Saarbrücken, Germany
`mario.albrecht@mpi-inf.mpg.de`
[2] School of Mathematics and Systems Engineering (MSI), Växjö University, Sweden
`andreas.kerren@vxu.se`
[3] Faculty of Computer Science, Dortmund University of Technology, Germany
`{karsten.klein,petra.mutzel,wolfgang.paul}@cs.tu-dortmund.de`
[4] Center for Bioinformatics, Eberhard Karls University Tübingen, Germany
`oliver.kohlbacher@uni-tuebingen.de`
[5] Leibniz Institute of Plant Genetics and Crop Plant Research (IPK) Gatersleben and Institute
for Computer Science, Martin-Luther-University Halle-Wittenberg, Germany
`schreibe@ipk-gatersleben.de`
[6] Clayton School of Information Technology, Monash University, Australia
`michael.wybrow@infotech.monash.edu.au`

Abstract. Much of the data generated and analyzed in the life sciences can be interpreted and represented by networks or graphs. Network analysis and visualization methods help in investigating them, and many universal as well as special-purpose tools and libraries are available for this task. However, the two fields of graph drawing and network biology are still largely disconnected. Hence, visualization of biological networks does typically not apply state-of-the-art graph drawing techniques, and graph drawing tools do not respect the drawing conventions of the life science community.

In this paper, we analyze some of the major problems arising in biological network visualization. We characterize these problems and formulate a series of open graph drawing problems. These use cases illustrate the need for efficient algorithms to present, explore, evaluate, and compare biological network data. For each use case, problems are discussed and possible solutions suggested.

1 Introduction

In recent years, the improvement of existing and development of novel high-throughput techniques have led to the generation of huge data sets in the life sciences. Since manual analysis of this data is costly and time-consuming, scientists are now turning towards computational methods that support data analysis.

The visualization and the visual analysis of biological networks are one of the key analysis techniques to cope with the enormous amount of data. In particular, the layout of networks should be in agreement with biological drawing conventions and should

---

* Corresponding author.

D. Eppstein and E.R. Gansner (Eds.): GD 2009, LNCS 5849, pp. 256–267, 2010.

draw the user's attention to relevant system properties that might remain hidden otherwise. While the approaches and the expertise of the bioinformatics, information visualization, and graph drawing communities may be ideally suited for solving these problems, little research has been performed to solve the special layout and visualization problems arising in the life sciences. Currently, most of the available software systems for the visual analysis of biological networks (e.g., CellDesigner [8], Cytoscape [5], VANTED [10], see also the review in [24]) provide only implementations of standard graph-drawing algorithms such as force-directed or layered approaches. Nevertheless, there are also some tools that offer specialist drawing algorithms more suitable for applications in the life sciences [2,3,6,7,13].

In general, visualization methods for the life sciences should allow for the layout and navigation of biological networks for both their static presentation as well as their interactive exploration. Such methods need to adhere to constraints that originate from recognized textbook and poster layouts (like [16,18]), from generally accepted drawing conventions within the life-science community as well as from standardization initiatives such as MIM (Molecular Interaction Maps) [14] and SBGN (Systems Biology Graphical Notation) [19].

In this paper, we want to identify graph drawing problems originating in applied bioinformatics and network biology. We start by presenting a characterization of common biological networks, describing their structure and semantics as well as the mapping of data onto network elements. Afterwards, we present a selection of use cases describing typical uses of biological network visualization. For each use case, we present the problem as well as its relevance and discuss existing or possible straightforward solutions as well as their drawbacks. The last section will present some conclusions. The focus of this paper is less on presenting novel methods for these use cases, but more on giving an overview about open problems to raise the awareness for the manifold tasks in information visualization and graph drawing related to biological networks.

2 The Nature of Biological Network Data

Biological networks are used to communicate many different types of data. These data can be encoded in the structure of the network as well as represented by the network layout, or as graphical or textual annotations. The data itself may be primary data (i.e., directly measured), secondary data (i.e., derived, inferred, or predicted), or a mixture of both. In this section, we discuss some common biological networks and the types of attributes used to annotate them. Throughout the rest of this article, graphs will play an important role when trying to represent biological networks. In fact, we will use the terms *graph* and *network* synonymously when talking about the representation of biological networks.

2.1 Types of Biological Networks

In the following subsection we will review various types of biological networks. This listing of networks is by no means complete and includes only those networks that are most central to research in systems biology.

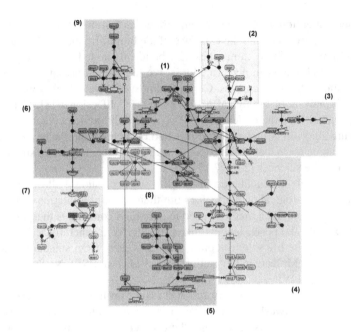

Fig. 1. A regulatory network representing the yeast cell cycle. The picture is taken from the CADLIVE homepage (http://www.cadlive.jp/) and was originally published in [15]. Li and Kurata used their implementation of a grid-layout algorithm.

Gene-regulatory and signal-transduction networks use sets of directed edges to convey a flow of information. While gene regulation (regulation of gene expression) occurs within a cell and represents a regulatory mechanism for the creation of gene products (RNA or proteins), signal transduction refers to any process that transports external or internal stimuli (often via signal cascades) to specific cellular parts where a cell response is triggered (e.g., gene regulation). While nodes in these networks represent molecular entities (genes, gene products, or other molecules), edges represent a flow of information (regulation or passing of a chemically encoded signal). Figure 1 gives an example layout of a graph representing some part of a gene regulatory network.

Protein Interaction Networks represent physical interactions of proteins with each other or with other binding partners such as DNA or RNA. The nodes in such networks represent proteins or sets of proteins. The time scale of protein interactions ranges from very short, transient processes (for instance, pairwise protein interactions and phosphorylation or glycosylation events) to very long lasting, permanent formation of protein assemblies (protein complexes) working as molecular machines. The interaction edges are normally undirected, but may be directed in case of specific interaction (such as activation) or of heterogeneous networks (e.g., protein-protein and protein-DNA/RNA interactions), resulting in mixed graphs. Each node and edge may be annotated with additional biological attributes like expression level, cellular localization, and the number of interaction partners. For an example layout of a protein network, see Figure 2.

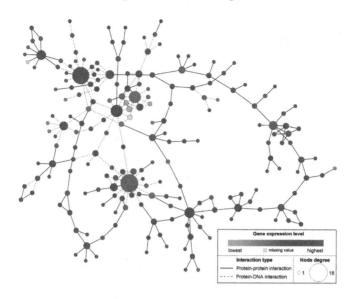

Fig. 2. Visual representation of the GAL4 protein interaction subnetwork in yeast. The protein nodes are colored by a shade gradient according to the expression value; green represents the lowest, red the highest value, and blue a missing value. The node size corresponds to the number of interactions. The shades and styles of the edges represent different interaction types; solid lines indicate protein-protein, and dashed lines protein-DNA interactions. The graph was drawn with Cytoscape [5] using its implementation of the spring-embedder algorithm.

Metabolic Networks describe how metabolites (chemical compounds) are converted into other metabolites. Such a network is a hypergraph that is usually represented as a bipartite graph $G = (V_1 \cup V_2, E)$. The node set is partitioned into the set V_1 of metabolite and enzyme nodes (enzymes catalyze the chemical reactions converting metabolites) and V_2 the set of reaction nodes. Large network posters (e.g., Nicholson's [18] and Michal's [16] pathway maps) are available, and several projects created graphical representations of metabolic networks and offer access to these graphs via web pages (e.g., Kyoto Encyclopedia of Genes and Genomes (KEGG) [21] or the BioCyc collection [11]). The availability of these representations has established a de facto standard for metabolic network drawings. These near-orthogonal drawings possess several characteristic features, i.e., the main direction of reaction pathways is accentuated, relevant subgraphs are placed close to the center of the drawing, substances and products of a reaction are clearly separated, and co-substances are placed out of the main path close to the reaction. Layout algorithms such as [12,22,23] obey established drawing styles of these networks. Figure 3 shows an example layout of a metabolic network.

2.2 The Attributes of Network Elements

The visual representation of primary and secondary data that has been mapped onto the elements of a biological network is an important research field. This is mainly due to the following two facts: firstly, manual analysis of primary and secondary data

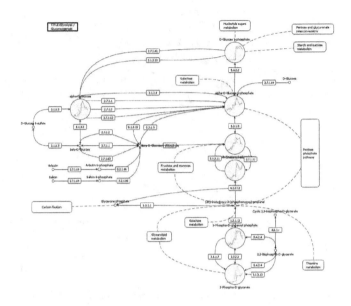

Fig. 3. Part of the glycolysis/gluconeogenesis pathway with additional data mapped onto some nodes. Circles encode metabolites, rectangles represent enzymes catalyzing the reaction, and rectangles with rounded corners denote other pathways. Solid and dashed lines represent reactions and connections to other pathways, respectively. The pathway data was derived from KEGG [21], and the graph was drawn with VANTED [10] in a style similar to the KEGG pathway picture.

has become virtually impossible as high-throughput analysis technologies have become widely available in all the life sciences. Secondly, in recent years, it has become evident that a deeper understanding of complex biological system such as cells, tissues or even whole organisms can only be achieved if the findings of life sciences such as proteomics or genomics are put into context.

The various types of primary data are defined by the different types of biochemical entities and experiments (e.g., time-series experiments or differential studies) and entities that are the subjects of analysis, namely, gene sequences, transcripts, expression levels, proteins, protein concentrations, metabolites, metabolite concentrations, or fluxes (of mass or information). The structure of the class of secondary data is, however, even more complex, and different categories of inferred, derived, or predicted information can be distinguished such as results of correlation analysis or comparisons of different biological states (e.g., healthy vs. diseased, before vs. after treatment with a drug, different organisms). The data belongs to different data types, namely:

- nominal data such as sequence names or categories
- ordinal data such as ontologies, rankings, or partly ordered information
- scalar data such as comparisons or ratios
- categorized spatial data such as data points that refer to biological entities from various parts or substructures of a cell.

The distinction between these categories is not clear in all cases. It should also be mentioned that primary as well as secondary data are subject to uncertainties (measurement errors, prediction confidences, etc.) and that the visualization of these uncertainties is often desired.

3 Use Cases and Related Visualization Problems

This section contains a collection of typical use cases that arise in constructing and viewing networks representing life-science data. Along with each use case, we present a formalized description of the graph-drawing, information-visualization, or visual-analytics problem behind the use case. Again, the list should not be interpreted as a closed set but rather as an attempt to stimulate further research and to demonstrate that there are many interesting and important visualization problems within the life sciences.

3.1 Visual Analysis of Data Correlation

Correlation graphs are frequently used as a means for visually expressing and exploring complex forms of correlation within data. As a means for visually exploring data sets within a life science context, the information contained within a data set is frequently mapped as annotation onto a graph that represents one or more types of pathways. Investigators are then interested in a graphical representation that highlights the inter-relation between the connectivity structure of pairs or subsets of nodes in the original network and their correlation. An example for an interesting correlation pattern would be a set of nodes that is closely connected within the underlying graph but exhibits only weak correlation in the data or vice versa. The represented connectivity structure should include only statistically significant correlations, for instance, significant up- or down-regulation of co-expressed genes or proteins. In particular, two or more nodes representing biological entities with multiple annotations may be considered to correlate if a minimum number of node annotations correspond with each other, e.g., regarding genotype, time value or number of the biological replicates.

One possible way to attack this problem would be to model the correlation data as a weighted graph. A weighted graph associates a label or weight with every edge in the graph. The weights are typically integers or real numbers, as this is usually a requirement induced by the algorithm applied to the weighted graph. We thus have a given graph $G_1 = (V, E_1)$ (called *network* in order to distinguish it from the correlation graph) with correlation data that induces a second graph $G_2 = (V, E_2)$ with edge weights on the same set of nodes V. This gives us a simultaneous embedding problem.

In our case, the two given graphs typically do not have too many edges in common. We seek either a layout of the union graph $G = (V, E_1 \cup E_2)$, in which the given network G_1 and the correlations are clearly displayed, or two disjoint layouts, in which the coordinates of the nodes in both layouts are identical. In the first case, a challenging task is to provide a layout which clearly emphasizes the two different edge sets E_1 and E_2. Often, a layout π_1 of the graph G_1 is given that has to be preserved as closely as possible. In this case, there is a trade-off between preserving the mental map (i.e., each user's individual mental representation of a graph and his or her landmarks) and emphasizing the correlation structure. Possible solutions may either fix the layout given in

π_1, or try to preserve the mental map by keeping the orthogonal relations, the topological embedding (i.e., each node of the graph is mapped to a point in the plane and each edge is associated with a curve segment between the points of its two end nodes), or the layout of a backbone (a subgraph of the given graph that serves as an abstraction).

3.2   Visual Comparison of Similar Biological Networks

Conservation of biochemical function during evolution results in structurally similar molecular subnetworks across different organisms and species. Uncovering relevant similarities and differences or comparing networks in different states (e.g., diseased vs. healthy), at different time points, or under various environmental conditions (for example, temperature, pressure or substrate concentrations) supports the life-science experts' knowledge-discovery process, for example, by identifying disease-specific patterns (biomarker discovery).

Given a set of graphs $G_1, \ldots, G_k$ with a high degree of similarity between each other, the task is to lay them out in a way so that the differences (or the similarities) are highlighted. This problem can be attacked via simultaneous embedding, which requires obtaining either a single layout of the union graph $G = G_1 \cup \ldots \cup G_k$ or k disjoint layouts of the graphs G_i ($i = 1, \ldots, k$) such that the coordinates of all nodes common to two or more subgraphs are the same. An alternative representation has been given in [4] where the third dimension has been used to stack the k layouts above each other. In these layouts, crossings between edges belonging to different graphs $G_i \neq G_j$ are either completely ignored or counted as less important than "real" crossings. In the context of biological networks and pathways, the stronger simultaneous embedding problem with fixed edges occurs, which forces not only the nodes but also the edges occurring in two or more graphs to be drawn identically. This guarantees that identical subnetworks have an identical layout. Sometimes, it may be important to keep a mental map of already given layouts of some of the graphs or their backbone structure. In any case, the layouts must obey the given biological constraints concerning the specific network type. Sometimes, the networks may be large, and it becomes desirable to hide some parts of the network and to highlight only the specific points of interest. Points of interest may be differences between the networks, but could also be important network structures such as the main pathways in a metabolic network. Here, one possibility would be to generate layouts in which the differences are all concentrated within only a few layout areas, which, of course, represents an abstraction and fails to preserve a mental map.

3.3   Integrated Representation of Multiple Overlapping Networks

The different types of biological networks describe different functional aspects of the whole cell, tissue, or organism in question. To get a deeper, system-wide understanding, these networks need to be combined. For example, the enzymes acting in metabolic networks are proteins and take part in protein interaction networks. It is thus becoming increasingly common to integrate these different types of networks into joint networks. A good joint layout of these networks should reveal the *interaction* between these networks, for example, how specific nodes of the gene regulatory network activate or inactivate whole subnetworks of the metabolic network. In order to simplify the identification of these subnetworks, mental map preservation on the level of the metabolic network is helpful.

We need a representation of combined networks in which the conventional layouts (there may be several different ones) of each of these networks need to be respected. Moreover, some groups of nodes in one network may belong to groups of nodes in another network. This mapping (which may be a $1 : 1$, $1 : n$, or $n : m$ mapping) needs to be displayed in the layout. We consider the case of integrating two networks, in which the involved mapping partners can be viewed as a cluster in a cluster graph $C = (G, T)$ of cluster depth 2. Using this approach, the problem may be attacked via the following formalized problem.

We are given two cluster graphs $C_1 = (G_1, T_1)$ and $C_2 = (G_2, T_2)$ with $G_i = (V_i, E_i)$ $(i = 1, 2)$ and cluster depth 2, and a mapping function $\Phi : \mathcal{C}_1 \to \mathcal{C}_2$, where $\mathcal{C}_i$ denote the clusters in G_i. Generate a layout $\pi(G)$ of the union graph $G = G_1 \cup G_2 \cup G[F]$, where F denotes the edges induced by the mapping, respecting the clusters as well as the conventional layouts of each of these networks. In the simpler case when the graph G_2 is highly disconnected (e.g., $E_2 = \emptyset$), we may prefer a solution in which the connected components of G_2 are integrated into a layout of G_1. In this case, we do not require to respect the root clusters in the problem mentioned above. Sometimes, the layout of G_1 may be given. If this is the case, there is often a trade-off between sticking to the given layout as closely as possible (or trying to preserve a user's mental map or the backbone of the original graph) and obtaining a better representation of G_2.

3.4 Visualization of Sub-cellular Localization

Cells consist of distinct compartments, subcellular locations, separated from each other by membranes. Examples for these are the cytosol, the nucleus, the mitochondria, or chloroplasts in plants. The membranes enclosing a compartment separate parts of the biological networks as well. Different partitions of the network will be localized in different subcellular locations and hence cannot interact with each other directly. It is thus essential for an understanding of the network's function to integrate that spatial information into the layout of the network. The required localization data may be either already contained in data sets derived from experiments, may be extracted from external sources, or may be predicted.

Given a network $G = (V, E)$ and additional localization annotation for the nodes in V, we search for a layout that reflects the topographical information of G and that conforms to the drawing conventions for that type of network. It should, of course, be at the same time aesthetically pleasing. Note that the subcellular localization does not just give a clustering of the nodes, for example, a specific relative position of the cellular compartments may be implicitly given by the biological morphology. If G reflects a flow of mass or information, the direction of the flow also needs to be displayed by, e.g., hierarchical layering of the nodes. A simple representation of a cross-sectional cut through a cell would be a stacking of layers as in [2]. In a layout that is inspired by biological morphology, subgraphs of the network should be arranged according to their subcellular localization in such a fashion that the physical structure of the compartments delimiting these subgraphs becomes evident [6,17]. In order to increase the user's acceptance of such layouts, it may be necessary to resemble the (manually generated) layouts from established projects like BioCarta [1]. Using a layered drawing approach has the inherent flaw that mass and information may only pass freely between members of neighboring layers.

3.5 Visualization of Multiple Attributes

Analyzing data in a life-science context often requires the consideration of multiple data attributes. One example for this would be the analysis of several time-series data sets in the context of a biological network. This is frequently done in order to better understand the dynamic behavior of a biological system. The combined representation of such time-series data and a corresponding network should allow investigators to gain new insights concerning the underlying system, such as co-regulated elements and their connection within the network. Such an analysis can, for example, be achieved by mapping the data onto the nodes of a network, see Figure 3 where time-series data from two series (day and night) were mapped on parts of the network (cp. enlarged nodes).

Mapping the given data onto nodes and/or edges is one possibility of solving the problem. It is also the simplest and most straight-forward one but has the general problem that simple use of small visualizations that replace, for example, the node representations is most of the time not sufficient, as a visual comparison of such small graphics in a large graph becomes very quickly infeasible. Furthermore, one has to consider that the attributes involved may belong to various types of data. Each data type needs specific requirements for its visual representation [9]. The challenge here lies in finding a harmonic combination of visual representations for mixtures of different data types.

A possible solution lies in the use of so-called preattentive features [9]. A preattentive feature is a visual property of a picture or drawing that can be very rapidly and accurately detected by the low-level human visual system. Therefore, users do not have to focus their attention onto one specific part of a drawing in order to understand its basic visual properties. Another possibility is to use an additional view separated from the original network view for displaying the attributes. This idea is based on standard coordinated and multiple view visualization techniques. Drawbacks of this method are the need to add connections between the views as well as the spatial distance between both. This is usually done using brushing-and-linking techniques that are used to highlight, select or delete subsets of elements by pointing to specific elements. When multiple views of the same data are used simultaneously, brushing is typically associated with linking, i.e., brushing of elements in one view affects the same elements in other views. First results that try to compare both approaches [25], i.e., multiple views and attribute integration, cannot be directly used in our complex case of biochemical networks.

3.6 Visualization of Flows and Paths in Networks

The qualitative and quantitative distribution of mass and signal flows (fluxes) within a biological network has to be analyzed under consideration of uncertainties in the data. The flow along certain paths may change over time (time-series of measurements) and the paths through the network may be numerous so that not all of them can be displayed. Investigators are, in such cases, primarily interested in the main paths through the network, i.e., those paths that possess a statistically significant flow and transport a considerable percentage of the overall flow through the entire network, and in the metabolites and reactions that are involved in these paths.

The given network, together with quantitative and qualitative information about the flow of mass or information (edge weights), may be a potentially very large one. For directed graphs such as metabolic networks, the layout must reflect the hierarchical nature of the flow, preserve layouts for subnetworks originating from textbook representations as closely as possibly, adhere to general drawing conventions, and, at the same time, focus on the relevant parts of the network, e.g., paths that at a certain point in time transport a large part of the flow. These main paths thus have to be visually emphasized (e.g., placed at the center of the layout and drawn as straight lines) and the distribution of the fluxes within the network have to be depicted, for example, by using different edge widths or colors.

If the dynamic change in the flow over time also needs to be visualized, smooth animations between layouts are required to preserve the user's mental map. In the past, this problem was mainly covered by software visualization techniques, especially by algorithm animation. In information visualization, solutions for showing flows and paths in networks are relatively rare and mostly limited to special cases or domains, to visualize communication flows in social networks [20].

3.7   Exploration of Hierarchical Networks

Biological networks often comprise several thousands nodes and edges. To support the exploration of such large and complex structures, the entire network is usually broken down in a hierarchical manner into pathways and subpathways. Investigators commonly focus on (sub)pathways in a region of interest and explore their relation to other pathways. However, due to the many connections between different pathways, an abstract overview-like picture of all pathways interconnections as well as an interactive navigation from a set of pathways to other connected or related pathways is often desired.

Given a huge biological network, methods for the biologically meaningful visualization of selected subsets $G_1, \ldots, G_k$ (e.g., pathways) and their interrelations, as well as techniques for the navigation within the network are needed. In order to allow the user to keep his orientation during exploration of the network, the layout changes resulting from a user interaction (e.g., selection of an additional pathway) should be small, and context information needs to be represented in an appropriate way. Expand-and-collapse mechanisms thus need to be incorporated into layout algorithms such that drawing conventions and the mental map are preserved. These operations could be restricted to certain levels of abstraction, e.g., by only collapsing/expanding semantically meaningful substructures like pathways. One of the main challenges is that layouts for such subnetworks as well as their relative position to each other may be pre-specified. This layout information needs to be preserved as closely as possible. As these networks are too large to be laid out nicely as a whole, an overview graph or a representation of the backbone could be defined by reduction or abstraction that covers the topologically or semantically relevant features of the network, thus helping the user in navigating through the network.

The subsets $G_1, \ldots, G_k$ do not need to be disjoint but may partially overlap. This poses an additional challenge for the visualization problem: Either the duplicate components are merged, which complicates the task of mental map preservation, or it has to be clearly emphasized that they represent the same biological entity.

4 Conclusions

Biological networks play a crucial role in systems biology. Many universal as well as special-purpose tools and libraries are available for laying out and drawing graphs in order to help visually investigating these networks. However, these tools either do not adhere to the special drawing conventions and recognized layouts in biology or are not adequate for handling large graphs.

The use cases presented here reveal graph drawing as well as information visualization problems arising in the biological domain. While we present possible solutions to these problems, we also consider this paper a challenge to the graph drawing community and people working on network visualization in systems biology as well as the life sciences. The problems described here are far from being solved for all practical scenarios and certainly merit further attention. Developing improved solutions will require custom state-of-the-art graph-drawing approaches, and more importantly, collaboration between researchers from graph drawing, information visualization, visual analytics, and the life sciences. We hope that this paper encourages such collaborations and presents interesting research directions in these fields.

Acknowledgements

We would like to thank Schloss Dagstuhl, the Leibniz Center for Informatics, for providing the congenial atmosphere for the seminar 08191 "Graph Drawing with Applications to Bioinformatics and Social Sciences," which led to the conception of this paper. We are also grateful to Hagen Blankenburg for Figure 2 as well as to Stephan Diehl, Aaron Quigley, and Idan Zohar for helpful and stimulating discussions during the Dagstuhl seminar. M.A. acknowledges financial support by NGFN and DFG (KFO 129/1-2 and MMCI Cluster of Excellence), and O.K. funding by DFG (SFB 685/B1).

References

1. BioCarta, http://biocarta.com/
2. Barsky, A., Gardy, J., Hancock, R., Munzner, T.: Cerebral: a cytoscape plugin for layout of and interaction with biological networks using subcellular localization annotation. Bioinformatics 23(8), 1040–1042 (2007)
3. Becker, M., Rojas, I.: A graph layout algorithm for drawing metabolic pathways. Bioinformatics 17(5), 461–467 (2001)
4. Brandes, U., Dwyer, T., Schreiber, F.: Visualizing related metabolic pathways in two and a half dimensions. In: Liotta, G. (ed.) GD 2003. LNCS, vol. 2912, pp. 111–122. Springer, Heidelberg (2004)
5. Cline, M., Smoot, M., Cerami, E., Kuchinsky, A., Landys, N., Workman, C., Christmas, R., Avila-Campilo, I., Creech, M., Gross, B., Hanspers, K., et al.: Integration of biological networks and gene expression data using Cytoscape. Nature Protocols 2(10), 2366–2382 (2007)

6. Demir, E., Babur, O., Dogrusöz, U., Gürsoy, A., Nisanci, G., Çetin Atalay, R., Ozturk, M.: PATIKA: an integrated visual environment for collaborative construction and analysis of cellular pathways. Bioinformatics 18(7), 996–1003 (2002)

7. Dogrusöz, U., Giral, E., Cetintas, A., Civril, A., Demir, E.: A compound graph layout algorithm for biological pathways. In: Pach, J. (ed.) GD 2004. LNCS, vol. 3383, pp. 442–447. Springer, Heidelberg (2005)

8. Funahashi, A., Morohashi, M., Kitano, H.: CellDesigner: a process diagram editor for gene-regulatory and biochemical networks. Biosilico 1(5), 159–162 (2003)

9. Görg, C., Pohl, M., Qeli, E., Xu, K.: Visual Representations. In: Kerren, A., Ebert, A., Meyer, J. (eds.) GI-Dagstuhl Research Seminar 2007. LNCS, vol. 4417, pp. 163–230. Springer, Heidelberg (2007)

10. Junker, B., Klukas, C., Schreiber, F.: VANTED: A system for advanced data analysis and visualization in the context of biological networks. BMC Bioinformatics 7, 109 (2006)

11. Karp, P., Ouzounis, C., Moore-Kochlacs, C., Goldovsky, L., Kaipa, P., Ahren, D., Tsoka, S., Darzentas, N., Kunin, V., Lopez-Bigas, N.: Expansion of the BioCyc collection of pathway/genome databases to 160 genomes. Nucleic Acids Research 33, 6083–6089 (2005)

12. Karp, P., Paley, S.: Automated drawing of metabolic pathways. In: Proc. International Conference on Bioinformatics and Genome Research, pp. 225–238 (1994)

13. Karp, P., Paley, S., Romero, P.: The pathway tools software. Bioinformatics 18(S1), S225–S232 (2002)

14. Kohn, K., Aladjem, M.: Circuit diagrams for biological networks. Molecular Systems Biology 2, e2006.0002 (2006)

15. Li, W., Kurata, H.: A grid layout algorithm for automatic drawing of biochemical networks. Bioinformatics 21(9), 2036–2042 (2005)

16. Michal, G.: Biochemical Pathways, 4th edn. (Poster). Roche (2005)

17. Nagasaki, M., Doi, A., Matsuno, H., Miyano, S.: Genomic Object Net: a platform for modeling and simulating biopathways. Applied Bioinformatics 2, 181–184 (2004)

18. Nicholson, D.: Metabolic Pathways Map (Poster). Sigma Chemical Co. (1997)

19. Novère, N.L., Hucka, M., Mi, H., Moodie, S., Schreiber, F., Sorokin, A., Demir, E., Wegner, K., Aladjem, M., Wimalaratne, S., Bergman, F.T., et al.: The Systems Biology Graphical Notation. Nature Biotechnology 27(8), 735–741 (2009)

20. Offenhuber, D., Donath, J.: Comment Flow: visualizing communication along network paths. Poster presented at IEEE InfoVis 2007 (2007)

21. Ogata, H., Goto, S., Sato, K., Fujibuchi, W., Bono, H., Kanehisa, M.: KEGG: Kyoto encyclopedia of genes and genomes. Nucleic Acids Research 27, 29–34 (1999)

22. Schreiber, F.: High quality visualization of biochemical pathways in BioPath. Silico Biology 2(2), 59–73 (2002)

23. Sirava, M., Schäfer, T., Eiglsperger, M., Kaufmann, M., Kohlbacher, O., Bornberg-Bauer, E., Lenhof, H.: BioMiner - modeling, analyzing, and visualizing biochemical pathways and networks. Bioinformatics 18(S2), 219–230 (2002)

24. Suderman, M., Hallett, M.: Tools for visually exploring biological networks. Bioinformatics 23(20), 2651–2659 (2007)

25. Yost, B., North, C.: Single complex glyphs versus multiple simple glyphs. In: CHI 2005 extended abstracts on human factors in computing systems, pp. 1889–1892. ACM, New York (2005)

A Novel Grid-Based Visualization Approach
for Metabolic Networks
with Advanced Focus&Context View

Markus Rohrschneider[1], Christian Heine[1], André Reichenbach[1],
Andreas Kerren[2], and Gerik Scheuermann[1]

[1] University of Leipzig, Department of Computer Science, Germany
[2] Växjö University, School of Mathematics and Systems Engineering, Sweden

Abstract. The universe of biochemical reactions in metabolic pathways
can be modeled as a complex network structure augmented with domain
specific annotations. Based on the functional properties of the involved
reactions, metabolic networks are often clustered into so-called pathways
inferred from expert knowledge. To support the domain expert in the
exploration and analysis process, we follow the well-known Table Lens
metaphor with the possibility to select multiple foci.

In this paper, we introduce a novel approach to generate an interactive layout of such a metabolic network taking its hierarchical structure
into account and present methods for navigation and exploration that
preserve the mental map. The layout places the network nodes on a fixed
rectilinear grid and routes the edges orthogonally between the node positions. Our approach supports bundled edge routes heuristically minimizing a given cost function based on the number of bends, the number
of edge crossings and the density of edges within a bundle.

1 Introduction

To fully comprehend and appreciate the existing knowledge on chemical processes in living organisms it is essential to develop suitable tools to explore and
navigate through vast amounts of information stored in biological databases. In
biochemistry, complex networks defined by interactions and relations between
different chemical compounds are considered as pathways, such as regulatory
pathways controlling gene activity or metabolic pathways comprising chemical
reactions for synthesis, transformation and degradation of organic substances in
biological systems.

In this work, we combine and apply information visualization techniques to
present the complete set of biochemical reactions of metabolic pathways in a
eucaryotic cell supplying means of exploration and navigation. Although the
emphasis of this paper is placed on biochemical network data, the presented
application is not limited to this area. Instead, it can handle any large graph
carrying arbitrary annotational information by mapping given data properties
to attributes being visualized by the software. To capture the complex chemical
interactions of such a reaction network, metabolic pathways may be modeled as

D. Eppstein and E.R. Gansner (Eds.): GD 2009, LNCS 5849, pp. 268–279, 2010.

Fig. 1. A hyperedge depicting a transaminase reaction, which converts amino acids into corresponding alpha-keto acids and vice versa. In this example, the enzyme Aspartate Aminotransferase converts the substrates L-Aspartate (amino acid) and 2-Ketoglutarate (alpha-keto acid) [input nodes] into the products Oxaloacetate (alpha-keto acid) and L-Glutamate (amino acid) [output nodes]. Many reactions are reversible, so the direction of the hyperedge simply gives a hint on the reaction's chemical equilibrium.

hypergraphs, where unlike a regular graph, each edge can connect an arbitrary number of nodes. In this hypergraph model, each substance is represented by a node of the graph, and each reaction by a (directed) hyperedge connecting the input node set—substrates—with the output node set—products— of the chemical reaction (see Fig. 1). To obtain a hierarchical graph, each metabolic pathway is represented by a node at the top level, where the pathway's reaction network constitutes the nested graph at the bottom level. The division into separate pathways, although based on expert knowledge, is somewhat arbitrary and may not be a strict partition of the graph. Nevertheless, we consider the clustering of the node and edge set as a partition to obtain a strictly confined hierarchy on the graph. Compound nodes and reactions belonging to more than one pathway are simply duplicated for the sake of simplicity of the resulting graph. This step has two benefits: layouting the graph will be a much simpler task, and we can use the hierarchy to explore the network in a top-down manner by examining the top-level graph at first and adding additional information on pathways of interest by expanding nodes.

2   Related Work

The visualization of large and complex biological networks is one of the key analysis techniques to cope with this enormous amount of data. Here, the layout of networks should be in agreement with biological drawing conventions and draw attention to relevant system properties that might remain hidden otherwise [14,13]. Further important issues are the preservation of the so-called mental map [1] when applying small changes to the graph and the possibility of clustering nodes. Depending on the concrete network drawing, there are further important visual representation and interaction techniques that play important roles, e.g., navigation in the complete network, focusing on parts of the network, or gradual differentiability of nodes with less importance (side metabolites) [12]. However, only little research has been done in the past to solve the special layout and visualization problems arising in this area. A lot of the most used software systems for the visual analysis of generic biological networks, i.e., different kinds of networks like regulatory networks or protein-protein interactions, only provide implementations of standard graph drawing algorithms, such as force-directed or hierarchical approaches [8].

Cytoscape [5] is one of the most popular tools for generic biochemical network visualization and supports a number of standard graph layout algorithms. Filtering functions are provided to reduce network complexity. For instance, the user can select nodes and edges according to their name and other attributes. This system allows a simple mapping of data attributes to visual elements of nodes and edges. VisANT [4] is another system designed to visualize generic biochemical networks. In addition to the features of Cytoscape, it provides statistical analysis tools, e.g., based on node degrees or the distribution of clustering coefficients. Their results are displayed in separate views, such as scatter plots.

Especially for metabolic networks, large and hand-drawn posters were produced in the past, for example, Nicholson's pathway map [9] or the widely-used metabolic pathway poster published by Roche Applied Science [18]. Other projects have created graphical representations of metabolic networks and offer them via web pages (e.g., the BioCyc collection [7]). The widespread pathway drawings of the Kyoto Encyclopedia of Genes and Genomes (KEGG) database [16], see also Section 3, were also produced by hand. These drawings are connected via links, but real interaction is not available. Because of their manual generation, they are well readable and can thus serve as an example in terms of quality and user conventions. Moreover, the availability of these representations has established a de facto standard for metabolic network drawings: it features near-orthogonal drawings where, for example, important paths are aligned or relevant subgraphs are placed close to the center of the drawing [14,13].

Newer approaches are based on a close interdisciplinary work between researchers in visualization and biochemistry. An example is the Caleydo framework [21] that extends the standard pathways of KEGG into 2.5D, similar to the report of Kerren [12] and the work of Brandes et al. [15], combined with brushing, highlighting, focus&context, and detail on demand. In this way, it supports the interactive exploration and navigation between several interconnected (but static!) networks.

Saraiya et al. [19] discussed the requirements of metabolic network visualization collected from interviews with biologists. They observed five requirements that are important for biologists working on pathway analysis, but still not completely realized in existing visualization systems (adapted from [10]):

1. automated construction and updating of pathways by searching literature databases;
2. overlaying information on pathways in a biologically relevant format;
3. linking pathways to multi-dimensional data from high-throughput experiments, such as microarrays;
4. overlooking multiple pathways simultaneously with interconnections between them; and
5. scaling pathways to higher levels of abstraction to analyze effects of complex molecular interactions at higher levels of biological organization.

Currently, our approach addresses several of the aforementioned requirements and improves the most previous work by using of interaction techniques from information visualization. Our new, interactive layouts are based on the KEGG data (Req. 1), and we provided the visualization with an intuitive focus&context view. In this way, we can handle, for example, the *complete* metabolism of a generalized eucaryotic cell (Req. 4) by following Shneiderman's mantra [20]: overview first, zoom and filter, details on demand. If the user explores the pathways interactively, the visualization approach preserves the mental map. To the best of our knowledge, no other system can provide that to this extent. Our system is also able to embed textual information into the drawings and to use glyphs/icons for the representation of lower-level subgraphs if needed, similar to the Pathway tools [3]. The integration of more complicated attributes as well as biological patterns regarding topological substructures are still missing. Here, other tools, such as BioPath [6], still have an advantage to be fully accepted by biologists.

The generation of the actual layout of the hierarchical pathway graph is motivated by the style of the "official" KEGG diagrams to be consistent with the domain experts expectations. The diagrams usually use the orthogonal style for drawing edges. To avoid overlapping labels, we ensure a minimum separation of the diagrams elements by using a regular grid based approach. Algorithms for orthogonal grid drawing have been widely studied; we cannot provide an extensive overview here and refer the reader to [8,26] for an introduction. These approaches often follow a topology-shape-metrics approach [25]: First, compute a planar embedding of the input graph, possibly planarizing it by augmenting vertices at crossings, second, compute an orthogonal representation of the embedding, and finally generate coordinates by compaction of the orthogonal representation. Usually, edges are not allowed to run simultaneously on the same grid segment, i.e., connection between two neighboring grid positions. The pathways of the KEGG database can be converted into graphs, but a planarization of them requires an enormous amount of augmented vertices. If edges are not allowed to run on the same grid segment, their layout dominates the area of the drawing resulting in poor resolution. Furthermore, as a pathway constitutes a semantic entity, they should be presented as a unit and without diagram elements from foreign pathways interfering. No existing orthogonal drawing algorithm was able to take these constraints into account, therefore we developed our own that does not planarize the graph but keeps track of edge crossings and heuristically minimizes them, allows "edge bundling" although it penalizes it, shows pathways as units and performs dynamic compactions based on the currently focused parts of the pathway hierarchy.

3   Network Data Source

The development of graph interaction techniques especially suited to fit biological problems makes it necessary to experiment with realistic datasets. To generate artificial graph data is of course possible, but it is hard to estimate the required complexity of such datasets to simulate realistic scenarios. The Kyoto Encyclopedia of Genes and Genomes (KEGG, [16]) System provides annotated pathway data facilitating the construction of metabolic pathway graphs of different sizes. KEGG is one of the major bioinformatics resources publicly accessible. It integrates genomic, chemical, i.e., molecular, and systemic functional information describing cellular processes and organism behavior. It provides a knowledge base for systematic analysis in bioinformatics research and the life sciences. We extracted the hypergraph structure including semantic information as discussed in [17]. The constructed graph covers the complete metabolism of a generalized eucaryotic cell and contains 4980 compound and 154 pathway nodes, 4943 reactions and 1248 inter-pathway edges.

4   Hierarchical Orthogonal Grid Layout

A *hypergraph* $H = (V, E)$ as an extension of the graph concept allows the elements of E called *hyperedges* to connect multiple vertices. Conceptually, a chemical reaction can be described as a hyperedge between compounds that are modeled as vertices. This requires a mark whether a vertex is a substrate or product. We model the data in the KEGG database as a hierarchy of one top-level graph that contains a vertex for each pathway and one hypergraph per pathway. If two pathways exchange compounds according to KEGG, both a regular edge exists between them in the top level graph as well as an edge between the two hypergraphs representing the two pathways.

The layout of the hierarchical KEGG hypergraph is generated by converting the hypergraphs of the hierarchy into their corresponding bipartite graphs and computing a layout of this graph hierarchy. The generated layout is orthogonal to match the style of the official KEGG diagrams. Furthermore, its vertices' positions lie on a grid to ensure both a minimum separation between labels and to make the algorithm both simpler and faster. The layout algorithm allows multi-edges but no loops and proceeds recursively—parents before their children. For each graph of the hierarchy, the layout consists of three phases named: VERTEX POSITION, EDGE ROUTING, and EDGE BUNDLING.

In the VERTEX POSITION phase, we try to find a unique integer position for each vertex that minimizes the stress: the amount of error that takes place by the projection of the "high-dimensional" graph-theoretic distances to the geometric distances between the vertices positions. As vertices and edges are laid out on a regular grid, the Manhattan distance is used as geometric distance. When an edge leads outside the graph, its hierarchy parent has already been laid out. Thus, the direction from which the edge enters the graph is known. For each of the four orientations, we temporarily add a *port* vertex to the graph and connect the edges to foreign graphs to that port. Unlike the other graph's vertices the position of ports is fixed on the boundary of the grid in the upcoming optimization phase.

We implemented a stress minimization algorithm inspired by Kamada and Kawai [22]: starting from an inital random integer positioning of vertices, we select a vertex with high local stress and find a continuous position for that vertex where its local stress becomes minimal using the Newton-Raphson method. We insert it then at the closest integer position not taken by any vertex. We compared this method with two different approaches. A *brute force* version picks a random vertex and tests all integer positions in a vincinity of its position and insert it at the best position. A *simulated annealing* [23] variant picks a random vertex and new position in the vincinity of the old position, but performs the insertion only on improvement and deterioration with decreasing likelihood. We found that the *brute force* method optimized quality as it is difficult to trap this method in a local minimum. The *simulated annealing* is very fast, but does not provide nearly the same quality. All heuristics terminate after a fixed number of iterations that is proportional to the number of vertices.

The EDGE ROUTING phase computes a combinatorial description of an edge routing along the edges of the regular grid. The vertices' positions are not altered by this phase. The combinatorical description is computed one edge at a time and after all edges have been processed once, an iterative process removes single edges and adds them again optimizing on the global cost of the layout. Given a combinatorial description of the current edge routing, we construct a *route graph* that consists of the original graph's vertices and the grid's edges as vertices and edges for valid transitions between these elements. Given this representation, we are able to compute the optimal routing of an edge by solving a single-pair-shortest-path instance on the *route graph*. The optimality is given by a cost function that takes the number of crossings, the number of bends, the length of an edge, and the "density" of edges on a grid segment into account. Note that the quality of the resulting configuration depends both on the original vertices' positions and the actual order of edge insertions. Good performance was achieved when inserting the edges in the order of increasing distances of their incident vertices. To reduce runtime and memory consumption, we use the A^* search algorithm [24] to solve the SPSP instance using the Manhattan distance as heuristic.

The EDGE BUNDLING phase shifts segments of edges' routes orthogonal to the grid segments they lie on to remove overlaps. It preserves the edges' relative ordering and straightens them in the process. This problem can be solved for each row and each column separately. We generate for each row and each column a directed acyclic graph that contains line segments as vertices and edges between these lines, if they are ordered in the combinatorial edge routing. Any topological numbering of this graph gives a displacement that avoids occlusions between edge routes of the same column/row, and using the topological numbering of minimum weight packs the edge bundles nicely together.

5 Graph Interaction

The graph interaction and exploration methods described in this section have all been implemented in our visualization software. The grid layout algorithm is the central component of the adapted Table Lens method to explore hierarchical graphs. We firstly present this technique with supplementary search and highlighting operations and explain later how the graphical user interface lets the user apply these methods to interact with the metabolic network graph.

5.1   Exploration Techniques

Two fundamental navigation operations on hierarchical graphs are node expansion to reveal the node's nested graph and collapse. For 2D graph representations, it is natural and desirable to present a flat graph at all times regardless the graph's expansion state. This means that the expansion of a node requires it to be hidden and replaced by its nested graph. The inverse operation replaces the nested graph by its parent. The well-known Table Lens metaphor [2] applied to hierarchical graph exploration fulfills this requirement. It is an established focus&context method to give an overview on large tabular datasets to examine obvious patterns and to provide detailed view on specific items at the same time. In our application, pathway nodes at the top level are placed in the center of a cell, edges are routed along the cell borders as intended result of the previously presented layout algorithm. When a node is expanded, the row and the column are enlarged in which the node is situated. Edges leading to and from one of the four ports (see Sec. 4 and Fig. 4 for example) of the pathway node are elongated while the remaining elements keep their relative position. This approach follows Ben Shneiderman's mantra of visual information-seeking: overview first, zoom and filter, details on demand [20]. Our application supports this concept in the following ways:

Overview first. The grid layout algorithm positions top-level nodes on a regular grid where each grid position can be regarded as a cell in a table. The user starts with examining the completely collapsed graph, i.e., only top-level nodes are visible. The application allows to display a node simply by showing the associated pathway's name as caption (see Fig. 2) or by creating an iconized view of the node's nested graph.

Zoom and Filter. We have implemented *semantic zooming* to display labels once a certain threshold is reached. Tool tips add additional information on each pathway node. If enabled, icons in top-level nodes depicting the nested graph give a quick hint on the pathway's size and complexity.

Details on Demand. The user can expand selected pathway nodes to explore the detailed network of chemical reactions. In contrast to the established Table Lens method, an arbitrary number of cells (pathways) can be enlarged (*multiple foci*) and examined in detail (see Fig. 3 and 4). Advanced selection and highlighting techniques facilitate and support the exploration process: selecting a pathway node highlights all objects belonging to that cluster. Selecting a specific reaction node highlights all edges to the associated substrate and product nodes. Selecting a compound node highlights all reactions this compound is involved including its connections to adjacent pathways.

5.2   Design of the Graphical User Interface

The GUI of the visualization software basically consists of three components, see Fig. 6 and 7.

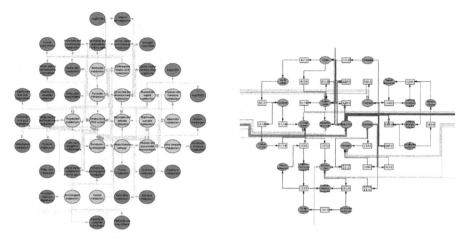

Fig. 2. The top-level graph of the "Carbohydrate Metabolism" (bright) and related pathways (dark). The highlighted nodes can be expanded to reveal the detailed reaction network.

Fig. 3. Detailed view of the expanded node "Citrate Cycle (TCA)". The highlighted compound node "Acetyl-CoA" plays a central role within this pathway and establishes several connections to adjacent pathways.

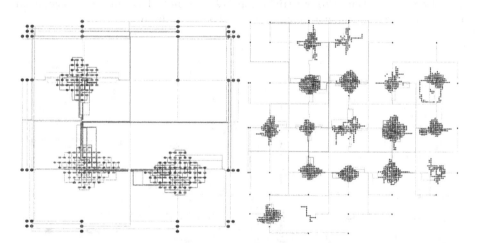

Fig. 4. Three expanded pathway nodes: "Citrate Cycle (TCA)", "Pentose Phosphate Pathway", and "Glycolysis / Gluconeogenesis"; the latter being highlighted in red with the connections to adjacent pathway nodes.

Fig. 5. Bottom-level graph. Reaction network of pathways associated with the carbohydrate metabolism. The search result for "Pyruvate" is highlighted including its incident edges.

The *Graph View* at the left hand side of the window renders the graph and provides an interface to interact with or edit the topology of the graph directly. Each graphical object can be individually selected, and applicable properties can

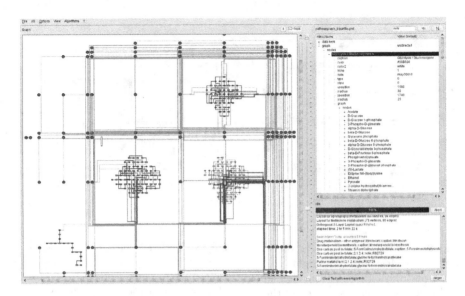

Fig. 6. GUI of the Visualization and Editing Tool. The top-level graph consisting of 154 pathway vertices with 4 expanded pathways. The node "Glycolysis/Gluconeogenesis" was selected in the Data Browser (right, top) resulting in highlighting all its compound and reaction nodes including connections to adjacent pathways.

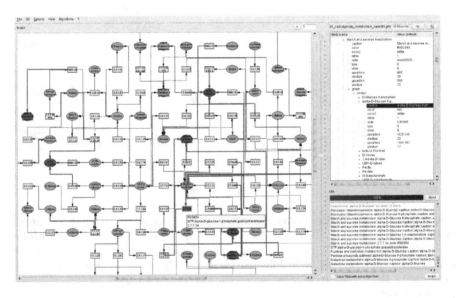

Fig. 7. A more detailed view of the bottom level graph. This portion of the graph displays the pathway *Starch and Sucrose metabolism*. The *Algorithm Info Area* (bottom, right) gives feedback on invoked algorithms and displays search results. In this scenario, a search for the term *alpha-D-Glucose* was performed and resulted in 13 matches being highlighted in the Graph View and in the hierarchical Data Browser view.

be assigned via a context menu. The integrated graph editing capability allows the user to manually construct pathway graphs or to modify a given layout either generated by the algorithm or loaded from file. Expanding or collapsing individual nodes can be performed by either double-clicking a node or selecting the operation via the associated entry in the context menu.

The *Data Browser* displays the hierarchical structure of the grap (explorer layout) and grants access to textual or numerical attributes of each graph element. Generic graph element properties, e.g., edge width, node size and shape, color or transparency, can be manipulated, and the effects will be directly displayed in the graph view. A simple search function among the textual attributes can be used as a filter to highlight and select a group of graph elements. This is an intuitive way to state queries like *"Select all pathways containing the compound Pyruvate"* (see Fig. 5). Highlighting elements matching a given search pattern is also propagated to the top-level.

The *Algorithm Info Area* at the bottom-right hand side displays textual output giving feedback on the progress of invoked graph or layout algorithms and to present search results, cf. Fig. 7.

6   Performance Results

Our KEGG import routine is suitable to construct pathway graphs of different size and complexity. To implement, test and demonstrate the discussed techniques, we constructed two graphs. Images 2 through 5 were created from 17 pathway files downloaded from the KEGG database covering the complete carbohydrate metabolism. Additional non-expandable pathway nodes were created when referenced in one of the input files. A graph with a total of 649 compound nodes, 50 pathway nodes, 814 reaction hyperedges and 149 regular inter-pathway edges was created. The portion of the graph containing the hyperedges and nodes was converted into a bipartite graph, where the previous hyperedges are displayed as rectangular nodes (yellow) labeled with the EC numbers of the catalyzing enzymes and the nodes as ellipses (green) labeled with the compound's chemical name, resulting in a total number of 1,513 nodes and 1,861 edges. This graph could easily be handled by the visualization software. On an Intel® Xeon® (2 GHz, 32 GB RAM) machine, response times of the graphical user interface were less than 0.2 sec for any operation discussed previously enabling a smooth interaction with the displayed graph. A second example was more complex. After converting the hypergraph portion into a bipartite subgraph, the graph describing the complete metabolism of a generalized eucaryotic cell had a total number of 10,067 nodes and 11,706 edges. Depending on the visible portion in the scrollable graph view area, any collapse/expand operation took up to 4 sec if the complete graph was visible, and up to 2 sec if one pathway was located in the visible area. The response times for scrolling, zooming, and highlighting elements for the worst-case scenario (all pathways expanded) were less than 0.75 sec if up to 1/4 of the graph's elements were visible, and less than 0.25 sec if the visible portion was 1/10 of the completely laid out graph.

The runtime of the grid layout algorithm heavily depends on the choice of parameters. For large graphs, the brute-force method testing all grid positions naturally takes longer compared to the simulated annealing method. The choice of the area ratio $a = 4 \cdot |V|$ generally produced more aesthetic layouts for cyclic and chain-like structures because of the larger space available to unfold those substructures, but resulted in increased runtime for the brute-force method.

7   Conclusion

The proposed software is able to layout and display complex graphs with a high number of elements. The development process was intensively accompanied by domain experts from biology and biochemistry. For metabolic pathway networks, not only the graph topology is relevant, a high number of additional attributes—textual annotations in our case—need to be visualized. Semantic zooming and focus&context methods are applied to accomplish this goal, instant highlighting of graph elements fitting the pattern of a string based search operation is an intuitive way to extract specific information on the dataset. The main benefit of the adapted Table Lens method is the preservation of the mental map. Many of the visualization tools lack this key feature. Even though node expansion and collapse produce very discrete and rather abrupt changes in the graph appearance, only the row and the column of the grid position are affected while the remaining elements keep their relative position. In combination with continuous zooming, it is a straightforward task to explore even large graphs. Highlighting individual or groups of edges greatly facilitates the tracking of routes. In the presented grid layout algorithm, vertex placement and edge routing are performed in two separate steps. This offers the opportunity to develop alternative node placement routines fitting the specific needs of pathway visualization in the future.

Acknowledgments. This work has been funded by the Volkswagen Stiftung under grant I/82 719.

References

1. Misue, K., Eades, P., Lai, W., Sugiyama, K.: Layout Adjustment and the Mental Map. Journal of Visual Languages and Computing 6(2), 183–210 (1995)
2. Rao, R., Card, S.K.: The table lens: merging graphical and symbolic representations in an interactive focus + context visualization for tabular information. In: Proceedings of the SIGCHI conference on Human factors in computing systems, pp. 318–322. ACM, New York (1994)
3. Karp, P.D., Paley, S.M., Romero, P.: The Pathway Tools software. Bioinformatics 18, 225–232 (2002)
4. Hu, Z., Mellor, J., Wu, J., DeLisi, C.: VisANT: an online visualization and analysis tool for biological interaction data. BMC Bioinformatics 5(1), e17 (2004)
5. Shannon, P., Markiel, A., Ozier, O., Baliga, N.S., Wang, J.T., Ramage, D., Amin, N., Schwikowski, B., Ideker, T.: Cytoscape: a software environment for integrated models of biomolecular interaction networks. Gen. Res. 13(11), 2498–2504 (2003)
6. Schreiber, F.: High Quality Visualization of Biochemical Pathways in BioPath. In Silico Biology 2(2), 59–73 (2002)

7. Karp, P.D., Ouzounis, C.A., Moore-Kochlacs, C., Goldovsky, L., Kaipa, P., Ahren, D., Tsoka, S., Darzentas, N., Kunin, V., Lopez-Bigas, N.: Expansion of the Bio-Cyc collection of pathway/genome databases to 160 genomes. Nucleic Acids Research 33(19), 6083–6089 (2005)
8. DiBattista, G., Eades, P., Tamassia, R., Tollis, I.G.: Graph Drawing: Algorithms for the Visualization of Graphs. Prentice Hall, New Jersey (1999)
9. Nicholson, D.E.: Metabolic Pathways Map (Poster). Sigma Chemical Co., St. Louis (1997)
10. Lungu, M., Xu, K.: Biomedical Information Visualization. In: Kerren, A., Ebert, A., Meyer, J. (eds.) GI-Dagstuhl Research Seminar 2007. LNCS, vol. 4417, pp. 311–342. Springer, Heidelberg (2007)
11. Kerren, A., Ebert, A., Meyer, J.: Human-Centered Visualization Environments. In: Kerren, A., Ebert, A., Meyer, J. (eds.) GI-Dagstuhl Research Seminar 2007. LNCS, vol. 4417, pp. 1–9. Springer, Heidelberg (2007)
12. Kerren, A.: Interactive Visualization and Automatic Analysis of Metabolic Networks – A Project Idea. Technical Report, Institute of Computer Graphics and Algorithms, Vienna University of Technology, Austria (2003)
13. Albrecht, M., Kerren, A., Klein, K., Kohlbacher, O., Mutzel, P., Paul, W., Schreiber, F., Wybrow, M.: On Open Problems in Biological Network Visualization. In: Proc. of the 17th International Symposium on Graph Drawing, Chicago, USA. Springer, Heidelberg (2009) (to appear)
14. Albrecht, M., Kerren, A., Klein, K., Kohlbacher, O., Mutzel, P., Paul, W., Schreiber, F., Wybrow, M.: A Graph-drawing Perspective to Some Open Problems in Molecular Biology. Technical report TR08-01-003, Lehrstuhl XI für Algorithm Engineering, Fakultät für Informatik, TU Dortmund, Germany (2008)
15. Brandes, U., Dwyer, T., Schreiber, F.: Visualizing Related Metabolic Pathways in Two and a Half Dimensions. In: Liotta, G. (ed.) GD 2003. LNCS, vol. 2912, pp. 111–122. Springer, Heidelberg (2004)
16. Kyoto Encyclopedia of Genes and Genomes, http://www.kegg.jp/kegg/
17. Klukas, C., Schreiber, F.: Dynamic exploration and editing of KEGG pathway diagrams. Bioinformatics 23(3), 344–350 (2007)
18. Michal, G.: Biochemical Pathways: Biochemie-Atlas. Spektrum Akademischer Verlag, Heidelberg (1999)
19. Saraiya, P., North, C., Duca, K.: Visualizing biological pathways: requirements analysis, systems evaluation and research agenda. Information Visualization 4(3), 191–205 (2005)
20. Shneiderman, B.: The Eyes Have It: A Task by Data Type Taxonomy for Information Visualizations. In: VL, pp. 336–343 (1996)
21. Streit, M., Kalkusch, M., Kashofer, K., Schmalstieg, D.: Navigation and Exploration of Interconnected Pathways. Eurographics / IEEE-VGTC Symposium on Visualization 27(3) (2008)
22. Kamada, T., Kawai, S.: An algorithm for drawing general undirected graphs. Inf. Process. Lett. 31(1), 7–15 (1989)
23. Kirkpatrick, S., Gelatt, C.D., Vecchi, M.P.: Optimization by simulated annealing. Science 220, 671–680 (1983)
24. Hart, P.E., Nilsson, N.J., Raphael, B.: A Formal Basis for the Heuristic Determination of Minimum Cost Paths. IEEE Transactions on Systems Science and Cybernetics 4(2), 100–107 (1968)
25. Batini, C., Nardelli, E., Tamassia, R.: A Layout Algorithm for Data Flow Diagrams. IEEE Trans. Software Eng. 12(4), 538–546 (1986)
26. Kaufmann, M., Wagner, D. (eds.): Drawing Graphs. LNCS, vol. 2025. Springer, Heidelberg (2001)

Small Drawings of Series-Parallel Graphs and Other Subclasses of Planar Graphs*

Therese Biedl

David R. Cheriton School of Computer Science, University of Waterloo,
Waterloo, Ontario N2L 3G1, Canada
biedl@uwaterloo.ca

Abstract. In this paper, we study small planar drawings of planar graphs. For arbitrary planar graphs, $\Theta(n^2)$ is the established upper and lower bound on the worst-case area. It is a long-standing open problem for what graphs smaller area can be achieved, with results known only for trees and outer-planar graphs. We show here that series-parallel can be drawn in $O(n^{3/2})$ area, but 2-outer-planar graphs and planar graphs of proper pathwidth 3 require $\Omega(n^2)$ area.

1 Introduction

A planar graph is a graph that can be drawn without crossing. It was established 20 years ago [15,20] that it has a straight-line drawing in area $O(n^2)$ with vertices placed at grid points. This is asymptotically optimal, since there are planar graphs that need $\Omega(n^2)$ area [14].

A number of other graph drawing models (e.g., poly-line drawings, orthogonal drawings, visibility representations) exist for planar graphs. In all these models, $O(n^2)$ area can be achieved for planar graphs, see for example [17,23]. On the other hand, $\Omega(n^2)$ area is needed, in all models, for the graph in [14]. This raises the natural question [5] whether $o(n^2)$ area is possible for subclasses of planar graphs.

Known results. Every **tree** has a straight-line drawing in $O(n \log n)$ area and in $O(n)$ area if the maximum degree is asymptotically smaller than n. See [7] for references and many other upper and lower bounds regarding drawings of trees.

It is quite easy (and appears to be folklore) to create straight-line drawings of **outer-planar graphs** that have area $O(nd)$, where d is the diameter of the dual tree of the graph. In an earlier paper [3], we showed that any outer-planar graph has a visibility representation (and hence a poly-line drawing) of area $O(n \log n)$. Since then, some work has been done on improving the bounds for straight-line drawings, with the best bounds now being $O(n^{1.48})$ [8] and $O(\Delta n \log n)$ [12].

Many drawing results are known for **series-parallel graphs**, see e.g. [1,6,16, 22]. However, the emphasis here was on displaying the series-parallel structure of the graph, and/or to use the structure to allow for additional constraints. All

---

* Research supported by NSERC. Part of the work was done while the author was on sabbatical leave at University of Passau.

D. Eppstein and E.R. Gansner (Eds.): GD 2009, LNCS 5849, pp. 280–291, 2010.

known algorithms bound the area by $O(n^2)$ area or worse. Quite recently, Frati proved a lower bound (for straight-line or poly-line drawings) of $\Omega(n \log n)$ for a series-parallel graph [13].

No graph drawing results specifically tailored to k-**outer-planar graphs** (for $k \geq 2$), or planar graphs with **small treewidth/pathwidth** appear to be known for 2-dimensional drawings. Planar graphs with small pathwidth play a critical role in drawings where the height is bounded by a constant [9], but not all graphs with small pathwidth have such a drawing.

While higher-dimensional drawings are not the focus of our paper, we would like to mention briefly that all graph classes considered in this paper can be drawn with linear area in 3D, because they are partial k-trees for constant k; see [10], and also [11] for some earlier 3D results for outer-planar graphs.

We would also like to note that all these graphs have small separators, hence all of them allow a non-planar two-dimensional orthogonal drawing in $O(n)$ area if the maximum degree is at most 4 [18].

Our Results. In this paper, we provide the following results:

- Every series-parallel graph has a visibility representation with $O(n^{3/2})$ area.
- A series-parallel graph for which at most f graphs are combined in parallel has a visibility representation with $O(fn \log n)$ area. We know $f \leq \Delta$.
- There are series-parallel graphs that require $\Omega(n^2)$ area in any poly-line drawing that respects the planar embedding.
- There are 2-outer-planar graphs that require $\Omega(n^2)$ area in any poly-line drawing. Moreover, these graphs have pathwidth 3.
- There are graphs of proper pathwidth 3 and maximum degree 4 that require $\Omega(n^2)$ area.

For algorithms, we restrict our attention to visibility representations, because any such drawing can be converted to a poly-line drawing with asymptotically the same area. Hence all our upper bounds also hold for poly-line drawings.

2   Background

Let $G = (V, E)$ be a graph with $n = n(G) = |V|$ vertices and $m = m(G) = |E|$ edges. Throughout this paper, we will assume that G is *simple* (has no loops or multiple edges) and *planar*, i.e., can be drawn without crossing. A planar drawing splits the plane into connected pieces; the unbounded piece is called the *outer-face*, all other pieces are called *interior faces*. An *outer-planar graph* is a planar graph that can be drawn such that all vertices are on the outer-face.

A *2-terminal series-parallel graph with terminals* s, t is a graph defined recursively with one of the following three rules: (a) An edge (s, t) is a 2-terminal series-parallel graph. (b) If G_i, $i = 1, 2$ are 2-terminal series-parallel graphs with terminals s_i and t_i, then in a *series composition* we identify t_1 with s_2 to obtain a 2-terminal series-parallel graph with terminals s_1 and t_2. (c) If G_i, $i = 1, \ldots, k$, are 2-terminal series-parallel graphs with terminals s_i and t_i, then in a *parallel composition* we identify $s_1, s_2, \ldots, s_k$ into one terminal s and $t_1, t_2, \ldots, t_k$ into one terminal t to obtain a 2-terminal series-parallel graph with terminals s and t. Here k is as large as possible, i.e., none of the graphs G_i is itself obtained via a parallel composition. The *fan-out* of a series-parallel graph is the maximum number of subgraphs k used in a parallel composition.

Given a 2-terminal series-parallel graph G, a *subgraph from the composition* is any of the subgraphs $G_1, \ldots, G_k$ that was used to create G, or recursively any subgraph from the composition of $G_1, \ldots, G_k$. Since we never consider any other subgraphs, we will say "subgraphs" instead of "subgraphs from the composition".

A *series-parallel graph*, or *SP-graph* for short, is a graph for which every biconnected component is a 2-terminal series-parallel graph. It is *maximal* if no edge can be added while maintaining a simple SP-graph. Any maximal series-parallel graph is a 2-terminal series-parallel graph where in any parallel composition there exists an edge between the terminals, and in any series composition each subgraph is either an edge or obtained from a parallel composition. We will only considering drawings of maximal series-parallel graph, since this makes no difference for asymptotic upper bounds on the area of graph drawings.

A *polyline-drawing* is an assignment of vertices to points and edges to a path of finitely many line segments connecting their endpoints. A *visibiliy representation* is an assignment of vertices to boxes[1] and edges to horizontal or vertical line segment connecting boxes of their endpoints. For a planar graph, such drawings should be planar, i.e., have no crossing. We also assume that all defining features have integral coordinates; in particular points of vertices and transition-points (*bends*) in the routes of edges have integral coordinates, and boxes of vertices have integral corner points. We allow boxes to be degenerate, i.e., to be line segments or points.

The *width* of a box is the number of vertical grid lines (*columns*) that are occupied by it. The *height* of a box is the number of horizontal grid lines (*rows*) that are occupied by it. A drawing whose minimum enclosing box has width w and height h is called a $w \times h$-drawing, and has *area* $w \cdot h$.

3   Visibility Representations of Series-Parallel Graphs

In this section, we study how to create a small visibility representation of a maximal SP-graph G. Our algorithm draws G and recursively all its subgraphs H. To ease putting drawings together, we put constraints on the drawing (see also Fig. 1):

- The visibility representation is what we call *flat*: every vertex is represented by a horizontal line segment.
- Vertex s is placed in the upper right corner of the bounding box.
- Vertex t is placed in the lower right corner of the bounding box.

With our construction we develop a recursive formula for the height: $h(m)$ is the maximum height of a drawing obtained with our algorithm over all maximal

---

[1] In this paper, the term "box" always refers to an axis-parallel box.

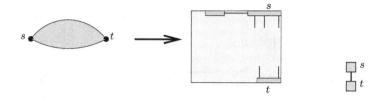

Fig. 1. Illustration of the invariant, and the base case $n = 2$

SP-graphs with m edges.(We have $m = 2n - 3$, but use m to simplify the computations.) In the base case ($m = 1$), simply place s atop t; see Fig. 1. The conditions are clearly satisfied, and we have $h(m) = 2$ for $m = 1$.

Modifying drawings. If $m \geq 2$, then we obtain the drawing by merging drawings of subgraphs together suitably. Before doing this, we sometimes modify them with an operation used earlier [3]. We say that in a drawing a vertex *spans the top (bottom) row* if its vertex box contains both the top (bottom) left point and the top (bottom) right point of the smallest enclosing box of the drawing. We can always achieve that terminal s spans the top row after adding a row; we call this *releasing terminal s*. Similarly we can also release terminal t after adding a row.

Lemma 1. *[3] Let $\Gamma(H)$ be a flat visibility representation of H of height $h \geq 2$ that satisfies the invariant. Then there exists a flat visibility representation $\Gamma'(H)$ of H of height $h + 1$ that satisfies the invariant, and vertex s spans the top row.*

Subgraphs from parallel compositions. Assume H is a subgraph of G which is obtained in a parallel composition from subgraphs $H_1, \ldots, H_k$, $k \geq 2$. After possible renaming, assume that $m_i = m(H_i)$ satisfies $m_1 \geq m_2 \geq \ldots \geq m_k$. Recursively obtain drawings of $H_1, \ldots, H_k$; the drawing of H_i has height at most $h(m_i)$. Combine them after releasing both terminals in all of $H_2, \ldots, H_k$ and adding rows so that all drawings have the same height. Place H_1 leftmost, and all other H_i to the right of it; this gives a drawing of H that satisfies the invariant, see Fig. 2. Since $m_2 \geq m_3 \geq \ldots \geq m_k$, the height of this drawing is

$$h(m) \leq \max\{h(m_1), h(m_2) + 2, \ldots, h(m_k) + 2\} = \max\{h(m_1), h(m_2) + 2\} \quad (1)$$

Subgraphs from series compositions. Now let H (with terminals s, t) be obtained from a series composition of graphs H_a and H_b with terminals s, x and x, t, respectively. Since we consider maximal SP-graphs, each of H_a and H_b is either an edge or obtained from a parallel composition. We distinguish cases.

Case (S1): One subgraph, say H_b, is an edge. Then we draw H_a recursively, extend the drawing of terminal s to the right, place t in the bottom row, and connect edge (x, t) horizontally. See Fig. 3. The case that H_a is an edge is symmetric. We have $h(m) = h(m - 1)$ in this case.

Case (S2): Both subgraphs have at least two edges. Assume that $m(H_b) \leq m(H_a)$; the other case is symmetric. Graph H_b was obtained from a parallel

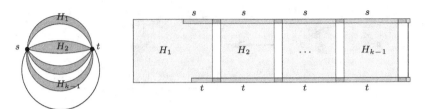

Fig. 2. Combining subgraphs in parallel

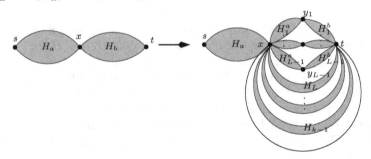

Fig. 3. A series composition when one subgraph is an edge

composition of subgraphs, say $H_1, \ldots, H_k$ such that $m(H_1) \geq \ldots \geq m(H_k)$. Note that H_k is the edge (x, t), which exists since the SP-graph is maximal.

Let L be an integer; we will discuss later how to choose L. For all $i <$ $\min\{L, k\}$, we break subgraph H_i up further. Graph H_i is not an edge (since $i < k$ and H_k is an edge), and so is obtained in a series composition of two subgraphs H_i^a and H_i^b with terminals x, y_i and y_i, t, respectively. See also Fig. 4. Set $m_\alpha^\beta = m(H_\alpha^\beta)$ for any strings α and β.

Fig. 4. Breaking down subgraph H_b

Recursively draw each of the subgraphs H_a, H_i^a, H_i^b (for $i = 1, \ldots,$ $\min\{k, L\}-1,$) and H_i (for $i = L, \ldots, k-1$.) Before we can combine these drawings, we need to release some terminals again (recall Lemma 1). We proceed as follows:

- The drawing of H_a is unchanged and has height $h(m_a)$.
- For $i = 1, \ldots, \min\{L, k\} - 1$, release terminal x in the drawing of H_i^a, and terminal t in the drawing of H_i^b. The drawings hence have height at most $h(m_i^a) + 1$ and $h(m_i^b) + 1$.
- For $i = L, \ldots, k - 1$, release both terminals in the drawing of H_i.

To explain how we put these drawings together, we distinguish two sub-cases:

Case (S2a): Assume first that $k \leq L$, and consider Fig. 5. We place H_a on the left, followed by $H_1^a, H_2^a, \ldots, H_{k-1}^a$. All these graphs share terminal x, and it spans the bottom row for $H_1^a, H_2^a, \ldots, H_{k-1}^a$, so this draws x as a horizontal segment. Now for $i = 1, \ldots, k-1$, rotate the drawing of H_i^b such that terminal t spans the bottom row and terminal y_i occupies the top left corner. We place these rotated drawings in order $H_{k-1}^b, H_{k-2}^b, \ldots, H_1^b$; then t is in the bottom row and can be connected to x with a horizontal edge.

We increase the heights of these drawings (by inserting rows, if needed) such that the two representations of y_i are in the same row, y_i is above the drawing of y_{i+1} (for $i < k-1$), and s is above all y_i's. Then all terminals can be represented as line segments and the invariant holds.

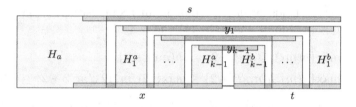

Fig. 5. Combining the subgraphs for a series composition. The case $k \leq L$.

Let h_i be the height of the drawing of H_i^a and H_i^b together in the final drawing. Then $h_{k-1} \leq \max\{h(m_{k-1}^a) + 1, h(m_{k-1}^b) + 1\} \leq h(m_{k-1}) + 1$. For $i < k - 1$, the height has been increased further to keep y_i above y_{i+1}, hence $h_i \leq \max\{h(m_i)+1, h_{i+1}+1\}$. Therefore, y_1 is at height $h_1 \leq \max\{h(m_1)+1, h(m_2) + 2, \ldots, h(m_{k-1}) + k - 1\}$, s is at least one higher, and the total height is

$$h(m) \leq \max\{h(m_a), h(m_1) + 2, h(m_2) + 3, \ldots, h(m_{k-1}) + k\} \qquad (2)$$

Case (S2b): Now we study the case $k > L$, where we treat the graphs $H_L, \ldots, H_{k-1}$ differently. Place $H_a, H_1^a, \ldots, H_{L-1}^a, H_{L-1}^b, \ldots, H_1^b$ exactly as before. Add rows until $H_L, \ldots, H_{k-1}$ all have the same height, say h_d, and place them *below* the segment of x. We may have to add some columns to x if it is not wide enough for the subgraphs. To make the two occurrences of t match up, we extend the drawings of $H_{L-1}^b, \ldots, H_1^b$ downwards and draw edge (x, t) vertically. See Fig. 6.

To obtain a formula for the resulting height, we hence need to add $h_d - 1$ to the formula of (2) (after replacing k by L in it.) Since h_d is the maximum height among $H_L, \ldots, H_{k-1}$, and $m_L \geq \ldots \geq m_k$, we have $h_d \leq h(m_L) + 2$ (recall that both terminals were released for $H_L, \ldots, H_{k-1}$), and therefore

$$h(m) \leq \max\{h(m_a), h(m_1) + 2, h(m_2) + 3, \ldots, h(m_{L-1}) + L\} + h(m_L) + 1 \quad (3)$$

Analysis. Now we show that the above algorithm indeed yields a small area.

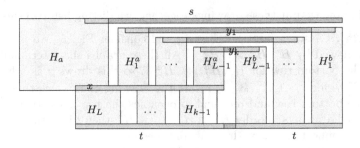

Fig. 6. Combining the subgraphs for a series composition. The case $k \geq L$.

Lemma 2. *For a suitable choice of L, we have $h(m) \leq 12\sqrt{m}$.*

Proof. This clearly holds for $m = 1$. For a parallel composition, we have $m_1 \geq m_i$ and hence $m_i \leq m/2$ for $i \geq 2$, so by (1) and $m \geq 2$

$$h(m) \leq \max\{h(m_1), h(m_2) + 2, \ldots, h(m_k) + 2\}$$
$$\leq \max\{h(m), h(m/2) + 2\} \leq \max\{12\sqrt{m}, 12\sqrt{m/2} + 2\} \leq 12\sqrt{m}.$$

In case (S1), we have $h(m) = h(m_a) \leq 12\sqrt{m_a} \leq 12\sqrt{m}$. In case (S2), we assumed $m_a \geq m_b$. Also, $m_b \geq 3$ (because H_1^a and H_1^b have each an edge, and (x, t) exists), and hence $m \geq 6$. We choose $L = 3\sqrt{m_a} + 1$.[2] Now for case (S2a), we have by (2)

$$h(m) \leq \max\{h(m_a), h(m_1) + 2, h(m_2) + 3, \ldots, h(m_{k-1}) + k\}$$
$$\leq \max\{h(m_a), h(m/2) + L\} \text{ since } m_i \leq m_b \leq m/2 \text{ and } k \leq L$$
$$\leq \max\{12\sqrt{m_a}, 12\sqrt{m/2} + 3\sqrt{m_a} + \frac{1}{\sqrt{6}}\sqrt{m}\} \text{ since } L = 3\sqrt{m_a} + 1 \text{ and } m \geq 6$$
$$\leq \max\{12, (\frac{12}{\sqrt{2}} + 3 + \frac{1}{\sqrt{6}})\}\sqrt{m} \leq 12\sqrt{m}$$

Finally we consider case (S2b). We have $m_1 \leq m_b \leq m_a$ and $m_i \leq m_1$, hence $m_i \leq m_b/2 \leq m_a/2$ for all $i \geq 2$. Recall that the height in case (S2b) is by (3)

$$h(m) \leq \max\{h(m_a), h(m_1) + 2, h(m_2) + 3, \ldots, h(m_{L-1}) + L\} + h(m_L) + 1$$
$$\leq \max\{h(m_a), h(m_a/2) + L - 2\} + h(m_L) + 3 \leq 12\sqrt{m_a} + 12\sqrt{m_L} + 3,$$

where the last inequality holds by induction and because $h(m_a/2) + L - 2 \leq 12\sqrt{\frac{m_a}{2}} + 3\sqrt{m_a} - 1 \leq 12\sqrt{m_a}$. But

$$(\sqrt{m_a} + \sqrt{m_L} + \frac{1}{4})^2 = m_a + m_L + \frac{1}{16} + 2\sqrt{m_a}\sqrt{m_L} + \frac{1}{2}\sqrt{m_a} + \frac{1}{2}\sqrt{m_L}$$
$$\leq m_a + m_L + \frac{1}{16}\sqrt{m_a m_L} + 2\sqrt{m_a m_L} + \frac{1}{2}\sqrt{m_a}m_L + \frac{1}{3}\sqrt{m_a}m_L$$

---

[2] Many thanks to Jason Schattman for helping with MAPLE to find small constants.

by $\sqrt{m_a} \geq \sqrt{3} \geq \frac{3}{2}$

$$\leq m_a + m_L + 3\sqrt{m_a}m_L = m_a + m_L + (L-1)m_L \text{ by } L = 3\sqrt{m_a} + 1$$

$$\leq m_a + m_L + m_1 + m_2 + \ldots + m_{L-1} \text{ by } m_i \geq m_L \text{ for } i < L$$

which is at most m. Putting it together, we get $h(m) \leq 12(\sqrt{m_a} + \sqrt{m_L} + \frac{1}{4}) \leq 12\sqrt{m}$ as desired. $\qquad\square$

Theorem 1. *Any series-parallel graph has a visibility representation with area* $O(n^{3/2})$.

Proof. By the previous lemma, the height is $O(\sqrt{m}) = O(\sqrt{n})$ by $m = 2n-3$. To analyze the width, notice that at the most we use one column for each edge. (Each vertex obtains at least one incident vertical edge in the base case, and hence does not contribute additional width.) Hence the width is at most $m \leq 2n - 3$, and the total area is $O(n^{3/2})$. $\qquad\square$

We get better bounds if case (S2b) does not happen, i.e., if the series-parallel graph has small fan-out.

Theorem 2. *Any series-parallel graph with fan-out f has a visibility representation of area* $O(fn \log n)$.

Proof. Assume first the graph is maximal. As in Theorem 1 the width is $O(n)$, so it suffices to show that $h(m) \leq 2 + f \log m$ for a maximal SP-graph with fanout f. We proceed by induction on the number of edges. In the base case $h(1) = 2 \leq 2 + f \log m$. In case of a parallel composition, by (1) we have $m_2 \leq m/2$ and height

$$h(m) \leq \max\{h(m_1), h(m_2) + 2\} \leq \max\{h(m_1), h(m/2) + 2\}$$
$$\leq \max\{2 + f \log m_1, 2 + f \log(m/2) + 2\} \leq 2 + f \log m$$

since $f \geq 2$. For case (S1), the height is $h(m) = h(m_a) \leq 2 + f \log m_a \leq 2 + f \log m$. In case (S2), we choose $L = f$, and hence always have $k \leq L$ and are in case (S2a). Here, the height is by (2)

$$h(m) \leq \max\{h(m_a), h(m_1) + 2, h(m_2) + 3, \ldots, h(m_{k-1}) + k\}$$
$$\leq \max\{h(m_a), h(m/2) + f\} \text{ since } m_i \leq m/2 \text{ and } k \leq f$$
$$\leq \max\{2 + f \log m_a, 2 + f \log(m/2) + f\} \leq 2 + f \log m.$$

If the graph is not maximal, then it can be made a maximal SP-graph by adding edges; this adds at most one to the fan-out f and hence the drawing of the super-graph has area $O(fn \log n)$. $\qquad\square$

Note in particular that a series-parallel graph with maximum degree Δ has fan-out at most Δ, so any series-parallel graph has a flat visibility representation of area $O(\Delta n \log n)$. Also, any outer-planar graph is an SP-graph with fan-out $f \leq 2$, so this theorem implies our earlier result [3], and in fact yields exactly the same visibility representation.

We note here that most algorithms for visibility representations of planar graphs (e.g. [23, 21]) are *uni-directional*, i.e., all edges are drawn as vertical line segments. Our visibility representations use two directions, but since all boxes of vertices have unit height, they can be made uni-directional at the cost of at most doubling the height. Details are omitted.

4   Lower Bounds

Series-parallel graphs. Most of the previously given lower bounds for planar drawings (see e.g. [14, 2, 19]) rely on an argument that we call the "stacked cycle argument", which we briefly review here because we will modify it later. Assume we have a planar graph G with a fixed planar embedding and outer-face. A set of disjoint cycles $C_1, \ldots, C_k$ is called *stacked cycles* if C_i is outside the region defined by C_{i-1} for all $i > 1$. The following is well-known:

Fact 1. *If G has k stacked cycles, then G needs at least a $2k \times 2k$-grid in any planar polyline drawing that reflects the planar embedding and outer-face.*

Therefore, to get a bound of $\Omega(n^2)$ on the area, construct graphs that consist of $n/3$ stacked triangles [14], or $\Omega(n)$ stacked cycles for some graph classes that do not allow stacked triangles [19]. The left graph in Fig. 7 is a series-parallel graph that has $n/3$ stacked cycles.

Theorem 3. *There exists a series-parallel graph that requires a $\frac{2}{3}n \times \frac{2}{3}n$-grid in any polyline drawing that respects the planar embedding and outer-face.*

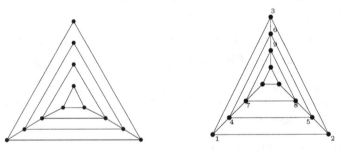

Fig. 7. Two graphs with $n/3$ stacked cycles

Note that our graph (contrary to the other lower bound graphs cited above) has many different planar embeddings, and using a different embedding one can easily construct drawings of it in area $O(n)$. Our algorithm (which changes the planar embedding) achieves area $O(n \log n)$ since the graph has fan-out 2. Frati [13] showed that another series-parallel graph (consisting of $K_{2,n}$ and a complete ternary tree) needs $\Omega(n \log n)$ in any poly-line drawing. Closing the gap between his lower bound and our upper bound of $O(n^{3/2})$ remains open.

k-outerplanar graphs. A k-outer-planar graph is defined as follows. Let G be a graph with a fixed planar embedding. G is called *1-outer-plane* if all vertices of G are on the outer-face (i.e., if G is outer-planar in this embedding.) G is called k-*outer-plane* if the graph that results from removing all vertices from the outer-face of G is $(k-1)$-outer-plane in the induced embedding. A graph G is called k-*outer-planar* if it is k-outer-plane in some planar embedding.

Clearly, k-outer-planar graphs generalize the concept of outer-planar graphs, and hence for small (constant) k are good candidates for $o(n^2)$ area. Also, by definition we cannot use a stacked cycle argument on them (a k-outer-planar graph has at most k stacked cycles.) Nevertheless, we can show an $\Omega(n^2)$ lower bound on the area even for 2-outer-planar graphs.

To show this, we modify the stacked-cycle argument. Let G be a graph with a fixed planar embedding, and let $C_1, \ldots, C_k$ be k cycles that are edge-disjoint and any two cycles have at most one vertex in common. We say that $C_1, \ldots, C_k$ are *1-fused stacked cycles* if C_i is outside the region defined by C_{i-1} except at the one vertex that they may have in common. See Fig. 8.

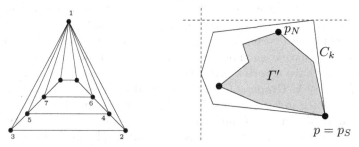

Fig. 8. A 2-outerplanar graph with $(n-1)/2$ 1-fused stacked cycles, and adding a 1-fused cycle around a drawing

Lemma 3. *Let G be a planar graph with a fixed planar embedding and outer-face, and assume G has k 1-fused stacked cycles $C_1, \ldots, C_k$. Then any poly-line drawing of G that respects the planar embedding and outer-face has width and height at least $k+1$.*

Proof. We proceed by induction on k. Clearly we need width and height 2 to draw the cycle C_1. For $k > 1$, let G' be the subgraph formed by the 1-fused stacked cycles $C_1, \ldots, C_{k-1}$.

Consider an arbitrary poly-line drawing Γ of G, and let Γ' be the induced drawing of G', which has width and height at least k by induction. Consider Fig. 8. The drawing of C_k in Γ must stay outside Γ', except at the point p where C_k and C_{k-1} have a vertex in common (if any.) Let p_N and p_S be points at a vertex or bend in the topmost and bottommost row of Γ'; by $k \geq 2$ they are distinct. So $p \neq p_N$ or $p \neq p_S$; assume the former. To go around p_N, the drawing of C_k in Γ must reach a point strictly higher than p_N, and hence uses at least one more row above Γ'. Similarly one shows that Γ has at least one more column than Γ'. □

Now we give a lower bound for 2-outerplanar graphs. The same graph also has small pathwidth (defined precisely below.)

Theorem 4. *There exists a 3-connected 2-outer-planar graph of pathwidth* 3 *that requires an $\frac{n+1}{2} \times \frac{n+1}{2}$-grid in any poly-line drawing that reflects the planar embedding and outer-face.*

Proof. (Sketch) Fig. 8 shows a graph that has $(n-1)/2$ 1-fused stacked cycles and hence needs an $(n+1)/2 \times (n+1)/2$-grid. Clearly it is 2-outerplanar and has pathwidth 3. $\qquad\square$

Since this graph is 3-connected, no other planar embedding is possible. It is possible to choose a different outer-face, but at least $(n-1)/4$ 1-fused stacked cycles will remain regardless of this choice, and hence an $\Omega(n^2)$ lower bound applies to any planar drawing of this graph.

Graphs of small (proper) pathwidth. The same graph can also serve as a lower-bound example for another restriction of planar graphs, namely, graphs of bounded treewidth, pathwidth, and proper pathwidth. See for example Bodlaender's overview [4] for exact definition of treewidth and applications of these graph classes. Graphs of treewidth 2 are exactly SP-graphs. Graphs of *pathwidth* k are those that have a vertex order $v_1, \ldots, v_n$ such that for any i, at most k vertices in $v_1, \ldots, v_i$ have a neighbour in $v_{i+1}, \ldots, v_n$. Graphs of *proper pathwidth* k are those that have a vertex order $v_1, \ldots, v_n$ such that for any edge (v_i, v_j), we have $|j - i| \le k$. Graphs of proper pathwidth k are a subset of graphs of pathwidth k, which in turn are a subset of graphs of treewidth k.

The labelling of vertices of the graph in Fig. 8 show that it has pathwidth at most 3. Many other previously given lower-bound graphs that consist of stacked cycles (see e.g. [2]) have constant pathwidth, even constant proper pathwidth, usually equal to the length of the stacked cycles. We give one more example that also has small maximum degree.

Theorem 5. *There exists a 3-connected graph of proper pathwidth* 3 *with maximum degree 4 that requires $\Omega(n^2)$ area in any poly-line drawing.*

Proof. The right graph in Fig. 7 shows an example with proper pathwidth at most 3 and maximum degree 4, and $n/3$ stacked cycles, hence needs a $\frac{2}{3}n \times \frac{2}{3}n$-grid in any polyline drawing. $\qquad\square$

Since planar partial 3-trees are also partial k-trees for any $k \ge 3$, our lower bounds holds for all partial k-trees with $k \ge 3$, hence destroying the hope that the linear-area layouts in 3D [10] could be replicated in 2D.

References

1. Bertolazzi, P., Cohen, R.F., Di Battista, G., Tamassia, R., Tollis, I.G.: How to draw a series-parallel digraph. Intl. J. Comput. Geom. Appl. 4, 385–402 (1994)
2. Biedl, T.: New lower bounds for orthogonal graph drawings. Journal of Graph Algorithms and Applications 2(7), 1–31 (1998)
3. Biedl, T.: Drawing outer-planar graphs in $O(n \log n)$ area. In: Goodrich, M.T., Kobourov, S.G. (eds.) GD 2002. LNCS, vol. 2528, pp. 54–65. Springer, Heidelberg (2002)

4. Bodlaender, H.: Treewidth: algorithmic techniques and results. In: Privara, I., Ružička, P. (eds.) MFCS 1997. LNCS, vol. 1295, pp. 19–36. Springer, Heidelberg (1997)
5. Brandenburg, F., Eppstein, D., Goodrich, M.T., Kobourov, S.G., Liotta, G., Mutzel, P.: Selected open problems in graph drawing. In: Liotta, G. (ed.) GD 2003. LNCS, vol. 2912, pp. 515–539. Springer, Heidelberg (2004)
6. Cohen, R., Di Battista, G., Tamassia, R., Tollis, I.: Dynamic graph drawings: Trees, series-parallel digraphs, and planar st-digraphs. SIAM J. Comput. 24(5), 970–1001 (1995)
7. Di Battista, G., Eades, P., Tamassia, R., Tollis, I.: Graph Drawing: Algorithms for Geometric Representations of Graphs. Prentice-Hall, Englewood Cliffs (1998)
8. Di Battista, G., Frati, F.: Small area drawings of outerplanar graphs. In: Healy, P., Nikolov, N.S. (eds.) GD 2005. LNCS, vol. 3843, pp. 89–100. Springer, Heidelberg (2006)
9. Dujmovic, V., Fellows, M., Kitching, M., Liotta, G., McCartin, C., Nishimura, N., Ragde, P., Rosamond, F., Whitesides, S., Wood, D.: On the parameterized complexity of layered graph drawing. Algorithmica 52, 267–292 (2008)
10. Dujmovic, V., Morin, P., Wood, D.K.: Layout of graphs with bounded tree-width. SIAM J. on Computing 34(3), 553–579 (2005)
11. Felsner, S., Liotta, G., Wismath, S.: Straight-line drawings on restricted integer grids in two and three dimensions. Journal of Graph Algorithms and Applications 7(4), 335–362 (2003)
12. Frati, F.: Straight-line drawings of outerplanar graphs in o(dn log n) area. In: Proceedings of the 19th Canadian Conference on Computational Geometry (CCCG 2007), pp. 225–228 (2007)
13. Frati, F.: A lower bound on the area requirements of series-parallel graphs. In: Broersma, H., Erlebach, T., Friedetzky, T., Paulusma, D. (eds.) WG 2008. LNCS, vol. 5344, pp. 159–170. Springer, Heidelberg (2008)
14. de Fraysseix, H., Pach, J., Pollack, R.: Small sets supporting fary embeddings of planar graphs. In: Twentieth Annual ACM Symposium on Theory of Computing, pp. 426–433 (1988)
15. de Fraysseix, H., Pach, J., Pollack, R.: How to draw a planar graph on a grid. Combinatorica 10, 41–51 (1990)
16. Hong, S.H., Eades, P., Quigley, A., Lee, S.H.: Drawing algorithms for series-parallel digraphs in two and three dimensions. In: Whitesides, S.H. (ed.) GD 1998. LNCS, vol. 1547, pp. 198–209. Springer, Heidelberg (1999)
17. Kant, G.: Drawing planar graphs using the canonical ordering. Algorithmica 16, 4–32 (1996)
18. Leiserson, C.: Area-efficient graph layouts (for VLSI). In: 21st IEEE Symposium on Foundations of Computer Science, pp. 270–281 (1980)
19. Miura, K., Nishizeki, T., Nakano, S.: Grid drawings of 4-connected plane graphs. Discrete Computational Geometry 26, 73–87 (2001)
20. Schnyder, W.: Embedding planar graphs on the grid. In: 1st Annual ACM-SIAM Symposium on Discrete Algorithms, pp. 138–148 (1990)
21. Tamassia, R., Tollis, I.: A unified approach to visibility representations of planar graphs. Discrete Computational Geometry 1, 321–341 (1986)
22. Tayu, S., Nomura, K., Ueno, S.: On the two-dimensional orthogonal drawing of series-parallel graphs. Discr. Appl. Mathematics 157(8), 1885–1895 (2009)
23. Wismath, S.: Characterizing bar line-of-sight graphs. In: 1st ACM Symposium on Computational Geometry, Baltimore, Maryland, USA, pp. 147–152 (1985)

Drawing Trees in a Streaming Model[*]

Carla Binucci[1], Ulrik Brandes[2], Giuseppe Di Battista[3],
Walter Didimo[1], Marco Gaertler[4], Pietro Palladino[1],
Maurizio Patrignani[3], Antonios Symvonis[5], and Katharina Zweig[6]

[1] Dipartimento di Ing. Elettronica e dell'Informazione, Università degli Studi di Perugia
[2] Fachbereich Informatik & Informationswissenschaft, Universität Konstanz
[3] Dipartimento di Informatica e Automazione, Università Roma Tre
[4] Fakultät für Informatik, Universität Karlsruhe (TH)
[5] Department of Mathematics, National Technical University of Athens
[6] Department of Biological Physics, Eötvös Loránd University

Abstract. We introduce a data stream model of computation for Graph Drawing, where a source produces a graph one edge at a time. When an edge is produced, it is immediately drawn and its drawing can not be altered. The drawing has an image persistence, that controls the lifetime of edges. If the persistence is k, an edge remains in the drawing for the time spent by the source to generate k edges, then it fades away. In this model we study the area requirement of planar straight-line grid drawings of trees, with different streaming orders, layout models, and quality criteria. We assess the output quality of the presented algorithms by computing the competitive ratio with respect to the best known offline algorithms.

1 Introduction

We consider the following model. A source produces a graph one edge at a time. When an edge is produced, it is immediately drawn (i.e., before the next edge is produced) and its drawing can not be altered. The drawing has an image persistence, that controls the lifetime of edges. If the persistence is infinite, edges are never removed from the drawing. Otherwise, suppose the persistence is k, an edge remains in the drawing for the time spent by the source to generate k edges, and then it fades away.

Studying this model, which we call *streamed graph drawing*, is motivated by the challenge of offering visualization facilities to streaming applications, where massive amounts of data, too large even to be stored, are produced and processed at a very high rate [12]. The data are available one element at a time and need to be processed quickly and with limited resources. Examples of application fields include computer network traffic analysis, logging of security data, stock exchange quotes' correlation, etc.

For the user of the visualization facility it is natural to associate any graphic change with a new datum coming from the stream. Hence, moving pieces of the drawing would be potentially ambiguous. On the other hand, the drawing should have a size as limited as possible.

---

[*] Work on this problem began at the BICI Workshop on Graph Drawing: Visualization of Large Graphs, held in Bertinoro, Italy, in March 2008.

D. Eppstein and E.R. Gansner (Eds.): GD 2009, LNCS 5849, pp. 292–303, 2010.
© Springer-Verlag Berlin Heidelberg 2010

Although streamed graph drawing is related to incremental and dynamic graph drawing, it is qualitatively different from both. In incremental graph drawing the layout is constructed step by step according to a precomputed vertex ordering that ensures invariants regarding, e.g., its shape [3,7]. In streamed graph drawing the order cannot be chosen. Dynamic graph drawing [4,11,13] usually refers to drawing sequences of graphs, where drawings of consecutive graphs should be similar. Insertions and/or deletions of vertices/edges are allowed and the current graph must be drawn without knowledge of future updates. However, the current layout is only weakly constrained by previous drawings. In streamed graph drawing modifications concern only single edges and previous layout decisions may not be altered.

While there is some work on computing properties of streamed graphs (see, e.g., [1,5,8]), very little has been done in the context of graph drawing. A result that applies to streamed graph drawing with infinite persistence is shown in [13] in what is called *no change scenario*. In that paper, a graph of maximum degree four is available one-vertex-at-time and it is drawn orthogonally and with a few crossings.

We consider both a finite persistence and an infinite persistence model. Our results in these models concern the area requirement for planar straight-line grid drawings of trees, where we assume that the tree is streamed in such a way that the subtree to be drawn is connected. Since a streamed graph drawing algorithm is a special case of an online algorithm, it is reasonable to assess its output quality in terms of its competitive ratio with respect to the best known offline algorithm. The obtained results are summarized in Table 1, where n is the number of vertices of the current graph. For each type of streaming order and for each class of trees investigated within each model, the table reports the competitive ratio of a (specific) drawing algorithm, and the corresponding lemma/theorem. The table puts in evidence the practical applicability of the finite persistence model. On the other hand, the results on the infinite persistence model show the intrinsic difficulty of the problem. In fact, in the paper we prove that a large family of algorithms for the infinite persistence model requires $\Omega(2^{\frac{n}{8}}/n)$ competitive ratio to draw binary trees (see Lemma 5).

Another way to interpret our results on the infinite persistence model is the following: All the area-efficient tree-drawing algorithms known in the literature have the capability to inspect the entire tree for exploiting some balancing consideration. In the infinite persistence model we ask the question of which is the achievable area bound if such an inspection can not be done.

This paper is organized as follows. In Sect. 2 we introduce the concept of streamed graph drawing. Area requirements for tree drawings in our two main models are derived in Sects. 3 and 4, and we conclude with directions for future work in Sect. 5.

2   Framework

Let $G = (V, E)$ be a simple undirected graph. A *straight-line grid drawing* $\Gamma = \Gamma(G)$ is a geometric representation of G such that each vertex is drawn as a distinct point of an integer-coordinate grid, and each edge is drawn as a straight-line segment between the points associated with its end-vertices. A drawing is *planar* if no two edges cross.

Table 1. Summary of the results: competitive ratios of the proposed algorithms

Finite persistence model (constant persistence k)

Streaming order	Graph class	Area (competitive ratio)	
Eulerian tour	tree	$O(k^2)$	Sect. 3, Theorem 2

Infinite persistence model (unbounded memory, n is the current graph size)

Streaming order	Graph class	Area (competitive ratio)	
connected	binary tree	$\Theta(2^n)$	Sect. 4.1, Lemma 4
	tree, max. degree d	$\Theta((d-1)^n)$	Sect. 4.1, Lemma 6
	tree	$\Omega(2^n/n)$	Sect. 4.1, Lemma 7
BFS, DFS	tree	$\Theta(n)$	Sect. 4.2, Lemma 8
layered	tree, max. degree d	$\Theta(dn)$	Sect. 4.3, Lemma 9

Since we only consider planar straight-line grid drawings we simply refer to them as *drawings* in the remainder.

Given a subset of edges $E' \subseteq E$, the *edge-induced (sub)graph* $G[E']$ contains exactly those vertices of V incident with edges in E', and the edges in E'. We study the problem of drawing a (potentially infinite) graph G described by a sequence of edges $(e_1, e_2, e_3, \dots)$, which we call a *stream of edges*, where e_i is known at time i. Throughout this paper, let $W_i^k = \{e_{i-k+1}, \dots, e_i\}$ denote a *window* of the stream of size k and let $E_i = \{e_1, \dots, e_i\}$ denote the *prefix* of the stream of length i. Observe that $E_i = W_i^i$.

Our goal is to design online drawing algorithms for streamed graphs. An online drawing algorithm incrementally constructs a drawing of the graph, by adding one edge at a time according to the order in which they appear in the stream. Once a vertex is placed, however, the decision must not be altered unless the vertex is removed.

Let Γ_0 be an initially empty drawing. We deal with two models.

Finite persistence model. At each time $i \geq 1$ and for some fixed parameter $k \geq 1$, called *persistence*, determine a drawing Γ_i of $G_i = G[W_i^k]$ by adding e_i to Γ_{i-1} and dropping (if $i > k$) e_{i-k} from Γ_i.

Infinite persistence model. At each time $i \geq 1$, determine a drawing Γ_i of $G_i = G[E_i]$ by adding e_i to Γ_{i-1}.

We relate the connectivity of the graph to the persistence of the drawing. If the persistence k is finite, a stream of edges is *connected* if $G[W_i^k]$ is connected for all $i \geq 1$. If the persistence is infinite, then a stream is *connected* if $G[E_i]$ is connected for all $i \geq 1$. In both models we assume that the stream of edges is connected. Also, in the finite persistence model we assume that our memory is bounded by $O(k)$.

Since streamed graph drawing algorithms are special online algorithms, an important assessment of quality is their competitive ratio. For a given online drawing algorithm A and some measure of quality, consider any stream of edges $S = (e_1, e_2, \dots)$. Denote by $A(S)$ the quality of A executed on S, and by $Opt(S)$ the quality achievable by an optimal offline algorithm, i.e. an algorithm that knows the streaming order in advance.

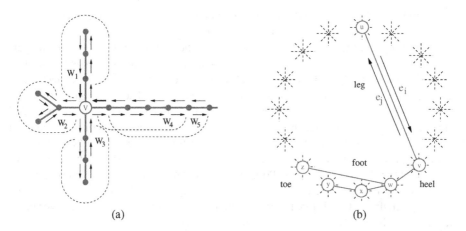

Fig. 1. (a) An Eulerian tour with a persistence $k = 5$. When W_5 is the current window, vertex v is disappeared from the drawing. (b) A leg of vertex u.

Where possible, we measure the effectiveness of A by evaluating its *competitive ratio*: $R_A = \max_S \frac{A(S)}{Opt(S)}$.

In the remainder of the paper we restrict our attention to the case where G is a tree, the goal is to determine a planar straight-line grid drawing, and the measure of quality is the area required by the drawing. Note that, $Opt(S) = \Theta(n)$ if $G[S]$ is a binary tree with n vertices [9] or if G[S] is a tree with n vertices and vertex degree bounded by $\sqrt{n}$ [10]. The best known area bound for general trees is $O(n \log n)$ [6,14]. In the next two sections we give upper and lower bounds on the area competitive ratio of streamed tree drawing algorithms under several streaming orders.

3   Finite Persistence Drawings of Trees

We consider the following scenario, corresponding to the intuitive notion of a user traversing an undirected tree: the edges of the stream are given according to an Eulerian tour of the tree where we suppose that the persistence k is much smaller than the number of the edges of the tree (the tree may be considered "infinite"). Each edge is traversed exactly twice: the first time in the *forward* direction and the second time in the *backward* direction. This corresponds to a DFS traversal where backtracks explicitly appear. Observe that window W_i^k contains in general both forward and backward edges and that $G[W_i^k]$ is always connected. Figure 1 shows an example of an Eulerian tour where several windows of size 5 are highlighted: window W_1 contains two forward and three backward edges; window W_5 contains all backward edges.

In this scenario a vertex may be encountered several times during the traversal. Consider edge $e_i = (u, v)$ and assume that the Eulerian tour moves from u to v. We say that e_i *leaves* u and *reaches* v. Also, if v was already a vertex of G_{i-1} (and hence is already drawn in Γ_{i-1}) then, we say that e_i *returns* to v. Otherwise, v has to be inserted into the drawing Γ_i of G_i. Observe that if a vertex v, reached at time i, is reached again at time j, with $j > i + k + 1$, and is not reached at any intermediate time, then v has (in general) two different representations in Γ_i and Γ_j.

The first algorithm presented in this section is the following. Consider m integer-coordinate points $p_0, p_1, \ldots, p_{m-1}$ in convex position. An easy strategy is to use such points clockwise in a greedy way. At each time i, we maintain an index $next_i$ such that point p_{next_i} is the first unused point in clockwise order. The first edge e_1 is drawn between points p_0 and p_1 and $next_2 = 2$. Suppose that edge $e_i = (u, v)$ has to be added to the drawing. If v is not present in Γ_{i-1}, assign to v the coordinates of p_{next_i} and set $next_{i+1} = (next_i + 1) \mod m$. We call this algorithm GREEDY-CLOCKWISE (GC).

Algorithm GC guarantees a non-intersecting drawing provided that two conditions are satisfied for all i: (**Condition 1**) Point p_{next_i} is not used in Γ_i by any vertex different from v. (**Condition 2**) Edge e_i does not cross any edge of Γ_i. Lemma 1 and Lemma 2 show that satisfying Condition 1 implies satisfying Condition 2. Let w be a vertex of Γ_i, we denote by $i(w)$ the time when vertex w entered Γ_i.

Lemma 1. *Let Γ_i be a drawing of G_i constructed by Algorithm GC and let v_1, v_2, and v_3 be three vertices of G_i such that $i(v_1) < i(v_2) < i(v_3)$ in Γ_i. If there is a sequence of forward edges from v_1 to v_3, then there is a sequence of forward edges from v_1 to v_2.*

Proof. Consider edges $e_{i(v_1)} = (v_0, v_1)$ and $e_j = (v_1, v_0)$ of the stream. The Eulerian tour implies that the vertices reached by a forward path from v_1 are those vertices incident to some edge e_h, with $i(v_1) < h < j$. Suppose for a contradiction that v_2 is not reached by a forward path from v_1. Since v_2 was drawn after v_1, this implies $i(v_2) > j$. It follows that also $i(v_3) > j$. Hence, v_3 can not be reached by a forward path from v_1. □

Lemma 2. *Let Γ_{i-1} be a drawing of G_{i-1} constructed by Algorithm GC and consider a vertex v that is not in G_{i-1} and should be added to G_{i-1} at time i. If Condition 1 is satisfied, then no crossing is introduced by drawing v at p_{next_i}.*

Proof. Let $e_i = (u, v)$. Draw v on p_{next_i}. Since Condition 1 is satisfied, then p_{next_i} is not used by any vertex. Suppose for contradiction that Γ_i has a crossing. It follows that there exists in Γ_i an edge (x, y), such that vertices x, u, y, v appear in this relative order in the clockwise direction. By Condition 1 and since the points are used in a greedy way, $i(x) < i(u) < i(y) < i(v)$. Because of edge (x, y), there is a forward path from x to y and hence by Lemma 1 there is a forward path from x to u. Analogously, because of edge (u, v), there is a forward path from u to v and hence by Lemma 1 there is a forward path from u to y. Hence, there is an undirected cycle in G_i involving x, u, and y. This is a contradiction since we are exploring a tree. □

Consider two edges $e_i = (u, v)$ and $e_j = (v, u)$, with $j > i$. Observe that $j - i$ is odd. Edges $e_i, e_{i+1}, \ldots, e_j$ are a *leg* of u. Vertices discovered at times $i, i+1, \ldots, j$, i.e., the $\frac{j-i+1}{2}$ distinct vertices incident to edges $e_{i+1}, \ldots, e_{j-1}$, are a *foot* of u. Node v is the *heel* of the foot and the last discovered vertex of the foot is the *toe*. Figure 1(b) shows the drawing of a leg (and provides a hint of why its vertices are called a foot). A foot is itself composed of smaller feet, where the smallest possible foot is when a leaf of the tree is reached, that is, when its heel and its toe are the same vertex (as for vertex y of Fig. 1(b)).

Consider the case when $j - i \leq k$. This implies that u is present in all the drawings $\Gamma_{i-1}, \ldots, \Gamma_{j+k}$. In this case we say that the foot is a *regular foot* (or *R-foot*). Otherwise, we say that it is an *extra-large foot* (or *XL-foot*).

Property 1. A regular foot has maximum size $\lceil \frac{k}{2} \rceil$.

Observe that in any drawing constructed by Algorithm GC the vertices of a regular foot are contiguously placed after its heel, the toe being the last in clockwise order.

Property 2. Let i be the time when an extra-large foot of v is entered by the Eulerian tour. Vertex v disappears from the drawing at time $i + k$.

Now, we exploit the above properties and lemmas to prove that, if k is the persistence of the drawing and if the tree has maximum degree d (where a binary tree has $d = 3$), then it suffices using $\lceil \frac{k}{2} \rceil \cdot (d - 1) + k + 1$ points in convex position to guarantee to GC that Condition 1 is satisfied. In order to prove this we need the following lemma.

Lemma 3. *Consider Algorithm GC on m points in convex position. Suppose that for each vertex v it holds that during the time elapsing from when v is discovered and when it disappears from the drawing at most $m - 1$ other vertices are discovered. Then Condition 1 holds at each time.*

Proof. Suppose, for a contradiction that there exists a vertex u, discovered at time i, for which Condition 1 does not hold because point p_{next_i} is used by vertex $w \neq u$. Since GC is greedy, after u has been inserted all the m points have been used. This implies that after w and before u, $m - 1$ vertices have discovered. Summing up, we have that w violates the condition of the statement.

Theorem 1. *Let S be a stream of edges produced by an Eulerian tour of a tree of degree at most d. Algorithm GC draws S with persistence k without crossings on $\lceil \frac{k}{2} \rceil \cdot (d - 1) + k + 1$ points in convex position. Also $R_{GC} = O(d^3 k^2)$.*

Proof. Due to Lemma 2 it suffices to show that Condition 1 holds at each time i. We exploit Lemma 3 to show that during the time elapsing from when a vertex v is discovered and when it disappears from the drawing at most $\lceil \frac{k}{2} \rceil \cdot (d - 1) + k$ other vertices are discovered. Suppose v is discovered by edge $e_i = (u, v)$. Three cases are possible: (i) v is a leaf; (ii) all feet of v are regular; (iii) v has an XL-foot. Case (i) is simple: we have that v disappears from the drawing at time $i + k + 1$. Hence, at most k vertices can be discovered before it disappears. In Case (ii) since each R-foot can have at most $\lceil \frac{k}{2} \rceil$ vertices (Property 1) and since at most $(d - 1)$ of them can be traversed, the maximum number of vertices that can be discovered after v enters the drawing and before it disappears is $\lceil \frac{k}{2} \rceil \cdot (d - 1) + k$ (see Fig. 1(a) for an example with $k = 5$). In Case (iii), because of Property 2, after the XL-foot is entered, at most k vertices can be discovered before v disappears. Hence, the worst case is that the XL-foot follows $d - 2$ R-feet. Overall, a maximum of $\lceil \frac{k}{2} \rceil \cdot (d - 2) + k$ vertices can be discovered before v disappears.

Regarding the competitive ratio, m grid points in convex position take $O(m^3)$ area [2], and therefore the area of the drawing of our online algorithm is $\Theta(d^3 k^3)$. Finally, any offline algorithm requires $\Omega(k)$ area for placing $O(k)$ vertices.   □

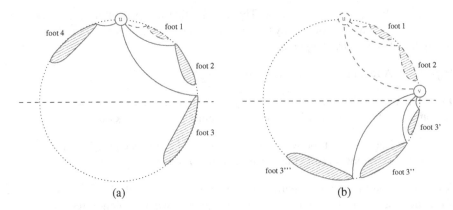

Fig. 2. (a) Feet 1, 2, and 3 are drawn by GC, foot 4 is drawn by GCC. (b) Foot 3 is an XL-foot of u. Its size is large enough to promote v as the oldest vertex in place of u.

Theorem 1 uses a number of points that is proportional to the maximum degree of the tree. In the following we introduce a second algorithm that uses a number of points that only depends on the persistence k.

Intuitively, the basic strategy is to alternate Algorithm GC with its mirrored version, called GREEDY-COUNTER-CLOCKWISE (GCC), where, at each step, $next_i$ is possibly decreased rather than increased. Namely, let $old(\Gamma_i)$ be the *oldest vertex of* Γ_i, that is, the vertex that appears in $\Gamma_i, \Gamma_{i-1}, \ldots, \Gamma_{i-j}$ with highest j. The decision of switching between GC and GCC (or vice versa) is taken each time you start to draw a new foot of $old(\Gamma_i)$. We begin with Algorithm GC and use points in the clockwise direction with respect to $old(\Gamma_i)$ until we have used enough of them to ensure that the points near to $old(\Gamma_i)$ in the counter-clockwise direction are available. At this point, we switch to Algorithm GCC, starting from the point immediately next to $old(\Gamma_i)$ in the counter-clockwise direction, and we use Algorithm GCC to draw the next feet of $old(\Gamma_i)$ until the last drawn foot of $old(\Gamma_i)$ has used enough points in the counter-clockwise direction to ensure that the points in clockwise direction are available. Figure 2(a) shows an example where three feet were drawn by GC and the fourth foot is drawn by GCC.

Formally, Algorithm SNOWPLOW (SP) works as follows. Let old_i be the index of the point of Γ_i where $old(\Gamma_i)$ is drawn. Suppose that edge $e_i = (u, v)$ has to be added to the drawing. If v is present in Γ_{i-1} then $\Gamma_i = \Gamma_{i-1}$. Otherwise, if $u \neq old_i$ or $u = old_i$ but $(next_i - old_i) \bmod m \leq \lceil \frac{k}{2} \rceil$, place v on p_{next_i} and set $next_{i+1} = (next_i + 1) \bmod m$. If $u = old_i$ and $(next_i - old_i) \bmod m > \lceil \frac{k}{2} \rceil$, then switch to GCC, that is, place v on point $p_{(old_i-1) \bmod m}$ and set $next_{i+1} = (old_i - 2) \bmod m$.

A critical step is when $old(\Gamma_i) \neq old(\Gamma_{i-1})$. This happens when an XL-foot is drawn either by GC or by GCC. In this case the heel of such a foot becomes the oldest vertex (see Fig. 2(b) for an example).

We show in the following that SP needs $2k - 1$ points in convex position to produce a non-crossing drawing of the stream of edges independently of the degree of the vertices.

Theorem 2. *Let S be a stream of edges produced by an Eulerian tour of a tree. Algorithm SP draws S with persistence k without crossings on $2k - 1$ points in convex position. Also $R_{SP} = O(k^2)$.*

Sketch of Proof: Suppose that Algorithm SP is in its GC phase. Assume, without loss of generality, that $p_{old_i} = p_0$, and denote by $P^+ = \{p_1, p_2, \ldots p_{\lceil \frac{k}{2} \rceil - 1}\}$ ($P^- = \{p_{-1}, p_{-2}, \ldots p_{-\lceil \frac{k}{2} \rceil + 1}\}$) the points after p_{old_i} in clockwise (counter-clockwise) order. Consider the case when p_{old_i} has a sequence of R-feet. In order to switch to the GCC phase at least $\lceil \frac{k}{2} \rceil$ points and at most $2\lceil \frac{k}{2} \rceil - 1$ points of P^+ are used. Since at least $\lceil \frac{k}{2} \rceil$ points are used of P^+, at least the same amount of time elapsed from when the current GC phase started. Hence, points in P^- are not used by any vertex. □

4 Infinite Persistence Drawings of Trees

We consider different scenarios depending on the ordering of the edges in the stream: (i) The edges come in an arbitrary order, with the only constraint that the connectivity is preserved, (ii) the edges come according to a DFS/BFS traversal, (iii) the edges come layer by layer. For each scenario different classes of trees are analyzed.

4.1 Arbitrary Order Scenario

In the arbitrary order scenario we first analyze the case of binary trees, then we give results for bounded degree trees and, eventually, for general trees. The following lemma deals with a very simple drawing strategy.

Lemma 4. *Let $S = (e_1, e_2, \ldots)$ be any stream of edges such that, at each time $i \geq 1$, G_i is a rooted binary tree. Suppose that the root of all G_i is one of the two end-vertices of e_1. There exists a drawing algorithm A for S in the infinite persistence model, such that the drawing of any G_i is downward with respect to the root and $R_A = \Theta(2^n)$.*

Sketch of Proof: Place the root at $(0, 0)$ and place its first child v_1 on $(0, 1)$ and its second child v_2 on $(1, 1)$. For every vertex v placed at (x, y) reserve in each subsequent row $z > y$ the points from $x_l = 2^{z-y} \cdot x$ to $x_r = 2^{z-y}(x + 1) - 1$ (see Fig. 3(a)). It is easy to see that an area of $O(n) \times O(2^n)$ is always sufficient for any stream of edges representing a binary tree. The bound is tight since the stream describing the path-like tree of Fig. 3(a) uses an area of $\Omega(n) \times \Omega(2^n)$. Since the best offline algorithm can draw a binary tree in linear area, the statement follows. □

The algorithm in the proof of Lemma 4 is such that whatever is the order in which the edges of a complete binary tree are given, it always computes the same drawing, up to a permutation of the vertex labels. We call such an algorithm a *predefined-location algorithm* for binary trees. Since the competitive ratio of the very simple algorithm in Lemma 4 is exponential, one can ask if there exists a better algorithm that uses a similar strategy. Unfortunately, the next result shows that this is not the case.

Lemma 5. *Let A be any predefined-location algorithm for binary trees in the infinite persistence model. Then $R_A = \Omega(2^{\frac{n}{8}}/n)$.*

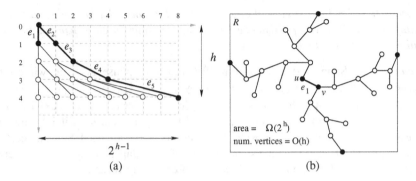

Fig. 3. (a) A drawing produced by the algorithm in the proof of Lemma 4. Bold edges represent an edge sequence that causes exponential area. (b) Schematic illustration of the proof of Lemma 5.

Proof. We show that there exists a sequence of edges such that the drawing computed by A for the binary tree induced by this sequence requires $\Omega(2^{\frac{n}{8}})$ area, where n is the number of vertices of the tree. Since there exists an offline algorithm that computes a drawing of optimal area $\Theta(n)$ for the binary tree, this implies the statement. Refer to Fig. 3(b). By hypothesis the algorithm always computes the same drawing for a rooted complete binary tree of depth h. Consider the bounding box R of such a drawing. Clearly, the area of R is $\Omega(2^h)$. Every path between two vertices of a complete binary tree of depth h consists of at most $2h + 1$ vertices and $2h$ edges (the first level has number 0). Independently of the position of the first edge $e_1 = (u, v)$ of the stream, we can define a subsequence of the stream with at most $8h$ edges that forces the algorithm to draw two paths, one consisting of at most $4h$ edges and going from the left side to the right side of R, and the other consisting of at most $4h$ edges and going from the bottom side to the top side of R, as shown in the figure. Therefore, for this subsequence of $n = 8h$ edges and vertices the algorithm constructs a drawing of area $\Omega(2^h)$.   □

If the stream of edges induces at each time a tree whose vertices have degree bounded by a constant d, then we can define a drawing algorithm similar to the one described in the proof of Lemma 4. Namely, when a new edge $e = (u, v)$ is processed and v is the k-th child of u, we set $y(v) = y(u) + 1$ and $x(v) = (d - 1) \cdot x(u) + k - 1$. Hence, using the same worst case analysis performed in the proof of Lemma 4, the drawing area used by this algorithm is $\Theta(n) \times \Theta((d-1)^n)$. Since there exists an offline drawing algorithm that takes $\Theta(n)$ area for bounded degree trees [10], we get the following result.

Lemma 6. *Let $S = (e_1, e_2, \dots)$ be any stream of edges such that, at each time $i \geq 1$, G_i is a tree with vertex degree at most d. There exists a drawing algorithm A for S in the infinite persistence model, such that $R_A = \Theta((d - 1)^n)$.*

The next result extends Lemma 6 to general trees. It proves that there exists an algorithm to draw any infinite tree in the infinite persistence model, under the hypothesis that the stream is connected. In this case we give only a lower bound of the area.

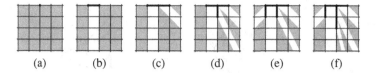

Fig. 4. Illustration of the algorithm described in the proof of Lemma 7

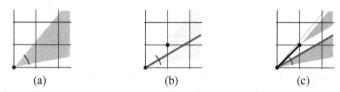

Fig. 5. Slicing the cone of a vertex and inserting its child: (a) initial configuration; (b) slicing the cone and finding the closest grid point; (c) inserting the edge.

Lemma 7. *There exists an algorithm A that draws in the infinite persistence model any stream of edges $S = (e_1, e_2, \dots)$ such that G_i is a tree of arbitrary vertex degree. The exponential competitive ratio of A is $R_A = \Omega(2^n/n)$.*

Sketch of Proof: A greedy drawing strategy is the following (see Fig. 4 and Fig. 5). For each vertex u already placed in the drawing, the algorithm reserves an infinite cone centered at u that does not intersect with any other cone. Each time a new edge $e = (u, v)$ is added to the drawing, the algorithm splits the cone of u into two halves, one of which will be the new cone of u and the other will be used to place v at the first available grid point inside it and to reserve a new (sub-)cone for v. Since all the cones assigned to vertices are infinite, it is always possible to add further edges.

The lower bound of the competitive ratio is obtained when using as input the family $G_n = (V_n, E_n)$ defined by $V_n = \{1, \dots, n\}$ and $E_n = \{(i, i+1) : 1 \leq i \leq n-2\} \cup \{(n-2, n)\}$. □

4.2 BFS and DFS Order Scenarios

If the edges in the stream are given according to some specific order, algorithms can be designed that improve the competitive ratio obtained in the case of the arbitrary order. We focus on orderings deriving from BFS or DFS traversals.

Lemma 8. *Let $S = (e_1, e_2, \dots)$ be a stream of edges such that G_i is a tree of any vertex degree, at each time $i \geq 1$, and the edges of the stream are given according to a BFS or to a DFS visit of the graph. There exists a drawing algorithm A for S in the infinite persistence model, such that $R_A = \Theta(n)$.*

Sketch of Proof: In the case of the BFS order, we place the first vertex at $(0,0)$ and all of its k children consecutively along the next row, starting at $(0,1)$. Processing the children of any vertex at (x, y) we place all its children on the leftmost position that is

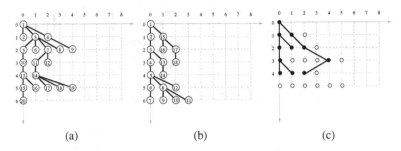

Fig. 6. Running examples of drawing algorithms for (a) BFS (b) DFS, and (c) layer ordering

not yet occupied, starting at $(0, y + 1)$ (see Fig. 6(a)). The required area is clearly in $O(n^2)$ for this algorithm. For offline algorithms the required area is bound from below by $\Omega(n)$ and thus the statement follows. It can be seen that the worst–case for the BFS order requires a quadratic area, which implies that the analysis is tight. When drawing a tree that comes in DFS order, every vertex can be placed at the leftmost unoccupied position below its parent. The area is $O(n^2)$ and the analysis of the worst case implies that this bound is tight.   □

4.3 Layer Order

This scenario is intermediate between the BFS order scenario and the arbitrary order one. In the *layer order* scenario edges come layer by layer, but the order of the edges in each layer is arbitrary. We prove the following.

Lemma 9. *Let $d > 0$ be a given integer constant and let $S = (e_1, e_2, \dots)$ be any stream of edges such that G_i is a tree of vertex degree at most d, at each time $i \geq 1$, and the edges of the stream are given according to a layer order. There exists a drawing algorithm A for S in the infinite persistence model, such that $R_A = \Theta(dn)$.*

Proof. Algorithm A works as follows. If $e_1 = (u, v)$ is the first edge of the stream, set $x(u) = 0$, $y(u) = 0$, $x(v) = 0$, $y(v) = 1$. Also, since u has at most other $d - 1$ adjacent vertices, reserve $(d - 1)$ consecutive grid points to the right of v. When the first vertex of a new level l ($l \geq 1$) enters the drawing, all vertices of the previous level $l - 1$ have been already drawn. Hence, if n_{l-1} is the number of vertices of level $l - 1$, reserve $(d-1)n_{l-1}$ consecutive grid points for the vertices of level l. Namely, denote by $u_1, u_2, \dots, u_{n_{l-1}}$ the vertices at level $l - 1$, from left to right. Use the leftmost $(d - 1)$ grid points at level l for arranging the children of u_1, the next $(d - 1)$ grid points for arranging the children of u_2, and so on. See Fig. 6(c) for an example. The width of the drawing increases at most linearly with the number of vertices of the tree. Indeed, if the width of the drawing is w, there is at least one level l of the drawing having a vertex with x-coordinate equal to w. This implies that level $l - 1$ contains $w/(d - 1)$ vertices. Also, since each level contains at least one vertex and since the height of the drawing is equal to the number of levels, the height of the drawing increases at most linearly with the number of vertices. Hence, the area is $O(dn) \times O(n)$. Also, it is easy to find an instance requiring such an area. The best offline algorithm takes $\Theta(n)$ area.   □

5 Open Problems

This paper opens many possible research directions, including the following: (i) Some of our algorithms have high competitive ratio, hence it is natural to investigate better solutions. (ii) Computing tighter lower bounds would allow us to have a more precise evaluation of streaming algorithms. (iii) It would be interesting to extend the study to larger classes of planar graphs or even to general graphs. (iv) Other persistence models can be considered. For example we could have drawings where the persistence is $O(\log n)$, where n is the size of the stream.

Acknowledgments

We thank Ioannis G. Tollis for interesting conversations.

References

1. Bar-Yossef, Z., Kumar, R., Sivakumar, D.: Reductions in streaming algorithms, with an application to counting triangles in graphs. In: Proc. SODA, pp. 623–632 (2002)
2. Bárány, I., Tokushige, N.: The minimum area of convex lattice n-gons. Combinatorica 24(2), 171–185 (2004)
3. Biedl, T., Kant, G.: A better heuristic for orthogonal graph drawings. Computational Geometry 9, 159–180 (1998)
4. Branke, J.: Dynamic graph drawing. In: Kaufmann, M., Wagner, D. (eds.) Drawing Graphs. LNCS, vol. 2025, pp. 228–246. Springer, Heidelberg (2001)
5. Buriol, L., Donato, D., Leonardi, S., Matzner, T.: Using data stream algorithms for computing properties of large graphs. In: Proc. Workshop on Massive Geometric Datasets (MASSIVE 2005), pp. 9–14 (2005)
6. Crescenzi, P., Di Battista, G., Piperno, A.: A note on optimal area algorithms for upward drawings of binary trees. Comput. Geom. Theory Appl. 2, 187–200 (1992)
7. de Fraysseix, H., Pach, J., Pollack, R.: How to draw a planar graph on a grid. Combinatorica 10, 41–51 (1990)
8. Feigenbaum, J., Kannan, S., McGregor, A., Suri, S., Zhang, J.: On graph problems in a semi-streaming model. In: Díaz, J., Karhumäki, J., Lepistö, A., Sannella, D. (eds.) ICALP 2004. LNCS, vol. 3142, pp. 531–543. Springer, Heidelberg (2004)
9. Garg, A., Rusu, A.: Straight-line drawings of binary trees with linear area and arbitrary aspect ratio. In: Goodrich, M.T., Kobourov, S.G. (eds.) GD 2002. LNCS, vol. 2528, pp. 320–331. Springer, Heidelberg (2002)
10. Garg, A., Rusu, A.: Straight-line drawings of general trees with linear area and arbitrary aspect ratio. In: Kumar, V., Gavrilova, M.L., Tan, C.J.K., L'Ecuyer, P. (eds.) ICCSA 2003. LNCS, vol. 2669, pp. 876–885. Springer, Heidelberg (2003)
11. Huang, M.L., Eades, P., Wang, J.: On-line animated visualization of huge graphs using a modified spring algorithm. J. Vis. Lang. Comput. 9(6), 623–645 (1998)
12. Muthukrishnan, S.: Data streams: Algorithms and applications. Foundations and Trends in Theoretical Computer Science 1(2), 117–236 (2005)
13. Papakostas, A., Tollis, I.G.: Interactive orthogonal graph drawing. IEEE Trans. Computers 47(11), 1297–1309 (1998)
14. Shiloach, Y.: Arrangements of Planar Graphs on the Planar Lattice. Ph.D. thesis, Weizmann Institute of Science (1976)

The Planar Slope Number of Planar Partial 3-Trees of Bounded Degree

Vít Jelínek[1,2], Eva Jelínková[1], Jan Kratochvíl[1,3], Bernard Lidický[1], Marek Tesař[1], and Tomáš Vyskočil[1,3]

[1] Department of Applied Mathematics, Charles University in Prague*
[2] Combinatorics Group, Reykjavík University
[3] Institute for Theoretical Computer Science, Charles University in Prague**
{jelinek,eva,honza,bernard,tesulo,whisky}@kam.mff.cuni.cz

Abstract. It is known that every planar graph has a planar embedding where edges are represented by non-crossing straight-line segments. We study the planar slope number, i.e., the minimum number of distinct edge-slopes in such a drawing of a planar graph with maximum degree Δ. We show that the planar slope number of every series-parallel graph of maximum degree three is three. We also show that the planar slope number of every planar partial 3-tree and also every plane partial 3-tree is at most $2^{\mathcal{O}(\Delta)}$. In particular, we answer the question of Dujmović et al. [Computational Geometry 38 (3), pp. 194–212 (2007)] whether there is a function f such that plane maximal outerplanar graphs can be drawn using at most $f(\Delta)$ slopes.

Keywords: graph drawing; planar graphs; slopes; planar slope number.

1 Introduction

The *slope number* of a graph G was introduced by Wade and Chu [10]. It is defined as the minimum number of distinct edge-slopes in a straight-line drawing of G. Clearly, the slope number of G is at most the number of edges of G, and it is at least half of the maximum degree Δ of G.

Dujmović et al. [2] asked whether there was a function f such that each graph with maximum degree Δ could be drawn using at most $f(\Delta)$ slopes. In general, the answer is *no* due to a result of Barát et al. [1]. Later, Pach and Pálvölgyi [9] and Dujmović et al. [3] proved that for every $\Delta \geq 5$, there are graphs of maximum degree Δ that need an arbitrarily large number of slopes.

On the other hand, Keszegh et al. [6] proved that every subcubic graph with at least one vertex of degree less than three can be drawn using at most four slopes; Mukkamala and Szegedy [8] extended this bound to every cubic graph. Dujmović et al. [3] give a number of bounds in terms of the maximum degree: for interval graphs, cocomparability graphs, or AT-free graphs. All the results mentioned so far are related to straight-line drawings which are not necessarily non-crossing.

* Supported by project 1M0021620838 of the Czech Ministry of Education.
** Supported by grant 1M0545 of the Czech Ministry of Education.

D. Eppstein and E.R. Gansner (Eds.): GD 2009, LNCS 5849, pp. 304–315, 2010.
© Springer-Verlag Berlin Heidelberg 2010

It is known that every planar graph G can be drawn so that edges of G are represented by non-crossing segments [5]. Hence, it is natural to examine the minimum number of slopes in a planar embedding of a planar graph.

In this paper, we make the (standard) distinction between *planar graphs*, which are graphs that admit a plane embedding, and *plane graphs*, which are graphs accompanied with a fixed prescribed combinatorial embedding, including a prescribed outer face. Accordingly, we distinguish between the *planar slope number* of a planar graph G, which is the smallest number of slopes needed to construct any straight-line plane embedding of G, as opposed to the *plane slope number* of a plane graph G, which is the smallest number of slopes needed to realize the prescribed combinatorial embedding of G as a straight-line plane embedding.

The research of slope parameters related to plane embedding was initiated by Dujmović et al. [2]. In [4], there are numerous results for the plane slope number of various classes of graphs. For instance, it is proved that every plane 3-tree can be drawn using at most $2n$ slopes, where n is its number of vertices. It is also shown that every 3-connected plane cubic graph can be drawn using three slopes, except for the three edges on the outer face.

In this paper, we study both the plane slope number and the planar slope number. The lower bounds of [1,9,3] for bounded-degree graphs do not apply to our case, because the constructed graphs with large slope numbers are not planar. Moreover, the upper bounds of [6,8] give drawings that contain crossings even for planar graphs.

For a fixed $k \in \mathbb{N}$, a *k-tree* is defined recursively as follows. A complete graph on k vertices is a k-tree. If G is a k-tree and K is a k-clique of G, then the graph formed by adding a new vertex to G and making it adjacent to all vertices of K is also a k-tree. A subgraph of a k-tree is called a *partial k-tree*.

A *two-terminal graph* (G, s, t) is a graph together with two distinct prescribed vertices $s, t \in V(G)$, known as *terminals*. The vertex s is called *source* and t is called *sink*. For a pair $(G_1, s_1, t_1), (G_2, s_2, t_2)$ of two-terminal graphs, a *serialization* is an operation that identifies t_1 with s_2, yielding a new two-terminal graph with terminals s_1 and t_2. Similarly, a *parallelization* is an operation which consists of identifying s_1 with s_2 into a single vertex s, and t_1 with t_2 into a single vertex t, thus yielding a two-terminal graph with terminals s and t. A two-terminal graph (G, s, t) is called *series-parallel graph* or *SP-graph* for short, if it either consists of a single edge connecting the vertices s and t, or if it can be obtained from smaller SP-graphs by serialization or parallelization.

We present several upper bounds on the plane and planar slope number in terms of the maximum degree Δ. The most general result of this paper is the following theorem, which deals with plane partial 3-trees.

Theorem 1. *The plane slope number of any plane partial 3-tree with maximum degree Δ is at most $2^{\mathcal{O}(\Delta)}$.*

Note that the above theorem implies that the planar slope number of any partial planar 3-tree is also at most $2^{\mathcal{O}(\Delta)}$. Since every outerplanar graph is also a partial 3-tree, the result above answers a question of Dujmović et al. [4], who asked whether a plane maximal outerplanar graph can be drawn using at most $f(\Delta)$ slopes.

In this extended abstract, we omit the proof of Theorem 1. Section 3 contains the proof of a weaker version of this result which deals with (non-partial) plane 3-trees.

In the special case of series-parallel graphs of maximum degree at most 3, we are able to prove an even better (in fact optimal) upper bound.

Theorem 2. *Any series-parallel graph with maximum degree at most 3 has planar slope number at most 3.*

Parts of the proof of Theorem 2 are in Section 2.

Let us introduce some basic terminology and notation that will be used throughout this paper. Let s be a segment in the plane. The smallest angle $\alpha \in [0, \pi)$ such that s can be made horizontal by a clockwise rotation by α, is called the *slope of s*. The *directed slope* of a directed segment is an angle $\alpha' \in [0, 2\pi)$ defined analogously.

A plane graph is called a *near triangulation* if all faces, except the outer face, are triangles.

2   Series-Parallel Graphs

In this section, we show the main ideas of the proof of Theorem 2.

We will in fact show that any series-parallel graph of maximum degree three can be embedded using the slopes from the set $S = \{0, \pi/4, -\pi/4\}$. This particular choice of S is purely aesthetic. Throughout this section, segments of slope $\pi/4$ (or 0, or $-\pi/4$) will be known as *increasing* (or *horizontal*, or *decreasing*, respectively).

First we give some useful definitions. For a pair of integers j and k, we say that a series-parallel graph (G, s, t) is a (j, k)-*graph* if G has maximum degree three, and furthermore, the vertex s has degree at most j and the vertex t has degree at most k.

Let us begin by a simple but useful lemma whose proof is omitted.

Lemma 1. *Let (G, s, t) be a $(1, 1)$-graph. Then G is either a single edge, a serialization of two edges, or a serialization of three graphs G_1, G_2 and G_3, where G_1 and G_3 consist of a single edge and G_2 is a $(2, 2)$-graph.*

We proceed with more terminology. An *up-triangle abc* is a right isosceles triangle whose hypotenuse ab is horizontal and whose vertex c is above the hypotenuse. We say that a series parallel graph (G, s, t) has an *up-triangle embedding* if it can be embedded inside an up-triangle abc using the slopes from S, in such a way that the two vertices s and t coincide with the two endpoints of the hypotenuse of abc, and all the remaining vertices are either inside or on the boundary of abc.

The concept of up-triangle embedding is motivated by the following lemma.

Lemma 2. *Every $(2, 2)$-graph has an up-triangle embedding.*

Proof. Let (G, s, t) be a $(2, 2)$-graph. We proceed by induction on the size of G. If G is a single edge, it obviously has an up-triangle embedding. If G is obtained by serialization or parallelization then there are a few cases to discuss. They are depicted in Fig. 1. □

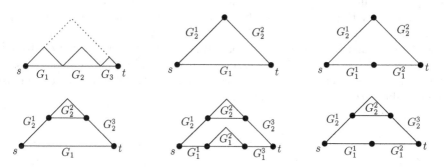

Fig. 1. Possible construction of a $(2, 2)$-graph G by serialization and parallelization of $(1, 1)$-graphs

To deal with $(3, 2)$-graphs, we need a more general concept than up-triangle embeddings. To this end, we introduce the following definitions.

An *up-spade* is a convex pentagon with vertices a, b, c, d, e in counterclockwise order, such that the segment ab is decreasing, the segment bc is horizontal, the segment cd is increasing, the segment ed is decreasing and the segment ae is increasing. We say that a series-parallel graph (G, s, t) has an *up-spade embedding* if it can be embedded into an up-spade $abcde$ using the slopes from S, in such a way that the vertex s coincides with the point a, the vertex t coincides either with the point b or with the point c, and all the remaining vertices of G are inside or on the boundary of the up-spade. Analogously, a *reverse up-spade embedding* is an embedding of a series-parallel graph (G, s, t) in which s coincides with b or c and t coincides with d.

Lemma 3. *Every $(3, 2)$-graph (G, s, t) has an up-spade embedding or an up-triangle embedding. Similarly, every $(2, 3)$-graph (G, s, t) has a reverse up-spade embedding or an up-triangle embedding.*

Proof. It suffices to prove just the first part of the lemma; the other part is symmetric. We again proceed by induction.

Let (G, s, t) be a $(3, 2)$-graph. If G is also a $(2, 2)$-graph, then G has an up-triangle embedding by Lemma 2. Assume that G is not a $(2, 2)$-graph. It is easy to see that in such case G has no up-triangle embedding, since it is impossible to embed three edges into an up-triangle in such a way that they meet in the endpoint of its hypotenuse.

Assume that G has been obtained by a serialization of a sequence of graphs $G_1, G_2, \ldots, G_k$, and that each of the graphs G_i is a single edge or a parallelization of smaller graphs. It follows that the graph G_2 is a single edge, because otherwise the two graphs G_1 and G_2 would share a vertex of degree at least 4. Let G_3^+ be the (possibly empty) serialization of $G_3, \ldots, G_k$. If G_3^+ is nonempty, it has an up-triangle embedding by Lemma 2. The graph G_1 has an up-spade embedding by induction. We may combine these embeddings as shown in Fig. 2 to obtain an up-spade embedding of G. If G_3^+ is empty, the construction is even simpler.

Assume now that G has been obtained by parallelization. Necessarily, it was a parallelization of a $(1,1)$-graph G_1 and a $(2,1)$-graph G_2. The graph G_2 can then be obtained by a serialization of a $(2,2)$-graph G_2^1 and a single edge G_2^2. The graph G_2^1 has an up-triangle embedding. Combining these embeddings, we obtain an up-spade embedding of G, as shown in Fig. 2. Note that we distinguish the possible structure of G_1 using Lemma 1.   □

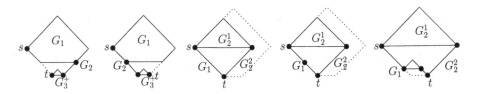

Fig. 2. Constructing an up-spade embedding of a $(3,2)$-graph by serialization and parallelization of smaller graphs

A similar case analysis can be done also for a $(3,3)$-graph. The serialization is easy by connecting two $(3,2)$-graphs while the parallelization takes a few cases. This finishes the proof of Theorem 2.

3   Planar 3-Trees

In this section, we outline a proof of a considerably weaker version of Theorem 1. Our current goal is to prove the following result.

Theorem 3. *There is a function g, such that every plane 3-tree with maximum degree Δ can be drawn using at most $g(\Delta)$ slopes.*

It is known that any plane 3-tree can be generated from a triangle by a sequence of vertex-insertions into inner faces. Here, a *vertex-insertion* is an operation that consists of creating a new vertex in the interior of a face, and then connecting the new vertex to all the three vertices of the surrounding face, thus subdividing the face into three new faces.

Throughout this section, we assume that Δ is a fixed integer.

For a partial plane 3-tree G we define the *level* of a vertex v as the smallest integer k such there is a set V_0 of k vertices of G with the property that v is on the outer face of the plane graph $G - V_0$. Let G be a partial plane 3-tree. An edge of G is called *balanced* if it connects two vertices of the same level of G. An edge that is not balanced is called *tilted*. Similarly, a face of G whose vertices all belong to the same level is called balanced, and any other face is called tilted. In a 3-tree, the level of a vertex v can also be equivalently defined as the length of the shortest path from v to a vertex on the outer face. However, this definition cannot be used for plane partial 3-trees.

Note that whenever we insert a new vertex v into an inner face of a 3-tree, the level of v is one higher than the minimum level of its three neighbors; note also that the level of all the remaining vertices of the 3-tree is not affected by the insertion of a new vertex.

Recall that a near triangulation is a plane graph whose every inner face is a triangle.

Let u, v be a pair of vertices forming an edge. A *bubble* over uv is an outerplanar plane near triangulation that contains the edge uv on the boundary of the outer face. The edge uv is called the *root* of the bubble. An *empty bubble* is a bubble that has no other edge apart from the root edge. A *double bubble* over uv is a union of two bubbles over uv which have only u and v in common and are attached to uv from its opposite sides. A *leg* is a graph L created from a path P by adding a double bubble over every edge of P. The path P is called the *spine of* L and the endpoints of P are also referred to as the endpoints of the leg. Note that a single vertex is also considered to form a leg.

A *tripod* is a union of three legs which share a common endpoint. A *spine* of a tripod is the union of the spines of its legs. Observe that a tripod is an outerplanar graph. The vertex that is shared by all the three legs of a tripod is called *the central vertex*.

Let G be a near triangulation, let Φ be an inner face of G. Let T be a tripod with three legs X, Y, Z and a central vertex c. An *insertion of tripod T into the face Φ* is the operation performed as follows. First, insert the central vertex c into the interior of Φ an connect it by edges to the three vertices of Φ. This subdivides Φ into three subfaces. Extend c into an embedding of the whole tripod T, by embedding a single leg of the tripod into the interior of each of the three subfaces. Next, connect every non-central vertex of the spine of the tripod to the two vertices of Φ that share a face with the corresponding leg. Finally, connect each non-spine vertex v of the tripod to the single vertex of Φ that shares a face with v. See Fig. 3. Observe that the graph obtained by a tripod insertion into Φ is again a near triangulation.

Lemma 4. *Let G be a graph. The following statements are equivalent:*

1. *G is a plane 3-tree, i.e., G can be created from a triangle by a sequence of vertex insertions into inner faces.*
2. *G can be created from a triangle by a sequence of tripod insertions into inner faces.*
3. *G can be created from a triangle by a sequence of tripod insertions into balanced inner faces.*

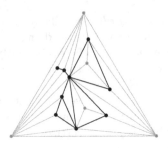

Fig. 3. An example of a tripod consisting of vertices of level 1 in a plane 3-tree

Proof. Clearly, (3) implies (2).

To observe that (2) implies (1), it suffices to notice that a tripod insertion into a face Φ can be simulated by a sequence of vertex insertions: first insert the central vertex of a tripod into Φ, then insert the vertices of the spine into the resulting subfaces, and then create each bubble by inserting vertices into the face that contains the root of the bubble and its subsequent subfaces.

To show that (1) implies (3), proceed by induction on the number of levels in G. If G only has vertices of level 0, then it consists of a single triangle and there is nothing to prove. Assume now that the G is a graph that contains vertices of $k > 0$ distinct levels, and assume that any 3-tree with fewer levels can be generated by a sequence of balanced tripod insertions by induction.

We will show that the vertices of level exactly k induce in G a subgraph whose every connected component is a tripod, and that each of these tripods is inserted inside a triangle whose vertices have level $k - 1$.

Let C be a connected component of the subgraph induced in G by the vertices of level k. Let $v_1, v_2, \ldots, v_m$ be the vertices of C, in the order in which they were inserted when G was created by a sequence of vertex insertions. Let Φ be the triangle into which the vertex v_1 was inserted, and let x, y and z be the vertices of Φ. Necessarily, all three of these vertices have level $k - 1$. Each of the vertices $v_2, \ldots, v_m$ must have been inserted into the interior of Φ, and each of them must have been inserted into a face that contained at least one of the three vertices of Φ.

Note that at each point after the insertion of v_1, there are exactly three faces inside Φ that contain a pair of vertices of Φ; each of these three faces is incident to an edge of Φ. Whenever a vertex v_i is inserted into such a face, the subgraph induced by vertices of level k grows by a single edge. These edges form a union of three paths that share the vertex v_1 as their common endpoint.

On the other hand, when a vertex v_i is inserted into a face formed by a single vertex of Φ and a pair of previously inserted vertices v_j, v_ℓ, then the graph induced by vertices of level k grows by two edges forming a triangular face with another edge whose endpoints have level k.

With these observations, it is easily checked (e.g., by induction on i) that for every $i \geq 1$, the subgraph of G induced by the vertices $v_1, \ldots, v_i$ is a tripod inserted into Φ. From this fact, it follows that the whole graph G could be created by a sequence of tripod insertions into balanced faces. $\square$

Note that when we insert a tripod into a balanced face, all the vertices of the tripod will have the same level (which will be one higher than the level of the face into which we insert the tripod). In particular, each balanced face we create by this insertion is an inner face of the tripod that we insert.

We will use the construction of plane 3-trees by tripod insertions as a main tool of our proof. Note that if G is a plane 3-tree of maximum degree at most Δ, then any tripod T used in the construction of G has fewer than 3Δ vertices. This is because every vertex of T is adjacent to a vertex of the triangular face Φ into which T was inserted, but each vertex of Φ has fewer than Δ neighbors on T. Let us say that a tripod T is Δ-*bounded* if it has maximum degree at most Δ and if it has at most 3Δ vertices. We conclude that any plane 3-tree of maximum degree Δ can be constructed by insertions of Δ-bounded tripods into balanced inner faces.

Let us give some technical definitions. Let α be a directed slope and let p be a point. We use the notation (p, α) to denote the ray starting in p with direction α.

Let G be a plane graph, let v be a vertex of G. We say that the vertex v has *visibility in direction α with respect to G*, if the ray starting in v and having direction α does not intersect the embedding of G in any point except v.

Assume now that G is a graph that has been obtained by inserting a tripod T in to a triangle Φ with vertex set x, y, z. Assume that we are given an embedding of the three vertices x, y, z as points in the plane, and we are also given a plane embedding $\mathcal{E}_T$ of the tripod T. We say that the embedding $\mathcal{E}_T$ is *compatible* with the embedding of x, y, z, if $\mathcal{E}_T$ is inside the convex hull of x, y, z, and it is possible to extend the plane embedding $\mathcal{E}_T \cup \{x, y, z\}$ into a plane straight-line embedding of the whole graph G.

Let us explain in more detail the main idea of the proof. As the principal step, we show that for every tripod T with at most 3Δ vertices, there is a finite set $\mathcal{F}_T$ of "permissible" embeddings of T, with the property that for any triangle x, y, z embedded in the plane, there exists an embedding from $\mathcal{F}_T$ whose appropriately scaled and translated copy is compatible with x, y, z. Since there are only finitely many tripods to consider, and since each considered tripod has only finitely many embeddings specified, all these embeddings together only define finitely many slopes, and finitely many (up to scaling) distinct triangular faces.

We thus have only finitely many pairs (Φ, T), where Φ is an embedding of a triangular face appearing in a permissible embedding of a tripod T', and T is a tripod. For each of these pairs we select a permissible embedding $\mathcal{E}_T$ of the tripod T that is compatible with Φ. Whenever we want to insert T into a scaled copy of the face Φ, we use the appropriately scaled copy of $\mathcal{E}_T$, so that the slope of a segment connecting a given vertex of Φ to a given vertex of $\mathcal{E}_T$ will only depend on the two vertices but not on the scaling of Φ.

As we know, any plane 3-tree G can be constructed as a sequence of tripod insertions into balanced faces. We construct the embedding of G recursively, so that whenever we need to insert a tripod T into an already embedded balanced triangle, we use the embedding selected by the procedure from the previous paragraph. The total number of slopes of all the balanced edges in the embedding of G can then be bounded by the total number of slopes appearing in the permissible embeddings of all tripods. The total number of slopes of tilted edges is bounded as well, which follows from the argument at the end of the last paragraph.

Let us now turn towards the technical details of the argument.

Lemma 5. *Let uv be a horizontal segment in the plane, let H be a halfplane containing uv on its boundary and extending above uv, and let $\varphi \in (0, \pi/2)$ be an angle. Let z be the point in H such that the segments uz and vz have slopes φ and $-\varphi$, respectively. There is a set $S \subseteq (-\varphi, \varphi)$ of 2Δ slopes such that every bubble B with root uv has a straight line drawing using only the slopes from S. Furthermore, all the vertices of this drawing except u and v are in the interior of the triangle uvz, and each vertex has visibility in any direction $\alpha \in (\varphi, \pi - \varphi)$.*

Proof (Sketch of proof of Lemma 5). Assume φ and B are given. To construct the drawing, first fix a sequence of slopes $0 < \varphi_0 < \varphi_1 < \varphi_2 < \ldots < \varphi_{\Delta-2} < \varphi$. In the first step, draw the vertices adjacent to u or v on a common line parallel to line uv, such that the absolute values of the slopes of the edges between uv and their neighbors belong to the sequence $\varphi_0, \ldots, \varphi_{\Delta-2}$ (see Fig. 4).

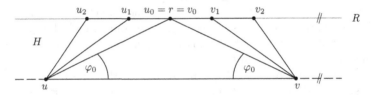

Fig. 4. Illustration of the proof of Lemma 5: drawing vertices adjacent to u and v

The rest of the bubble B can be expressed as a union of smaller bubbles, each of them rooted at a horizontal edge that has been drawn in the first step. We recursively apply the same drawing procedure to draw each of these smaller bubbles, each of them inside its own triangle similar to uvz, as illustrated in Fig. 5. □

Now that we can draw isolated bubbles, we may describe how to combine these drawings into a drawing of the whole leg of a tripod. Simply speaking, the procedure concatenates the drawings from Lemma 5 (appropriately rotated) on a single prescribed ray R.

Leg Drawing Procedure (LDP):
Input: A leg L with the central vertex u already drawn. A ray R with origin in u.
Output: Drawing of the leg L.

1. Assume that the spine of leg L contains vertices $u = u_0, u_1, \ldots, u_k$ such that $u_i u_{i-1}$ is an edge for $0 < i \leq k$.

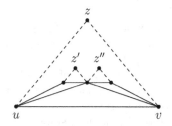

Fig. 5. Drawing a bubble in a triangle

2. Draw vertices u_i on R such that $|u_{i-1} - u| < |u_i - u|$ for $0 < i \leq k$.
3. Fix an angle $\varphi \in (0, \pi/2)$.
4. Use the drawing from Lemma 5, rotated and reflected if necessary, on both bubbles rooted at the edge $u_{i-1}u_i$ for $0 < i \leq k$. All the drawings use the same value of φ, and hence the same set S of slopes.

It is again not difficult to check that this procedure generates a correct plane straight-line drawing of a given leg. A careful analysis allows us to conclude that the drawing will use at most 2Δ slopes, and contain at most 4Δ distinct triangular faces, up to scaling and translation. These slopes and face-types only depend on the slope of the ray R and the choice of φ. Since by the choice of φ we can force each bubble to be embedded inside an arbitrarily "flat" isosceles triangle, we can easily argue that any vertex of the spine has visibility in any direction that differs from the undirected slope of R by more than φ. Moreover, a vertex $v \in L$ that does not belong to the spine has visibility in those directions that differ from the slope of R by more than φ and are directed towards the half-plane of R containing v. Finally, the central vertex u has visibility in any direction that differs from the directed slope of R by more than φ.

Finally we describe a procedure that combines the drawing of the individual legs into the drawing of the whole tripod. Let ε denote the value $\frac{\pi}{100}$ (any sufficiently small integral fraction of π is suitable here). The procedure expects a triangle Φ whose vertices are three points a, b, c. It then selects the position of a central vertex u, as well as the slopes of three rays R_1, R_2, R_3 emanating from u, and then draws the three legs of a given tripod on these rays by using LDP. The slopes of the legs are chosen in such a way that the resulting embedding of the tripod is compatible with Φ.

Furthermore, the slopes of the three rays are rounded to an integral multiple of ε. This rounding ensures, that the slopes of the spines of the legs can only take finitely many values (namely, at most $\frac{2\pi}{\varepsilon}$). It will follow that the procedure can only generate (up to scaling) a finite number of tripod embeddings, for all possible triangles Φ.

Tripod Drawing Procedure (TDP):
Input: A triangle $\Phi = \{a, b, c\}$ and a tripod T, where Φ is already drawn.
Output: Drawing of all vertices and edges of the tripod T inside Φ.

1. Fix an angle $\varphi \in (0, \frac{\pi}{4} - \varepsilon)$
2. Let u be the central vertex of T and let L_i for $i \in \{1, 2, 3\}$ be the legs of T.
3. Draw u to the intersection of the axes of the inner angles of Φ.
4. Process leg L_i for $i \in \{1, 2, 3\}$:
   (a) Let e be the endpoint of the spine of L_i different from u.
   (b) Let x, y be the vertices of Φ adjacent to the end e of L_i.
   (c) Let o be the axis of the angle cud.
   (d) Let R_i be a ray originating at u of slope o rounded to integral multiple of ε.
   (e) Use LDP to draw L_i on R_i. Scale the result so that it fits inside Ψ.

It is not difficult to check that the tripod-drawing procedure produces a correct straight-line embedding of any tripod inside any triangle Φ. Moreover, there is a set of slopes S of size at most $4\Delta\pi\varepsilon^{-1}$ and a set of triangles τ of size at most $8\Delta\pi\varepsilon^{-1}$, such that for any tripod T and any triangle Φ, the resulting embedding of T only uses the slopes from the set S and all its inner faces are scaled copies of triangles from τ.

The visibility properties and the "flatness" of the leg embedding guarantee that the resulting tripod embedding is compatible with Φ.

Let us now consider the slopes of the 'tilted' segments, i.e., those segments that connect a vertex of Φ with a vertex of the tripod embedded inside Φ by TDP. Assume that Φ is fixed. For each Δ-bounded tripod T, there are at most 9Δ segments connecting a vertex of Φ with a vertex of T. The number of Δ-bounded tripods is clearly finite (in fact, an upper bound of the form $2^{\mathcal{O}(\Delta)}$ can be obtained without much difficulty). We may now easily see that, for a fixed Φ, the total number of slopes of the segments that connect a vertex of Φ with a vertex of a Δ-bounded tripod is bounded.

Of course, for different triangles Φ, different slopes of this type arise. However, this is not an issue for us, because to generate a plane embedding of a plane 3-tree of maximum degree at most Δ, it is sufficient to insert Δ-bounded tripods into faces of previously inserted tripods. Thus, there are only finitely many triangles Φ for which we ever need to perform the tripod drawing procedure. Thus, by repeated calls of TDP, we may construct an embedding of any plane 3-tree with maximum degree Δ, while using at most $g(\Delta)$ slopes. More careful analysis of these arguments reveals a bound of the form $g(\Delta) = 2^{\mathcal{O}(\Delta)}$.

4   Conclusion and Open Problems

We have presented an upper bound of $2^{\mathcal{O}(\Delta)}$ for the planar slope number of planar partial 3-trees of maximum degree Δ. It is not obvious to us if the used methods can be generalized to a larger class of graphs, such as planar partial k-trees of bounded degree.

Let us remark that our proof of Theorem 1 actually implies a slightly stronger statement: for any Δ there is a set of slopes $S = S(\Delta)$ of size $2^{\mathcal{O}(\Delta)}$, such that all partial plane 3-trees of maximum degree Δ can be drawn using the slopes of S. This implies, for instance, that there is a constant $\varepsilon = \varepsilon(\Delta) > 0$ such that in our drawing of a partial plane 3-tree of maximum degree Δ, any two edges sharing a vertex have slopes differing by at least ε. Our method, however, is not necessarily suitable for obtaining good bounds on ε.

In view of the results of Keszegh et al. [6] and Mukkamala and Szegedy [8] for the slope number of (sub)cubic planar graphs, it would also be interesting to find analogous bounds for the planar slope number.

The main open problem is to determine whether the planar slope number of a planar graph can be bounded from above by a function of its maximum degree.

This paper does not address lower bounds for the planar slope number in terms of Δ; this might be another direction worth pursuing.

References

1. Barát, J., Matoušek, J., Wood, D.R.: Bounded-Degree Graphs have Arbitrarily Large Geometric Thickness. The Electronic Journal of Combinatorics 13, R3 (2006)
2. Dujmović, V., Suderman, M., Wood, D.R.: Really straight graph drawings. In: Pach, J. (ed.) GD 2004. LNCS, vol. 3383, pp. 122–132. Springer, Heidelberg (2005)
3. Dujmović, V., Suderman, M., Wood, D.R.: Graph drawings with few slopes. Computational Geometry 38(3), 181–193 (2007)
4. Dujmović, V., Eppstein, D., Suderman, M., Wood, D.R.: Drawings of planar graphs with few slopes and segments. Comp. Geom.-Theor. Appl. 38(3), 194–212 (2007)
5. Fáry, I.: On straight-line representation of planar graphs. Acta Univ. Szeged. Sect. Sci. Math. 11, 229–233 (1948)
6. Keszegh, B., Pach, J., Pálvölgyi, D., Tóth, G.: Drawing cubic graphs with at most five slopes. In: Kaufmann, M., Wagner, D. (eds.) GD 2006. LNCS, vol. 4372, pp. 114–125. Springer, Heidelberg (2007)
7. Kratochvíl, J., Thomas, R.: A note on planar partial 3-trees (in preparation)
8. Mukkamala, P., Szegedy, M.: Geometric representation of cubic graphs with four directions. Computational Geometry: Theory and Applications 42(9), 842–851 (2009)
9. Pach, J., Pálvölgyi, D.: Bounded-degree graphs can have arbitrarily large slope numbers. Electr. J. Combin. 13(1), N1 (2006)
10. Wade, G.A., Chu, J.-H.: Drawability of complete graphs using a minimal slope set. Comput. J. 37, 139–142 (1994)

Drawing Planar 3-Trees with Given Face-Areas⋆

Therese Biedl and Lesvia Elena Ruiz Velázquez

David R. Cheriton School of Computer Science, University of Waterloo,
Waterloo, ON N2L 3G1, Canada
{biedl,leruizve}@uwaterloo.ca

Abstract. We study straight-line drawings of planar graphs such that
each interior face has a prescribed area. It was known that such drawings
exist for all planar graphs with maximum degree 3. We show here that
such drawings exist for all planar partial 3-trees, i.e., subgraphs of a
triangulated planar graph obtained by repeatedly inserting a vertex in
one triangle and connecting it to all vertices of the triangle. Moreover,
vertices have rational coordinates if the face-areas are rational, and we
can bound the resolution. We also give some negative results for other
graph classes.

1 Introduction

A planar graph is a graph that can be drawn without crossing. Fáry, Stein and
Wagner [3,9,12] proved independently that every planar graph has a drawing
such that all edges are drawn as straight-line segments. Sometimes additional
constraints are imposed on the drawings. The most famous one is to have integer
coordinates while keeping the area small; it was shown in 1990 that this is always
possible in $O(n^2)$ area [4,8]. Another restriction might be to ask whether all edge
lengths are integral; this exists if the graph is 3-regular [5], but is open in general.

In this paper, we consider drawings with prescribed face areas. This has applications in cartograms, where faces (i.e., countries in a map) should be proportional to some property of the country, such as population. Ringel [7] showed
that such drawings do not exist for all planar graphs. Thomassen [10] showed
that they do exist for planar graphs with maximum degree 3. Quite a few results
are known for drawings with prescribed face areas that are not straight-line, but
instead use orthogonal paths, preferably with few bends [11,1,2].

We show that every planar partial 3-tree, for any given set of face areas, admits
a planar straight-line drawing that respects the face areas. It is quite easy to
show that such drawings exist; our main contribution is that the coordinates are
rational (presuming the face-areas are). This has not been studied before when
drawing planar graphs with prescribed face-areas. Furthermore, we can bound
the resolution in terms of the number of vertices (albeit not polynomially).

It remains open whether Thomassen's proof could be modified to yield rational
coordinates for all planar graphs of maximum degree 3; we provide some evidence
why this seems unlikely. We also show that some planar partial 4-tree cannot

---

⋆ Research supported by NSERC.

D. Eppstein and E.R. Gansner (Eds.): GD 2009, LNCS 5849, pp. 316–322, 2010.
© Springer-Verlag Berlin Heidelberg 2010

be realized at all, and another planar partial 4-tree can be realized only with irrational coordinates.

2   Background

Let $G = (V, E)$ be a graph with n vertices and m edges that is *simple* (has no loops or multiple edges) and *planar* (can be drawn without crossing.) A planar drawing of G splits the plane into connected pieces; the unbounded piece is called the *outer-face*, all other pieces are called *interior faces*. We assume that one combinatorial drawing (characterized by the clockwise order of edges around each vertex and choice of the outer-face) has been fixed for G.

A *planar straight-line drawing* of G is an assignment of vertices to distinct points in the plane such that no two (induced) straight-line segments of edges cross, and the fixed order of edges and outer-face are respected.

Let A be a function that assigns non-negative rationals[1] to interior faces of G. We say that a planar straight-line drawing of G *respects the given face areas* if every interior face f of G is drawn with area $const \cdot A(f)$, where the constant is the same for all faces. If $A \equiv 1$, then the drawing is called an *equifacial drawing*.

A graph G is a *k-tree* if it has a vertex order $v_1, \ldots, v_n$ such that for $i > k$ vertex v_i has exactly k *predecessors*, i.e., earlier neighbours, and they form a clique. A *partial k-tree* is a subgraph of a k-tree. Partial k-trees are the same as graphs of treewidth at most k; such graphs have received huge attention in the last few years due to the ability to solve many NP-hard problems in polynomial time on graphs of constant treewidth.

Assume G is a planar 3-tree. Then vertex v_i (for $i > 3$) has three predecessors and they form a triangle. Hence we can think of G as being built up by starting with a triangle, and repeatedly picking a face f (which is necessarily a triangle) and subdividing f into three triangles by inserting a new vertex in it. One can show that the first triangle in this process can be presumed to be the outer-face.

A planar partial 3-tree is a graph G' that is planar and is the subgraph of a 3-tree G [6]. Planar partial 3-trees include outerplanar graphs, series-parallel graphs, Halin graphs and IO-graphs.

3   Drawing Planar Partial 3-Trees

We now show that every planar partial 3-tree can be drawn with given face areas. A vital ingredient is how to draw K_4 by placing one point inside a triangle.

Lemma 1. *Let T be a triangle with area a and vertices v_0, v_1, v_2 in counterclockwise order. For any non-negative value $a_0 + a_1 + a_2 = a$, there exists a point v^* inside T such that triangle $\{v_{i+1}, v_{i-1}, v^*\}$ has area a_i, for $i = 0, 1, 2$ and addition modulo 3.*

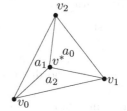

---
[1] Irrational face areas could be allowed, but would force irrational coordinates.

Proof. Let (x_0, y_0), (x_1, y_1), (x_2, y_2), (x^*, y^*) be the coordinates of v_0, v_1, v_2, v^*, respectively. The signed area formula expresses the area of a triangle via determinants; the result is positive if the vertices are counterclockwise around the triangle and negative otherwise. In particular, for a_i to be the area of a triangle $\{v_{i+1}, v_{i-1}, v^*\}$ (for $i = 0, 1, 2$ and addition modulo 3), we must have

$$2 \cdot a_i = \begin{vmatrix} x_{i+1} & y_{i+1} & 1 \\ x_{i-1} & y_{i-1} & 1 \\ x^* & y^* & 1 \end{vmatrix}$$
$$= (x_{i-1} \cdot y^* - x^* \cdot y_{i-1}) - (x_{i-1} \cdot y_{i+1} - x_{i+1} \cdot y_{i-1}) + (x^* \cdot y_{i+1} - x_{i+1} \cdot y^*)$$

Since the triangle defined by v_0, v_1, v_2 has area $a = a_1 + a_2 + a_3$, we also know

$$2 \cdot a = \begin{vmatrix} x_0 & y_0 & 1 \\ x_1 & y_1 & 1 \\ x_2 & y_2 & 1 \end{vmatrix} = (x_1 \cdot y_2 - x_2 \cdot y_1) - (x_1 \cdot y_0 - x_0 \cdot y_1) + (x_2 \cdot y_0 - x_0 \cdot y_2)$$

Combining these equations yields, after sufficient manipulation, that

$$x^* = \frac{a_1 \cdot x_1 + a_2 \cdot x_2 + a_3 \cdot x_3}{a_1 + a_2 + a_3} \quad \text{and} \quad y^* = \frac{a_1 \cdot y_1 + a_2 \cdot y_2 + a_3 \cdot y_3}{a_1 + a_2 + a_3} \tag{1}$$

Since $2a_i$ is non-negative, the signed-area formula guarantees that v^* lies to the left of the directed segments $v_0 v_1$, $v_1 v_2$, and $v_2 v_0$, and hence inside T. $\square$

Lemma 2. *Every planar 3-tree can be drawn respecting prescribed face areas.*

Proof. Assume $v_1, \ldots, v_n$ is the vertex-order that defined the 3-tree G, with $\{v_1, v_2, v_3\}$ the outer-face. We proceed by induction on n. The base case is $n = 3$, where this is obvious. If $n \geq 4$, then consider the K_4 formed by v_n and its neighbours. In $G - v_n$, these neighbours form a triangle T that is an interior face. Draw $G - v_n$ recursively, requiring as area for T the sum of the area of the faces around v_n. Then, by Lemma 1, v_n can be added inside T suitably. $\square$

Lemma 3. *Every planar partial 3-tree can be drawn respecting prescribed face areas.*

Proof. Recall that a planar partial 3-tree can be augmented into a planar 3-tree G by adding edges. Each time an edge is added, it divides a face f_i into two faces f_i^1 and f_i^2. Let a_i be the prescribed area for f_i, then we choose area a_i^j for face f_i^j such that $a_i^1 + a_i^2 = a_i$, e.g. $a_i^1 = a_i^2 = \frac{a_i}{2}$. By Lemma 2, G can be drawn respecting the prescribed face areas. Deleting all added edges then gives the desired drawing. $\square$

In our construction, we are interested not only in whether such a drawing exists, but what bounds can be imposed on the resulting coordinates. If all areas are rationals, then Equation (1) shows immediately that the newly placed vertex v^* has rational coordinates if the coordinates of T are rational. Hence, using induction and starting in the base case with a triangle with rational coordinates, one can immediately show that all coordinates of all vertices are rational. We summarize:

Theorem 1. *Let G be a planar partial 3-tree and A be an assignment of non-negative rationals to interior faces of G. Then G has a straight-line drawing such that each interior face f of G has area $A(f)$ and all coordinates are rationals.*

We can also give bounds on the required resolution.

Theorem 2. *Any planar 3-tree G has an equifacial straight-line drawing with integer coordinates and width and height at most $\prod_{k=1}^{n}(2k+1)$.*

Proof. We show that G has an equifacial straight-line drawing with rational coordinates in $[0, 1]$ with common denominator at most $\prod_{k=1}^{n}(2k+1)$; the result then follows after scaling. Let $v_1, \ldots, v_n$ be a vertex order of G with v_1, v_2, v_3 the outer-face. The drawing is the one from Theorem 1; we assume that v_1, v_2, v_3 are at the triangle $T = \{(1,0), (0,1), (0,0)\}$ (this can be enforced in the base case of Lemma 2.) Since G is triangulated, it has $2n - 5$ faces; so each interior face is drawn with area $a = 1/(4n - 10)$ since T has area $1/2$. We show the bound on the denominator only for x-coordinates; y-coordinates are proved similarly.

We need some notations. Recall that we can view graph G as being obtained by inserting vertex v_j into the triangle T_j spanned by the three predecessors of v_j. Let G_j be the subgraph of G induced by all vertices on or inside T_j. Since T_j was a face in the graph induced by $\{v_1, \ldots, v_{j-1}\}$, all vertices in G_j are either v_j, or one of its three predecessor, or a vertex in $\{v_{j+1}, \ldots, v_n\}$ and so G_j has at most $n - j + 4$ vertices. Let f_j be the number of interior faces in G_j; we have $f_j \leq 2(n - j + 4) - 5 = 2n - 2j + 3$. Also note that T_j contains exactly these f_j faces and they all have area $1/(4n - 10)$, so T_j has area $f_j/(4n - 10)$.

We will show by induction on i that vertex v_i has x-coordinate

$$x_i = \frac{integer}{\prod_{4 \leq j \leq i} f_j} \tag{2}$$

for some integer that we will not analyze further to keep notation simple. Nothing is to show for $i = 1, 2, 3$, since x_i is an integer by choice of the points for the outer-face triangle. For $i \geq 4$, let $v_{i_0}, v_{i_1}, v_{i_2}$ be the three predecessors of v_i.

For $k = 0, 1, 2$, Equation (2) holds for x_{i_k} by $i_k \leq i - 1$ and induction, and expanding with integers $f_{i_k+1}, \ldots, f_{i-1}$ yields

$$x_{i_k} = \frac{integer}{\prod_{4 \leq j \leq i_k} f_j} = \frac{integer}{\prod_{4 \leq j \leq i-1} f_j}$$

Equation (1) states that $x_i = (a_0 x_{i_0} + a_1 x_{i_1} + a_2 x_{i_2})/(a_0 + a_1 + a_2)$, where a_0, a_1, a_2 are the areas of faces incident to v_i. For $k = 0, 1, 2$, each a_k is the sum of faces in some subgraph, and therefore an integer multiple of $1/(4n - 10)$. Furthermore, $a_0 + a_1 + a_2$ is exactly the area of triangle T_i spanned by $v_{i_1}, v_{i_2}, v_{i_3}$, which we argued earlier is $f_i/(4n - 10)$. Hence, as desired,

$$x_i = \frac{a_0 x_{i_0} + a_1 x_{i_1} + a_2 x_{i_2}}{a_0 + a_1 + a_2} = \frac{\sum_{k=0}^{2} \frac{integer}{4n-10} \frac{integer}{\prod_{4 \leq j \leq i-1} f_j}}{\frac{f_i}{4n-10}} = \frac{integer}{\prod_{4 \leq j \leq i} f_j}.$$

Since $f_4, \ldots, f_n$ are integers, by Equation (2) all x_i's have common denominator

$$\prod_{4 \leq j \leq n} f_j \leq \prod_{4 \leq j \leq n} (2n - 2j + 3) = \prod_{k=1}^{n-3}(2k + 1) \qquad \square$$

We mention without proof that we can obtain similar (but uglier-looking) bounds for arbitrary integer face areas, by replacing 'f_j' by 'the sum of the f_j largest face areas in G'. We also did experiments to see whether our bounds are tight. We computed (using Maple) the coordinates in Theorem 2 for the planar 3-tree $v_1, \ldots, v_n$ where v_i has predecessors $v_{i-1}, v_{i-2}, v_{i-3}$ for $i \geq 4$; note that this graph has $f_i = 2n - 2i + 3$ and hence is a good candidate to obtain the bound in Theorem 2. Figure 1 shows the least common denominator for various values of n; they are smaller than the upper bound but are clearly growing in exponential fashion as well.

n	LCD in drawing	upper bound
10	$5.0 \cdot 10^3$	$2.0 \cdot 10^6$
50	$3.1 \cdot 10^{34}$	$2.8 \cdot 10^{75}$
100	$1.0 \cdot 10^{82}$	$1.7 \cdot 10^{183}$
500	$1.0 \cdot 10^{427}$	$2.0 \cdot 10^{1271}$
1000	$2.8 \cdot 10^{852}$	$4.8 \cdot 10^{2853}$

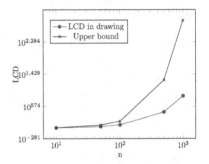

Fig. 1. Lower and upper bounds on the resolution in the drawing

4   Negative Results

In this section, we give some examples of graphs where no realization with rational coordinates is possible, hence providing counter-example to some possible conjectured generalizations of Theorem 1.

The first example is the octahedron where all face areas are 1 except for two non-adjacent, non-opposite faces, which have area 3. As shown by Ringel [7], any drawing that respects these areas must have some complex coordinates. (Ringel's result was actually for the graph G_1 obtained from the octahedron by subdividing two triangles further; the resulting graph then has no equifacial drawing.) Note that both the octahedron and G_1 are planar partial 4-trees, so not all partial 4-trees have equifacial drawings.

The second example is the octahedron where all face areas are 1 except that the three faces adjacent to the outer-face have area 3. (Alternatively, one could ask for an equifacial drawing of graph G_2 in Figure 2.) Assume, after possible linear transformation, that the vertices in the outer-face are at $(0, 0), (0, 13)$ and $(2, 0)$. Computing the signed area of all the faces one can show that the vertices

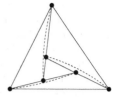

Fig. 2. Graphs G_1, G_2 and G_3

not on the outerface are at $(\frac{10}{3} + \frac{2\sqrt{3}}{13}, 5 - \sqrt{3})$, $(\frac{10}{3} - \frac{2\sqrt{3}}{13}, 3)$ and $(\frac{6}{13}, 5 + \sqrt{3})$. Thus even if a partial 4-tree has an equifacial drawing, it may not have one with rational coordinates.

The third example is again the octahedron, with three face areas prescribed to be 0, which forces some edges to be aligned as shown in Figure 2. If all other interior faces have area 1/8, and the outer-face is at $(1,0), (0,1), (0,0)$, then similar computations show that some of the coordinates of the other three vertices are $(3 \pm \sqrt{5})/8$. Let G_3 be the graph obtained from the octahedron by deleting the edges that are dashed in Figure 2. Graph G_3 is a crucial ingredient in Thomassen's proof [10] that every planar of maximum degree 3 graph has a straight-line drawing with given face areas: in one case he splits the input graph into G_3 and three subgraphs inside three interior faces of G_3, draws G_3 with the edges aligned as in Figure 2, and recursively draws and pastes the subgraphs. Since we showed that G_3 cannot always be drawn with rational coordinates, then Thomassen's proof, as is, does not give rational coordinates. It remains an open problem whether Thomassen's proof could be modified to show that any planar graph with maximum degree 3 has a drawing respecting given rational face-areas that has rational coordinates.

References

1. de Berg, M., Mumford, E., Speckmann, B.: On rectilinear duals for vertex-weighted plane graphs. In: Healy, P., Nikolov, N.S. (eds.) GD 2005. LNCS, vol. 3843, pp. 61–72. Springer, Heidelberg (2006)
2. Eppstein, D., Mumford, E., Speckmann, B., Verbeek, K.: Area-universal rectangular layouts. In: Proceedings of the 25th Annual Symposium on Computational Geometry. SCG 2009, Aarhus, Denmark, June 08-10, pp. 267–276. ACM, New York (2009)
3. Fáry, I.: On straight line representation of planar graphs. Acta. Sci. Math. Szeged 11, 229–233 (1948)
4. de Fraysseix, H., Pach, J., Pollack, R.: How to draw a planar graph on a grid. Combinatorica 10, 41–51 (1990)
5. Geelen, J., Guo, A., McKinnon, D.: Straight line embeddings of cubic planar graphs with integer edge lengths. Journal of Graph Theory 58(3), 270–274 (2008)
6. Kratochvil, J., Thomas, R.: Manuscript (in preparation)
7. Ringel, G.: Equiareal graphs. In: Contemporary methods in graph theory, in honour of Prof. Dr. K. Wagner, pp. 503–505 (1990)

8. Schnyder, W.: Embedding planar graphs on the grid. In: 1st Annual ACM-SIAM Symposium on Discrete Algorithms, pp. 138–148 (1990)
9. Stein, S.: Convex maps. Amer. Math. Soc. 2, 464–466 (1951)
10. Thomassen, C.: Plane cubic graphs with prescribed face areas. Combinatorics, Probability & Computing 1, 371–381 (1992)
11. van Kreveld, M., Speckmann, B.: On rectangular cartograms. In: Albers, S., Radzik, T. (eds.) ESA 2004. LNCS, vol. 3221, pp. 724–735. Springer, Heidelberg (2004)
12. Wagner, K.: Bemerkungen zum Vierfarbenproblem. Jahresbericht der Deutschen Mathematiker-Vereinigung 46, 26–32 (1936)

3D Visibility Representations by Regular Polygons

Jan Štola

Department of Applied Mathematics, Charles University
Malostranské nám. 25, Prague, Czech Republic
Jan.Stola@mff.cuni.cz

Abstract. We study 3D visibility representations of complete graphs where vertices are represented by equal regular polygons lying in planes parallel to the xy-plane. Edges correspond to the z-parallel visibility among these polygons.

We improve the upper bound on the maximum size of a complete graph with a 3D visibility representation by regular n-gons from $2^{O(n)}$ to $O(n^4)$.

1 Introduction

In this paper we study 3D visibility drawings that represent vertices by two-dimensional sets placed in planes parallel to the xy-plane. Two vertices are connected by an edge if and only if they can see each other in the direction that is orthogonal to their planes, i.e., parallel to the z-axis.

This type of representation was introduced as a generalization of the 2D visibility drawing. The 2D rectangle visibility drawing received a wide attention because of its connection to VLSI routing and circuit board layout [7,8].

The representation of vertices by rectangles remains popular also in the 3D visibility drawing. A lot of papers are focused on the maximum size of a complete graph with a 3D visibility representation by rectangles. Rote and Zelle provide a representation of K_{22} (see [6]). On the other hand, Bose et al. [4] showed that no complete graph with more than 102 vertices has such a representation. This result was then improved to 55 by Fekete et al. [3] and recently by Štola [5] to 50.

If the vertices are represented by unit squares then the largest complete graph with this type of representation is K_7 according to [3]. This is the only exact result known about representations by equal regular n-gons. Only estimates are known for $n \neq 4$. Babilon et al. [2] show that K_{14} can be represented by equal triangles. They also present a lower bound $\lfloor \frac{n+1}{2} \rfloor + 2$ on the maximum size of a complete graph with a 3D visibility representation by equal regular n-gons. Štola [1] then moved this bound to $n+1$. The first upper bound 2^{2^n} was given by Babilon et al. [2]. This doubly-exponential estimate was improved by Štola [1] to an exponential $\binom{6n-3}{3n-1} - 3 \approx 2^{6n}$. The main result of this paper is another significant improvement of this bound. We present a polynomial upper bound $O(n^4)$.

D. Eppstein and E.R. Gansner (Eds.): GD 2009, LNCS 5849, pp. 323–333, 2010.

2   Preliminaries

Let P be a regular n-gon inscribed in a unit circle (with the center c). Let $v_0, v_1, \ldots, v_n = v_0$ be the vertices of P, $s_0 = \overline{v_0 v_1}, \ldots, s_{n-1} = \overline{v_{n-1} v_n}, s_n = s_0$ the sides of P, m_i the center of s_i and p_i the half-line $\overrightarrow{cm_i}$. If P_i is a copy of P (shifted by a vector $\boldsymbol{w_i}$) then we denote its vertices by v_j^i and the sides by s_j^i.

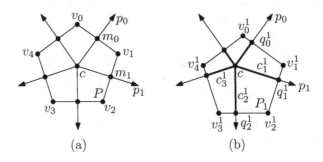

(a)   (b)

Fig. 1.

The distance of v_j and p_j is $\sin(\pi/n)$, similarly $\mathrm{dist}(v_j, p_{j-1}) = \sin(\pi/n)$ and $\mathrm{dist}(s_j, c) = \cos(\pi/n)$. Hence, if $|\boldsymbol{w_i}| < \sin(\pi/n)$ then v_j^i (the shifted copy of v_j) remains in the angle $\widehat{m_{j-1}cm_j}$. If in addition $|\boldsymbol{w_i}| < \cos(\pi/n)$ then s_j^i intersects p_j.

Definition 1. *Let $\{P_i, P_i = P + \boldsymbol{w_i}\}$ be the set of shifted copies of a regular n-gon P (inscribed in a unit circle). We say that this set is a* short-distance set *if $\forall i : |\boldsymbol{w_i}| < \min(\sin(\pi/n), \cos(\pi/n))$.*

The definition of a short-distance set requires a reference polygon P that is close to every polygon from the set. If the polygons $P_i = P + \boldsymbol{w_i}$ are far from P but close to each other, i.e., $\forall i, j : |\boldsymbol{w_i} - \boldsymbol{w_j}| < \min(\sin(\pi/n), \cos(\pi/n))$ then they also form a short-distance set because we can take any P_i as a reference polygon in this case.

For a polygon P_i from a short-distance set we can define $q_j^i = p_j \cap s_j^i$ and $c_j^i = \mathrm{dist}(c, q_j^i)$ (see Figure 1b). We call the n-tuple $(c_j^i)_{j=1}^n$ the *coordinates* of P_i.

Every polygon can be reconstructed from its coordinates (see Figure 2). If H_j^i is the half-plane with its boundary line h_j^i such that $c \in H_j^i$, $h_j^i \perp p_j$ and $\mathrm{dist}(h_j^i, c) = c_j^i$ then $P_i = \bigcap_{j=1}^n H_j^i$. Therefore the intersection $P_i \cap P_k = \bigcap_{j=1}^n (H_j^i \cap H_j^k)$ can be described by coordinates $(\min(c_j^i, c_j^k))_{j=1}^n$.

We assume in the sequel that P is a regular n-gon inscribed in a unit circle and $\{P_i = P + \boldsymbol{w_i}, i = 1, \ldots, m\}$ is a 3D visibility representation of a complete graph K_m. We assume that the z-coordinate of P_i is i but we use it to identify polygons that can block visibility between other polygons only. Otherwise, we ignore the z-coordinate and work with the polygons as if they were in the same

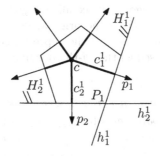

Fig. 2. Reconstruction of the polygon P_1 from its coordinates

xy-parallel plane. Formally, these operations represent operations over orthogonal projections of the relevant objects (points, lines, polygons) into a common xy-parallel plane and the projection of the results (for example, intersection points) into individual planes of the polygons.

Lemma 1. *Polygons P_i and P_k can see each other if and only if there exists l such that $\forall j, i < j < k : (c_l^j < \min(c_l^i, c_l^k) \text{ or } c_{l+1}^j < \min(c_{l+1}^i, c_{l+1}^k))$.*

Proof. $Q = P_i \cap P_k$ is a polygon given by coordinates $(\min(c_j^i, c_j^k))_{j=1}^n$. Let Q_l be the intersection of Q with the angle $\widehat{m_l c m_{l+1}}$ and q_l be the (only) vertex of Q in $\widehat{m_l c m_{l+1}}$.

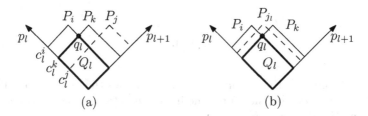

Fig. 3.

If $c_l^j < \min(c_l^i, c_l^k)$ or $c_{l+1}^j < \min(c_{l+1}^i, c_{l+1}^k)$ then P_j doesn't block the visibility of P_i and P_k in the neighborhood of q_l, see Figure 3a. Hence, if for a fixed l this condition holds for all polygons P_j between P_i and P_k then P_i and P_k can see each other in the neighborhood of q_l.

On the other hand, if $\forall l \; \exists j_l : i < j_l < k$, $c_l^{j_l} \geq \min(c_l^i, c_l^k)$ and $c_{l+1}^{j_l} \geq \min(c_{l+1}^i, c_{l+1}^k)$ then P_{j_l} blocks the visibility of P_i and P_k in the angle $\widehat{m_l c m_{l+1}}$, see Figure 3b. Therefore P_i cannot see P_k. □

Lemma 1 describes a sufficient and necessary condition for the visibility between two polygons from a short-distance set. If we shift the polygon P_i by a sufficiently small vector then we don't break any of the strict inequalities in Lemma 1. In other words, the shifted polygon can see all polygons that the original polygon can see. Therefore we can replace the original polygon P_i by the shifted one without breaking the completeness of the represented graph. This observation allows us to assume in the sequel that j-th coordinates of polygons are distinct, i.e., $\forall i, j, k, i \neq k : c_j^i \neq c_j^k$.

Lemma 2. *Let P_i be a regular n-gon with coordinates $(c_j^i)_{j=1}^n$ and $P_k = P_i + \boldsymbol{w}$ a shifted copy of P_i with coordinates $(c_j^k)_{j=1}^n$. If n is even then there are exactly $n/2$ adjacent coordinates with $\mathrm{sgn}(c_j^k - c_j^i) = 1$ and $n/2$ adjacent coordinates with the opposite signum. If n is odd then there are $\lfloor n/2 \rfloor$ or $\lceil n/2 \rceil$ adjacent coordinates with $\mathrm{sgn}(c_j^k - c_j^i) = 1$ and the rest with the opposite signum.*

Proof. The length of the orthogonal projection of $\boldsymbol{w}$ into a line containing p_j is $|c_j^k - c_j^i|$. The difference $c_j^k - c_j^i$ is positive (resp. negative) if this projection of $\boldsymbol{w}$ has the same (resp. the opposite) orientation as p_j.

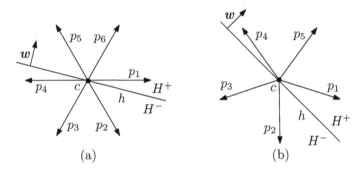

Fig. 4.

Let h be a line such that $h \perp \boldsymbol{w}$ and $c \in h$. h divides the plane into half-planes H^+ and H^-. Let H^+ be the half-plane in the direction of the vector $\boldsymbol{w}$. p_j lies in H^+ resp. H^- if $c_j^k > c_j^i$ resp. $c_j^i > c_j^k$.

If n is even then exactly $n/2$ adjacent half-lines from $(p_j)_{j=1}^n$ lie in H^+ and $n/2$ adjacent half-lines lie in H^-, see Figure 4a. If n is odd then $\lfloor n/2 \rfloor$ or $\lceil n/2 \rceil$ adjacent half-lines lie in H^+ and the rest of them lie in H^-, see Figure 4b. $\quad\square$

The next lemma shows that every 3D visibility representation of a complete graph contains a large short-distance subset. The following sections focus on these subsets.

Lemma 3. *Let $\{P_i = P + \boldsymbol{w}_i, i = 1, \ldots, m\}$ be a set of regular n-gons. If $\{P_i\}$ is a 3D visibility representation of a complete graph K_m then $\{P_i\}$ contains a short-distance subset with at least $\lceil m/16n^2 \rceil$ polygons.*

Proof. Every two polygons P_j, P_k from the representation have to intersect (to see each other). Polygons $\{P_i\}$ are shifted copies of P (a polygon inscribed into a unit circle). Hence, P_j can intersect P_k only if the distance of their centers is at most 2. Therefore the set C of centers of polygons from $\{P_i\}$ has the diameter at most 2.

Let S be a square that contains all points from C and whose side-length is 2. We can divide this square into $4n \times 4n = 16n^2$ sub-squares with the side-length $1/2n$. At least one of these sub-squares must contain at least $\lceil m/16n^2 \rceil$ points of C. We claim that the polygons with the center in this sub-square form a short-distance set.

It is sufficient to show that two points in one sub-square have the distance lower than $\min(\sin(\pi/n), \cos(\pi/n))$. For $x \in (0, \pi/3\rangle$ we have $\frac{x}{\sqrt{2}\pi} < \min(\sin x, \cos x)$. Hence, for $n \geq 3$ we have $\frac{1}{\sqrt{2}n} < \min(\sin(\pi/n), \cos(\pi/n))$ and $\frac{1}{\sqrt{2}n}$ is the maximum distance of two points in one sub-square. $\qquad\square$

3   Regular $2k$-gons

The goal of this section is a polynomial upper bound on the maximum size of a complete graph with a 3D visibility representation by regular $2k$-gons. We start with a lemma that points out an important forbidden configuration of three polygons.

Lemma 4. *Let $\{P_1, P_2, P_3\}$ be a short-distance set of regular $2k$-gons. If $\{P_1, P_2, P_3\}$ is a 3D visibility representation of a complete graph K_3 then it cannot happen that $c_1^1 < c_1^2 < c_1^3$ and $c_2^1 > c_2^2 > c_2^3$ (where $(c_j^i)_{j=1}^n$ are coordinates of P_i).*

Proof. If $c_1^1 < c_1^2 < c_1^3$ and $c_2^1 > c_2^2 > c_2^3$ then $c_l^1 > c_l^2 > c_l^3$ for $l \in \{2, \ldots, k+1\}$ and $c_l^1 < c_l^2 < c_l^3$ for $l \in \{k+2, \ldots, 2k\} \cup \{1\}$ by Lemma 2. Therefore, $c_l^2 > \min(c_l^1, c_l^3)$ for $l \in \{1, \ldots, 2k\}$ and P_1 cannot see P_3 according to Lemma 1 but this is a contradiction. $\qquad\square$

The following lemma shows that if the sequence $(c_1^i)_i$ of the first coordinates is monotone then the size of the representation is small.

Lemma 5. *Let $\{P_i, i = 1, \ldots, m\}$ be a short-distance set of regular $2k$-gons. If $\{P_i\}$ is a 3D visibility representation of a complete graph K_m and $(c_1^i)_{i=1}^m$ is a monotone sequence (where $(c_j^i)_{j=1}^n$ are coordinates of P_i) then $m \leq k+1$.*

Proof. We assume that the sequence $(c_1^i)_{i=1}^m$ is increasing. The proof for a decreasing sequence is similar. Let $I = \{\{i, j\} : i < j, c_2^i > c_2^j\}$, i.e., the pairs of polygons whose boundaries intersect in $\overline{m_1 c m_2}$. We claim that $I = \emptyset$ or $\bigcap I \neq \emptyset$.

We proceed by contradiction. Let's assume that $I \neq \emptyset$ and $\bigcap I = \emptyset$. At first we show that there must be (at least) two disjoint pairs in I. Let's assume that there aren't two disjoint pairs in I. If $\{a, \overline{a} : a < \overline{a}\} \in I$ then there exist $B = \{b, \overline{b} : b < \overline{b}\}$ and $C = \{c, \overline{c} : c < \overline{c}\}$ in I such that $a \notin B$ and $\overline{a} \notin C$ (because $a, \overline{a} \notin \bigcap I$). Moreover $\overline{a} \in B$ and $a \in C$ because the pairs $\{a, \overline{a}\}$ and B (resp. C) are not disjoint. If $\overline{a} = b$ then $c_1^a < c_1^{\overline{a}} = c_1^b < c_1^{\overline{b}}$ and $c_2^a > c_2^{\overline{a}} = c_2^b > c_2^{\overline{b}}$ which is in contradiction with Lemma 4. Therefore $\overline{a} = \overline{b}$ and $B = \{b, \overline{a}\}$.

An analogous argument shows that $a = c$ and $C = \{a, \bar{c}\}$. The pairs B and C are not disjoint according to our assumption. This can happen only if $\bar{c} = b$ but then $c_1^a < c_1^{\bar{c}} = c_1^b < c_1^{\bar{a}}$ and $c_2^a > c_2^{\bar{c}} = c_2^b > c_2^{\bar{a}}$ which is in contradiction with Lemma 4 again. This means that there must be two disjoint pairs in I.

Let $\{a, \bar{a} : a < \bar{a}\}$ and $\{b, \bar{b} : b < \bar{b}\}$ be disjoint pairs in I. We can assume without loss of generality that $a < b$.

Let's assume that $\bar{a} < \bar{b}$ (see Figure 5):

$$a < \bar{a} < \bar{b}, a < b < \bar{b}, (c_1^i)_i \text{ increasing} \Rightarrow c_1^a < c_1^{\bar{a}} < c_1^{\bar{b}}, c_1^a < c_1^b < c_1^{\bar{b}}$$

$$\{a, \bar{a} : a < \bar{a}\}, \{b, \bar{b} : b < \bar{b}\} \in I \Rightarrow c_2^a > c_2^{\bar{a}}, c_2^b > c_2^{\bar{b}}$$

$$c_1^b < c_1^{\bar{b}}, c_2^b > c_2^{\bar{b}} \Rightarrow c_l^b > c_l^{\bar{b}}, l \in \{2, \ldots, k+1\} \text{ by Lemma 2}$$

$$c_1^a < c_1^{\bar{a}}, c_2^a > c_2^{\bar{a}} \Rightarrow c_l^a < c_l^{\bar{a}}, l \in \{k+2, \ldots, 2k\} \cup \{1\} \text{ by Lemma 2}$$

$$c_1^{\bar{a}} < c_1^{\bar{b}} \Rightarrow c_{k+1}^{\bar{b}} < c_{k+1}^{\bar{a}} \text{ by Lemma 2}$$

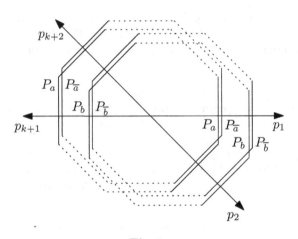

Fig. 5.

We can see that $c_1^a < c_1^b$ and $c_l^b > c_l^{\bar{b}}, l \in \{2, \ldots, k+1\}$. Therefore $c_l^b > \min(c_l^a, c_l^{\bar{b}})$, $l \in \{1, \ldots, k+1\}$. Similarly, $c_{k+1}^{\bar{b}} < c_{k+1}^{\bar{a}}$ and $c_l^a < c_l^{\bar{a}}$, $l \in \{k+2, \ldots, 2k\} \cup \{1\}$, i.e., $c_l^{\bar{a}} > \min(c_l^a, c_l^{\bar{b}})$, $l \in \{k+1, \ldots, 2k\} \cup \{1\}$. Hence, P_a cannot see $P_{\bar{b}}$ according to Lemma 1 but this cannot happen because $\{P_i\}$ is a representation of a complete graph. Therefore, it cannot be $\bar{a} < \bar{b}$.

If $\bar{b} < \bar{a}$ then $a < b < \bar{b} < \bar{a}$ and $c_1^a < c_1^b < c_1^{\bar{b}} < c_1^{\bar{a}}$ because $(c_1^i)_i$ is increasing. $c_2^{\bar{a}} < c_2^a$ and $c_2^{\bar{b}} < c_2^b$ because $\{a, \bar{a} : a < \bar{a}\}, \{b, \bar{b} : b < \bar{b}\} \in I$. If $c_2^{\bar{a}} < c_2^{\bar{b}}$ then $P_b, P_{\bar{b}}$ and $P_{\bar{a}}$ are in contradiction with Lemma 4. Similarly, if $c_2^b < c_2^a$ then P_a, P_b and $P_{\bar{b}}$ are in contradiction with Lemma 4. Therefore it must be $c_2^{\bar{b}} < c_2^{\bar{a}} < c_2^a < c_2^b$ but this means that the disjoint pairs $\{a, \bar{b} : a < \bar{b}\}$, $\{b, \bar{a} : b < \bar{a}\}$ satisfy assumptions of the previous paragraph and we again have a contradiction with the completeness of the represented graph.

We know that $\bar{a} \neq \bar{b}$ because $\{a, \bar{a}\}$ and $\{b, \bar{b}\}$ are disjoint. On the other hand, both possibilities $\bar{a} < \bar{b}$ and $\bar{b} < \bar{a}$ lead to a contradiction. Hence, the original assumption that $I \neq \emptyset$ and $\bigcap I = \emptyset$ cannot be satisfied. It must be either $I = \emptyset$ or $\bigcap I \neq \emptyset$.

If $I = \emptyset$ then $(c_2^i)_i$ is increasing. If $I \neq \emptyset$ then there exists $a \in \bigcap I$. This means that if $i < j$ and $c_i^2 > c_j^2$ then $i = a$ or $j = a$. In other words, the sequence $(c_2^i)_{i \in \{1,\ldots,m\} \setminus \{a\}}$ is increasing.

We can repeat this proof with $c_2, c_3, \ldots, c_k$ subsequently and show that there is a set A such that $|A| \leq k$ and $(c_{k+1}^i)_{i \in \{1,\ldots,m\} \setminus A}$ is increasing. On the other hand, this sequence is also decreasing by Lemma 2 because $(c_1^i)_{i \in \{1,\ldots,m\} \setminus A}$ is increasing. Therefore the sequence $(c_{k+1}^i)_{i \in \{1,\ldots,m\} \setminus A}$ has length at most 1 and $1 \geq |\{1, \ldots, m\} \setminus A| \geq m - k$. $\qquad\square$

Now we are ready to prove the main theorem of this section.

Theorem 1. *If $\{P_i, i = 1, \ldots, m\}$ is a 3D visibility representation of a complete graph K_m by regular n-gons (where $n = 2k$) then $m \leq 4n^2(n + 2)^2$.*

Proof. The set $\{P_i\}$ contains a short-distance subset $\{P_i'\}$ with at least $\lceil m/16n^2 \rceil$ polygons according to Lemma 3. Let $(c_j^i)_{j=1}^n$ be coordinates of P_i'. If $\lceil m/16n^2 \rceil \geq (k + 1)^2 + 1$ then due to Erdős-Szekeres theorem [9] the sequence $(c_1^i)_{i=1}^{\lceil m/16n^2 \rceil}$ contains a monotone subsequence of length $k + 2$ which is in contradiction with Lemma 5. Therefore $m/16n^2 \leq \lceil m/16n^2 \rceil \leq (k + 1)^2$. $\qquad\square$

4   Regular $(2k + 1)$-gons

We focus on regular $(2k+1)$-gons in this section. We prove a theorem analogous to Theorem 1. Unfortunately, Lemma 4 doesn't hold for $(2k + 1)$-gons. We have to use a more complicated version.

Lemma 6. *Let $\{P_1, P_2, P_3, P_4\}$ be a short-distance set of regular $(2k + 1)$-gons. If $\{P_i\}$ is a 3D visibility representation of a complete graph K_4 then it cannot happen that $c_1^1 < c_1^2 < c_1^3 < c_1^4$ and $c_2^1 > c_2^2 > c_2^3 > c_2^4$ (where $(c_j^i)_{j=1}^n$ are coordinates of P_i).*

Proof. If $c_1^1 < c_1^2 < c_1^3 < c_1^4$ and $c_2^1 > c_2^2 > c_2^3 > c_2^4$ then $c_l^1 > c_l^2 > c_l^3 > c_l^4$ for $l \in \{2, \ldots, k + 1\}$ and $c_l^1 < c_l^2 < c_l^3 < c_l^4$ for $l \in \{k + 3, \ldots, 2k + 1\} \cup \{1\}$ by Lemma 2. In other words, $c_l^2 > \min(c_l^1, c_l^3)$ and $c_l^3 > \min(c_l^2, c_l^4)$ for $l \in \{1, \ldots, 2k + 1\} \setminus \{k + 2\}$.

P_1 and P_3 can see each other. Therefore, $c_{k+2}^2 < \min(c_{k+2}^1, c_{k+2}^3)$ according to Lemma 1. Similarly, $c_{k+2}^3 < \min(c_{k+2}^2, c_{k+2}^4)$ because P_2 and P_4 can see each other. But this is a contradiction because the first inequality gives us $c_{k+2}^2 < c_{k+2}^3$ while $c_{k+2}^3 < c_{k+2}^2$ by the second inequality. $\qquad\square$

We need the following consequence of Lemma 6 several times in the sequel.

Corollary 1. *Let* $\{P_1, P_2, P_3, P_4\}$ *be a short-distance set of regular* $(2k+1)$-*gons. If* $\{P_i\}$ *is a 3D visibility representation of a complete graph* K_4 *then it cannot happen that* $c_1^1 < c_1^2 < c_1^3 < c_1^4$ *and* $c_{k+1}^1 < c_{k+1}^2 < c_{k+1}^3 < c_{k+1}^4$ *(or* $c_{k+2}^1 < c_{k+2}^2 < c_{k+2}^3 < c_{k+2}^4$*).*

Proof. If $c_1^1 < c_1^2 < c_1^3 < c_1^4$ and $c_{k+1}^1 < c_{k+1}^2 < c_{k+1}^3 < c_{k+1}^4$ then $c_{k+2}^1 > c_{k+2}^2 > c_{k+2}^3 > c_{k+2}^4$ by Lemma 1 but this is in contradiction with Lemma 6 for coordinates $k+1$ and $k+2$ (Lemma 6 holds for any pair of adjacent coordinates).

Similarly, if $c_1^1 < c_1^2 < c_1^3 < c_1^4$ and $c_{k+2}^1 < c_{k+2}^2 < c_{k+2}^3 < c_{k+2}^4$ then $c_{k+1}^1 > c_{k+1}^2 > c_{k+1}^3 > c_{k+1}^4$ by Lemma 1 and we have a contradiction again.   □

The next lemma is an analogy of Lemma 5. The proof of this lemma is more complicated because the representations by $(2k+1)$-gons are more complicated but the main ideas of both proofs (of Lemma 5 and Lemma 7) are the same.

Lemma 7. *Let* $\{P_i, i = 1, \ldots, m\}$ *be a short-distance set of regular* $(2k+1)$-*gons. There exists* $c > 0$ *independent of* k *such that if* $\{P_i\}$ *is a 3D visibility representation of a complete graph* K_m *and* $(c_1^i)_{i=1}^m$ *is a monotone sequence (where* $(c_j^i)_{j=1}^n$ *are coordinates of* P_i*) then* $m \leq ck$.

Proof. We assume that the sequence $(c_1^i)_{i=1}^m$ is increasing. The proof for a decreasing sequence is similar. Let $I = \{\{i, j\} : i < j, c_2^i > c_2^j\}$. We claim that there exists $n_0 \in \mathbb{N}$ (independent of k) such that I doesn't contain n_0 pairwise disjoint pairs.

Let's assume that $J \subseteq I : \forall A, B \in J, A \neq B \Rightarrow A \cap B = \emptyset$. Consider a complete graph on the vertex set J. We color the edge $\{\{a, \bar{a} : a < \bar{a}\}, \{b, \bar{b} : b < \bar{b}\} : a < b\}$ by

- color 1 when $\bar{a} < \bar{b}$ and $c_{k+2}^{\bar{a}} < \min(c_{k+2}^a, c_{k+2}^{\bar{b}})$
- color 2 when $\bar{a} < \bar{b}$ and $c_{k+2}^{\bar{a}} > \min(c_{k+2}^a, c_{k+2}^{\bar{b}})$
- color 3 when $\bar{b} < \bar{a}, c_2^a < c_2^b, c_2^{\bar{b}} < c_2^{\bar{a}}$ and $c_{k+2}^{\bar{b}} < \min(c_{k+2}^a, c_{k+2}^{\bar{a}})$
- color 4 when $\bar{b} < \bar{a}, c_2^a < c_2^b, c_2^{\bar{b}} < c_2^{\bar{a}}$ and $c_{k+2}^{\bar{b}} > \min(c_{k+2}^a, c_{k+2}^{\bar{a}})$
- color 5 when $\bar{b} < \bar{a}, c_2^a < c_2^b$ and $c_2^{\bar{a}} < c_2^{\bar{b}}$
- color 6 when $\bar{b} < \bar{a}, c_2^b < c_2^a$ and $c_2^{\bar{b}} < c_2^{\bar{a}}$
- color 7 when $\bar{b} < \bar{a}, c_2^b < c_2^a$ and $c_2^{\bar{a}} < c_2^{\bar{b}}$

If $\{\{a, \bar{a} : a < \bar{a}\}, \{b, \bar{b} : b < \bar{b}\} : a < b\}$ has the 7th color then $c_1^a < c_1^b < c_1^{\bar{b}} < c_1^{\bar{a}}$ because $a < b < \bar{b} < \bar{a}$ and $(c_1^i)_i$ is increasing. $c_2^{\bar{b}} < c_2^b$ because $\{b, \bar{b} : b < \bar{b}\} \in I$. Therefore $c_2^{\bar{a}} < c_2^{\bar{b}} < c_2^b < c_2^a$ and $P_a, P_b, P_{\bar{b}}, P_{\bar{a}}$ are in contradiction with Lemma 6. Hence, the 7th the color is not used and every edge of K_J has one of the first six colors.

According to Ramsey's theorem [10,11] there exists n_0 such that if $|J| \geq n_0$ then K_J contains a monochromatic subgraph K_S, $S = \{\{a, \bar{a} : a < \bar{a}\}, \{b, \bar{b} : b < \bar{b}\}, \{c, \bar{c} : c < \bar{c}\}, \{d, \bar{d} : d < \bar{d}\} : a < b < c < d\}$.

If K_S has color 1 then $c_{k+2}^{\bar{a}} < c_{k+2}^{\bar{b}} < c_{k+2}^{\bar{c}} < c_{k+2}^{\bar{d}}$, $\bar{a} < \bar{b} < \bar{c} < \bar{d}$ and $c_1^{\bar{a}} < c_1^{\bar{b}} < c_1^{\bar{c}} < c_1^{\bar{d}}$ (because $(c_1^i)_i$ is increasing). This is in contradiction with Corollary 1.

If K_S has color 2 then we have:

$$a < b < \bar{b}, a < \bar{a} < \bar{b}, (c_1^i)_i \text{ increasing} \Rightarrow c_1^a < c_1^b < c_1^{\bar{b}}, c_1^a < c_1^{\bar{a}} < c_1^{\bar{b}}$$

$$\{a, \bar{a} : a < \bar{a}\}, \{b, \bar{b} : b < \bar{b}\} \in I \Rightarrow c_2^{\bar{a}} < c_2^a, c_2^{\bar{b}} < c_2^b$$

$$c_1^b < c_1^{\bar{b}}, c_2^{\bar{b}} < c_2^b \Rightarrow c_l^{\bar{b}} < c_l^b, l \in \{2, \ldots, k+1\} \text{ by Lemma 2}$$

$$c_1^a < c_1^{\bar{a}}, c_2^{\bar{a}} < c_2^a \Rightarrow c_l^a < c_l^{\bar{a}}, l \in \{k+3, \ldots, 2k+1\} \cup \{1\} \text{ by Lemma 2}$$

We can see that $c_1^a < c_1^b$ and $c_l^{\bar{b}} < c_l^b, l \in \{2, \ldots, k+1\}$. Hence, $c_l^b > \min(c_l^a, c_l^{\bar{b}})$ for $l \in \{1, \ldots, k+1\}$. Similarly, $c_l^a < c_l^{\bar{a}}, l \in \{k+3, \ldots, 2k+1\} \cup \{1\}$ and $c_{k+2}^{\bar{a}} > \min(c_{k+2}^a, c_{k+2}^{\bar{b}})$. Therefore $c_l^{\bar{a}} > \min(c_l^a, c_l^{\bar{b}})$ for $l \in \{k+2, \ldots, 2k+1\} \cup \{1\}$. If $c_{k+1}^{\bar{a}} > \min(c_{k+1}^a, c_{k+1}^{\bar{b}})$ then P_a cannot see $P_{\bar{b}}$ according to Lemma 1. It must be $c_{k+1}^{\bar{a}} < \min(c_{k+1}^a, c_{k+1}^{\bar{b}})$, namely $c_{k+1}^{\bar{a}} < c_{k+1}^{\bar{b}}$. The same argument shows that also $c_{k+1}^{\bar{b}} < c_{k+1}^{\bar{c}} < c_{k+1}^{\bar{d}}$. On the other hand, $c_1^{\bar{a}} < c_1^{\bar{b}} < c_1^{\bar{c}} < c_1^{\bar{d}}$ (because $\bar{a} < \bar{b} < \bar{c} < \bar{d}$) which is in contradiction with Corollary 1.

If K_S has color 3 then $c_{k+2}^{\bar{d}} < c_{k+2}^{\bar{c}} < c_{k+2}^{\bar{b}} < c_{k+2}^{\bar{a}}, \bar{d} < \bar{c} < \bar{b} < \bar{a}$ and $c_1^{\bar{d}} < c_1^{\bar{c}} < c_1^{\bar{b}} < c_1^{\bar{a}}$ (because $(c_1^i)_i$ is increasing) and we have a contradiction again.

If K_S has color 4 then we proceed in a similar way as with the second color. We have:

$$c_2^a < c_2^b, c_2^{\bar{b}} < c_2^{\bar{a}}$$

$$a < b < \bar{b} < \bar{a}, (c_1^i)_i \text{ increasing} \Rightarrow c_1^a < c_1^b < c_1^{\bar{b}} < c_1^{\bar{a}}$$

$$\{a, \bar{a} : a < \bar{a}\} \in I \Rightarrow c_2^{\bar{a}} < c_2^a$$

$$c_1^b < c_1^{\bar{a}}, c_2^{\bar{a}} < c_2^a < c_2^b \Rightarrow c_l^{\bar{a}} < c_l^b, l \in \{2, \ldots, k+1\} \text{ by Lemma 2}$$

$$c_1^a < c_1^{\bar{b}}, c_2^{\bar{b}} < c_2^{\bar{a}} < c_2^a \Rightarrow c_l^a < c_l^{\bar{b}}, l \in \{k+3, \ldots, 2k+1\} \cup \{1\} \text{ by Lemma 2}$$

We can see that $c_1^a < c_1^b$ and $c_l^{\bar{a}} < c_l^b, l \in \{2, \ldots, k+1\}$. Hence, $c_l^b > \min(c_l^a, c_l^{\bar{a}})$ for $l \in \{1, \ldots, k+1\}$. Similarly, $c_l^a < c_l^{\bar{b}}, l \in \{k+3, \ldots, 2k+1\} \cup \{1\}$ and $c_{k+2}^{\bar{b}} > \min(c_{k+2}^a, c_{k+2}^{\bar{a}})$. Therefore, $c_l^{\bar{b}} > \min(c_l^a, c_l^{\bar{a}})$ for $l \in \{k+2, \ldots, 2k+1\} \cup \{1\}$. If $c_{k+1}^{\bar{b}} > \min(c_{k+1}^a, c_{k+1}^{\bar{a}})$ then P_a cannot see $P_{\bar{a}}$ according to Lemma 1. It must be $c_{k+1}^{\bar{b}} < \min(c_{k+1}^a, c_{k+1}^{\bar{a}})$, namely $c_{k+1}^{\bar{b}} < c_{k+1}^{\bar{a}}$. The same argument shows that also $c_{k+1}^{\bar{d}} < c_{k+1}^{\bar{c}} < c_{k+1}^{\bar{b}}$. On the other hand, $c_1^{\bar{d}} < c_1^{\bar{c}} < c_1^{\bar{b}} < c_1^{\bar{a}}$ (because $\bar{d} < \bar{c} < \bar{b} < \bar{a}$) which is in contradiction with Corollary 1.

If K_S has color 5 then $c_2^{\bar{a}} < c_2^{\bar{b}} < c_2^{\bar{c}} < c_2^{\bar{d}}, \bar{d} < \bar{c} < \bar{b} < \bar{a}$ and $c_1^{\bar{d}} < c_1^{\bar{c}} < c_1^{\bar{b}} < c_1^{\bar{a}}$ (because $(c_1^i)_i$ is increasing). This is in contradiction with Lemma 6.

If K_S has color 6 then $c_2^{\bar{d}} < c_2^{\bar{c}} < c_2^{\bar{b}} < c_2^{\bar{a}}, a < b < c < d$ and $c_1^a < c_1^b < c_1^c < c_1^d$ (because $(c_1^i)_i$ is increasing) and we have a contradiction with Lemma 6 again.

We can see that K_J cannot contain a monochromatic subgraph K_S. Therefore $|J| \le n_0 - 1$, i.e., I doesn't contain n_0 pairwise disjoint pairs.

Let $J_{max} \subseteq I$ be a maximal subset of pairwise disjoint pairs. We know that $|\bigcup J_{max}| = 2|J_{max}| \le 2(n_0 - 1)$. For any $A \in I$ there exists $B \in J_{max}$ such that $A \cap B \ne \emptyset$. Hence, the sequence $(c_2^i)_{i \in \{1, \ldots, m\} \setminus \bigcup J_{max}}$ is increasing.

We can repeat this proof with $c_2, c_3, \ldots, c_k$ subsequently and show that there is a set J' such that $|J'| \leq 2(n_0 - 1)k$ and $(c_{k+1}^i)_{i \in \{1,\ldots,m\} \setminus J'}$ is increasing. The sequence $(c_1^i)_{i \in \{1,\ldots,m\} \setminus J'}$ is also increasing. Therefore, its length is less than 4 by Corollary 1, i.e., $4 > |\{1, \ldots, m\} \setminus J'| \geq m - 2(n_0 - 1)k$. $\qquad \square$

Lemma 7 allows us to prove an analogy of Theorem 1 for regular $(2k + 1)$-gons.

Theorem 2. *There exists $c > 0$ such that if $\{P_i, i = 1, \ldots, m\}$ is a 3D visibility representation of a complete graph K_m by regular n-gons (where $n = 2k + 1$) then $m \leq cn^4$.*

Proof. The proof is the same as the proof of Theorem 1 (using Lemma 7 instead of Lemma 5). $\qquad \square$

If we combine Theorem 1 and Theorem 2 then we obtain the following result.

Theorem 3. *If $s(n)$ is the maximum size of a complete graph with a 3D visibility representation by equal regular n-gons then $s(n) = O(n^4)$.*

Proof. Theorem 1 if n is even and Theorem 2 if n is odd. $\qquad \square$

5   Conclusion

We show that the maximum size of a complete graph with a 3D visibility representation by regular n-gons is $O(n^4)$. This result is a significant improvement of the previously known exponential bound $\binom{6n-3}{3n-1} - 3 \approx 2^{6n}$ from [1]. We don't attempt to minimize constants in this estimate because there still remains a big gap between the lower bound $\Omega(n)$ and our upper bound $O(n^4)$.

References

1. Štola, J.: 3D Visibility Representations of Complete Graphs. In: Liotta, G. (ed.) GD 2003. LNCS, vol. 2912, pp. 226–237. Springer, Heidelberg (2004)
2. Babilon, R., Nyklová, H., Pangrác, O., Vondrák, J.: Visibility Representations of Complete Graphs. In: Kratochvíl, J. (ed.) GD 1999. LNCS, vol. 1731, pp. 333–340. Springer, Heidelberg (1999)
3. Fekete, S.P., Houle, M.E., Whitesides, S.: New Results on a Visibility Representation of Graphs in 3D. In: Brandenburg, F.J. (ed.) GD 1995. LNCS, vol. 1027, pp. 234–241. Springer, Heidelberg (1996)
4. Bose, P., Everett, H., Fekete, S.P., Lubiw, A., Meijer, H., Romanik., K., Shermer, T., Whitesides, S.: On a visibility representation for graphs in three dimensions. In: Di Battista, G., Eades, P., de Frayseix, H., Rosenstiehl, P., Tamassia, R. (eds.) GD 1993, pp. 38–39 (1993)
5. Štola, J.: Unimaximal Sequences of Pairs in Rectangle Visibility Drawing. In: Tollis, I.G., Patrignani, M. (eds.) GD 2008. LNCS, vol. 5417, pp. 61–66. Springer, Heidelberg (2009)
6. Fekete, S.P., Meijer, H.: Rectangle and box visibility graphs in 3D. Int. J. Comput. Geom. Appl. 97, 1–28 (1997)

7. Tamassia, R., Tollis, I.G.: A unifed approach to visibility representations of planar graphs. Discrete and Computational Geometry 1, 321–341 (1986)
8. Dean, A.M., Hutchinson, J.P.: Rectangle-Visibility Representations of Bipartite Graphs. In: Tamassia, R., Tollis, I.G. (eds.) GD 1994. LNCS, vol. 894, pp. 159–166. Springer, Heidelberg (1995)
9. Erdős, P., Szekeres, G.: A combinatorial problem in geometry. Compositio Math. 2, 463–470 (1935)
10. Ramsey, F.P.: On a problem of formal logic. Proc. London Math. Soc. 30, 264–286 (1930)
11. Graham, R., Rothschild, B., Spencer, J.: Ramsey Theory, 2nd edn. Wiley-Interscience Series in Discrete Mathematics. Wiley, New York (1990)

Complexity of Some Geometric and Topological Problems

Marcus Schaefer[*]

DePaul University, Chicago, IL 60604, USA
mschaefer@cs.depaul.edu

Abstract. We show that recognizing intersection graphs of convex sets has the same complexity as deciding truth in the existential theory of the reals. Comparing this to similar results on the rectilinear crossing number and intersection graphs of line segments, we argue that there is a need to recognize this level of complexity as its own class.

1 Introduction

We show that determining whether a graph can be realized as an intersection graph of convex sets in the plane has the same complexity as deciding the truth of existential first-order sentences over the real numbers. This connection between geometry and logic is not uncommon: Kratochvíl and Matoušek [11], for example, showed that recognizing intersection graphs of line segments also has the same complexity as the existential theory of the reals (we include a slightly simplified proof of that result), and there are several other geometric problems that share the same complexity. We therefore suggest the introduction of a new complexity class $\exists \mathbb{R}$, which captures the complexity of deciding the truth of the existential theory of the reals.

Remark 1. In the formal definition of $\exists \mathbb{R}$ we will not allow equality. If we define $\exists_{=} \mathbb{R}$ like $\exists \mathbb{R}$, but with equality allowed, we obtain a complexity class for which there is a name in the Blum-Shub-Smale model of computing over the reals: $\mathrm{BP}(\mathbf{NP}^0_{\mathbb{R}})$ [3]; this class has not played a major role in that model so far (as reflected by the complexity of the notation). Somewhat surprisingly, $\exists \mathbb{R} = \exists_{=} \mathbb{R}$ [21], even though algebraically the two classes differ, e.g. $x^2 = 2$ defines an irrational point, which is not possible without equality.

The first combinatorial problem shown complete for $\exists \mathbb{R}$ was stretchability of simple pseudoline arrangements, a result due to Mnëv as a byproduct of his universality theorem [14,18,23]. There have been several other problems classified as $\exists \mathbb{R}$-complete since, including the algorithmic Steinitz problem [2], intersection graphs of line segments [11], and straight-line realizability of abstract topological graphs [13]. Very often, however, $\exists \mathbb{R}$-completeness is not claimed explicitly;

---

[*] Some of this work was done in the beautiful library at Oberwolfach during the seminar on Discrete Geometry in September 2008.

D. Eppstein and E.R. Gansner (Eds.): GD 2009, LNCS 5849, pp. 334–344, 2010.

for example, in the case of the rectilinear crossing number, Bienstock gave a reduction from stretchability to the rectilinear crossing number problem. Since the problem can easily be shown to lie in $\exists\mathbb{R}$ (see Section 3) computing the rectilinear crossing number is $\exists\mathbb{R}$-complete. So—in a sense—the complexity of the problem is known precisely, but it is not unusual to see the complexity question for the rectilinear crossing number listed as an open problem [15]. There is some good reason for that: we do not know how to capture $\exists\mathbb{R}$ well with respect to classical complexity classes: we know that it contains **NP** (this follows easily from the definition of $\exists\mathbb{R}$; also, Shor gave a direct proof that stretchability is **NP**-hard [23]) and is itself contained in **PSPACE**, a remarkable improvement on Tarski's original decision procedure for the theory of reals by Canny [4]. So, in a sense, we do *not* know the complexity of the rectilinear crossing number problem, since we can only position it between **NP** and **PSPACE**. We should approach this situation in the same spirit as we do **NP**-completeness: **NP**-completeness of a problem does not exclude the possibility that the problems is in **P** or **EXP**-complete, but proving it **NP**-complete focuses that question on the real issue, away from the particular problem, and towards the study of the structural aspects of **NP**-completeness as a whole. Something similar should be possible for $\exists\mathbb{R}$-completeness. Knowing that a problem is $\exists\mathbb{R}$-complete does not tell us more than that it is **NP**-hard and in **PSPACE** in terms of classical complexity, but it does tell us where to start the attack: by studying the structure of $\exists\mathbb{R}$-complete problems; so asking, like [15], whether the rectilinear crossing number can be decided in **NP** is really asking whether $\exists\mathbb{R}$ lies in **NP**. And that puts a different perspective on the problem. A solution will likely not come out of graph drawing or graph theory, but out of a better understanding of real algebraic geometry and logic; what satisfiability is for **NP**, the existential theory of the reals is for $\exists\mathbb{R}$.

To justify our claim of the importance of $\exists\mathbb{R}$ and the necessity of a new complexity class, we need to find natural $\exists\mathbb{R}$-complete problems. In this note we give three examples: two known (one implicitly), one new. Plus one bonus problem in topological inference. This work is part of a more comprehensive project in which we survey many other problems as candidates for $\exists\mathbb{R}$-completeness including several other new results, including graph and linkage realizability and the complexity of finding Brouwer fixed points and Nash equilibria [19].

2   Background

The existential theory of the reals is the set of true sentences of the form

$$(\exists x_1, \ldots, x_n) \; \varphi(x_1, \ldots, x_n),$$

where φ is a quantifier-free Boolean formula (without negation) over the signature $(0, 1, +, *, <)$ interpreted over the universe of real numbers. It was first shown by Tarski that this theory is decidable; it is now known to be decidable in **PSPACE** by a result of Canny [4].

By disallowing negation, we restrict ourselves to strict inequalities, which is the version of the problem relevant to the examples presented in the current note; let us call the set of true sentences of this theory STRICT INEQ. With this we define the complexity class $\exists\mathbb{R}$ as the closure of STRICT INEQ under polynomial-time reductions. A problem is $\exists\mathbb{R}$-*complete* if it belongs to $\exists\mathbb{R}$ and every problem in $\exists\mathbb{R}$ can be reduced to it by a polynomial-time reduction. Note that **NP** $\subseteq \exists\mathbb{R}$, since we can express satisfiability of a Boolean formula in $\exists\mathbb{R}$. For example, $(x \vee \neg y \vee z) \wedge (\neg x \vee y \vee z) \wedge (\neg x \vee \neg y \vee \neg z)$ is equivalent to

$$(\exists x, y, z)[\ (-\varepsilon < x < 2) \wedge (-\varepsilon < y < 2) \wedge (-\varepsilon < z < 2)$$
$$\wedge\ (x(1-y)z) + ((1-x)yz) + ((1-x)(1-y)(1-z)) < \varepsilon],$$

if we choose $\varepsilon = 1/8(1 + 4m) = 1/104$ where m is the number of clauses, so $m = 3$ in the example.

A *pseudoline* is a simple closed curve in the projective plane that is homeomorphic to a straight line. An *arrangement of pseudolines* is a collection of pseudolines so that each pair of pseudolines cross at most once (and do not touch). An arrangement is *simple* if no more than two pseudolines pass through a point. Two arrangements are *equivalent* if there is a homeomorphism of the projective plane turning one into the other. An arrangement of pseudolines is *simply stretchable* if it is equivalent to a simple arrangement of straight lines. (So being simply stretchable means the original arrangement is simple and stretchable.)

Remark 2. If one wants to avoid the reference to the projective plane, one can define pseudolines in the plane as simple *x-monotone* curves, that is curves that cross every vertical line exactly once. If one takes this route, one needs to require that in an arrangement of pseudolines every pair of pseudolines crosses *exactly* once (as opposed to at most once).

Mnëv showed that STRICT INEQ reduces to SIMPLE STRETCHABILITY; since the reverse is also true, SIMPLE STRETCHABILITY is $\exists\mathbb{R}$-complete. Shor later simplified the reduction [23]. From this it immediately follows that SIMPLE STRETCHABILITY is **NP**-hard, since $\exists\mathbb{R}$-hardness implies **NP**-hardness as we saw above. (Shor [23] also gave a direct proof.)

$\exists\mathbb{R}$-hard problems typically require large representations; Goodman, Pollack and Sturmfels [8] showed that there are stretchable arrangements of n pseudolines whose coordinate representation requires 2^{cn} bits for some constant $c > 0$. (Equivalently, if we want to draw the endpoints on a grid, it must have size at least $2^{2^{c'n}}$ for some $c' > 0$.) Typically, reductions from an $\exists\mathbb{R}$-hard problem A to another problem B are *geometric* in the sense that if we are given a geometric representation of B, we can derive a geometric representation of A which is of at most polynomial size in the bit-size of the original representation. For example, this is the case for Bienstock's reduction from simple stretchability to rectilinear crossing number. We can then conclude (as Bienstock did) that there are graphs for which any straight-line drawing with optimal rectilinear crossing number requires 2^{cn} bits of storage. All other reductions in this note are also geometric, so geometric representations of these problem will require exponential precision.

3   Rectilinear Crossing Number

The *rectilinear crossing number* of G, lin-cr(G), is the smallest number of crossings in a straight-line drawing of G, that is, a drawing in which every edge is represented by a straight-line segment and at most two edges intersect in a point. The problem is **NP**-hard by Garey and Johnson's original proof that computing the crossing number is **NP**-hard [7] and it remains **NP**-hard even if the graph is cubic and 3-connected [9,16]. Bienstock gave an easy and elegant reduction that shows that SIMPLE STRETCHABILITY reduces to deciding whether lin-cr$(G) \leq k$, even if G is restricted to be cubic [1].

Theorem 1 (Bienstock [1]). *Computing the rectilinear crossing number of a (cubic) graph is $\exists\mathbb{R}$-complete. There are graphs for which the coordinates of the vertices in an* lin-cr*-optimal drawing of the graph require exponential precision (in the size of the graph).*

Proof. $\exists\mathbb{R}$-hardness follows from Bienstock's reduction as does the claim about exponential precision, so we only have to show that determining whether lin-cr$(G) \leq k$ lies in $\exists\mathbb{R}$; the only, small, difficulty is that we do not know which edges of the graph cross, so we need to guess a subset of pairs of edges of size at most k using real numbers.

Using quantifier-free formulas, we can define $\overline{colinear}(x_1, y_1, x_2, y_2, x_3, y_3)$ to express that the three points $(x_i, y_i)_{i \in [3]}$ are not colinear and a predicate $\overline{cross}(x_1, y_1, x_2, y_2, x_1', y_1', x_2', y_2')$ expressing that the two line segments determined by $(x_1, y_1), (x_2, y_2)$ and $(x_1', y_1'), (x_2', y_2')$ do not have a point in common (details in the full paper).

For a fixed k and $m = |E(G)|$, we can write a predicate $atmost_k(z_1, \ldots, z_{m^2})$ which guarantees that at most k of the z_i are greater than zero:

$$\bigwedge_{i \in [m^2]} \left((-1/2m^4 < z_i < 0) \vee (1 + 1/2m^2 < z_i) \right) \wedge \sum_{i \in [m^2]} z_i < k + 1.$$

Since lin-cr$(G) \leq \binom{m}{2}$, we can assume that $k \leq \binom{m}{2}$; so the sum of the negative z_i is at least $-1/2m^2$. If more than k of the z_i are positive, their sum is at least $k + 1 + (k+1)/2m^2$, but then the total sum is at least $k + 1$. On the other hand, given any subset of the z_i of size at most k, we can assign each z_i in the set the value $1 + 2/3m^2$ and every other z_i gets the value $-2/3m^4$, so that $\sum_{i \in [m^2]} z_i \leq k + 2/3 < k + 1$, showing that any subset of the z_i can be realized by $atmost_k$.

With these predicates, we can express lin-cr$(G) \leq k$; to simplify the formula, suppose that $V(G) = [n], E(G) = [m]$, and we have two functions $h, t : E \to V$ so that $h(e) = x$ and $t(e) = y$ if $e \in E$ is an edge between $x, y \in V$. We use $z_{(i-1)m+j} > 0$ to indicate that edges i and j are allowed to cross. Now lin-cr$(G) \leq k$ if and only if

$(\exists x_1, y_1, \ldots, x_n, y_n, z_1, \ldots, z_m)$ [$atmost_k(z_1, \ldots, z_{m^2})$

$\quad \wedge \quad \bigwedge_{i<j<k\in[n]} \overline{colinear}(x_i, y_i, x_j, y_j, x_k, y_k)$

$\quad \wedge \quad \bigwedge_{\substack{i<j\in[m], \\ \text{not adjacent}}} (z_{(i-1)m+j} > 0) \vee \overline{cross}(x_{h(i)}, y_{h(i)}, x_{t(i)}, y_{t(i)}, x_{h(j)}, y_{h(j)}, x_{t(j)}, y_{t(j)}))].$

4   Intersection Graphs of Segments

$G = (V, E)$ is an *intersection graph of line segments* if for each $v \in V$ there is a line segment ℓ_v in the plane so that $uv \in E$ if and only if ℓ_u and ℓ_v intersect.

Theorem 2 (Kratochvíl, Matoušek [11]). *Recognizing intersection graphs of line segments is* $\exists\mathbb{R}$*-complete. There are graphs for which the coordinates of the endpoints of the line segments in any intersection representation of the graph require exponential precision (in the size of the graph).*

Remark 3. Kratochvíl and Pergel showed that the recognition of intersection graphs of line segments remains **NP**-hard if the graphs have girth at least k for any fixed k [12]. Can this be extended to $\exists\mathbb{R}$-completeness?

We give a slightly simplified proof of Theorem 2; the argument will also be used in Theorem 3. We write $[n]$ for $\{1, \ldots, n\}$.

Lemma 1. *Suppose we have Jordan curves* ℓ, $(\ell_i)_{i\in[n]}$, $(s_i^j)_{i\in[n-1], j\in[3]}$, *and* $(c_i)_{i\in[4n]}$ *in the plane so that*

   (i) ℓ *crosses* ℓ_i, $i \in [n]$, *and* s_i^2, $i \in [n-1]$,
  (ii) c_i *crosses* c_{i+1} *(c_1 for $i = 4n$) exactly once,* $i \in [4n]$,
 (iii) ℓ_i *crosses* c_{2i} *and* $c_{4n-2i+2}$, $i \in [n]$,
 (iv) *both* s_i^1 *and* s_i^3 *cross* s_i^2, $i \in [n-1]$,
  (v) s_i^1 *crosses* c_{2i+1} *and* s_i^3 *crosses* $c_{4n-2i+1}$, $i \in [n-1]$,
 (vi) *the only other crossings among these curves are between pairs of* ℓ_i.

Then the curves ℓ_i *cross* ℓ *either in order* $\ell_1, \ldots, \ell_n$ *or in the reverse of that order. The conclusion remains true if instead of (i) we only require that (i')* ℓ *crosses* ℓ_i, $i \in [n]$, *and (i'')* s_i^2, $i \in [n-1]$, *may cross* ℓ, *but it does lie in the same connected component of* $\mathbb{R}^2 - \cup_{i\in[4n]}c_i$ *as* ℓ.

We call the collection of curves $(s_i^j)_{i\in[n-1], j\in[3]}$, and $(c_i)_{i\in[4n]}$ and the way they cross each other and the curves ℓ and $(\ell_i)_{i\in[n]}$ the *ordering gadget* for ℓ with respect to $(\ell_i)_{i\in[n]}$. The intended drawing of the curves of the lemma is shown in Figure 1, but there are other drawings.

Proof. The set $\cup_{i\in[4n]} c_i$ contains a (unique) closed Jordan curve C. C separates the plane into two faces; without loss of generality (since we are dealing with

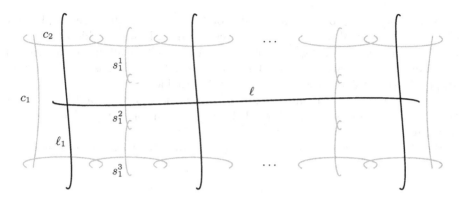

Fig. 1. An ordering gadget

curves), we may assume that ℓ lies in the inner face. Since ℓ crosses every s_i^2, and these curves do not cross any c_j, all the s_i^2 also lie in the inner face of C (indeed, (ii') is sufficient to draw this conclusion). Within each $S_i := c_{2i+1} \cup s_i^1 \cup s_i^2 \cup s_i^3 \cup c_{4n-2i+1}$, $i \in [n-1]$ choose a Jordan arc s_i with endpoints on C. The s_i are chords of C that lie in the inner face of C (since s_i^2 does); moreover, the s_i do not intersect each other (since any two S_i are disjoint) or any of the ℓ_i (since S_i and ℓ_j are disjoint for all $i, j \in [n]$). Now the ends of the s_i and ℓ_i along C are in order $\ell_1, s_1, \ell_2, \ldots, s_{n-1}, \ell_n, \ell_n, s_{n-1}, \ldots, s_1, \ell_1$ (up to cyclic shifts). Since every ℓ_i has to cross ℓ and has to do so within C, it must do so in order $\ell_1, \ldots, \ell_n$ or its reverse.

Proof (Theorem 2). It is easy to see that the problem lies in $\exists \mathbb{R}$. To show $\exists \mathbb{R}$-hardness, we reduce from **SIMPLE STRETCHABILITY**. Suppose we are given a simple arrangement $\mathcal{A}$ of pseudolines. Remark 2 allows us to think of the arrangement as a set of simple, x-monotone curves.

Add a triangle T formed by three pairwise crossing curves so that all crossings of curves in $\mathcal{A}$ lie within the region enclosed by T and one edge of T crosses all curves in $\mathcal{A}$ (for example, choose a vertical line segment to the left of all crossings in $\mathcal{A}$ that is long enough to cross every curve in $\mathcal{A}$). We can choose T so that we know the order of crossings of curves belonging to $\mathcal{A}$ with the curves of T.

Cut off the pseudolines just beyond the boundary of T and let $\mathcal{C}$ contain all the resulting curves together with the three curves from T. For each curve $\ell \in \mathcal{C}$ add the ordering gadget—as constructed in Lemma 1—with respect to all remaining curves in $\mathcal{C}$. Also, require that curves c_{2i} and $c_{4n-2i+2}$ for ℓ cross the corresponding two c-curves of ℓ_i (see Figure 2). Let $G_{\mathcal{A}}$ be the resulting intersection graph of all curves.

In any (curvilinear) drawing of $G_{\mathcal{A}}$, the order of crossings along each curve from $\mathcal{A}$ and T with curves from that set is as in the original arrangement or reversed by Lemma 1, since we added ordering gadgets for each of those curves. In particular, the crossings with T are first and last along each curve from $\mathcal{A}$, and therefore all crossings between curves of $\mathcal{A}$ occur within the region enclosed by

T. Now the crossings of $\mathcal{A}$ with the edge of T that crosses all curves in $\mathcal{A}$ occur either in the original order or in the reversed order. However, this means that the order of crossings along edges realizing $\mathcal{A}$ must either all be in the original order, or all of them are reversed. Hence, if $G_{\mathcal{A}}$ can be realized as an intersection graph of straight-line segments, then $\mathcal{A}$ is stretchable. ̄

It is easy to see that if the original arrangement $\mathcal{A}$ is stretchable, then so is the extended arrangement (the intended drawing of the ordering gadget is shown in Figure 1).

Finally, the reduction is geometric, so the claim about exponential precision follows.

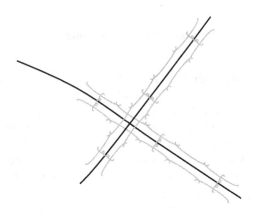

Fig. 2. Two arrangement lines crossing, with gadgets

5   Intersection Graphs of Convex Sets

$G = (V, E)$ is an *intersection graph of convex sets* if for every $v \in V$ there is a convex set C_v in the plane so that $uv \in E$ if and only if C_u and C_v intersect. We say two regions in the plane *intersect* if they share a common point. The problem is known to be in **PSPACE** and **NP**-hard [11].

Theorem 3. *Recognizing intersection graphs of convex sets is $\exists\mathbb{R}$-complete. There are graphs for which any realization as intersection graphs of convex polygons requires exponential precision in writing down the coordinates of the vertices of the polygon.*[1]

For the $\exists\mathbb{R}$-hardness proof we carefully adapt the reduction from SIMPLE STRETCHABILITY to SEG we saw in Theorem 2, and we begin by restating Lemma 1 for convex sets.

---

[1] The result on exponential precision has been independently obtained by Martin Pergel[10].

Lemma 2. *Suppose we have convex sets* L, $(L_i)_{i\in[n]}$, $(S_i^j)_{i\in[n-1],j\in[3]}$, *and* $(C_i)_{i\in[4n]}$ *in the plane so that*

(i) L *intersects* L_i, $i \in [n]$, *and* S_i^2, $i \in [n-1]$,
(ii) C_i *intersects* C_{i+1} *(C_1 for $i = 4n$), $i \in [4n]$,
(iii) L_i *intersects* C_{2i} *and* $C_{4n-2i+2}$, $i \in [n]$,
(iv) *both* S_i^1 *and* S_i^3 *intersect* S_i^2, $i \in [n-1]$,
(v) S_i^1 *intersects* C_{2i+1} *and* S_i^3 *intersects* $C_{4n-2i+1}$, $i \in [n-1]$,
(vi) *the only other intersections among these regions are between pairs of* L_i.

Moreover, suppose we have Jordan curves ℓ *in* L *and* ℓ_i *in* L_i, $i \in [n]$ *so that every* ℓ_i *crosses* ℓ. *Then the order of the intersections along* ℓ *is either* $\ell_1,\ldots,\ell_n$ *or the reverse of that order.*

We call the collection of convex sets $(S_i^j)_{i\in[n],j\in[3]}$, and $(C_i)_{i\in[8n-4]}$ and the way they intersect each other and the sets L and $(L_i)_{i\in[n]}$ the *ordering gadget* for L with respect to $(\ell_i)_{i\in[n]}$. The intended drawing of the convex sets is similar to the one shown in Figure 1 with line segments replaced by convex sets.

Proof. Pick vertices $v_i \in C_i \cap C_{i+1}$, $i \in [4n-1]$, and $v_{4n} \in C_{4n} \cap C_1$, and let c_i be a straight-line segment in C_i connecting v_i to v_{i+1} (v_1 for $i = 4n$). Then the c_i form a cycle C without crossings (since any two non-adjacent segments of C belong to disjoint convex sets). Now we can extend ℓ_i in $L_i \cup C_{2i}$ so it connects to v_{2i-1} and in $L_i \cup C_{4n-2i+2}$ so it connects to $v_{4n-2i+1}$ without crossing C. Pick vertices $t_i^1 \in S_i^1 \cap S_i^2$ and $t_i^2 \in S_i^2 \cap S_i^3$. We can connect t_i^1 by a curve s_i^1 in $S_i^1 \cup C_{2i+1}$ to v_{2i} and t_i^2 by a curve s_i^3 in $S_i^3 \cup C_{4n-2i+1}$ to v_{4n-2i} without crossing any of the curves we have already constructed; finally, we can connect t_i^1 to t_i^2 within S_i^2 by a curve s_i^2 not crossing any other curve except, possibly, ℓ. Now extend the curves we have constructed slightly, so that shared endpoints become crossing points. The resulting curves fulfill Lemma 1 with condition (i') in place of (i): (i') is true, since L intersects S_i^2, $i \in [n-1]$, and none of these sets intersect C, so they must all lie on the same side of C. Now Lemma 1 allows us to conclude that ℓ is crossed by $(\ell_i)_{i\in[n]}$ in order $\ell_1,\ldots,\ell_n$ or the reverse of that order.

Proof (Theorem 3). It it easy to see that the problem lies in $\exists\mathbb{R}$. Suppose we are given a simple arrangement $\mathcal{A}$ of pseudolines. As earlier, we think of the arrangement as a set of simple, x-monotone curves.

Let D be a disk which contains all the crossings of the pseudolines in its interior. Cut all the pseudolines at the boundary of D and let their order of intersection with the boundary be $A_1,\ldots,A_n,A_1,\ldots,A_n$. Add sets $(B_i)_{i\in[2n]}$, required to intersect cyclically: B_i with B_{i+1} and B_{2n} with B_1, with no other intersections, and sets $(H_i)_{i\in[2n]}$, so that H_i intersects B_i, and A_i if $i \leq n$ and A_{i-n} otherwise. Now for each of the A-, B- and H-sets add the ordering gadget described in Lemma 2. Call the resulting intersection graph $G_\mathcal{A}$. (We will only make use of the ordering gadgets for $(A_i)_{i\in[n]}$, but we need to add them in such a way that they allow for the intersections with the other sets as well.)

If $\mathcal{A}$ is stretchable, then the intersection graph we specified is realizable by convex sets (actually by line segments).

So suppose there is a drawing of convex sets realizing $G_{\mathcal{A}}$. Pick a vertex $u_i \in B_i \cap B_{i+1}$ for $i \in [2n-1]$, and $u_{2n} \in B_{2n} \cap B_1$, and let b_i be a straight-line segment between u_i and u_{i+1} (u_1 for $i = 2n$). Then the b_i form a cycle B (without crossings). None of the A_i intersect any of the B_j so all A_i must be on the same side of B. For each A_i, $i \in [n]$, pick a straight-line segment ℓ_i that starts in $A_i \cap H_i$ and ends in $A_i \cap H_{i+n}$. We claim that any two ℓ_i cross each other: each ℓ_i can be extended through H_i and H_{i+n} to connect to the cycle B. But then since two ℓ_i connect to alternating endpoints along B and both curves are on the same side of the cycle, the curves must cross; since the H_i do not intersect each other, that crossing must occur along the straight-line segments ℓ_i.

Now Lemma 2 implies that the order of crossings along each ℓ_i is either the original order or the reversed order; however, since the order of intersection with D is fixed by the cycle B, either all those orders are in the original order, or they are all reversed. But then, in either case, $\mathcal{A}$ is stretchable.

The claim about exponential precision again follows because the reduction we gave is geometric.

6 Topological Inference

Topological inference problems ask whether a specification of topological relationships can be realized by regions. The problems vary by what type of relationships (e.g. "contained in" and "disjoint with") and predicates (e.g. "connected", "convex") are available and what types of regions belong to the universe of discourse (2-dimensional, 3-dimensional, closed, regular, connected). For the current discussion we will restrict our universe to regular regions in the plane, not necessarily connected, where a region is *regular* if it is the closure of its interior.

There is a standard set of topological relationships, called RCC8, from the region connection calculus, that, in some sense, cover all possibilities of how two regions can be related to each other; the relations are, "disconnected" (DC), "externally connected" (EC), "equal" (EQ), "partially overlapping" (PO), "tangential proper part" (TPP), "tangential proper part inverse" (TPPi), "non-tangential proper part" (NTPP), and "non-tangential proper part inverse" (NTPPi), for details see [6,17]. Other relations can be defined from the basic relations, for example "proper part" (PP) is the disjunction of TPP and NTPP.

In the language of RCC8 a relationship between two regions is a *constraint* and the conjunction of several constraints a *constraint network* (we do not need to allow negation, since the 8 relations are exhaustive (at least one of them has to hold). These are special cases of *topological expressions*, that is, Boolean formulas involving the 8 relations (typically excluding negation, since it is not necessary). We say a topological expression is *realizable*, if it is realized by regular regions in the plane.

The problem of determining whether a constraint network (or a topological expression in general) is realizable lies in **NP** and is **NP**-complete for the full set of relations, though there are tractable fragments [17, Section 6]. If we restrict the universe to connected regions, the problem remains **NP**-complete, as shown in [20,22].

If we add the predicate "convex" to the signature of topological expressions, then the problem becomes ∃ℝ-complete.

Theorem 4 (Davis, Gotts, Cohn [5]). *RCC8 with convexity is ∃ℝ-complete, this remains true even if the signature is restricted to EC, PP and "convex" or PO, DC and "convex". In the second case the result remains true if the constraint network contains a constraint for every pair of regions (the constraint network is fully specified).*

Davis, Gotts, Cohn [5] only show the first part (EC, PP, and "convex"), we show the second part (restriction to PO, DC and "convex") here.

Proof. PO, DC and "convex" are enough to express that a graph G is the intersection graph of convex regions in the plane (we require every region to be convex, so we are not bothered by the disconnected regions contained in the universe), which, together with Theorem 3 suffices to establish ∃ℝ-hardness. Note that we specify for every pair of regions whether they overlap (PO) or are disjoint (DC), so the resulting constraint network is fully specified. Davis, Gotts, Cohn [5] show that the problem lies in ∃ℝ.

Remark 4. If we restrict the universe of discourse to connected sets, then it is not immediately obvious that the realizability problem (with convexity) remains in ∃ℝ: the issue at stake is that in this case the realizability problem without convexity is equivalent to the string graph problem, for which membership in **NP** is not trivial [22,20].

References

1. Bienstock, D.: Some provably hard crossing number problems. Discrete Comput. Geom. 6(5), 443–459 (1991)
2. Björner, A., Vergnas, M.L., Sturmfels, B., White, N., Ziegler, G.M.: Oriented matroids, 2nd edn. Encyclopedia of Mathematics and its Applications, vol. 46. Cambridge University Press, Cambridge (1999)
3. Bürgisser, P., Cucker, F.: Counting complexity classes for numeric computations. II. Algebraic and semialgebraic sets. J. Complexity 22(2), 147–191 (2006)
4. Canny, J.: Some algebraic and geometric computations in pspace. In: STOC 1988: Proceedings of the twentieth annual ACM symposium on Theory of computing, pp. 460–469. ACM, New York (1988)
5. Davis, E., Gotts, N.M., Cohn, A.G.: Constraint networks of topological relations and convexity. Constraints 4(3), 241–280 (1999)
6. Egenhofer, M.J.: Reasoning about binary topological relations. In: Günther, O., Schek, H.-J. (eds.) SSD 1991. LNCS, vol. 525, pp. 143–160. Springer, Heidelberg (1991)

7. Garey, M.R., Johnson, D.S.: Crossing number is NP-complete. SIAM Journal on Algebraic and Discrete Methods 4(3), 312–316 (1983)

8. Goodman, J.E., Pollack, R., Sturmfels, B.: Coordinate representation of order types requires exponential storage. In: STOC 1989: Proceedings of the twenty-first annual ACM symposium on Theory of computing, Seattle, Washington, United States, pp. 405–410. ACM, New York (1989)

9. Hliněný, P.: Crossing number is hard for cubic graphs. J. Combin. Theory Ser. B 96(4), 455–471 (2006)

10. Kratochvíl, J.: Personal communication

11. Kratochvíl, J., Matoušek, J.: Intersection graphs of segments. J. Combin. Theory Ser. B 62(2), 289–315 (1994)

12. Kratochvíl, J., Pergel, M.: Geometric intersection graphs: do short cycles help (extended abstract). In: Lin, G. (ed.) COCOON 2007. LNCS, vol. 4598, pp. 118–128. Springer, Heidelberg (2007)

13. Kyncl, J.: The complexity of several realizability problems for abstract topological graphs (unpublished manuscript) (based on Graph Drawing 2007 paper)

14. Mněv, N.E.: The universality theorems on the classification problem of configuration varieties and convex polytopes varieties. In: Topology and geometry—Rohlin Seminar. Lecture Notes in Math., vol. 1346, pp. 527–543. Springer, Berlin (1988)

15. Pach, J., Tóth, G.: Thirteen problems on crossing numbers. Geombinatorics 9(4), 194–207 (2000)

16. Pelsmajer, M.J., Schaefer, M., Štefankovič, D.: Crossing number of graphs with rotation systems. In: Hong, S.-H., Nishizeki, T., Quan, W. (eds.) GD 2007. LNCS, vol. 4875, pp. 3–12. Springer, Heidelberg (2008)

17. Renz, J.: Qualitative spatial reasoning with topological information. In: Renz, J. (ed.) Qualitative Spatial Reasoning with Topological Information. LNCS (LNAI), vol. 2293, p. 207. Springer, Heidelberg (2002)

18. Richter-Gebert, J.: Mněv's universality theorem revisited. Sém. Lothar. Combin. 34 (1995)

19. Schaefer, M.: The real logic of drawing graphs (unpublished manuscript)

20. Schaefer, M., Sedgwick, E., Štefankovič, D.: Recognizing string graphs in NP. J. Comput. System Sci. 67(2), 365–380 (2003); Special issue on STOC 2002, Montreal, QC

21. Schaefer, M., Štefankovič, D.: Fixed points, nash equilibria, and the existential theory of the reals (unpublished manuscript)

22. Schaefer, M., Štefankovič, D.: Decidability of string graphs. J. Comput. System Sci. 68(2), 319–334 (2004); Special issue on STOC 2001, Crete, Greece

23. Shor, P.W.: Stretchability of pseudolines is NP-hard. In: Applied geometry and discrete mathematics. DIMACS Ser. Discrete Math. Theoret. Comput. Sci., vol. 4, pp. 531–554. Amer. Math. Soc., Providence (1991)

On Planar Supports for Hypergraphs[*]

Kevin Buchin[1], Marc van Kreveld[2], Henk Meijer[3],
Bettina Speckmann[1], and Kevin Verbeek[1]

[1] Dep. of Mathematics and Computer Science, TU Eindhoven, The Netherlands
{k.a.buchin,b.speckmann,k.a.b.verbeek}@tue.nl
[2] Dep. of Computer Science, Utrecht University, The Netherlands
marc@cs.uu.nl
[3] Roosevelt Academy, Middelburg, The Netherlands
h.meijer@roac.nl

Abstract. A graph G is a *support* for a hypergraph $H = (V, \mathcal{S})$ if the vertices of G correspond to the vertices of H such that for each hyperedge $S_i \in \mathcal{S}$ the subgraph of G induced by S_i is connected. G is a *planar support* if it is a support and planar. Johnson and Pollak [9] proved that it is NP-complete to decide if a given hypergraph has a planar support. In contrast, there are polynomial time algorithms to test whether a given hypergraph has a planar support that is a path, cycle, or tree. In this paper we present an algorithm which tests in polynomial time if a given hypergraph has a planar support that is a tree where the maximal degree of each vertex is bounded. Our algorithm is constructive and computes a support if it exists. Furthermore, we prove that it is already NP-hard to decide if a hypergraph has a 3-outerplanar support.

1 Introduction

A *hypergraph* $H = (V, \mathcal{S})$ is a generalization of a graph, where V is a set of elements or vertices and $\mathcal{S}$ is a set of non-empty subsets of V, called *hyperedges* [3]. The set $\mathcal{S}$ of hyperedges is a subset of the powerset of V. Hypergraphs are not as common as graphs, but there are several application areas were they occur. For example, there is a natural correspondence between hypergraphs and database schemata in relational databases, with vertices corresponding to attributes and hyperedges to relations (e.g., see [2]). Further applications include VLSI design [13], computational biology [12], and social networks [5].

There is no single "standard" method of drawing hypergraphs, comparable to the point-and-arc drawings for graphs. In this paper we focus on a set of decision problems which are motivated by *subdivision drawings* of hypergraphs as proposed by Kaufmann et al. [10]. In a subdivision drawing each vertex corresponds uniquely to a face of a planar subdivision and, for each hyperedge, the union of the faces corresponding to the vertices incident to that hyperedge is connected. For example, vertex-based Venn diagrams [9] and concrete Euler diagrams [7] are both subdivision drawings.

---

[*] K. Buchin, B. Speckmann, and K. Verbeek were supported by the Netherlands' Organisation for Scientific Research (NWO) under project no. 639.022.707.

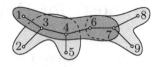

Fig. 1. Tree support for $H = (V, \mathcal{S})$ with $V = \{1, \ldots, 9\}$ and $\mathcal{S} = \{(2,3,4,5), (1,3,4,6,7), (6,7,8,9)\}$

A graph G is a *support* for a hypergraph $H = (V, \mathcal{S})$ if the vertices of G correspond to the vertices of H such that for each hyperedge $S_i \in \mathcal{S}$ the subgraph of G induced by S_i is connected (see Fig. 1). We say that S_i is connected in G. G is a *planar support* if it is a support and planar. Intuitively, a planar support is a subgraph of the dual graph of a subdivision drawing of H. Subdivisions and their dual graphs have been studied extensively and there are several methods that can turn a planar support into a dual subdivision. Hence we focus on finding planar supports for hypergraphs which can then easily be turned into subdivision drawings.

Johnson and Pollak [9] proved that it is NP-complete to decide if a given hypergraph has a planar support. In contrast, there are polynomial time algorithms that decide whether a given hypergraph has a planar support that is either a path, a cycle, or a tree. We discuss these results in some detail in Section 2. Path or cycle supports naturally lend themselves to the creation of pleasing and easily readable subdivision drawings which are *simple* and, in the case of path supports, *compact* [10]. However, not many hypergraphs admit a path or a cycle support. Tree supports, on the other hand, can have vertices of arbitrarily high degree and hence may not result in easily interpretable subdivision drawings. Therefore we consider tree supports of bounded vertex tree. For example, a binary tree support can be interpreted as the dual graph of a triangulation of a (convex) polygon and as such can be used to create a simple and compact subdivision drawing where each face of the subdivision is a triangle. In Section 3 we give an $O(kn^3)$ time constructive algorithm based on a flow formulation that solves the following decision problem: given a hypergraph H together with degrees d_i for each element i of the base set, is there a tree support for H such that each vertex i of the tree has degree at most d_i? Additionally, in Section 4 we strengthen the result by Johnson and Pollak by proving that it is even NP-complete to decide if a hypergraph has a 3-outerplanar support.

Notation and Definitions. Our input is a hypergraph $H = (V, \mathcal{S})$ with n vertices and k hyperedges. We denote the total input size by $N := \sum_i |S_i|$. In the remainder of the paper we interpret H as a set system $\mathcal{S} = \{S_1, \ldots, S_k\}$ on a base set $V = \{1, \ldots, n\}$ of n elements. Two elements h and j of V are *equivalent* with respect to $\mathcal{S}$ if every set $S_i \in \mathcal{S}$ contains either none or both of h and j. To simplify the discussion we assume that no two elements of V are equivalent. We also assume that each element of the base set occurs in at least one set (hence $N \geq n$) and that the elements within each set are sorted. The vertices of a planar support G correspond to the elements of V. We often directly identify a vertex with "its" element and use the same name to refer to both. Furthermore, for each hypergraph $H = (V, \mathcal{S})$ we consider a graph $G(H)$ on V. Two elements u and v of V are connected by an edge in $G(H)$ if there is a hyperedge $S_i \in \mathcal{S}$ that contains both u and v. We now define the connected components of H as the connected components of $G(H)$. Finally, a graph G is k-outerplanar if for $k = 1$, G is outerplanar and for $k > 1$, G has a planar embedding such that if all vertices on the exterior face are deleted, the connected components of the remaining graph are all $(k - 1)$-outerplanar.

2 Path, Cycle, and Tree Supports

In this section we summarize previous work on path, cycle and tree supports. For all three classes of graphs one can decide whether a given hypergraph has such a support in linear time.

Path support. Korach and Stern [11] observed that the decision problem for path supports is equivalent to finding a permutation π of $\{1, \ldots, n\}$ such that, for every set S_i, the elements of S_i are consecutive in π. This problem in turn is directly related to the *consecutive ones property*: a matrix of zeroes and ones is said to have the consecutive ones property if there is a permutation of its columns such that the ones in each row appear consecutively. Let M be a matrix with n columns and m rows such that entry (i, j) is 1 if $j \in S_i$, and 0 otherwise. H has a path support if and only if M has the consecutive ones property (see Fig. 2). There are algorithms [4,8] that can test the consecutive ones property and produce a corresponding permutation in $O(m + n + r)$ time, where $m \times n$ is the size of M, and r is the number of ones in M. Hence using such an algorithm a path support for a given hypergraph can be found in $O(N)$ time.

Cycle support. Finding a cycle support for a hypergraph H can be reduced to finding a path support for an auxiliary hypergraph H'. For a cycle support, a set S_i is connected if and only if its complement S_i^c is connected. For some $j \in V$, let H' be the hypergraph obtained by replacing the sets S_i for which $j \in S_i$ with S_i^c. As no set of H' contains j, H has a cycle support if and only if H' has a path support. By choosing j as the element that occurs in the minimum number of sets, one can reduce the problem of finding a cycle support for H to finding a path support for a hypergraph H' of size $O(N)$. This can be found in $O(N)$ time as described above. Finding a cycle support is also directly related to testing matrices for the *circular ones property* [16].

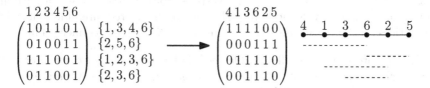

Fig. 2. Finding a path support via the consecutive ones property

Tree support. Johnson and Pollak [9] argued that one can efficiently decide whether a hypergraph has a tree support by considering its dual. The dual of a hypergraph $H = (V, \mathcal{S})$ is the hypergraph H^*, such that each hyperedge of H corresponds to a vertex of H^*, and each vertex $v \in V$ of H corresponds to a hyperedge of H^* that contains all hyperedges of H (vertices of H^*) that contain v. The dual of a hypergraph with a tree support is an acyclic hypergraph [2], and acyclicity can be tested in linear time [15].

Korach and Stern [11] considered the following generalization of finding a tree support: assume that for a hypergraph H a real weight is given for every pair of different numbers in the vertex set V, i.e., for each potential edge in the tree. They showed that the tree support with minimum total edge weight (if it exists), can be found in polynomial time.

3 Bounded-Degree Tree Supports

We describe an algorithm that solves the following decision problem: given a hypergraph $H = (V, \mathcal{S})$ together with degrees d_i for each element i of the base set V, is there a tree support for H such that each vertex i of the tree has degree at most d_i? Our algorithm is constructive and computes a support if it exists. To simplify the discussion we assume that $V \in \mathcal{S}$. This enforces that any support is connected and does not influence the outcome of the decision problem.

To construct a bounded-degree tree support we need to know what our choices are when connecting vertices. Consider the sets $S_1 = \{1, 2, 3\}$ and $S_2 = \{2, 3, 4\}$, all tree supports are shown in Fig. 3. Each support has an edge connecting 2 to 3, but 1 and 4 can be connected to either 2 or 3. So it appears that the intersection $\{2, 3\}$ of S_1 and S_2 must be connected in any tree support. Korach and Stern proved this observation in [11], for completeness we include a short proof.

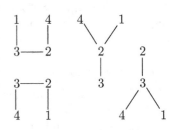

Fig. 3. All tree supports

Observation 1. *The intersection $A \cap B$ of two sets $A, B \in \mathcal{S}$ must be connected in every tree support.*

Proof. Since $A \cap B$ is always connected if it contains zero or one elements, we assume that $|A \cap B| \geq 2$. Let T be a tree support for H. So A and B are both connected in T. Let $x \in A \cap B$ and $y \in A \cap B$. Since A is connected in T, there is a path in T from x to y using only vertices from A. Also there is a path in T from x to y using only vertices from B. Since paths in trees are unique it follows all vertices on the path from x to y are in $A \cap B$. So $A \cap B$ is connected in T.   □

Let $\mathcal{S}^*$ denote the set of all possible sets that can be obtained by intersecting any number of sets from $\mathcal{S}$. Clearly $\mathcal{S}^*$ is closed under intersection and $\mathcal{S} \subseteq \mathcal{S}^*$. Observation 1 implies that H has a (bounded-degree) tree support if and only if $H^* = (V, \mathcal{S}^*)$ does. We now define the *intersection structure* $\mathcal{I}$ as follows. $\mathcal{I}$

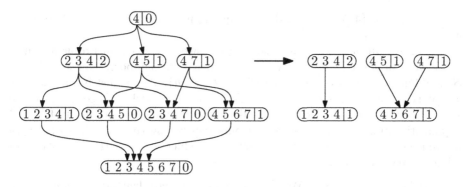

Fig. 4. The intersection structure for $\{\{1,2,3,4\},\{2,3,4,5\},\{4,5,6,7\},\{2,3,4,7\}\}$ (with the demands next to the sets) and the corresponding connectivity structure

is a directed acyclic graph whose vertices are the sets in $\mathcal{S}^*$. $\mathcal{I}$ has a directed edge (S_1, S_2) if and only if $S_1 \subset S_2$ and for no set $S_3 \in \mathcal{S}^*$, we have $S_1 \subset S_3 \subset S_2$. That is, edges are directed from smaller to larger sets and represent direct containment—$\mathcal{I}$ does not contain transitive edges (see Fig. 4 (left)).

The minimum number of edges of any support of a hypergraph H can be deduced directly from its intersection structure. Let B and $A_1, \ldots, A_h$ be vertices of $\mathcal{I}$ such that (A_j, B), $1 \leq j \leq h$, are incoming edges of B in $\mathcal{I}$ and there are no further incoming edges of B. We call the sets A_j the *children* of B, and B is a *parent* of each A_j. Let us assume that the sets A_j are connected in a support G of H and that G has the fewest edges among all supports with that property. Let c be the number of connected components implied by the sets A_j, i.e., the number of connected components of the hypergraph $(B, \{A_1, \ldots, A_h\})$. To connect B we need to add at least $c - 1$ additional edges to G—the *demand* of B (see Fig. 4 (left)). The sum of the demands of all sets in $\mathcal{S}^*$ is the *total demand*.

Lemma 1. *The total demand of the sets in $\mathcal{S}^*$ equals the minimum number of edges required for any support of H.*

Proof. By definition, the demand of a set B is the number of edges required to connect B, given that its children in $\mathcal{I}$ are connected. It remains to argue that no edge of a support G can simultaneously connect two sets B and B'. Assume that $|B'| \leq |B|$. The statement is obviously true if $B \cap B' = \emptyset$. If $B' \subset B$ then B' is part of a single connected component of B and hence no edge that is used to connect B connects two elements of B'. Finally, if $B \cap B' = A \neq \emptyset$, then, because $\mathcal{S}^*$ is closed under intersection, A must be a vertex of $\mathcal{I}$ as well. If an edge e of G is used to connect simultaneously both B and B', then e must connect two elements of A. But then e counts towards the demand of A. □

Recall that we assume that the base set V is an element of $\mathcal{S}$. Then, by Lemma 1, a hypergraph H has a tree support if and only if the total demand equals $n - 1$. $\mathcal{I}$ also indicates between which vertices the edges of a support should be. Consider the example in Fig. 4. The set $\{4, 5, 6, 7\}$ has a demand of 1. Since the connected components are $\{4, 5, 7\}$ and $\{6\}$, the support must contain an edge between 6 and either 4, 5 or 7.

$\mathcal{I}$ contains all necessary information to answer our decision problem, but it can have exponential complexity even if H has a tree support. Consider the set $\mathcal{S}$ of all but one subsets of size $n-1$ of $V = \{1, \ldots n\}$. There must be one element j that is contained in each set of $\mathcal{S}$. The star graph with j as center is a tree support for $H = (V, \mathcal{S})$. However $\mathcal{S}^*$ is nearly the complete powerset of V and exponential in size. Hence we restrict ourselves to the *connectivity structure*, a limited version of the intersection structure for which we prove that it still carries all necessary information.

Connectivity structure. We say that sets with zero demand are *implied*. We remove all sets with zero demand from $\mathcal{S}^*$ and call the resulting set $\mathcal{S}^-$. The connectivity structure $\mathcal{C}$ is built on $\mathcal{S}^-$ in the same manner as the intersection structure on $\mathcal{S}^*$ (see Fig. 4 (right)). The demand of a set in $\mathcal{C}$ equals its demand in $\mathcal{I}$. If H has a tree support then $\mathcal{S}^-$ contains at most $n-1$ sets. One can easily construct examples where also in this case $\mathcal{C}$ has $\Omega(n^2)$ edges.

Clearly we do not want to compute $\mathcal{S}^-$ and the connectivity structure by first constructing $\mathcal{S}^*$ and the intersection structure and pruning sets with zero demand. Instead we incrementally compute a graph that is the connectivity structure if H has a tree support. Let $\mathcal{S} = \{S_1, \ldots, S_k\}$ with $S_1 = V = \{1, \ldots n\}$. We incrementally compute the connectivity structures $\mathcal{C}_i$ $(1 \leq i \leq k)$ for the sets $S_1, \ldots, S_i$. To compute $\mathcal{C}_{i+1}$ from $\mathcal{C}_i$, we first compute all intersections between the new set S_{i+1} and all sets in $\mathcal{C}_i$. We then add those intersections which are not implied to the connectivity structure, starting with the smallest set by inclusion (see Fig. 5). If as a result any previous sets become implied, then we remove them. If at any point the total demand exceeds $n-1$, then we directly stop and conclude that the hypergraph has no tree support. We argue in the lemmas below that this approach is indeed correct.

The graph computed by this incremental construction might conceivably be missing sets since the intersection of a new set with a (removed) implied set might not be implied itself and hence should have been included. However, we can argue that for hypergraphs with a tree support this incremental approach indeed computes the correct connectivity structure (Lemma 2). But, if a hypergraph has no tree support, then the algorithm computes a total demand greater than $n-1$. Equivalently, if the total demand determined by the algorithm is $n-1$, then the hypergraph has a tree support (Lemma 3).

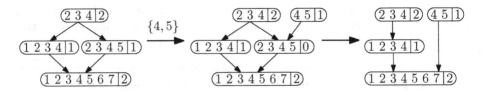

Fig. 5. Incremental construction of the connectivity structure

Lemma 2. *The incremental approach described above correctly computes the connectivity structure $\mathcal{C}$ if the hypergraph H has a tree support.*

Proof. We could use a similar approach to compute the complete intersection structure. So it remains to argue that removing implied sets in an intermediate stage does not influence the final result for hypergraphs with a tree support.

Assume that we have removed an implied set S from $\mathcal{C}_i$. Hence there must be sets $A_1, \ldots, A_h$ in $\mathcal{C}_i$ that imply that S is connected. Note that $A_j \subset S$ for all $1 \leq j \leq h$. Let S' be a new set that is added to $\mathcal{C}_i$. We have to argue that $S' \cap S$ is implied if H has a tree support. In fact we show that $S' \cap S$ is implied by the sets $A'_j = S' \cap A_j$. Assume for contradiction that this is not the case and hence the sets A'_j form at least two connected components in $S' \cap S$. Because $S' \cap S$ must be connected, these connected components are directly connected by edges in a tree support. However, because the sets A_j imply the connectedness of S, these connected components are also connected in a different manner in the tree support, introducing a cycle, which contradicts the assumption that H has a tree support. Since the total demand of a hypergraph with a tree support is $n - 1$, the algorithm does not terminate early for such hypergraphs. $\square$

Lemma 3. *If the total demand during the incremental construction is $n - 1$ then H has a tree support.*

Proof (by induction). Base case: $S_1 = \{1, \ldots n\}$ which has a demand of $n - 1$ and clearly has a tree support. Now assume that the sets $S_1, \ldots, S_i$ have a tree support T. In the inductive step we add the set S_{i+1} to $\mathcal{C}_i$, that is, we add the non-implied intersections of S_{i+1} with the sets in $\mathcal{C}_i$ starting with the smallest by inclusion. Let S be one of these intersections. After S has been added to $\mathcal{C}_i$, it has exactly one parent P. If it had two or more parents then it would be the non-implied intersection of at least two sets in $\mathcal{C}_i$ and as such already be contained in $\mathcal{C}_i$. If S had no parent then its demand would have to be zero for the total demand not to exceed $n - 1$. Hence S would be implied.

Let P be the parent of S and let $A_1, \ldots, A_h$ be the children of P before adding S. Assume that $(P, \{A_1, \ldots, A_h\})$ had c connected components before S was added and that S connects x of these components into one connected component. Then the demand of P becomes $c - x$. Since the total demand remains $n - 1$, the demand of S becomes $x - 1$. Since all children of S are former children of P none of the demand of S can be subsumed by its children. Let $B_1, \ldots, B_x$ be the connected components of $(P, \{A_1, \ldots, A_h\})$ that were connected by S. We change the tree support T as follows. Disconnect the connected components of P in T. Let $B'_j = S \cap B_j$ for $1 \leq j \leq x$. Note that all B'_j are connected in T, because these intersections have already been added. Now use the $x - 1$ edges covering the demand of S to connect the B'_j into a tree. Finally connect S with the remaining connected components of $(P, \{A_1, \ldots, A_h\})$, using the $c - x$ edges covering the demand of P. By construction the new tree still connects all sets of $\mathcal{C}_i$ as well as all intersections already added and S. $\square$

Lemma 3 directly implies that if H does not have a tree support then the total demand necessarily exceeds $n - 1$ at some point during the construction.

Flow formulation. Using the connectivity structure $\mathcal{C}$ we can formulate our decision problem as a flow problem. To simplify matters we add some additional sets to $\mathcal{C}$. Let S be a vertex of $\mathcal{C}$ and let $A_1, \ldots, A_h$ be children of S such that $A_1, \ldots, A_h$ form a (maximal) single connected component C of S. We say that C is a *connection set* (or *c-set* for short) and add C to $\mathcal{C}$ in between $A_1, \ldots, A_h$ and S. By construction all c-sets have zero demand. We also add all singleton sets. The resulting graph $\mathcal{C}^*$ is called the *augmented connectivity structure*. Every set in $\mathcal{C}^*$ is either a singleton set, a c-set, or a *normal* set. Normal sets now have the property that all their children are disjoint, hence the demand of a normal set is the number of its children minus one. Let c_S be the number of connected components of a set S in $\mathcal{C}$. The number of c-sets we add to $\mathcal{C}^*$ is $k_c \leq \sum_S c_S \leq (n-1) + \sum_S (c_S - 1) = 2n - 2$. So $\mathcal{C}^*$ has $O(n)$ vertices as well.

We construct a flow network $\mathcal{F}$ from $\mathcal{C}^*$ as follows. We add a source and connect it to the singletons with edges whose capacities are the maximal degree of each element. That is, the edge from the source to $\{i\}$ has capacity d_i. The capacities of the remaining edges are unbounded. Every incoming edge of a normal set requires at least one unit of flow, that is, we have a lower bound for the flow on these edges. The source produces $2n - 2$ units of flow which is consumed by the normal sets, each normal set consumes twice its demand (see Fig. 6 (left)). Intuitively the units of flow correspond to the degrees of the vertices in the tree support. Consider a normal set S and its children $A_1, \ldots, A_h$. Since these children are disjoint in $\mathcal{C}^*$ we need at least one unit of flow from each A_i to connect S. Also, S has to consume exactly $2h - 2$ units of flow. Observation 2 follows from a simple inductive argument.

Observation 2 (Tamura and Tamura [14]). *For a given degree sequence $(d_1, \ldots, d_h)$ with $d_i \geq 1$ for all i, a tree exists whose vertices have precisely these degrees if and only if $\sum_{j=1}^{h} d_j = 2h - 2$.*

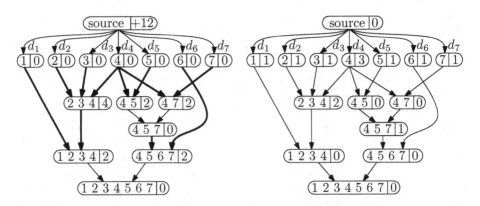

Fig. 6. The original (left) and the new (right) flow network. Thick edges denote a lower bound on the flow. In the flow networks the production/consumption (left) or the capacity to the sink (right) is shown instead of the demand.

Lemma 4. *Every tree support T that respects the degree bounds corresponds to a feasible flow F.*

Proof. As argued in the proof of Lemma 1 each edge $e = \{u, v\}$ of T can be mapped uniquely to a normal set S of C^*. Let $A_1, \ldots, A_h$ be the children of S. We have $u \in A_i$ and $v \in A_j$ for some $i \neq j$, $1 \leq i, j \leq h$. We choose an arbitrary path from the source to S through $\{u\}$ and A_i and add a unit of flow to every edge on this path. We do the same for $\{v\}$ and A_j. Repeating this procedure for every edge of T constructs a flow F. It remains to argue that F is indeed feasible. Consider again a normal set S and one of its children A_i. Since S is connected in T there is at least one edge of T which is mapped to S and contains an element of A_i. Hence the edge from A_i to S has at least one unit of flow. By the fact that $h - 1$ edges are mapped to S, the number of paths from the source to S is exactly $2h - 2$, so S consumes the correct amount of flow. Finally, we add flow exactly once for each edge in T and so the flow from the source to a singleton $\{i\}$ is at most d_i. $\qquad\square$

Before we explain how to construct a valid tree support from a feasible flow, we first discuss how to compute such a feasible flow using a standard construction (see [1], Sections 6.2 and 6.7): we transform the flow network $\mathcal{F}$ to a max-flow network $\mathcal{F}'$. We remove the lower bounds and the production/consumption restrictions. We add a sink to $\mathcal{F}$ and add edges from all sets to this sink. For a set S let δ_S denote its demand. As capacity of the edge from a set S to the sink we take the number of outgoing edges to a normal set (edges that had a lower bound of one), and if S is a normal set, we further add $\delta_S - 1$ to the capacity (Fig. 6 (right)). The sum of the capacities of the incoming edges of the sink is now exactly $2n - 2$. The following is shown in [1].

Lemma 5. *A feasible flow exists for $\mathcal{F}$ if and only if the max-flow of $\mathcal{F}'$ is exactly $2n - 2$.*

Tree construction. We now describe how to construct a tree support T from a feasible flow for $\mathcal{F}$. Although the flow tells us the degrees for each vertex in T, we need to use the entire flow to build T correctly. We handle the sets of $\mathcal{F}$ in topological order. Consider a normal set S with children $A_1, \ldots, A_h$. We can use the flow consumed by S to connect the sets A_j in T. To know which vertices need to be connected in T, we need to know from which singletons the flow to a set A_j originates. We maintain this information using a list L_i for each set S_i. The list L_i contains a vertex number for every unit of outgoing flow. Initially the list L_j for a singleton $\{j\}$ contains a number of copies of j corresponding to the amount of outgoing flow. Now let S again be a normal set S with children $A_1, \ldots, A_h$. First we build a tree on the components A_j depending on the flow towards S (note that the ingoing flow of S might exceed $2h - 2$, but in that case we can make choices as long as we use one unit of flow from each child). If we want to connect A_i to A_j, then we simply take the first elements x of L_i and y of L_j and add an edge (x, y) to T (Fig. 7). Then we remove x and y from L_i and L_j, respectively. After the tree is built for S, we take k_j elements from each list

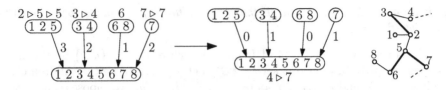

Fig. 7. A step in the tree construction algorithm. The lists are shown above/below the sets. The thick edges have been added to the tree support.

L_j of A_j, where k_j is the remaining flow from A_j to S, and merge them into the list L of S. In case S is a c-set, we perform only this final step.

Lemma 6. *The method described above correctly constructs a tree support T, which respects the degree bounds, from a feasible flow.*

Proof. We have to show three things: (i) every set S_i is connected in T, (ii) T is a tree and (iii) the degree bounds are respected. The algorithm adds exactly $n - 1$ edges to T, so (ii) follows from (i) since V is an element of S and hence T is necessarily connected. When we handle a set S_i, we make sure that it is connected in T. Since we never remove edges from T, (i) holds. Finally, when we add an edge incident to a vertex x to T, we remove it from a list. Note that vertex numbers are added to the singleton lists, but after that they are only moved from list to list. So the degree of x can be at most the size of L_x, which is properly bounded in a feasible flow.   □

Theorem 1. *Given a hypergraph H together with degrees d_i for each element i of the base set, we can construct a tree support T for H such that each vertex i of T has degree at most d_i in $O(kn^3)$ time—if such a tree support exists.*

Proof. The first step is to compute the connectivity structure. To add a set S_{i+1} to $\mathcal{C}_i$, we compute and add all intersections between S_{i+1} and the $O(n)$ sets in $\mathcal{C}_i$ ($O(n^2)$ time). Then we compute the direct containment graph on the resulting $O(n)$ sets which can easily be done in $O(n^3)$ time. Next we compute the demand of every set. We can easily find the connected components formed by the children of a set in $O(n^2)$ time, so this takes $O(n^3)$ time. Finally we remove the implied sets to obtain $\mathcal{C}_{i+1}$. Hence we can compute the connectivity structure in $O(kn^3)$ time. Then we can augment the connectivity structure and construct the max-flow network in $O(n^3)$ time. We compute the max-flow using the Ford-Fulkerson algorithm [6], which runs in $O(|E|f^*)$ time, where $|E|$ is the number of edges in the flow network and f^* is the maximum flow. As f^* is $O(n)$, this takes at most $O(n^3)$ time. Finally we construct the tree in $O(n^2)$ time.   □

4   Hardness for 3-Outerplanar Graphs

We show that for any instance of 3-SAT (or of SAT), we can reduce it to an instance of finding a planar support for a hypergraph such that there is a planar support if and only if the 3-SAT instance is satisfiable. The planar support—if one exists—will be 3-outerplanar (or we can assume it to be so without limiting any options in the planar support).

Let $a, b, c, \ldots$ be the variables used in a 3-SAT instance. We represent each variable, say b, by six elements $b_1, \ldots, b_6$ and some sets. Many of these sets have size two, and so any planar support must include an edge that connects the vertices of these two elements. The sets for b are $\{b_1, b_2\}$, $\{b_2, b_3\}$, $\{b_1, b_3\}$, $\{b_2, b_4\}$, $\{b_3, b_5\}$, $\{b_4, b_5\}$, $\{b_5, b_6\}$, and $\{b_4, b_6\}$, see Fig. 8. We connect the variable elements into some sequence (in any order; we assume it is $a, b, c, \ldots$) by extra elements $a', b', c', \ldots$ and $a'', b'', c'', \ldots$, and use sets $\{a', a_1\}$, $\{a_1, b'\}$, $\{b', b_1\}$, etc., and $\{a'', a_6\}$, $\{a_6, b''\}$, $\{b'', b_6\}$, etc. We also use extra elements $a''', b''', c''', \ldots$ and sets $\{a', a'''\}$, $\{a''', a''\}$, $\{b', b'''\}$, $\{b''', b''\}$, etc., to separate the variables from each other. Next, we use sets $\{a''', a_2\}$, $\{a''', a_4\}$, $\{b''', a_3\}$, and $\{b''', a_5\}$ for each variable, and four more sets $\{a', a_2\}$, $\{a'', a_4\}$, $\{z', z_3\}$, and $\{z'', z_5\}$. We add one more set, namely $\{a''', z'''\}$ (see Fig. 8), which ensures that no edge between any of $a', a_1, b', b_1, c', c_1, \ldots$ and any of $a'', a_6, b'', b_6, c'', c_6, \ldots$ can exist in the planar support. All sets of cardinality two imply a 3-connected planar graph as a support, so its embedding is fixed up to the choice of the outer face.

A 3-SAT clause $(a \vee \bar{c} \vee x)$ is represented by a set

$$\{a_1, b_1, c_1, \ldots, a', b', c', \ldots, a_6, b_6, c_6, \ldots, a'', b'', c'', \ldots, a_2, a_5, c_3, c_4, x_2, x_5\} \ .$$

In Fig. 8, these are all vertices of the top row, all vertices of the bottom row, the subscript-2 and subscript-5 vertices of the variables that occur as a literal in the clause, and the subscript-3 and subscript-4 vertices of the variables that occur negated as a literal in the clause. Their connection in the fixed part of the planar support is shown in grey in the figure.

The only way to extend the fixed part of the planar support so that the set of a clause has a connected support is to use at least one of the edges (a_2, a_5), (c_3, c_4), or (x_2, x_5). The only choices of edges in the support that can help to give sets planar support are ones like (a_2, a_5) and (a_3, a_4) (dotted in Fig. 8).

For any variable, it is easy to see that we can only take the edge (a_2, a_5) or (a_3, a_4), and not both, otherwise the support graph is not planar. This corresponds to the variable assignment of a to TRUE (take edge (a_2, a_5)) or FALSE (take edge (a_3, a_4)). Hence, the 3-SAT instance has a variable assignment that

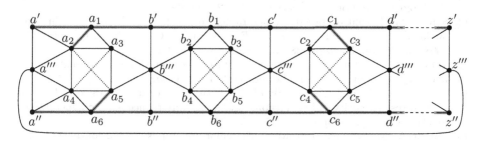

Fig. 8. Construction of a hypergraph and its planar support from a 3-SAT instance

makes it true if and only if the constructed hypergraph has a planar support. It is easy to see that the planar support is 3-outerplanar. Hence, if a planar support exists, then a 3-outerplanar support exists.

Theorem 2. *It is NP-complete to decide if a hypergraph has a 3-outerplanar support.*

Acknowledgements. We would like to thank the anonymous referees for pointing us to some related work.

References

1. Ahuja, R.K., Magnanti, T.L., Orlin, J.B.: Network Flows: Theory, Algorithms, and Applications. Prentice Hall, Englewood Cliffs (1993)
2. Beeri, C., Fagin, R., Maier, D., Yannakakis, M.: On the desirability of acyclic database schemes. Journal of the ACM 30, 479–513 (1983)
3. Berge, C.: Graphs and Hypergraphs. North-Holland, Amsterdam (1973)
4. Booth, K., Lueker, G.: Testing for the consecutive ones property, interval graphs, and planarity using pq-tree algorithms. Journal of Computer and System Sciences 13, 335–379 (1976)
5. Brinkmeier, M., Werner, J., Recknagel, S.: Communities in graphs and hypergraphs. In: 16th ACM Conference on Information and Knowledge Management, pp. 869–872 (2007)
6. Cormen, T., Leiserson, C., Rivest, R., Stein, C.: Introduction to Algorithms, 2nd edn. MIT Press, Cambridge (2001)
7. Flower, J., Howse, J.: Generating Euler diagrams. In: Hegarty, M., Meyer, B., Narayanan, N.H. (eds.) Diagrams 2002. LNCS (LNAI), vol. 2317, pp. 61–75. Springer, Heidelberg (2002)
8. Hsu, W.L.: A simple test for the consecutive ones property. Journal of Algorithms 43(1), 1–16 (2002)
9. Johnson, D., Pollak, H.: Hypergraph planarity and the complexity of drawing Venn diagrams. Journal of Graph Theory 11(3), 309–325 (1987)
10. Kaufmann, M., van Kreveld, M., Speckmann, B.: Subdivision drawings of hypergraphs. In: Tollis, I.G., Patrignani, M. (eds.) GD 2008. LNCS, vol. 5417, pp. 396–407. Springer, Heidelberg (2009)
11. Korach, E., Stern, M.: The clustering matroid and the optimal clustering tree. Mathematical Programming, Series B 98, 385–414 (2003)
12. Lundgren, J.R.: Food webs, competition graphs, competition-common enemy graphs and niche graphs. Applications of Combinatorics and Graph Theory to the Biological and Social Sciences 17, 221–243 (1989)
13. Sander, G.: Layout of directed hypergraphs with orthogonal hyperedges. In: Kreowski, H.-J., Montanari, U., Orejas, F., Rozenberg, G., Taentzer, G. (eds.) GD 2004. LNCS, vol. 3393, pp. 381–386. Springer, Heidelberg (2005)
14. Tamura, A., Tamura, Y.: Degree constrained tree embedding into points in the plane. Information Processing Letters 44, 211–214 (1992)
15. Tarjan, R.E., Yannakakis, M.: Simple linear-time algorithms to test chordality of graphs, test acyclicity of hypergraphs, and selectively reduce acyclic hypergraphs. SIAM Journal on Computing 13, 566–579 (1984)
16. Tucker, A.: Matrix characterizations of circular-arc graphs. Pacific Journal of Mathematics 39(2), 535–545 (1971)

DAGmaps and ε-Visibility Representations of DAGs

Vassilis Tsiaras and Ioannis G. Tollis

Institute of Computer Science, Foundation for Research and Technology-Hellas,
Department of Computer Science, University of Crete,
Vassilika Vouton, P.O. Box 1385, Heraklion, GR-71110 Greece
{tsiaras,tollis}@ics.forth.gr

Abstract. DAGmaps are space filling visualizations of DAGs that generalize treemaps. Deciding whether or not a DAG admits a DAGmap is NP-complete. Recently we defined a special case called one-dimensional DAGmap where the admissibility is decided in linear time. However there is no complete characterization of the class of DAGs that admit a one-dimensional DAGmap. In this paper we prove that a DAG admits a one-dimensional DAGmap if and only if it admits a directed ε-visibility representation. Then we give a characterization of the DAGs that admit directed ε-visibility representations. Finally we show that a DAGmap defines a directed three-dimensional ε-visibility representation of a DAG.

Keywords: DAGmap, Treemap, DAG, Visibility.

1 Introduction

Among the many alternative ways to visualize a tree, space filling visualizations, such as treemaps, have become very popular due to their efficiency, their scalability, and their easiness of navigation and user interaction [1]. Space filling techniques make optimal use of the available space and have the capacity to show thousands of items legibly. On the other hand, the node-link representations do not make optimal use of the available space since most of the pixels are used for background. Recently, we investigated space filling visualizations for hierarchies that are modeled by Directed Acyclic Graphs (DAG). We assumed that the available space is a rectangle and we defined the constraints for a visualization that extends the treemap techniques [3,10,1,8] to DAGs and where the vertices and edges of a DAG are drawn as rectangles [13]. In [13] we use the term "DAGmap" to describe space filling visualization according to the constraints and we show that there are DAGs that admit and DAGs that do not admit DAGmap drawings. Moreover deciding whether or not a DAG admits a DAGmap drawing is NP-complete. In the special cases, of Two Terminal Series Parallel digraphs [14] and of layered planar DAGs the admissibility question can be answered in linear time with respect to input size [13].

D. Eppstein and E.R. Gansner (Eds.): GD 2009, LNCS 5849, pp. 357–368, 2010.

A visibility representation of a graph G maps vertices of G to sets in Euclidean space and the edges are expressed as visibility relations between these sets. In a (two-dimensional) visibility representation of a graph G, the vertices are drawn as horizontal segments in the plane and the edges are represented by pairs of vertically visible segments [11,5,15,7,4]. Recently, interest has developed in investigating visibility representations in three-dimensions where vertices are represented by disjoint axis-aligned closed rectangles lying in planes parallel to the xy-plane and edges correspond to z-parallel visibility among these rectangles [2]. If graph G is directed then for every edge (u, v) the rectangle of v is below the rectangle of u. Note that in order to be consistent with the downward representation of DAGs we draw the visibility representation downwards whereas in the literature it is drawn upwards [5].

In a DAGmap as well as in a visibility representation of a DAG the vertices are represented by closed rectangles and the edges are closed sets which have non-empty intersection with the source and destination vertex rectangles. In this paper we show that a DAGmap (resp. treemap) determines a directed three-dimensional visibility representation of a DAG (resp. tree). Additionally we show that there is a one-to-one correspondence between a one-dimensional DAGmap and a directed ε-visibility representation of a DAG. Using this correspondence we show that the class of DAGs that admit a one-dimensional DAGmap is the class of downward planar digraphs that admit an embedding such that all source and sink vertices appear on the boundary of the external face. Additionally we propose an admissibility and drawing algorithm that runs in O(n) time.

2   Preliminaries

Let $G = (V, E)$ be a directed acyclic graph (DAG) with $n = |V|$ vertices and $m = |E|$ edges. A path of length k from a vertex u to a vertex w is a sequence $v_0, v_1, v_2, \ldots, v_k$ of vertices such that $u = v_0$, $w = v_k$, and $(v_{i-1}, v_i) \in E$ for $i = 1, 2, \ldots, k$. There is always a zero-length path from u to u. If there is a path p from u to w, we say that w is reachable from u via p.

A topological numbering of G is an assignment of numbers to the vertices of G, such that for every edge (u, v) of G, the number assigned to v is greater than the one assigned to u (i.e., $number(v) > number(u)$). If the edges of G have nonnegative weights assigned to them, then the number assigned to v is greater than or equal to the number assigned to u plus the weight of (u, v) (i.e., $number(v) \geq number(u) + weight(u, v)$). The numbering is optimal if the range of numbers assigned to vertices is minimized.

If $e = (u, v) \in E$ is a directed edge, we say that e is incident from u (or outgoing from u) and incident to v (or incoming to v); vertex u is the origin of e and vertex v is the destination of e. The origin of e is denoted by $orig(e)$ and the destination of e by $dest(e)$. For every vertex $u \in V$, $\Gamma^+(u) = \{e \in E \mid orig(e) = u\}$ and $\Gamma^-(u) = \{e \in E \mid dest(e) = u\}$ are the sets of edges incident from and to vertex u, respectively.

A *drawing* $\mathcal{G}$ of a graph (digraph) G is a function which maps each vertex v to a distinct point $\mathcal{G}(v)$ and each edge (u, v) to a simple open Jordan curve $\mathcal{G}(u, v)$, with endpoints $\mathcal{G}(u)$ and $\mathcal{G}(v)$. A drawing is *planar* if no two distinct edges intersect. A graph is planar if it admits a planar drawing. A (planar) *embedding* $\widehat{G}$ of G is an equivalence class of planar drawings and is described by the circular order of the neighbors of each vertex. An *embedded graph* is a graph with a specified embedding.

An st-graph is an acyclic digraph with a single source s and a single sink t. A planar st-graph is an st-graph that is planar and embedded with vertices s and t on the boundary of the external face.

Let $\mathcal{S}$ be a set of horizontal non-overlapping segments in the plane. Two segments σ, σ' of $\mathcal{S}$ are said to be visible if they can be joined by a vertical segment not intersecting any other segment of $\mathcal{S}$. Furthermore, σ and σ' are called *ε-visible* if they can be joined by a vertical band of nonzero width that does not intersect any other segment of $\mathcal{S}$.

Definition 1. *A directed (weak) w-visibility representation for a DAG G consists of mapping each vertex v of G into a horizontal segment $\sigma(v)$ (called vertex-segment), and each edge $(u, v) \in E$ into a vertical segment $\sigma(u, v)$ (called edge-segment), so that, the vertex-segments do not overlap, and for each edge $(u, v) \in E$ the corresponding edge-segment $\sigma(u, v)$ has its top endpoint on $\sigma(u)$, its bottom endpoint on $\sigma(v)$, and it does not cross any other vertex-segment $\sigma(a), a \neq u, v$.*

Definition 2. *A directed ε-visibility representation for a DAG G is a directed w-visibility representation with the additional property that two vertex-segments are directed ε-visible if and only if the vertex that corresponds to the bottom vertex-segment is adjacent to the vertex that corresponds to the top vertex-segment.*

Now consider an arrangement of closed, non-overlapping rectangles in $\mathbb{R}^3$ such that the planes determined by the rectangles are perpendicular to the z-axis, and the sides of the rectangles are parallel to the x-or y-axes. Two rectangles R_i and R_j are ε-visible if and only if between the two rectangles there is a closed cylinder C of positive height and radius such that the ends of C are contained in R_i and R_j, the axis of C is parallel to the z-axis, and the intersection of C with any other rectangle in the arrangement is empty [2].

Definition 3. *[9] A directed three-dimensional ε-visibility representation for a DAG G consists of mapping each vertex v of G into a rectangle R_v (called vertex-rectangle), and each edge $(u, v) \in E$ into a vertical closed cylinder C of positive length and radius (called edge-cylinder), so that, the vertex-rectangles do not overlap, and for each edge $(u, v) \in E$ the corresponding edge-cylinder C has its top base on R_u, its bottom base on R_v, and does not intersect any other vertex-rectangle $R_w, w \neq u, v$. Additionally, two vertex-rectangles are ε-visible if and only if the vertex that corresponds to the bottom vertex-rectangle is adjacent to the vertex that corresponds to the top vertex-rectangle.*

In the following R denotes the initial rectangle, R_u denotes the drawing region of a vertex $u \in V$ and R_e denotes the drawing region of an edge $e \in E$.

Definition 4 (DAGmap drawing [12,13]). *A DAGmap drawing of a DAG*
$G = (V, E)$ *is a space filling visualization of G that satisfies the following drawing
constraints:*

B1. *Every vertex is drawn as a rectangle (R_u is a rectangle for every $u \in V$).*
B2. *The union of the rectangles of the sources of G form a partition of the initial
    drawing rectangle ($R = \cup_{s \in S} R_s$ and $\forall s_1, s_2 \in S$ with $s_1 \neq s_2$ area($R_{s_1} \cap
    R_{s_2}$) = 0, where $S \subset V$ is the set of sources of G).*
B3. *Every edge is drawn as a rectangle that has positive area ($\forall e = (u, v) \in E$,
    R_e is a rectangle and area(R_e) > 0).*
B4. *The rectangle of every non-source vertex $u \in V$ is equal to the union of the
    rectangles of edges incident to u ($R_u = \cup_{e \in \Gamma^-(u)} R_e$).*
B5. *The rectangles of edges incident from a non-sink vertex $u \in V$ form a
    partition of the rectangle of u ($R_u = \cup_{e \in \Gamma^+(u)} R_e$ and $\forall e_1, e_2 \in \Gamma^+(u)$ with
    $e_1 \neq e_2$ area($R_{e_1} \cap R_{e_2}$) = 0).*

Theorem 1. *[12] In a DAGmap drawing of DAG $G = (V, E)$, if for some pair
of edges $e_1, e_2 \in E$ with $e_1 \neq e_2$, it holds that that orig(e_1) is not reachable from
dest(e_2) and orig(e_2) is not reachable from dest(e_1), then the rectangles R_{e_1} and
R_{e_2} do not overlap (i.e., area($R_{e_1} \cap R_{e_2}$) = 0).*

Proposition 1. *[13] In a DAGmap drawing of a DAG G the following hold:
For every pair of vertices $u, v \in V$ if there is no path from u to v and from v to
u then their rectangles R_u, R_v do not overlap (area($R_u \cap R_v$) = 0).*

3   One-Dimensional DAGmaps and Directed ε-Visibility Representations

One-dimensional DAGmaps were introduced in [13]. They are constructed by
partitioning the space only along the vertical direction. We will show that
one-dimensional DAGmaps are related to directed ε-visibility representations
of DAGs.

Definition 5. *A DAGmap is called one-dimensional if the initial drawing rectangle is sliced in one dimension either the vertical or the horizontal. See Figure
1 for an example.*

Since the height of all the rectangles is constant and equal to the height of
the initial drawing rectangle, the combinatorial properties of the problem are
unaffected if instead of the vertex and edge rectangles R_q we consider their
projections on the horizontal axis. These projections are intervals I_q.

From the vertex rectangles R_q (resp. intervals I_q) of a one-dimensional DAGmap
we can construct a directed three-dimensional (resp. two-dimensional) ε-visibility
representation by assigning to rectangles (resp. intervals) a z-coordinate. The
construction is described in Theorem 2. A directed three-dimensional ε-visibility
representation of the DAG of Fig. 1(a) is shown in Fig. 2(a). The corresponding
directed (two-dimensional) ε-visibility representation of this DAG is shown in Fig.
2(b). The segments of Fig. 2(b) are the projections of the rectangles of Fig. 2(a)
onto the xz-plane.

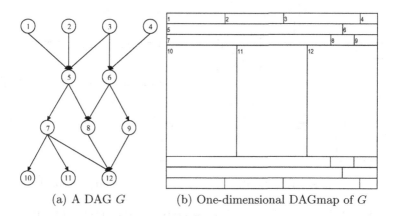

(a) A DAG G (b) One-dimensional DAGmap of G

Fig. 1. An example of a one-dimensional DAGmap drawing of a DAG G. The hierarchy structure is visualized via nesting [8] along the vertical direction.

Theorem 2. *If a DAG $G = (V, E)$ admits a one-dimensional DAGmap then it admits a directed ε-visibility representation.*

Sketch of Proof. Suppose that G admits a one-dimensional DAGmap. From the one-dimensional DAGmap we will construct a directed ε-visibility representation as follows: We compute an optimal topological numbering Y of G, such that only integer numbers are used and the sources are assigned the number 0. We also compute the longest path length, h, in the DAG. Each interval I_u, $u \in V$ of the one-dimensional DAGmap is shifted along the vertical direction and is drawn on the horizontal line with equation $y = y(u) = h - Y(u) + \epsilon \cdot j(u)$, where ϵ is a small positive number (e.g. $0 < \epsilon < \frac{1}{1000 \cdot |V|}$) and $j = j(u) \in \{0, 1, \cdots, |V| - 1\}$ is a unique vertex index. The shifted intervals I_u, $u \in V$ become the vertex-segments $\sigma(u)$ of the ε-visibility representation. The vertex-segments do not overlap since they are drawn on different horizontal lines. Next we add the edge-segments. For each $e = (u, v) \in E$, an edge-segment $\sigma(u, v) = \{(\mu_{uv}, y) \mid y(u) \geq y \geq y(v)\}$ is created, where μ_{uv} is the horizontal coordinate of the middle of the interval I_e. This construction is always possible since $length(I_e) > 0$.

Note that the above construction can also be done using st-numbering instead of an optimal topological numbering. The advantage in this case is that the term $\epsilon \cdot j(u)$ is not needed since the vertex rectangles have distinct y coordinates and therefore are disjoint. The disadvantages are: a) Figures 2, 3 and 4 require more space and b) some proofs become slightly longer since in order to apply Proposition 1 we need further checks to show that there is no path between two vertices in the case when two vertices have the same optimal topological numbering.

Due to space limitation we briefly outline the rest of the proof. First we show that the conditions of w-visibility are satisfied and then we prove that if

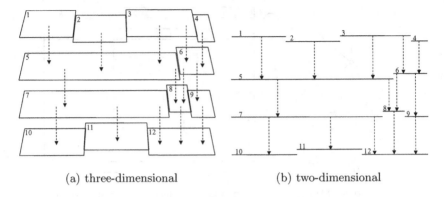

(a) three-dimensional   (b) two-dimensional

Fig. 2. Directed ε-visibility representation of the DAG of Fig. 1(a)

two vertex segments $\sigma(u)$ and $\sigma(v)$, such that $Y(u) < Y(v)$, are ε-visible then $(u, v) \in E$. Therefore the conditions of ε-visibility are satisfied.   □

4 Characterization of Directed ε-Visibility Representations

Theorem 2 reveals an interesting relationship between one-dimensional DAGmaps and ε-visibility representations. Namely a one-dimensional DAGmap defines an ε-visibility representation of a DAG. The converse of this theorem is even more interesting because a) the problem of visibility representation of a DAG has been thoroughly studied and b) it allows us to characterize the class of DAGs that admit a one-dimensional DAGmap. Before we give the converse of Theorem 2 we will characterize the class of DAGs that admit a directed ε-visibility representation. This minor result seems to be lacking from the literature. We should mention here that a complete characterization of the class of (undirected) graphs that admit an ε-visibility representation was given by Tamassia and Tollis [11]. A complete characterization of the class of digraphs that admit a (weak) w-visibility representation was given by Di Battista and Tamassia [5].

Theorem 3. *[5] A digraph G admits a directed w-visibility representation if and only if G is a subgraph of a planar st-graph.*

For directed ε-visibility representation the following lemma holds.

Lemma 1. *[12] If a DAG G admits a directed ε-visibility representation, then there exists a planar embedding $\widehat{G}$ of G such that all source and sink vertices appear on the boundary of the external face.*

Let C_w and C_ε be the classes of DAGs that admit a directed w-visibility and a directed ε-visibility representation respectively. From the definition of directed

ε-visibility representation we have that $C_\varepsilon \subset C_w$. From Lemma 1 and Theorem 3 it follows that C_ε is properly included in C_w.

Lemma 2. *[11] For every vertex v of a planar st-graph G, the incoming (outgoing) edges appear consecutively around v.*

Now we will present an algorithm, which is based on the Algorithm Tessellation [4], that computes a directed ε-visibility representation of a planar st-graph. To describe the algorithm we need to introduce some definitions. Let G be a planar st-graph and F be its set of faces (recall that G is embedded). We conventionally assume that F contains two representatives for the external face: the *left external face* s^*, which is incident with the edges on the left boundary of G, and the *right external face* t^*, which is incident with the edges on the right boundary of G. For each element o of $V \cup E$ we define $orig(o)$, $dest(o)$, $left(o)$, and $right(o)$ as follow:

1) If $o = v \in V$, we define $orig(v) = dest(v) = v$. Also, with reference to Lemma 2 we denote by $left(v)$ (resp. $right(v)$) the face that separate the incoming from the outgoing edges of a vertex $v \neq s, t$ in the clockwise direction (resp. counter-clockwise direction). For $v = s$ or $v = t$, we conventionally define $left(v) = s^*$ and $right(v) = t^*$.
2) If $o = e \in E$, we denote by $left(e)$ (resp. $right(e)$) the face on the left (resp. right) side of e. Also, $orig(e)$ (resp. $dest(e)$) denotes the origin (resp. destination) vertex of e.

We define a digraph G^*, associated with planar st-graph G, as follows: The vertex set of G^* is the set of faces of G. For every edge $e \neq (s, t)$ of G, G^* has an edge $e^* = (f, g)$ where $f = left(e)$ and $g = right(e)$.

Theorem 4. *Let G be a planar st-graph with n vertices. Algorithm 1 constructs a directed ε-visibility representation in $O(n)$ time.*

Sketch of Proof. For any pair of vertices $u, v \in V$, the vertex segments $\sigma(u)$ and $\sigma(v)$ do not overlap since they have distinct y coordinates.

For each edge $e = (u, v) \in E$, the corresponding maximal visibility band $b(e)$ has its top side on $\sigma(u)$ since $y_T(e) = y(u)$ and $x_L(u) \leq x_L(e) < x_R(e) \leq x_R(u)$, and its bottom side on $\sigma(v)$ since $y_B(e) = y(v)$ and $x_L(v) \leq x_L(e) < x_R(e) \leq x_R(v)$. Then we can choose a vertical band $b'(e) \subset b(e)$ of non-zero width that has its top side on $\sigma(u)$ its bottom side on $\sigma(v)$ and does not intersect with any other vertex-segment $\sigma(w)$.

Finally since the topological numbering X is optimal the vertex-segment of a non-sink vertex u is covered by the bottom sides of the maximal visibility bands of edges incident from u. Similarly the vertex-segment of a non-source vertex v is covered by the top sides of the maximal visibility bands of edges incident to v. Therefore vertex-segment $\sigma(v)$ is ε-visible from vertex-segment $\sigma(u)$ only if G has an edge (u, v).

The $O(n)$ time bound follows easily since each step of the algorithm can be accomplished in $O(n)$ time. $\square$

Algorithm 1. Directed ε-visibility representation

Input: a planar st-graph $G = (V, E)$
Output: a) a directed ε-visibility representation $\mathcal{G}$ of G
b) visibility bands of maximal width that their internal points do
not intersect with any vertex-segment of $\mathcal{G}$ and that each one
having its top and bottom sides on two vertex-segments
$s(u)$ and $s(v)$ respectively if and only if $(u, v) \in E$.

1. Construct the planar st-graph G^*.
2. Compute an optimal topological numbering Y of G such that only integer numbers
   are used.
3. Compute an optimal topological numbering X of G^*.
4. Let ϵ be a very small positive number e.g. $0 < \epsilon < \frac{1}{1000 \cdot |V|}$
5. $j = 0$;
6. For each vertex $u \in V$, let the coordinates of segment $\sigma(u)$ be:
   $x_L(u) = X(left(u)); \quad x_R(u) = X(right(u));$
   $y(u) = Y(t) - Y(u) + \epsilon \cdot j; \quad //\text{perturb slightly by adding } \epsilon \cdot j$
   $j = j + 1;$
7. For each edge $e \in E$, let the coordinates of the corresponding maximal
   visibility band $b(e)$ be:
   $x_L(e) = X(left(e)); \quad x_R(e) = X(right(e));$
   $y_T(e) = y(orig(e)); \quad y_B(e) = y(dest(e)).$

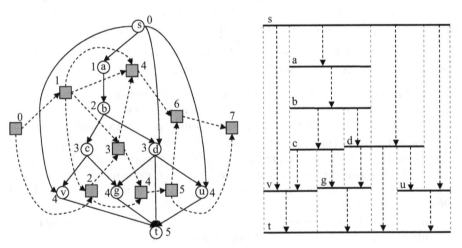

(a) Planar st-graphs G and G^*

(b) ε-visibility representation of G and
maximal visibility bands

Fig. 3. Example of a run of Algorithm 1. Planar st-graphs G and G^* are labeled by
topological numbering Y and X, respectively.

Let $G = (V, E)$ be a DAG and $G' = (V', E')$ be the DAG that is formed by augmenting DAG G with two vertices s' and t' and edges from s' to all sources of G and edges from all sinks of G to t' (i.e. $V' = V \cup \{s', t'\}$ and $E' = E \cup \{(s', s) \mid s$ is a source of $G\} \cup \{(t, t') \mid t$ is a sink of $G\}$). Note that in order to test for st-planarity and find an appropriate planar embedding of G' we add edge (s', t'). This edge constrains the embedding such that vertices s' and t' appear on the boundary of the same face, say the external face. The following theorem holds.

Theorem 5. *[12] DAG G admits a directed ε-visibility representation if and only if DAG G' is a planar st-graph.*

Corollary 1. *A DAG G admits a directed ε-visibility representation if and only if there exists a planar embedding $\widehat{G}$ of G such that all source and sink vertices appear on the boundary of the external face.*

5   Characterization of One-Dimensional DAGmaps

We are now in a position to prove the converse of Theorem 2

Theorem 6. *If a DAG $G = (V, E)$ admits a directed ε-visibility representation then it admits a one-dimensional DAGmap.*

Sketch of Proof. We compute a directed ε-visibility representation $\mathcal{G}'$ of G' using Algorithm 1 and from $\mathcal{G}'$ a directed ε-visibility representation $\mathcal{G}$ of G by deleting segments $\sigma(s')$ and $\sigma(t')$. From the arrangement of the vertex-segments of $\mathcal{G}$ we will construct a one-dimensional DAGmap drawing of G. To each vertex-segment $\sigma(u), u \in V$ we correspond an interval I_u by taking its projection on the horizontal axis. For each edge $e = (u, v) \in E$ there is exactly one vertical band b of maximal width that has its bottom side on $\sigma(u)$ its top side on $\sigma(v)$ and does not ε-intersect any other segment. The coordinates of b are calculated by Algorithm 1. The edge rectangle I_e is equal to the projection of b on the horizontal axis.

We show that intervals $\{I_u \mid u \in V\}$ and $\{I_e \mid e \in E\}$ satisfy the DAGmap drawing constraints. Drawing constraints B1, B3 are clearly satisfied. Constraint B2 is satisfied due to the optimality of the topological numbering X of G^*. Constraints B4 and B5 are satisfied when the ε-visibility representation of G is produced by Algorithm 1 due to the optimality of the topological numbering X of G^*.   □

Combining Theorems 2 and 6 we have the following theorem:

Theorem 7. *A DAG $G = (V, E)$ admits a one-dimensional DAGmap if and only if it admits a directed ε-visibility representation.*

Corollary 2. *The class of DAGs that admit a one-dimensional DAGmap are the planar st-graphs that admit a planar embedding such that all source and sink vertices appear on the boundary of the external face.*

Algorithm 2. One-dimensional DAGmap drawing

Input: DAG $G = (V, E)$ and drawing rectangle $R = (x_{left}, y_{bottom}, x_{right}, y_{top})$
Output: one-dimensional DAGmap drawing of G if G is admits such a drawing
            or error message otherwise.

1. From DAG $G = (V, E)$ we construct an st-digraph $G' = (V', E')$, where $V' = V \cup \{s', t'\}$ and $E' = E \cup \{(s', u) \mid u$ is a source of $G \} \cup \{(u, t') \mid u$ is a sink of $G \} \cup (s', t')$.
2. If G' is not st-planar return "DAG G does not admit a one-dimensional DAGmap."
3. Else find a planar embedding of G' such that s' and t' appear on the boundary of the external face.
4. Remove edge (s', t') from G'
5. Call Algorithm 1 with input G' to compute the horizontal coordinates of vertex-segments and maximal visibility bands.
6. Use these coordinates to fill the coordinates of vertex and edge rectangles of G.

Algorithm 2 recognizes whether or not a DAG admits a one-dimensional DAGmap and in the first case it constructs a one-dimensional DAGmap drawing. All steps of this algorithm can be computed in $O(n)$ time. Therefore we have the following theorem.

Theorem 8. *Algorithm 2 computes a one-dimensional DAGmap of a DAG or returns an error message in time $O(n)$ time.*

6   DAGmaps and Three Dimensional ε-Visibility Representations

A treemap determines a three-dimensional ε-visibility representation of a tree T if vertex rectangles are placed in three dimensional space such that their x and y coordinates are unaltered and their z coordinates are equal to the tree height minus the distance of the corresponding vertices from the root (plus tiny perturbations in order to keep the rectangle disjoint) (see Fig. 4). The above discussion is extended to DAGmaps as described by Theorem 9.

Theorem 9. *When a DAG G admits a DAGmap then it admits a directed three-dimensional ε-visibility representation which can be constructed by shifting the vertex rectangles along the vertical direction in such a way that their z-coordinates are determined by an optimal topological numbering of G (plus a tiny perturbation).*

The proof of Theorem 9 is similar to the proof of Theorem 2 and is omitted in this version. It is interesting to investigate if the converse of the above theorem holds. If yes, then DAGmap admissibility would be equivalent to three-dimensional visibility representation and results derived for the former problem would be valid for the second and vise versa. However the converse of Theorem 9 does not hold as the counter-example in Fig. 5 shows.

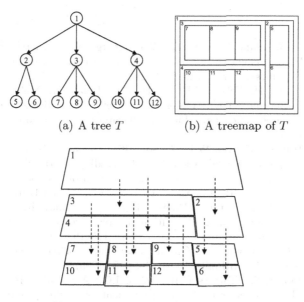

(a) A tree T (b) A treemap of T

(c) Three-dimensional ε-visibility of T

Fig. 4. A tree T, a treemap of T and a three-dimensional ε-visibility representation of T that is constructed from the treemap of T

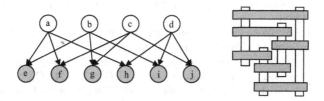

Fig. 5. The DAG of this figure does not admit a DAGmap [13]. However it admits a three-dimensional directed ε-visibility representation.

7 Discussion

In this paper we show that there is a one-to-one correspondence between a one-dimensional DAGmap and an ε-visibility representation of a DAG. Based on this correspondence we give a characterization of the class of DAGs that admit a one-dimensional DAGmap. They are those that admit a planar embedding such that all source and sink vertices appear on the boundary of the external face. Additionally we propose linear time, $O(n)$, testing and drawing algorithms for one-dimensional DAGmaps. Our next steps are a) to implement Algorithm 2 using SPQR trees [6] to find a planar embedding if one exists and b) to investigate under what restrictions the converse of Theorem 9 holds.

References

1. Bederson, B.B., Shneiderman, B., Wattenberg, M.: Ordered and quantum treemaps: Making effective use of 2D space to display hierarchies. ACM Transactions on Graphics 21(4), 833–854 (2002)
2. Bose, P., Everett, H., Fekete, S.P., Houle, M.E., Lubiw, A., Meijer, H., Romanik, K., Rote, G., Shermer, T.C., Whitesides, S., Zelle, C.: A visibility representation for graphs in three dimensions. Journal of Graph Algorithms and Applications 2, 1–16 (1998)
3. Bruls, M., Huizing, K., van Wijk, J.: Squarified treemaps. In: de Leeuw, W., van Liere, R. (eds.) Data Visualization 2000, Proceedings of the Second Joint Eurographics and IEEE TCVG Symposium on Visualization, pp. 33–42 (2000)
4. Di Battista, G., Eades, P., Tamassia, R., Tollis, I.G.: Graph Drawing: Algorithms for the Visualization of Graphs. Prentice Hall, Upper Saddle River (1998)
5. Di Battista, G., Tamassia, R.: Algorithms for plane representations of acyclic digraphs. Theoretical Computer Science 61(2-3), 175–198 (1988)
6. Di Battista, G., Tamassia, R.: On-line maintenance of triconnected components with SPQR-trees. Algorithmica 15(4), 302–318 (1996)
7. Kant, G., Liotta, G., Tamassia, R., Tollis, I.G.: A visibility representation for graphs in three dimensions. Area requirement of visibility representations of trees 62, 81–88 (1997)
8. Lü, H., Fogarty, J.: Cascaded treemaps: examining the visibility and stability of structure in treemaps. In: ACM Proceedings of graphics interface 2008, pp. 259–266 (2008)
9. Romanik, K.: Directed VR-representable graphs have unbounded dimension. In: Tamassia, R., Tollis, I.G. (eds.) GD 1994. LNCS, vol. 894, pp. 177–181. Springer, Heidelberg (1995)
10. Shneiderman, B.: Tree visualization with tree-maps: 2-d space-filling approach. ACM Transactions on Graphics 11(1), 92–99 (1992)
11. Tamassia, R., Tollis, I.G.: A unified approach to visibility representations of planar graphs. Discrete and Computational Geometry 1(1), 321–341 (1986)
12. Tsiaras, V.: Algorithms for the Analysis and Visualization of Biomedical Networks. Ph.D. thesis, Computer Science Department, University of Crete (2009) (submitted)
13. Tsiaras, V., Triantafilou, S., Tollis, I.G.: DAGmaps: Space Filling Visualization of Directed Acyclic Graphs. Journal of Graph Algorithms and Applications 13(3), 319–347 (2009)
14. Valdes, J., Tarjan, R.E., Lawler, E.L.: The recognition of series parallel digraphs. SIAM Journal on Computing 11(2), 298–313 (1982)
15. Wismath, S.K.: Characterizing bar line-of-sight graphs. In: SCG 1985: Proceedings of the first annual symposium on Computational geometry, pp. 147–152. ACM, New York (1985)

Drawing Directed Graphs Clockwise

Christian Pich

Chair of Systems Design, ETH Zürich*
cpich@ethz.ch

Abstract. We present a method for clockwise drawings of directed cyclic graphs. It is based on the eigenvalue decomposition of a skew-symmetric matrix associated with the graph and draws edges clockwise around the center instead of downwards, as in the traditional hierarchical drawing style. The method does not require preprocessing for cycle removal or layering, which often involves computationally hard problems. We describe an efficient algorithm which produces optimal solutions, and we present some application examples.

1 Introduction

Directed graphs are usually drawn with the desire to have edges pointing in the same direction, say, downwards, assuming that there is a general trend or direction of flow in the graph. The most popular and thoroughly researched drawing method is the Sugiyama framework [15], which works well for directed graphs with no or only few cycles. After preprocessing, in which some edges are temporarily removed or reversed, the graph is acyclic, which allows all nodes to be assigned to layered in such a way that all edges point in the same direction.

Instead of discrete levels, nodes may also be assigned continuous vertical coordinates. Carmel et al. [2] minimize a hierarchy energy in which every edge in a directed graph induces a target height difference between the two incident nodes; an iterative optimization process computes coordinates which attain these height differences as well as possible. Sometimes, however, it is not appropriate to assume that there is an overall linear trend of direction; cycles may not just be considered as "noise", but as essential information which should be highlighted and conveyed in a drawing.

An alternative to the traditional style of hierarchical layouts are *recurrent hierarchies* [15], which have long gone unnoticed until recently. Such drawings are read clockwise with respect to a distinguished point of origin. For constructing a drawing, a cyclic order on all nodes has to be found in which as many edges as possible point forward.

Sugiyama and Misue introduced a set of modifications of force-directed algorithms to get a cyclic orientation [14]. They use a concentric force field which rotates around the center and takes edges along, and report about satisfactory

---

* Part of this work was done while the author was at the University of Konstanz, Department of Computer & Information Science.

D. Eppstein and E.R. Gansner (Eds.): GD 2009, LNCS 5849, pp. 369–380, 2010.
© Springer-Verlag Berlin Heidelberg 2010

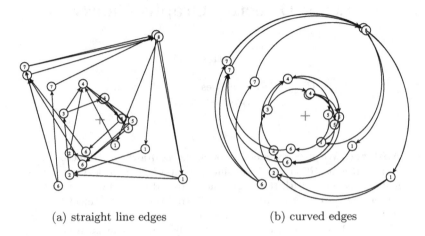

| (a) straight line edges | (b) curved edges |

Fig. 1. Clockwise drawings of the graph on the cover page of [8]. The crosses indicate the location of the origin, relative to which the configuration is oriented. The labels represent the layers in the original drawing.

experimental results for small example graphs. This method is intuitive and works for small graphs, but is, like many other force-directed methods, susceptible to local minima, sensitive to the choice of initial configurations, and not very scalable.

Bachmaier et al. extend the traditional Sugiyama approach by a cyclic level assignment [1]. The lowest level is considered to be on top of the highest level; this modification renders some of the involved optimization problems $\mathcal{NP}$-complete; for the combinatorial background of cyclic arrangements for directed graphs, see also [3,11]. The assignment of levels to nodes in such a cyclic setting is done with various heuristics.

We describe a novel approach for drawing directed graphs in a cyclic style, which does not require a discrete leveling, and gives direct and, in a sense to be specified later, optimal solutions; see Fig. 1 for an example. Positions are given by eigenvectors of a matrix associated with the graph, which is technically similar in style to other spectral layout methods [10], but conceptually different. We give the mathematical background and present a drawing algorithm which is efficient and easy to implement, together with some application results.

2   Preliminaries

In the following, let $G = (V, E)$ be a directed connected graph with directed edges $(u, v) \in E \subseteq V \times V$. The cardinalities of node and edge sets are denoted by $n = |V|, m = |E|$. When $(u, v) \in E$ we say that u *precedes* v and v *succeeds* u. The sets of predecessors and successors of a node $v \in V$ are denoted $N^-(v)$ and $N^+(v)$. Node coordinates are written as column vectors of the form $x = (x_v)_{v \in V} \in \mathbb{R}^n$, and the norms of vectors and matrices are denoted by $\|x\| = (\sum_{v \in V} x_v^2)^{1/2}$ and $\|A\| = (\sum_{i=1}^{n} \sum_{j=1}^{n} a_{ij}^2)^{1/2}$.

3 Skew-Symmetry

Let $A = A(G) = (a_{uv})_{u,v \in V}$ denote the *adjacency matrix* of G, with entries

$$a_{uv} = \begin{cases} 1 & \text{if } (u,v) \in E \\ 0 & \text{otherwise.} \end{cases} \tag{1}$$

We will assume in the following that between every pair of nodes $u, v \in V$ there is at most one directed edge, and that there are no self-edges (v, v).

From the adjacency matrix, which is asymmetric in general, a *skew-symmetric* matrix is derived. A square matrix $S = (s_{uv})_{u,v \in V}$ is skew-symmetric if and only if $s_{uv} = -s_{vu}$ for all $u, v \in V$, or equivalently, $S = -S^T$.

The *skew-symmetric adjacency matrix* $S(G)$ of a directed graph $G = (V, E)$ is connected to its adjacency matrix $A(G)$ by

$$S = S(G) = A(G) - A(G)^T \tag{2}$$

with entries

$$s_{uv} = \begin{cases} 1 & \text{if } (u,v) \in E, (v,u) \notin E \\ -1 & \text{if } (v,u) \in E, (u,v) \notin E \\ 0 & \text{otherwise} \end{cases} \tag{3}$$

for all $u, v \in V$.

We will now use the eigenvalues and eigenvectors of S to obtain positions for every node and thus a drawing of G. Without loss of generality, the *eigenvalue decomposition* of S may be written in the form

$$S = U \Phi U^T , \tag{4}$$

where $U \in \mathbb{R}^{n \times n}$ is an orthogonal matrix whose columns are real unit length eigenvectors $u_1, \ldots, u_n \in \mathbb{R}^n$, $\|u_i\| = 1$ for all $i \in \{1, \ldots, n\}$, and $\Phi \in \mathbb{C}^{n \times n}$ is a diagonal matrix of complex eigenvalues.

Since S is skew-symmetric, the complex eigenvalues of S are purely imaginary and occur in conjugated complex pairs

$$\pm \sqrt{-1}\phi_1, \pm \sqrt{-1}\phi_2, \ldots, \pm \sqrt{-1}\phi_{\lfloor n/2 \rfloor} \tag{5}$$

with an additional singleton zero eigenvalue if n is odd. We will refer to a pair of eigenvalues $\pm \sqrt{-1}\phi_i$ as *the eigenvalue* ϕ_i. Without loss of generality we assume that the eigenvalues in Φ are ordered non-increasingly by their absolute magnitude, $\phi_1 \geq \cdots \geq \phi_{\lfloor n/2 \rfloor} \geq 0$. With the i-th eigenvalue ϕ_i $(1 \leq i \leq \lfloor n/2 \rfloor)$ a pair of eigenvectors u_{2i-1}, u_{2i} is associated, which span a two-dimensional space frequently called *(i-th) bimension*.

Through orthogonal transformation, (4) can be brought into a slightly different form known as the *Gower decomposition* [5]

$$
\begin{bmatrix} u_1 & u_2 & \cdots & u_{n-1} & u_n \end{bmatrix}
\begin{bmatrix} 0 & \phi_1 & & & \\ -\phi_1 & 0 & & & \\ & & \ddots & & \\ & & & 0 & \phi_{\lfloor \frac{n}{2} \rfloor} \\ & & & -\phi_{\lfloor \frac{n}{2} \rfloor} & 0 \end{bmatrix}
\begin{bmatrix} u_1^T \\ u_2^T \\ \vdots \\ u_{n-1}^T \\ u_n^T \end{bmatrix}
\tag{6}
$$

which allows S to be written as a sum of $\lfloor n/2 \rfloor$ elementary rank-2 matrices

$$
S = \sum_{i=1}^{\lfloor n/2 \rfloor} \phi_i \left(u_{2i} u_{2i-1}^T - u_{2i-1} u_{2i}^T \right),
\tag{7}
$$

all of which are skew-symmetric.

An intuitive interpretation of the decomposition (7) is that each of the (at most) $\lfloor n/2 \rfloor$ summands explains a share of the directional information expressed by S; the magnitude of the eigenvalue ϕ_i is equal to the share of the ith bimension. Note that a pair of eigenvectors u_{2i-1}, u_{2i} may be replaced by any orthogonal pair of vectors spanning the same two-dimensional space.

4   Clockwise Drawings

Each of the $\lfloor n/2 \rfloor$ bimensions of $S(G)$ may be used to obtain a two-dimensional drawing of a graph $G = (V, E)$. Since ϕ_1 is the largest eigenvalue, the information expressed by the edges of G is best captured in two-dimensions by using the corresponding eigenvectors u_1 and u_2 as follows.

Positions for every node $v \in V$ are simply obtained by setting

$$
x = \sqrt{\phi_1} u_1, y = \sqrt{\phi_1} u_2
\tag{8}
$$

and using the entries x_v, y_v as the coordinates of v in two-dimensional Euclidean space. In such a configuration, the particular skew-symmetry s_{uv} between two nodes u and v, which comes from the orientation (or absence) of the edge (u, v), is fitted by

$$
s_{uv} \approx x_u y_v - x_v y_u.
\tag{9}
$$

This quantity is proportional to the *signed* area of the triangle of the positions (x_u, y_u) and (x_v, y_v) subtended by the origin, since

$$
x_u y_u - (x_v y_v/2 + x_u y_v/2 + (x_u - x_v)(y_v - y_u)/2) = (x_u y_v - x_v y_u)/2
$$

as illustrated in Fig. 2.

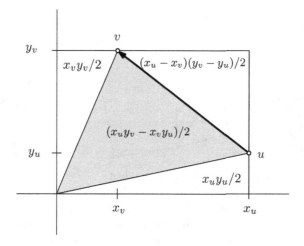

Fig. 2. The signed area of the triangle of the edge pointing from (x_u, y_u) to (x_v, y_v) subtended by the origin represents the amount and direction of the skew-symmetry between u and v

It can be shown that setting the positions (x_v, y_v) for each node $v \in V$ as in (8) minimizes the objective function

$$\sum_{(u,v)\in E} (x_u y_v - x_v y_u - 1)^2 + \sum_{(u,v),(v,u)\notin E} (x_u y_v - x_v y_u)^2 \qquad (10)$$

among all two-dimensional layouts $x, y \in \mathbb{R}^n$ [6]. Intuitively, minimizing (10) tries to represent all directed edges with a correspondingly oriented triangle having *positive unit area*, while all non-adjacent pairs of nodes should be located on a line through the origin, forming a triangle with area zero; this may be interpreted as a global repulsion energy for non-adjacent node pairs.

Unlike distance-based layout methods, the origin and the relation of nodes to the origin are crucial for reading the clockwise drawing. The angle of a node's position is determined largely by the angle of its predecessors and successors. Since two eigenvectors spanning a bimension share the same eigenvalue, axes are not meaningful, and the configuration may be freely rotated around the center without modifying triangle areas and signs. Furthermore, it is not determined whether the bimension blocks in the block-diagonal matrix in (6) are of the form

$$\begin{pmatrix} 0 & \phi_i \\ -\phi_i & 0 \end{pmatrix} \qquad \text{or} \qquad \begin{pmatrix} 0 & -\phi_i \\ \phi_i & 0 \end{pmatrix}$$

so that the orientation is made clockwise or counterclockwise, as desired, by reflecting it on any line through the origin. Note that some edges may point against the desired orientation because the associated triangle areas are negative. Depending on the context, these edges may be visually highlighted, or re-oriented by letting them follow the opposite, longer way around the origin.

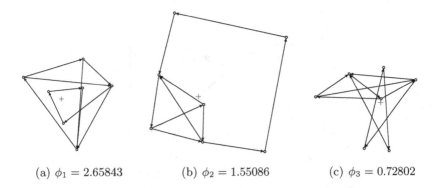

(a) $\phi_1 = 2.65843$ (b) $\phi_2 = 1.55086$ (c) $\phi_3 = 0.72802$

Fig. 3. Drawings of an example graph ($n = 7, m = 10$) in all three possible bimensions. Note that $\phi_1^2 + \phi_2^2 + \phi_3^2 = m$, and the bimensions account for about $\phi_1^2/m \approx 70.6\%$, $\phi_2^2/m \approx 24.1\%$ and $\phi_3^2/m \approx 5.3\%$ of the skew-symmetry information.

Although the bimension for the largest eigenvalue is the best one can do with two dimensions in the sense of the criterion (10), drawings in other bimensions may also be helpful, since they visualize additional, less dominant parts of the directional information. A small graph and the layout in all possible bimensions is given in Fig. 3. The second bimension explains as much as possible of the skew-symmetry remaining after removing the contribution of the first bimension from the sum in (7), and so on.

5   Implementation

There are some dedicated algorithms for computing the complete spectral decomposition (4) of a skew-symmetric matrix [12,16]. Fortunately, we need only the largest eigenvalue and the two associated eigenvectors, and a simple power iteration is thus sufficient.

Since $S = -S^T$ and hence $-SS^T = S^2$, we can use the fact that the eigenvectors of S are identical to the eigenvectors of the symmetric matrix $SS^T = -S^2$, and power-iterate with SS^T, which is convenient to handle computationally.

An initial non-zero vector, which may be chosen randomly, is iteratively multiplied with SS^T by carrying out the multiplication step

$$x \leftarrow \frac{SS^T x}{\|SS^T x\|}. \tag{11}$$

over and over again; in general, x will converge to the desired eigenvector [4]. Instead of materializing the matrix SS^T, which would require $\mathcal{O}(n^3)$ real multiplication operations, the step (11) can be split into

$$\hat{x} \leftarrow \frac{S^T x}{\|S^T x\|} \tag{12}$$

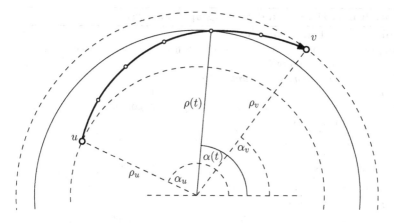

Fig. 4. Drawing curved edges using splines with control points

$$x \leftarrow \frac{S\hat{x}}{\|S\hat{x}\|} \tag{13}$$

which encompasses only $\mathcal{O}(n^2)$ operations, assuming a constant number of iterative steps to achieve convergence. Furthermore, sparsity of a graph G, i.e., when $S(G)$ has $o(n^2)$ non-zero entries, may be exploited to obtain a power iteration with linear time per step. The power iteration process in (12) and (13) becomes

$$\hat{x} \leftarrow \frac{Ax - A^T x}{\|Ax - A^T x\|} \tag{14}$$

$$x \leftarrow \frac{A^T \hat{x} - A\hat{x}}{\|A^T \hat{x} - A\hat{x}\|} \tag{15}$$

which is just a linear scan over all edges, since only positions of adjacent nodes need to be accumulated; this is reminiscent of hubs and authorities [9], where the eigenvectors of AA^T and $A^T A$ are computed.

The required second eigenvector of SS^T is computed similarly, but with orthogonalizing against the first eigenvector after each step. Pseudo-code of an algorithm with running time and space complexity in $\mathcal{O}(n + m)$ per iteration step is given in Alg. 1.

To avoid unnecessary crossings by straight lines, edges may be drawn as clockwise curves around the origin, e.g., using splines. The corresponding control points are determined in a linear interpolation between the angles α_u, α_v, and the radii ρ_u, ρ_v of the nodes u, v by

$$\rho(t) = (1 - t) \cdot \rho_u + t \cdot \rho_v \tag{16}$$

$$\alpha(t) = (1 - t) \cdot \alpha_u + t \cdot \alpha_v \tag{17}$$

Algorithm 1. Drawing a directed graph clockwise

Input: Directed graph $G = (V, E)$
Output: Coordinate vectors $x, y \in \mathbb{R}^n$ with positions for every $v \in V$
$x \leftarrow$ random, $y \leftarrow$ random
while x *and* y *change significantly* **do**

$\quad$ $x \leftarrow x/\|x\|, y \leftarrow y/\|y\|$ $\qquad\qquad$ // normalize
$\quad$ $y = y - x^T y \cdot x$ $\qquad\qquad\qquad$ // orthogonalize
$\quad$ **foreach** $v \in V$ **do**

$\qquad$ $\hat{x}_v \leftarrow \displaystyle\sum_{u \in N^-(v)} x_u - \sum_{w \in N^+(v)} x_w$ $\qquad$ // $\hat{x} \leftarrow (A - A^T) \cdot x$

$\qquad$ $\hat{y}_v \leftarrow \displaystyle\sum_{u \in N^-(v)} y_u - \sum_{w \in N^+(v)} y_w$ $\qquad$ // $\hat{y} \leftarrow (A - A^T) \cdot y$

$\quad$ **foreach** $v \in V$ **do**

$\qquad$ $x_v \leftarrow \displaystyle\sum_{w \in N^+(v)} \hat{x}_w - \sum_{u \in N^-(v)} \hat{x}_u$ $\qquad$ // $x \leftarrow (A^T - A) \cdot \hat{x}$

$\qquad$ $y_v \leftarrow \displaystyle\sum_{w \in N^+(v)} \hat{y}_w - \sum_{u \in N^-(v)} \hat{y}_u$ $\qquad$ // $y \leftarrow (A^T - A) \cdot \hat{y}$

$\quad$ $\phi \leftarrow \sqrt{\|x\|}$ $\qquad\qquad\qquad\qquad$ // estimate for largest eigenvalue
$x \leftarrow x/\phi^{3/2}, y \leftarrow y/\phi^{3/2}$ $\qquad\qquad$ // scale eigenvectors to have length $\sqrt{\phi}$

---

where $0 \le t \le 1$; when k control points are used, $t \in \{0, \frac{1}{k}, \frac{2}{k}, \ldots, \frac{k-1}{k}, 1\}$. Note that when $|\alpha_u - \alpha_v| > \pi$, this interpolation results in the edge (u, v) winding around the center with an angle greater than π; the shorter counterpart of that curve is obtained by adding 2π to the smaller of α_u, α_v.

6   An Application

A special class of directed graphs is called *tournaments* [7,13]. A tournament $G = (V, E)$ on n nodes is an orientation of the complete undirected graph on n nodes. Tournaments are a model for round-robin competitions in which everybody competes with everybody else, and every competition $\{u, v\}$ for $u, v \in V, u \ne v$ has a winner u and a loser v, say, which is represented by the orientation (u, v).

Here we use a variant of tournaments, in which the underlying undirected graph is almost complete, but some edges are allowed to be missing because there are situations in which no winner can be determined. The method of clockwise drawing is applied to results of international football leagues in England, Germany, Italy, and Spain, in the seasons ending in 2006, 2007, and 2008. In every season, between every possible pair of teams two matches are carried out, each team being the home team once. The tournament graph contains an edge $(u, v) \in E$ when u dominates v, i.e., u has won more matches against v than v against u; ties are not considered.

Fig. 5 shows drawings of all 12 tournaments, as given by the positions in the bimension of the largest eigenvalue. A cyclic structure is displayed in some of

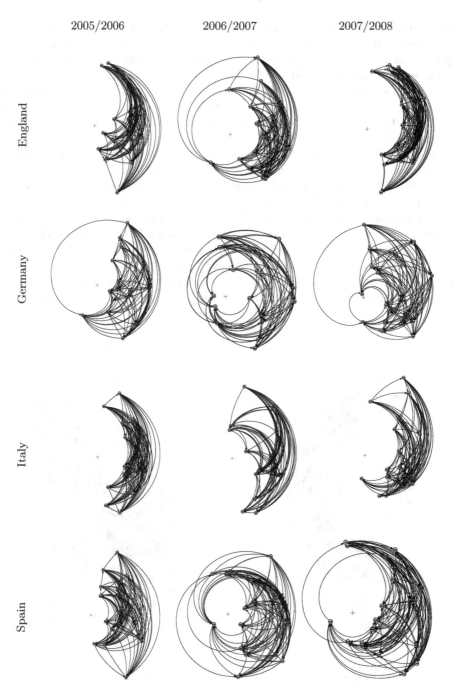

Fig. 5. Clockwise drawings (bimension of the largest eigenvalue) of the tournament graphs in European football leagues in three consecutive seasons. The span of edges around the center is emphasized by edges curving around the center.

the configurations, such as England 2006/2007, Germany 2006/2007, and Spain 2006/2007, 2007/2008, which leads to the conjecture that these seasons were quite balanced, with no clear dominator. In these tournaments, some otherwise weak teams, which are dominated by most others, won against otherwise strong teams. For example, in the 2006/2007 season of the English Premier League, West Ham United (node on the lower left) closed the season on rank 15 of 20 teams, but dominated the champions Manchester United and fourth-ranked Arsenal FC.

In contrast, it is interesting to observe that the drawings of some other tournaments appear to be rather non-cyclic, especially England 2005/2006 and 2007/2008, all three seasons in Italy, and Spain 2005/2006. Since all nodes are on the same side of a line through the origin, the signed triangle areas do not allow for cyclic node triples in this bimension. Thus, most of the dominance structure in the skew-symmetric adjacency matrix is intrinsically rather non-cyclic, and suggests that the classical hierarchical approach is actually more appropriate than the cyclic one. In the context of football matches, there is a clear tendency for strong teams to consistently dominate weaker teams and weak teams to be consistently dominated by stronger teams, with no or only few exceptions.

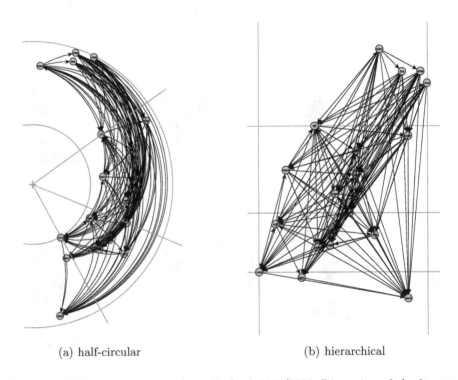

(a) half-circular   (b) hierarchical

Fig. 6. Football tournament graph in England 2007/2008 (bimension of the largest eigenvalue). The graph exhibits a substantially hierarchical structure, which justifies the transformation from a polar into a cartesian domain.

In fact, a polar transformation easily transforms the half-circular arrangement into the traditional hierarchical drawing style. A natural ranking is given by the total order of angles of all nodes with respect to the origin, as they are given by their angular representation. The coordinates for node v after this polar transformation are given by

$$\rho_v = \sqrt{x_v^2 + y_v^2}, \quad \alpha_v = \operatorname{atan2}(y_v, x_v) \tag{18}$$

where $\operatorname{atan2}(\cdot, \cdot) \colon \mathbb{R}^2 \to [0, 2\pi]$ denotes the two-argument inverse of the tangent function implemented in most modern programming languages. ρ_v represents the transformed clockwise rank of v and α_v the amount of skew-symmetry of v with all other nodes. An example of such a half-clockwise configuration and its polar transform is given in Fig. 6.

7   Conclusion

The decomposition of the skew-symmetric adjacency matrix yields a method for drawing directed graphs in a cyclic fashion and provides direct and unique solutions. The drawing area is oriented either clockwise or counterclockwise around a distingiuished center point; if necessary, the sense of rotation is inverted by reflecting one axis.

The algorithm is easy to implement because it requires only essential array operations and no sophisticated data structures. Since no cycle removal or level assignment is required, some of the computationally hard problems are avoided. The sparsity of the skew-symmetric adjacency matrix can be used to obtain a power iteration algorithm which runs in linear time per step and requires linear space.

When discrete levels or radial level assignments are required, they may be obtained from the continuous coordinates by a quantization scheme. The clockwise configurations can be combined with the force-directed methods in [1,14]. A straightforward extension would be to use non-uniform edge lengths. For strongly connected graphs, all distances are finite, and the analysis is also applicable to the corresponding skew-symmetric distance matrix.

While there is no space here for a detailed discussion of quantitative measures to characterize the cyclicity of a dominance relation, we would like to point out that clockwise oriented drawings and the distribution of the involved eigenvalues are useful for testing hypotheses about the cyclic or hierarchical structure of directed graphs.

Beyond the graph drawing application, we expect that the presented method is also useful for generating initial solutions to heuristic methods for general $\mathcal{NP}$-hard arrangement problems [3,11].

References

1. Bachmaier, C., Brandenburg, F.J., Brunner, W., Lovász, G.: Cyclic leveling of directed graphs. In: Tollis, I.G., Patrignani, M. (eds.) GD 2008. LNCS, vol. 5417, pp. 348–359. Springer, Heidelberg (2009)

2. Carmel, L., Harel, D., Koren, Y.: Combining hierarchy and energy for drawing directed graphs. IEEE Transactions on Visualization and Computer Graphics 10(1), 46–57 (2004)
3. Ganapathy, M.K., Lodha, S.P.: On minimum circular arrangement. In: Diekert, V., Habib, M. (eds.) STACS 2004. LNCS, vol. 2996, pp. 394–405. Springer, Heidelberg (2004)
4. Golub, G.H., van Loan, C.F.: Matrix Computations, 3rd edn. The Johns Hopkins University Press, Baltimore (1996)
5. Gower, J.C.: The analysis of asymmetry and orthogonality. In: Recent Developments in Statistics, pp. 109–123 (1977)
6. Gower, J.C., Constantine, A.G.: Graphical representation of asymmetric matrices. Applied Statistics 27, 297–304 (1978)
7. Harary, F., Moser, L.: The theory of round robin tournaments. Amer. Math. Monthly 73, 231–246 (1966)
8. Kaufmann, M., Wagner, D. (eds.): Drawing Graphs. LNCS, vol. 2025. Springer, Heidelberg (2001)
9. Kleinberg, J.M.: Authoritative sources in a hyperlinked environment. Journal of the ACM 46(5), 604–632 (1999)
10. Koren, Y.: On spectral graph drawing. In: Warnow, T.J., Zhu, B. (eds.) COCOON 2003. LNCS, vol. 2697, pp. 496–508. Springer, Heidelberg (2003)
11. Liberatore, V.: Circular arrangements and cyclic broadcast scheduling. Journal of Algorithms 51(2), 185–215 (2004)
12. Paardekooper, M.H.C.: An eigenvalue algorithm for skew-symmetric matrices. Numerische Mathematik 17(3), 189–202 (1971)
13. Reid, K.B., Beineke, L.W.: Tournaments. In: Beineke, L.W., Wilson, R.J. (eds.) Selected Topics in Graph Theory, pp. 169–204. Academic Press, London (1978)
14. Sugiyama, K., Misue, K.: A simple and unified method for drawing graphs: Magnetic-spring algorithm. In: Tamassia, R., Tollis, I.G. (eds.) GD 1994. LNCS, vol. 894, pp. 364–375. Springer, Heidelberg (1995)
15. Sugiyama, K., Tagawa, S., Toda, M.: Methods for visual understanding of hierarchical system structures. IEEE Transactions on Systems, Man, and Cybernetics 11(2), 109–125 (1981)
16. Ward, R.C., Gray, L.C.: Ward and Leonard C. Gray. Eigensystem compuation for skew-symmetric matrices and a class of symmetric matrices. ACM Transactions on Mathematical Software 4(3), 278–285 (1978)

An Improved Algorithm for the Metro-line Crossing Minimization Problem

Martin Nöllenburg

Fakultät für Informatik, Universität Karlsruhe (TH) and
Karlsruhe Institute of Technology (KIT), Karlsruhe, Germany
noellenburg@iti.uka.de

Abstract. In the metro-line crossing minimization problem, we are given a plane graph $G = (V, E)$ and a set $\mathcal{L}$ of simple paths (or *lines*) that *cover* G, that is, every edge $e \in E$ belongs to at least one path in $\mathcal{L}$. The problem is to draw all paths in $\mathcal{L}$ along the edges of G such that the number of crossings between paths is minimized. This crossing minimization problem arises, for example, when drawing metro maps, in which multiple transport lines share parts of their routes.

We present a new line-layout algorithm with $O(|\mathcal{L}|^2 \cdot |V|)$ running time that improves the best previous algorithms for two variants of the metro-line crossing minimization problem in unrestricted plane graphs. For the first variant, in which the so-called *periphery condition* holds and terminus side assignments are given in the input, Asquith et al. [1] gave an $O(|\mathcal{L}|^3 \cdot |E|^{2.5})$-time algorithm. For the second variant, in which all lines are paths between degree-1 vertices of G, Argyriou et al. [2] gave an $O((|E| + |\mathcal{L}|^2) \cdot |E|)$-time algorithm.

1   Introduction

Schematic *metro maps* are effective and popular visualizations of public transport networks all over the world; see Ovenden's comprehensive collection of metro maps [3]. Several methods for automatically drawing metro maps have been suggested in recent years [4, 5, 6]. These methods, however, focus on drawing the *underlying graph*, that is, the graph that represents stations as vertices and direct links between two stations as edges. This graph represents the infrastructure of the transport network, for example, railway tracks or roads. A schematic layout of the underlying graph, whether created manually or by one of the existing methods mentioned above, does not necessarily yield a proper metro map yet. The reason is that most real-world networks contain many different transport lines whose routes partially overlap, that is, some edges of the underlying graph are shared by multiple transport lines. In practice, each transport line is therefore drawn in a distinct color along the edges of its path in the underlying graph. Consequently, edges that belong to several lines consist in fact of a bundle of colored parallel curves. As an example, Fig. 1 shows a detail of the metro map of Cologne.

D. Eppstein and E.R. Gansner (Eds.): GD 2009, LNCS 5849, pp. 381–392, 2010.

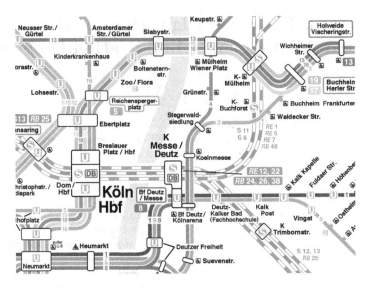

Fig. 1. Detail of the metro map of Cologne

An immediate consequence of such a visualization is that there are situations in which two lines in the—otherwise plane—network cross. Some line crossings are mandatory, induced by the prescribed network topology, others depend on the line orders in each vertex and can be avoided by choosing the right orders. Hence, the *metro-line crossing minimization* (MLCM) problem arises as a secondary problem in the metro-map layout process: find an ordering of the parallel lines along each edge of the underlying graph such that as few pairs of lines as possible cross each other in the final layout. Additionally, the relative order of lines traversing a vertex in the same direction must not change within this vertex, that is, we do not allow to hide line crossings "below" the area occupied by the representation of a vertex. Note that the MLCM problem is independent of the actual layout of the underlying graph. The combinatorial embedding of the underlying graph, which is usually defined by its geographic input embedding, is all one needs to define the orderings of the parallel lines. Hence, algorithms for MLCM can be used both for reducing line crossings in existing layouts and, as a second step in combination with layout methods for the underlying graph, for creating metro maps from scratch.

Although we present our results in terms of the classic problem of visualizing transportation networks, we note that the metro map metaphor has also been used as a means to visualize potentially much larger networks in other fields, for example, metabolic pathways [7]. Actually, the MLCM problem appears whenever multiple parallel edges in a graph need to be drawn separately along a common geometric path with the minimum number of crossings among them.

Benkert et al. [8] introduced the general MLCM problem. Subsequently, MLCM was considered in several variants and for different classes of underlying graphs [9, 2, 1], which are discussed in detail in Section 3. One important

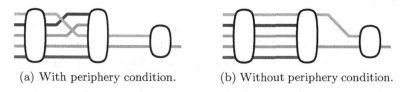

(a) With periphery condition.   (b) Without periphery condition.

Fig. 2. Layout of a terminus (middle vertex) with three terminating lines. The layout in (b) introduces a gap between the continuing lines.

variant, posed as an open problem by Benkert et al. [8] and addressed by Bekos et al. [9] and by Asquith et al. [1], restricts the positions of each line's start and end point (called *termini*) to be left- or rightmost in the ordered sequence of lines along the underlying edges leading to its termini. This restriction is called the *periphery condition* and prevents gaps between continuing lines, see Fig. 2. Gaps between parallel lines disrupt the uniform appearance of the underlying edge and hence are to be avoided in order to improve readability. Apart from avoiding gaps, an exposed outer position for terminating lines also allows for better highlighting or labeling of the termini. Often the final destination of a train or bus in a transport network is used to indicate its direction and hence termini and their names should be prominent features that are easy to recognize in a metro map. Many of the real-world maps in Ovenden's collection [3] adhere to the periphery condition, as does the metro map of Cologne in Fig. 1. Bekos et al. [9] showed that the MLCM problem is NP-hard under the periphery condition if each terminus can lie on either side of the respective final edge. On the other hand, Asquith et al. [1] showed that the MLCM problem under the periphery condition can be solved efficiently for general plane graphs if the terminus side assignment is given as part of the input.

In this paper, we investigate the MLCM problem under the periphery condition with terminus side assignments and present a new algorithm in Section 4 that solves this problem in $O(|\mathcal{L}|^2 \cdot |V|)$ time for a graph $G = (V, E)$ and a set of lines $\mathcal{L}$. The algorithm has two phases. First, for each pair of lines that share a common subpath, we determine their required relative order at the end of their common subpath. Then, in a second step, we iteratively insert one line at a time into the layout such that the relative orders computed in the first phase are respected and no unnecessary line crossings are created. Our algorithm improves the algorithm of Asquith et al. [1] for the same problem, which has a running time of $O(|\mathcal{L}|^3 \cdot |E|^{2.5})$. Our algorithm can also be used to solve a closely related problem considered by Argyriou et al. [2], where all lines must be paths connecting two degree-1 vertices in G. Hence, it also improves the algorithm of Argyriou et al., which has a running time of $O((|E| + |\mathcal{L}|^2) \cdot |E|)$. These are the only two variants of MLCM that are known to be efficiently solvable, and our algorithm is to the best of our knowledge currently the fastest method to solve both of them for general plane underlying graphs.

2 Model

The input to the MLCM problem is a *metro graph* $(G, \mathcal{L})$, where $G = (V, E)$ is a planar embedded graph and $\mathcal{L}$ is a *line cover* of G, that is, a set of simple paths (or *lines*) that cover G. Note that existing edge crossings in the input graph can easily be modeled as dummy vertices. For notational convenience, we consider each undirected edge $\{u, v\} \in E$ as a pair of directed edges uv and vu. Both notations refer to the same single edge just from two different perspectives.

The vertices v_0 and v_k of a line $\ell = (v_0, v_1, \ldots, v_k) \in \mathcal{L}$ of length $|\ell| = k$ are called the *termini* of ℓ, the vertices $v_1, \ldots, v_{k-1}$ are called *intermediate vertices* of ℓ. An edge uv is included in a line ℓ, in short $uv \in \ell$, if u and v are consecutive vertices in ℓ. We denote as $\mathcal{L}_{uv} = \mathcal{L}_{vu} = \{\ell \in \mathcal{L} \mid uv \in \ell\}$ the set of all lines that include an edge uv. The *total edge size* of $\mathcal{L}$ is defined as $N_{\mathcal{L}} = \sum_{\ell \in \mathcal{L}} |\ell| = \sum_{uv \in E} |\mathcal{L}_{uv}|$. Note that $N_{\mathcal{L}} \in O(|\mathcal{L}| \cdot |V|)$ since $|\ell| \leq |V|$ for each line $\ell \in \mathcal{L}$.

Each vertex u has a cyclic sequence of $\sum_{uv \in E} |\mathcal{L}_{uv}|$ consecutive *ports*, one for each line of each incident edge uv. Each port is a point on the boundary of the geometric representation of u, at which the individual lines in $\bigcup_{uv \in E} \mathcal{L}_{uv}$ enter (or leave) u. We are interested in the order in which the lines in $\mathcal{L}_{uv}$ connect to the consecutive subsequence of ports of u (and of v) that correspond to the lines along edge uv. So for each edge $uv \in E$, we define two *line orders* $<^u_{uv}$ and $<^v_{uv}$ of $\mathcal{L}_{uv}$ in the endpoints of uv. For two lines ℓ_1 and ℓ_2 in $\mathcal{L}_{uv}$ we write $\ell_1 <^u_{uv} \ell_2$ (or $\ell_1 <^v_{uv} \ell_2$) if ℓ_1 is right of ℓ_2 at the endpoint u (or v) with respect to the direction of uv. Note that the orders are reversed if we use the oppositely directed edge vu instead of uv, that is, $\ell_1 <^u_{uv} \ell_2$ if and only if $\ell_2 <^u_{vu} \ell_1$. The sorted sequence of the lines in $\mathcal{L}_{uv}$ with respect to $<^u_{uv}$ is denoted as s^u_{uv}; analogously s^v_{uv} is the sorted sequence of lines with respect to $<^v_{uv}$. Again, the sequences s^u_{vu} and s^v_{vu} are the reversed sequences of s^u_{uv} and s^v_{uv}.

A *line crossing* is a crossing between two lines ℓ_1 and ℓ_2 along a shared edge uv. The two lines cross on uv if $\ell_1 <^u_{uv} \ell_2$ and $\ell_2 <^v_{uv} \ell_1$ or vice versa. Abstracting from geometry, the number of line crossings along an edge uv is thus equal to the number of inversions in the sequences s^u_{uv} and s^v_{uv}.

In order to avoid confusion for the map viewer, it is not allowed to hide line crossings "below" a vertex. To that end we define a line order $<^v_{uv}$ to be *compatible* with the vertex v if the following holds. Apart from uv, let $vw_1, vw_2, \ldots, vw_k$ be the other edges incident to v in counterclockwise order starting from uv. We consider the sequence s^v_{uv} and the concatenated sequence $s' = \prod_{i=1}^{k} s^v_{vw_i}$. Then $<^v_{uv}$ is compatible with v if s^v_{uv} is a subsequence of s'. In other words, the lines that enter v through the edge uv and leave v through the edges $vw_1, vw_2, \ldots, vw_k$ do not change their relative order. We say that a vertex v is *admissible* if the line orders for all incident edges are compatible with v.

3 MLCM Variants and Previous Work

In this section we present four different variants of the MLCM problem that have been considered in the literature so far. Previous results and the results obtained in this paper are summarized in Table 1.

Table 1. Overview of results for the MLCM problem and its variants. Algorithmic results are given by their running time.

problem	graph class	restrictions	result	reference
MLCM	single edge uv	–	$O(|\mathcal{L}_{uv}|^2)$	[8]
MLCM-P	path	–	NP-hard	[9]
	plane graph	–	ILP + MLCM-PA	[1]
MLCM-PA	path	2-side model	$O(|\mathcal{L}| \cdot |V|)$	[9]
	left-to-right tree	2-side model	$O(|\mathcal{L}| \cdot |V|)$	[9]
	plane graph	–	$O(|\mathcal{L}|^3 \cdot |E|^{2.5})$	[1]
	plane graph	2-side model	$O(|V| \cdot (|E| + |\mathcal{L}|))$	[2]
	plane graph	–	$O(|\mathcal{L}|^2 \cdot |V|)$	Theorem 1
MLCM-T1	left-to-right tree	2-side model	$O(|\mathcal{L}| \cdot |V|)$	[9]
	plane graph	2-side model	$O(|V| \cdot (|E| + |\mathcal{L}|))$	[2]
	plane graph	–	$O((|E| + |\mathcal{L}|^2) \cdot |E|)$	[2]
	plane graph	–	$O(|\mathcal{L}|^2 \cdot |V|)$	Corollary 1

The original *metro-line crossing minimization* problem as introduced by Benkert et al. [8] is as follows.

Problem 1 (MLCM). Given a metro graph $(G = (V, E), \mathcal{L})$, find for each edge $uv \in E$ two line orders $<^u_{uv}$ and $<^v_{uv}$ of the lines in $\mathcal{L}_{uv}$ such that the number of line crossings is minimal and all vertices are admissible.

A solution to MLCM is denoted as a *line layout*. Benkert et al. [8] gave a quadratic-time algorithm to solve MLCM for a single edge of G. Their algorithm does not extend to larger subgraphs and it is a remaining open problem whether MLCM is NP-hard in its general form.

We have already introduced the periphery condition, which additionally requires that each line terminates in an outer or *peripheral* position in each of its two termini (recall Fig. 2). Formally, this means that for each vertex v and each edge uv all lines in $\mathcal{L}_{uv}$, for which v is a terminus, must be placed in the beginning or in the end of the sequence s^v_{uv}. In other words, no terminating line can lie between two continuing lines in the order $<^v_{uv}$. We denote the following variant as MLCM *with periphery condition* (MLCM-P).

Problem 2 (MLCM-P). Given a metro graph $(G = (V, E), \mathcal{L})$, find for each edge $uv \in E$ two line orders $<^u_{uv}$ and $<^v_{uv}$ of the lines in $\mathcal{L}_{uv}$ such that the number of line crossings is minimal, all vertices are admissible, and each terminating line is placed at a peripheral position in each of its two termini.

Bekos et al. [9] showed that MLCM-P is NP-hard, even if G is a path, and Asquith et al. [1] formulated an integer linear program (ILP) to solve MLCM-P. Still, Problem 2 gives rise to a closely related (but computationally feasible) variant that additionally specifies in the input fixed terminus sides for each line. We denote this variant as MLCM *with periphery condition and terminus side assignments* (MLCM-PA).

Problem 3 (MLCM-PA). Given a metro graph $(G = (V, E), \mathcal{L})$ and terminus side assignments for all lines in $\mathcal{L}$, find for each edge $uv \in E$ two line orders $<_{uv}^u$ and $<_{uv}^v$ of the lines in $\mathcal{L}_{uv}$ such that the number of line crossings is minimal, all vertices are admissible, and each terminating line is placed at a peripheral position on the specified side of each of its two termini.

Problem 3 occurs in situations, in which, for example, the physical location of the tracks or the bus stop of the terminating line in a terminus yields this information. Alternatively, the optimal terminus side assignments can be obtained from the ILP formulation of Asquith et al. [1]. Asquith et al. also presented an $O(|\mathcal{L}|^3 \cdot |E|^{2.5})$-time algorithm to solve MLCM-PA for general plane graphs. Bekos et al. [9] gave two algorithms to solve MLCM-PA in the restricted *2-side model* for paths and for a special class of left-to-right directed trees with bounded vertex degree in $O(|\mathcal{L}| \cdot |V|)$ time, respectively. In the 2-side model, all vertices are drawn as rectangles and all lines are drawn as x-monotone paths that pass through vertices from the left to the right side. Argyriou et al. [2] recently presented an algorithm to solve MLCM-PA in the 2-side model for general plane graphs in $O(|V| \cdot (|E| + |\mathcal{L}|))$ time.

Another interesting MLCM variant restricts the lines in $\mathcal{L}$ to terminate at degree-1 vertices only, that is, all termini in $(G, \mathcal{L})$ are leaves of G.

Problem 4 (MLCM-T1). Given a metro graph $(G = (V, E), \mathcal{L})$ in which the degree of any terminus v of any path in $\mathcal{L}$ equals 1, find for each edge $uv \in E$ two line orders $<_{uv}^u$ and $<_{uv}^v$ of the lines in $\mathcal{L}_{uv}$ such that the number of line crossings is minimal and all vertices are admissible.

Problem 4 is of practical interest since in many real-world networks transport lines lead from one terminus station in the outskirts of a city through the city center to another terminus station in the outskirts. This is exactly the situation in which lines terminate at leaves of the underlying graph. Argyriou et al. [2] presented an algorithm to solve MLCM-T1 in general plane graphs in $O((|E| + |\mathcal{L}|^2) \cdot |E|)$ time. For MLCM-T1 in the previously mentioned 2-side model, they improved the running time to $O((|E| + |\mathcal{L}|) \cdot |V|)$.

We observe that a line layout for an MLCM-T1 instance trivially satisfies the periphery condition. Since each terminus v is a degree-1 vertex in G, there cannot be any continuing lines in v, and any position in the line order at v is peripheral by definition. Furthermore, there is no need to distinguish two different sides for the assignment of the terminus positions: not being separated by a continuing line, the two sides of the edge leading to v coincide. Hence, we can reduce any MLCM-T1 instance to an equivalent MLCM-PA instance by assigning all lines that terminate at the same leaf v to the same terminus side. This actually means that there is no restriction to the line order in v at all, and we indeed model the general setting of MLCM-T1. Obviously, the reduction takes only linear time. This is summarized in the following lemma.

Lemma 1. *An instance of MLCM-T1 can be reduced to an equivalent instance of MLCM-PA in linear time.*

4   An Improved Algorithm for MLCM-PA

In this section we present our main result, an $O(|\mathcal{L}|^2 \cdot |V|)$-time algorithm for MLCM-PA and MLCM-T1 in general plane graphs. We first show a simple lemma about the line crossings in an optimal layout for an MLCM-PA instance. We define a line crossing of two lines in a metro graph $(G, \mathcal{L})$ to be *unavoidable*, if it is present in any line layout of $(G, \mathcal{L})$.

Lemma 2. *Given a metro graph $(G, \mathcal{L})$ and terminus side assignments for all lines in $\mathcal{L}$, all line crossings in a crossing-minimal line layout are unavoidable crossings.*

Proof. By definition every unavoidable crossing is present in any crossing-minimal line layout. We want to show that the opposite is also true: every line crossing in a crossing-minimal line layout is unavoidable.

So let ℓ_1 and ℓ_2 be two lines that cross in a crossing-minimal line layout along an edge uv. By $P = (w_0, \ldots, w_i = u, w_{i+1} = v, \ldots, w_k)$, $0 \le i < k$, we denote the maximal common subpath of ℓ_1 and ℓ_2 that contains uv. First of all note that the crossing along uv is the only crossing of ℓ_1 and ℓ_2 along P; any two consecutive crossings of two lines along a common subpath could be removed by routing the upper line just below the lower line along the edges between the two crossings—this contradicts the optimality of the line layout and has been observed by Asquith et al. [1] before.

We can assume that $\ell_1 <^u_{uv} \ell_2$ and $\ell_2 <^v_{uv} \ell_1$. Since there is a single crossing between ℓ_1 and ℓ_2 along P, this implies that $\ell_1 <^{w_0}_{w_0 w_1} \ell_2$ and $\ell_2 <^{w_k}_{w_{k-1} w_k} \ell_1$. This inversion of ℓ_1 and ℓ_2 in the line orders of vertices w_0 and w_k is either enforced by the combinatorial embedding of G as the line orders $<^{w_0}_{w_0 w_1}$ and $<^{w_k}_{w_{k-1} w_k}$ must be compatible with w_0 and w_k (if the respective line continues beyond w_0 or w_k) or by the given terminus side assignment (if the respective line terminates at w_0 or w_k). The only case where the relative order of ℓ_1 and ℓ_2 is not fixed by the compatibility requirements or the terminus side assignments is if both lines terminate at the same vertex, say w_0, and are assigned to the same terminus side. In that case, however, they can always be reordered in $<^{w_0}_{w_0 w_1}$ such that they reflect their relative order in $<^{w_k}_{w_{k-1} w_k}$ and the crossing would disappear. This contradicts the optimality of the layout.

We conclude that the crossing of ℓ_1 and ℓ_2 is unavoidable: the relative order of ℓ_1 and ℓ_2 at one end of P is the inverse of their order at the other end of P due to the given terminus side assignments or the compatibility requirements for the embedding of G.   □

Lemma 2 implies that there is a line layout that realizes exactly the unavoidable crossings and, consequently, that any such layout is optimal. Algorithm 1 constructs such a line layout. It first computes all maximal common subpaths of all pairs of lines to determine their relative orders as induced by the topology or the terminus side assignments. In a second phase all lines are iteratively inserted into the line orders of their edges and the final line layout is fixed.

Algorithm 1. MLCM-PA line layout

Input: metro graph $(G, \mathcal{L})$, terminus side assignments for all $\ell \in \mathcal{L}$
Output: line orders $<_{uv}^u, <_{uv}^v$ for all edges $uv \in E$

/* Phase 1   */
foreach $(\ell_1, \ell_2) \in \mathcal{L} \times \mathcal{L}$, $\ell_1 \neq \ell_2$, $\ell_1 = (v_0, v_1, \ldots, v_k)$ **do**
  compute set $\Lambda(\ell_1, \ell_2)$ of all maximal common subpaths of ℓ_1 and ℓ_2
  foreach $(v_i, v_{i+1}, \ldots, v_j) \in \Lambda(\ell_1, \ell_2)$ **do**
    if ℓ_2 leaves ℓ_1 towards the left or terminates left of ℓ_1 in v_j **then**
      for $l = i$ to $j - 1$ **do**
        $\quad$ side$(\ell_1, \ell_2, v_l v_{l+1}) \leftarrow$ left
    else
      for $l = i$ to $j - 1$ **do**
        $\quad$ side$(\ell_1, \ell_2, v_l v_{l+1}) \leftarrow$ right

/* Phase 2   */
foreach $\ell = (v_0, v_1, \ldots, v_k) \in \mathcal{L}$ **do**
  for $i = 0$ to $k - 1$ **do**
    insert ℓ into $<_{v_i v_{i+1}}^{v_i}$
    insert ℓ into $<_{v_i v_{i+1}}^{v_{i+1}}$

Theorem 1. *Given a metro graph $(G, \mathcal{L})$ and terminus side assignments for all lines in $\mathcal{L}$, Algorithm 1 computes a crossing-minimal line layout under the periphery condition in $O(|\mathcal{L}| \cdot N_{\mathcal{L}})$ time.*

Proof. In Phase 1 of Algorithm 1 we compute the value of a binary variable side(ℓ_1, ℓ_2, uv) for each triple of two lines ℓ_1 and ℓ_2 and an edge uv such that uv is a common edge of ℓ_1 and ℓ_2. This value represents the side to which line ℓ_2 tends with respect to ℓ_1 on edge uv. So if side$(\ell_1, \ell_2, uv) = $ *left* (*right*), we know that at the end of the maximal common subpath of ℓ_1 and ℓ_2 that contains uv the line ℓ_2 must be placed left (right) of ℓ_1.

In order to compute the set $\Lambda(\ell_1, \ell_2)$ of maximal common subpaths of ℓ_1 and ℓ_2 we walk along $\ell_1 = (v_0, \ldots, v_k)$ and check for each edge $v_i v_{i+1}$ whether ℓ_2 shares that edge with ℓ_1. If this is the case, we either open a new subpath or extend the current subpath. Otherwise we close the current subpath if there is one. We assume that the input $(G, \mathcal{L})$ contains a Boolean edge-line array of size $|E| \times |\mathcal{L}|$ so that we can check whether a line uses an edge in constant time.

For each subpath $\lambda = (v_i, v_{i+1}, \ldots, v_j) \in \Lambda(\ell_1, \ell_2)$ we need to determine whether ℓ_2 *tends* left- or rightward along λ with respect to ℓ_1, that is, whether at the end of λ the line ℓ_2 must be left or right of ℓ_1. There are three cases to consider.

(1) If $v_j = v_k$, that is, ℓ_1 terminates in v_j, and ℓ_2 does not terminate in v_j, then ℓ_2 tends leftward (rightward) if ℓ_1 is assigned a right (left) terminus position, respectively.

(2) If $v_j = v_k$ and ℓ_2 also terminates in v_j, then either ℓ_1 and ℓ_2 are assigned to different terminus sides and ℓ_2 tends to its assigned side, or both are assigned to the same side. In the latter case, ℓ_2 shall stay on the same side of ℓ_1 as in the first vertex v_i of λ. So if ℓ_2 enters v_i to the left of ℓ_1, then ℓ_2 also tends leftward along λ; otherwise it tends rightward.

(3) If $v_j \neq v_k$ then ℓ_2 tends leftward if either ℓ_2 is assigned to terminate on the left in v_j or ℓ_2 continues along an edge $v_j w$ that is left of ℓ_1 in the embedding of the underlying graph G; otherwise ℓ_2 tends rightward.

In all three cases the value of $\text{side}(\ell_1, \ell_2, uv)$ is either an immediate consequence of the lines' terminus assignments or can be determined by a constant-time query for the relative order of three incident edges in the embedding of G.

Summarizing the above, Phase 1 takes $O(|\mathcal{L}| \cdot N_{\mathcal{L}})$ time and space since we check for each edge of each line if any of the other lines in $\mathcal{L}$ share the edge; if this is the case we assign the leftward/rightward value to the corresponding variable.

In Phase 2 the actual line layout is computed by iteratively fixing the course of each line. We show the correctness of the algorithm by maintaining two invariants during Phase 2.

Invariant 1: There are no invalid intra-vertex crossings, that is, for each vertex u and each edge uv the line order $<^u_{uv}$ is compatible with u.

Invariant 2: All line crossings are unavoidable crossings with respect to the input embedding of G and the given terminus side assignments.

Inserting the first line as the only line into the empty line orders clearly satisfies both invariants. So assume that we already have a partial line layout that satisfies the invariants and that we want to insert the next line $\ell = (v_0, v_1, \ldots, v_k)$ into this partial layout.

We start by inserting ℓ into the order $<^{v_0}_{v_0 v_1}$. Let's assume ℓ is assigned to a left terminus in v_0 with respect to the first edge $v_0 v_1$ (for a right terminus the insertion is analogous). If ℓ is currently the only line with a left terminus on this edge, we insert ℓ as the last edge into $<^{v_0}_{v_0 v_1}$. Otherwise we scan the lines with a left terminus in $<^{v_0}_{v_0 v_1}$, starting with the largest (or leftmost) element, for the first line ℓ' for which $\text{side}(\ell, \ell', v_0 v_1) = right$. We insert ℓ into $<^{v_0}_{v_0 v_1}$ immediately after (or left) of ℓ'. This first insertion does not create any intra-vertex crossings, so Invariant 1 is clearly satisfied. Furthermore, if there are multiple lines terminating along $v_0 v_1$ on the same side as ℓ then ℓ is inserted exactly between those lines that tend leftward and those lines that tend rightward with respect to ℓ. Hence all those lines are already on the correct side of ℓ and no line crossings are created; Invariant 2 is satisfied.

Next, we consider inserting ℓ into the order $<^{v_i}_{v_i v_{i+1}}$ for $i > 0$ such that Invariant 1 is satisfied. If one of the neighboring lines in the previous line order $<^{v_i}_{v_{i-1} v_i}$ also continues along $v_i v_{i+1}$, then ℓ simply keeps its position directly next to that line. Since the previous layout did not contain any invalid intra-vertex crossings and ℓ follows a previous line, Invariant 1 is still satisfied. This case is illustrated in Figure 3a, where the red line ℓ_1 follows the neighboring black line through the

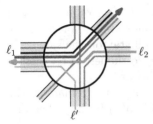
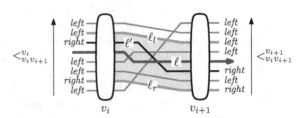

(a) Lines ℓ_1 and ℓ_2 are inserted so that Invariant 1 is maintained.

(b) Line ℓ is inserted into $<^{v_{i+1}}_{v_i v_{i+1}}$ so that Invariant 2 is maintained. The values side$(\ell, \cdot, v_i v_{i+1})$ are indicated for all lines.

Fig. 3. Insertion of lines into an existing partial line layout

vertex. Otherwise, if ℓ is the only line continuing along $v_i v_{i+1}$, we scan $<^{v_i}_{v_i v_{i+1}}$, starting with the smallest (rightmost) element, for the first line ℓ', whose previous edge wv_i is left of ℓ in the embedding of G or that terminates in v_i with a left terminus along $v_i v_{i+1}$. We insert ℓ immediately before ℓ' in $<^{v_i}_{v_i v_{i+1}}$. This is illustrated in Figure 3a by the blue line ℓ_2 which is inserted immediately before the yellow line ℓ'. If no line ℓ' is found then ℓ becomes the largest (leftmost) element in $<^{v_i}_{v_i v_{i+1}}$. The chosen position for ℓ ensures that $<^{v_i}_{v_i v_{i+1}}$ remains compatible with v_i and that Invariant 1 is satisfied.

It remains to determine the position of ℓ in the order $<^{v_{i+1}}_{v_i v_{i+1}}$. Figure 3b illustrates the situation. We scan the already determined line order $<^{v_i}_{v_i v_{i+1}}$ for the smallest (rightmost) line ℓ_l left of ℓ for which side$(\ell, \ell_l, v_i v_{i+1}) = left$ and for the largest (leftmost) line ℓ_r right of ℓ, for which side$(\ell, \ell_r, v_i v_{i+1}) = right$. Note that it is possible that one or both lines ℓ_l and ℓ_r do not exist. If they exist, these two lines ℓ_l and ℓ_r are the closest lines to ℓ that are already on the correct side. Since Invariant 2 holds for the previous partial layout, ℓ_l and ℓ_r do not cross each other along $v_i v_{i+1}$, that is, $\ell_r <^{v_i}_{v_i v_{i+1}} \ell_l$ and $\ell_r <^{v_{i+1}}_{v_i v_{i+1}} \ell_l$. Obviously, ℓ may not cross either of them and we must insert ℓ between ℓ_r and ℓ_l in $<^{v_{i+1}}_{v_i v_{i+1}}$ (otherwise Invariant 2 will be violated). More precisely, we insert ℓ immediately left of the largest (leftmost) line ℓ' in the interval $[\ell_r, \ell_l]$ of $<^{v_{i+1}}_{v_i v_{i+1}}$ for which side$(\ell, \ell', v_i v_{i+1}) = right$, see Figure 3b. If ℓ_r (ℓ_l) does not exist we symbolically assign $\ell_r = -\infty$ ($\ell_l = \infty$) so that the interval $[\ell_r, \ell_l]$ may become unbounded. The position of ℓ is determined as before. If there is no line ℓ' then ℓ becomes the rightmost line in $<^{v_{i+1}}_{v_i v_{i+1}}$.

We claim that in the assigned position ℓ crosses only lines that were to its left and tend to the right or lines that were to its right and tend to the left—crossings that are unavoidable. Assume to the contrary that ℓ crosses a line $\hat{\ell}$ that was to its left and also tends to the left. Since we insert ℓ immediately to the left of ℓ', the two lines $\hat{\ell}$ and ℓ' also cross each other. This is a contradiction to Invariant 2 for the previous partial layout, though, since $\hat{\ell}$ crosses ℓ' from left to right but eventually needs to cross ℓ' again from right to left in order to reach its leftward destination. If there is no line ℓ' then ℓ is the rightmost line in $<^{v_{i+1}}_{v_i v_{i+1}}$ by definition and cannot cross $\hat{\ell}$. Similarly, assume that ℓ crosses a line $\tilde{\ell}$ that was to its right and also tends to the right. Then $\ell_l <_{v_i v_{i+1}}^{v_{i+1}} \tilde{\ell}$ since otherwise we would have placed ℓ left of $\tilde{\ell}$ in the interval $[\ell_r, \ell_l]$. But this means that $\tilde{\ell}$ crosses ℓ_l from right to left, which again violates Invariant 2 for the previous partial layout: there must be a second crossing, where $\tilde{\ell}$ crosses ℓ_l from left to right in order to reach its rightward destination. If $\ell_l = \infty$ we would have placed ℓ left of $\tilde{\ell}$ which is also a contradiction. So Invariant 2 holds for the selected position of ℓ.

Finally, we show that Invariant 1 holds for the position of ℓ in $<_{v_i v_{i+1}}^{v_{i+1}}$. The first possibility for a violation is a line $\hat{\ell}$ with side$(\ell, \hat{\ell}, v_i v_{i+1}) = $ *left* that is still to the right of ℓ but does not continue further along $v_{i+1} v_{i+2}$. By definition $\hat{\ell}$ can only be right of ℓ if $\hat{\ell} <_{v_i v_{i+1}}^{v_{i+1}} \ell'$. But then Invariant 1 would have been violated before by $\hat{\ell}$ and ℓ'. The other possibility for a violation of Invariant 1 is a line $\tilde{\ell}$ with side$(\ell, \tilde{\ell}, v_i v_{i+1}) = $ *right* that is still to the left of ℓ but does not continue further along $v_{i+1} v_{i+2}$. By definition this can only be the case if $\ell_l <_{v_i v_{i+1}}^{v_{i+1}} \tilde{\ell}$. But this means that Invariant 1 would have been violated before by $\tilde{\ell}$ and ℓ'.

Since both invariants hold at the end of Algorithm 1, we have proven its correctness. By Invariant 1 all vertices are admissible, and by Invariant 2 the final line layout realizes exactly the unavoidable crossings and is thus crossing-minimal by Lemma 2. The running time of Phase 1 is $O(|\mathcal{L}| \cdot N_{\mathcal{L}})$. The running time of Phase 2 is again $O(|\mathcal{L}| \cdot N_{\mathcal{L}})$ since there are $2N_{\mathcal{L}}$ insertion operations, each of which determines a position for the current line by scanning the line orders of size $O(|\mathcal{L}|)$ of the current edge.   □

We note that the size of a solution for MLCM-PA is $\Omega(N_{\mathcal{L}})$ and thus the running time of our algorithm is only a factor of $|\mathcal{L}|$ away from the output size. Since for the total edge size $N_{\mathcal{L}}$ we have $N_{\mathcal{L}} \in O(|\mathcal{L}| \cdot |V|)$, the running time of Algorithm 1 can also be expressed as $O(|\mathcal{L}|^2 \cdot |V|)$.

By Lemma 1, we can reduce any instance of MLCM-T1 to an equivalent instance of MLCM-PA in linear time. We thus obtain the following corollary.

Corollary 1. *Given a metro graph* $(G, \mathcal{L})$ *in which the degree of any terminus* v *of any line in* $\mathcal{L}$ *equals 1, we can use Algorithm 1 to compute a crossing-minimal line layout in* $O(|\mathcal{L}| \cdot N_{\mathcal{L}})$ *time.*

5   Conclusions

In this paper we have presented a new algorithm that improves the best previous algorithms for both the MLCM-PA and the MLCM-T1 problem. The running time of the new algorithm is $O(|\mathcal{L}| \cdot N_{\mathcal{L}})$, where $N_{\mathcal{L}} \in O(|\mathcal{L}| \cdot |V|)$.

We conclude with two observations about practical MLCM instances as found, for example, in Ovenden's book [3]. First, the number of lines $|\mathcal{L}|$ in a transport network is usually much smaller than the size of the underlying graph G. Since the output size is already $\Omega(N_{\mathcal{L}})$, our algorithm runs in linear time if the number of lines is constant. Second, many lines in practice indeed terminate at degree-1 vertices of the underlying graph as modeled in the MLCM-T1 variant. Still, most networks also have some lines that start or end in non-leaf vertices. We therefore suggest to use the ILP formulation of Asquith et al. [1] (or a simple exhaustive-search algorithm) to determine an optimal terminus side assignment for those lines. We can then transform the original MLCM-P instance together with the additional terminus side assignments into an MLCM-PA instance that can be solved efficiently with our algorithm.

There are a few remaining open problems in MLCM. First of all, it is still an unsolved question whether the general MLCM problem (without periphery condition) is NP-hard for general plane graphs or even for paths. Another interesting open question is whether the NP-hard problem MLCM-P is fixed-parameter tractable for a suitable small parameter, such as the maximum multiplicity of the edges. Furthermore, no approximation algorithms for MLCM-P are known so far.

Acknowledgments. We thank Joachim Gudmundsson, Damian Merrick, and Thomas Wolle for initial discussions about the problem during a visit in Sydney.

References

1. Asquith, M., Gudmundsson, J., Merrick, D.: An ILP for the metro-line crossing problem. In: Harland, J., Manyem, P. (eds.) Proc. 14th Computing: The Australasian Theory Symp. (CATS 2008). CRPIT, vol. 77, pp. 49–56. Australian Comput. Soc. (2008)
2. Argyriou, E., Bekos, M.A., Kaufmann, M., Symvonis, A.: Two polynomial time algorithms for the metro-line crossing minimization problem. In: Tollis, I.G., Patrignani, M. (eds.) GD 2008. LNCS, vol. 5417, pp. 336–347. Springer, Heidelberg (2009)
3. Ovenden, M.: Metro Maps of the World. Capital Transport Publishing (2003)
4. Stott, J.M., Rodgers, P.: Metro map layout using multicriteria optimization. In: Proc. 8th Internat. Conf. Information Visualisation (IV 2004), pp. 355–362. IEEE, Los Alamitos (2004)
5. Hong, S.H., Merrick, D., do Nascimento, H.A.D.: Automatic visualization of metro maps. J. Visual Languages and Computing 17, 203–224 (2006)
6. Nöllenburg, M., Wolff, A.: A mixed-integer program for drawing high-quality metro maps. In: Healy, P., Nikolov, N.S. (eds.) GD 2005. LNCS, vol. 3843, pp. 321–333. Springer, Heidelberg (2006)
7. Hahn, W.C., Weinberg, R.A.: A subway map of cancer pathways (2002); Poster in Nature Reviews Cancer
8. Benkert, M., Nöllenburg, M., Uno, T., Wolff, A.: Minimizing intra-edge crossings in wiring diagrams and public transportation maps. In: Kaufmann, M., Wagner, D. (eds.) GD 2006. LNCS, vol. 4372, pp. 270–281. Springer, Heidelberg (2007)
9. Bekos, M.A., Kaufmann, M., Potika, K., Symvonis, A.: Line crossing minimization on metro maps. In: Hong, S.-H., Nishizeki, T., Quan, W. (eds.) GD 2007. LNCS, vol. 4875, pp. 231–242. Springer, Heidelberg (2008)

Layout with Circular and Other Non-linear Constraints Using Procrustes Projection

Tim Dwyer and George Robertson

Microsoft Research,
Redmond, USA
{timdwyer,ggr}@microsoft.com

Abstract. Recent work on constrained graph layout has involved projection of simple two-variable linear equality and inequality constraints in the context of majorization or gradient-projection based optimization. While useful classes of containment, alignment and rectangular non-overlap constraints could be built using this framework, a severe limitation was that the layout used an axis-separation approach such that all constraints had to be axis aligned. In this paper we use techniques from Procrustes Analysis to extend the gradient-projection approach to useful types of non-linear constraints. The constraints require subgraphs to be locally fixed into various geometries—such as circular cycles or local layout obtained by a combinatorial algorithm (e.g. orthogonal or layered-directed)—but then allow these sub-graph geometries to be integrated into a larger layout through translation, rotation and scaling.

1 Introduction

Our past work has explored methods for incorporating various types of constraints over node positions and edge routing into force-directed layout. A key component in achieving stable incremental constraint satisfaction in the context of such layout has been *gradient-projection* techniques. Optimization of a goal function subject to constraints using gradient projection involves finding a gradient related descent vector which is then *projected* against the constraints to obtain a descent vector that is feasible with respect to those constraints. Projection, as described in Section 3, involves solving a constrained least-squares problem.

Recent work has focused on interactive applications of such constraint-based layout. For example, a diagram authoring tool [11] and on-line exploration of large graphs [8]. To achieve interactive responsiveness in such applications the projection step needs to be efficient and for certain classes of constraints we have been able to find methods of projection that compare favourably in running time to the basic unconstrained layout. In [6] we gave a simple active-set algorithm for projection subject to *orthogonal ordering* constraints; i.e. a partial ordering of nodes in either the horizontal or vertical axes of the drawing. In [7] we gave an algorithm for more general *separation constraints*: linear equality or inequality constraints over pairs of either x- or y-position variables.

D. Eppstein and E.R. Gansner (Eds.): GD 2009, LNCS 5849, pp. 393–404, 2010.

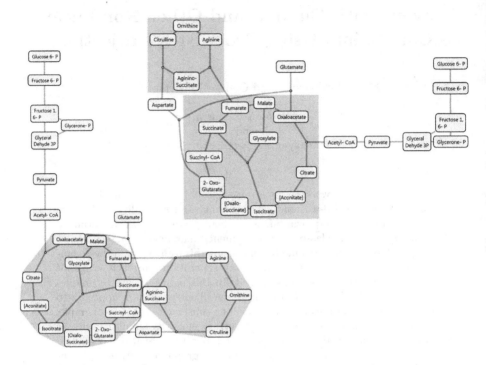

Fig. 1. A metabolic pathway network with two cycles arranged in two ways using differ-ent (user defined) constraints. In both cases Procrustes projection (see Section 4) is used to keep the cycles circular and groups are created around the two cyclic components. Constraints prevent members of these groups from overlapping with other parts of the graph. In the lower-left drawing the non-overlap constraint is based on the convex-hulls of the groups, projected apart as described in Section 3.2. The upper-right drawing is arranged with rectangular group boundaries using separation constraints (see Section 3.1). Various horizontal and vertical alignment constraints (using equality separation constraints) have been added interactively by the user to customize the layouts.

Most recently, following position-based dynamics approaches used success-fully in computer game animation, we showed that a simple class of nonlinear constraint could also be projected in a cyclical Gauss-Seidel scheme [5]. The con-straints were simple equalities or inequalities over Euclidean distance between pairs of nodes. Although simple, we were able to compose these constraints into more complex rigid structures. In particular we demonstrated wheel-like constructions to draw directed-graph cycles in a reorientable, but fixed radius circle. Such circular constraints are useful for achieving the kind of drawing con-ventions commonly seen, for example, in biology textbooks, for drawing cycles in metabolic pathways.

Although projecting cycles in this way was successful it led us to an investiga-tion to see whether a closed-form solution to the projection of such circular con-straints was possible. Also, we wanted circular constraints with variable as well as fixed radii. In this paper we show that the technique of *Procrustes analysis*—more commonly used by statisticians to fit experimental observations to a model—efficiently solves this exact problem and further more, can be used to obtain a projection of any rigid shape with minimal translation, rotation and scaling.

2   Related Work

A survey of graph-drawing literature—particularly regarding circular layout style—reveals a number of scenarios where the Procrustes projection described in this paper could provide a concrete improvement to either the quality of the drawings or the stability of the layout method.

Six and Tollis [19] give a multi-stage force-directed approach for layout of circular subgraphs in a non-circular arrangement of the larger graph. At first the subgraphs are replaced with single nodes and this abridged graph is arranged using a typical force-directed technique. Then a circular ordering of the subgraphs is found to minimize internal edge-crossings. The radius of each circle is fixed based on the number of nodes and an orientation is found by what sounds like a brute-force search. Finally, another relaxation step is applied using an ad-hoc local search method over node angles.

Becker and Rojas [2] discuss a technique for drawing the cycles in metabolic pathways as circles. They do not give many algorithmic details but the brief description of their two-stage force-directed approach suggests that it is similar in spirit to Six and Tollis.

Baur and Brandes [1] investigate techniques for circular ordering of nodes in subgraphs to minimize crossings between both edges internal to the subgraph, and edges linking the subgraph to other circular subgraphs in so called "Micro/Macro" graphs, i.e. graphs with one level of semantic grouping. They do not consider the problem of orienting the circular "micro" graphs in the context of the larger "macro" graph layout and in many of their examples it is clear that a little rotation of the circles would significantly reduce edge length.

Friedrich and Eades [14] give a complicated (and unproven) algebraic expression for finding an affine transformation of a graph to transition between different layouts such that squared displacement of the transformed graph from the target graph is minimized. This is exactly a Procrustes problem although Friedrich and Eades also allow shear transformations. Shearing is forbidden by the orthogonal Procrustes model described in Section 4 since shearing does not preserve the "shape" of the model and can collapse the dimensionality: e.g. transform a 2-d shape to a line [4, pg. 430]. In addition the Procrustes formulation that follows is easier to describe, implement and debug and does not suffer from potential singularities that may be a problem in the formulation in [14] (Friedrich and Eades do not explain how to handle zero value denominators in their expression).

3   Constraint Projection

A key ingredient to the constraint-based layout described in this paper is the idea of solving a *projection* problem. Projecting the variables $x = (x_1, \ldots, x_n)$

with starting or desired positions $d = (d_1, \ldots, d_n)$ against a set of constraints that define a feasible region S means finding the point x in S *closest to d*.

$$\arg\min_{x \in S} \sum_{i=1}^{n} (x_i - d_i)^2 \tag{1}$$

While we have in the past considered different ways to project against certain classes of constraints using specially developed solver techniques, this is the first paper where we have combined different projection methods for different classes of constraints in a single unifying framework. Before introducing the new type of Procrustes constraint projection in Section 4, we briefly review the two other types of constraint projection that will be used in combination.

3.1   Separation Constraint Projection

A separation constraint is an equality or inequality between a pair of (exclusively) horizontal or vertical node positions. For example, $u_x + g \leq v_x$ requires that nodes u and v be separated horizontally by at least g. In [7] and also [11] we give gradient projection techniques for layout using only such horizontal and vertical separation constraints. They are useful for many drawing conventions involving constraints that are aligned with the page or screen axes such as rectangular node and cluster non-overlap constraints, constraints requiring the end node of a directed edge be strictly above the start node, or for persistent horizontal or vertical alignments.

Efficient scan-line techniques for generating horizontal or vertical non-overlap constraints have been developed, see [9]. We also have fast techniques for finding the projection of all separation constraints in a given axis, see [7].

Although separation constraints are useful there are many drawing conventions requiring non-linear constraints, or linear constraints that are not axis aligned. In [10] and [12] we experimented with augmentation of the goal function to simulate other types of constraint. Simply adding terms to the goal function, however, does not provide the strict "rigidity" of real constraints. Increasing the weighting of such terms to reduce "stretchiness" usually overwhelms the underlying layout goal function or can lead to instability.

3.2   Euclidean Distance Projection

A Euclidean distance constraint of the form $|\mathbf{pq}| \geq d$ requires a minimum distance d between the positions of two nodes p and q. If such a constraint is violated the projection, i.e. feasible positions $\mathbf{p}'$ and $\mathbf{q}'$ that minimize the squared displacement from $\mathbf{p}$ and $\mathbf{q}$, are trivially computed as $\mathbf{p}' = \mathbf{p} - \frac{w_q}{w_p+w_q}\mathbf{r}$, $\mathbf{q}' = \mathbf{q} + \frac{w_p}{w_p+w_q}\mathbf{r}$ where $\mathbf{r} = |\mathbf{pq}|^{-1}(d - |\mathbf{pq}|)\mathbf{pq}$. The "weights" w_p and w_q for p and q are by default 1. However, for a constraint involving a cluster of n nodes it is useful to take the weight as n.

In [5] Euclidean distance constraints (including equality constraints) were the only type of constraint and the above calculation was the only type of projection used. Complex constraints like rigid circles were built with a wheel-like frame of

Euclidean distance equality constraints. The Procrustes projection technique described in Section 4 makes this usage redundant. However, this type of Euclidean distance projection is still useful in the framework described in Section 5, for preventing overlap between the convex hulls of node/cluster boundaries. That is, given two overlapping convex hulls we can minimally project them apart by chosing the displacement vector $\mathbf{r}$ (above) from the minimum penetration depth vector, computed from the Minkowski Difference of the two hulls, see Figure 2. The time to compute this vector is proportional to the sum of vertices in the two hulls. We use a binary space partition tree to quickly identify potentially overlapping hulls (rather than computing Minkowski Differences for all pairs). Figure 1 shows a graph with non-overlapping cluster boundaries projected apart using this technique.

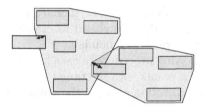

Fig. 2. To prevent overlap between convex hull cluster boundaries and nodes or other cluster boundaries we project apart overlapping boundaries using the minimum penetration depth vector

4   Procrustes Projection

Procrustes analysis is a technique for fitting an observed data configuration to an expected model using only linear transformations. Borg and Groenen [4] give a comprehensive overview and introduction to Procrustes methods, although the techniques have been known to statisticians since the 1950s. For a statistical technique, it is rather colourfully named after the character in Greek mythology of the same name. Procrustes was a keeper of an inn who "fit" his victims to an iron bed using drastic means.

The problem that we consider in this paper is projecting a set of n 2-d points X onto a target constrained configuration Y with a shape that is rigid but which can be scaled by a factor s, translated by a vector t or rotated by an orthogonal matrix T such that the sum of squared distances from the transformed Y to the original X is minimized. That is, we want to find s, t and T that minimize:

$$\sum_{i=1}^{n} (X_i - (sY_iT + t))^2 \tag{2}$$

subject to $T'T = I$, i.e. only orthogonal rotation.

The optimal translation vector t is optained by differentiating (2) with respect to t and setting the derivative equal to 0 (see [4]):

$$t = \frac{1}{n} \sum_{i=1}^{n} (X_i - sY_iT) \tag{3}$$

The optimal scale s is obtained similarly by substituting (3) for t in (2), differentiating with respect to s and setting this derivative to 0 giving:

$$s = \frac{\operatorname{tr} X'YT}{\operatorname{tr} Y'Y} \tag{4}$$

where tr is the matrix trace of the 2×2 result of the inner products. Note that this assumes that Y is centered on the origin (or has been centered by subtracting the barycenter of Y from all of its elements).

Substituting (3) and (4) into (2) we see that the optimal rotation T is invariant to scale or translation. Conveniently, it can be shown (see Appendix) that $T = QP'$, where P and Q are found from the singular value decomposition $X'Y = P\Phi Q'$, is exactly the optimal rotation. The singular value decomposition of the 2×2 matrix $X'Y$ can be obtained in closed form using the quadratic formula to find roots of the characteristic polynomial.

To summarize, the following procedure takes a matrix X of n points (i.e. node positions), a matrix Y of n points with the target configuration (centered on the origin), and returns the projection of X on the optimally transformed Y:

procedure *ProjectXonY(X, Y)*
   $C \leftarrow X'Y$
   $(P, \Phi, Q') \leftarrow SingularValueDecomposition(C)$
   $T \leftarrow QP'$
   $s \leftarrow (\operatorname{tr} CT)/(\operatorname{tr} Y'Y)$
   $t \leftarrow \frac{1}{n}\sum_{i=1}^{n}(X_i - sY_iT)$
return $sT'Y' + \mathbf{1}'t$

Procedure *ProjectXonY* runs in $O(n)$ time since the most expensive operation is computing the inner-product of $n \times 2$ matrices.

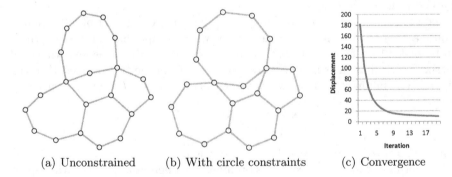

(a) Unconstrained   (b) With circle constraints   (c) Convergence

Fig. 3. Circular constraints applied using cyclical Gauss-Seidel Procrustes projection can even be interlocking. Provided a feasible solution exists, and the starting layout is reasonably untangled (e.g. the unconstrained layout on the left) cyclical projection rapidly converges (see Section 5.1).

4.1 Choosing the Target Configuration

The target configuration matrix Y can be any shape centered at the origin. For example, to require that n nodes be equally spaced in a given order around a circle we simply chose Y as the vector of n points $(y_1, \ldots, y_n)$ where $y_i = (\cos i\theta, \sin i\theta)$ taking $\theta = \frac{2\pi}{n}$. Figures 1, 3 and 6 show the results of using such circular constraints—in combination with other constraints—in the constrained layout scheme described in Section 5.

The target configuration can equally easily be the result of a complete layout algorithm applied to the subgraph. Figure 6(c) demonstrates this by taking the target configuration Y as the result of a layered layout algorithm applied to subgraphs with tree structure.

5 Combining Constraints in an Incremental Layout Step

The procedure *FeasibleLayoutStep* summarizes the operations in a single iteration of layout for a graph $G = (V, E)$, with nodes initially positioned horizontally and vertically at V_x and V_y respectively, a set C of Procrustes or any other constraints that we know how to project and horizontal and vertical separation constraints C_h and C_v respectively. The last parameter α controls the size of the unconstrained descent step, see below.

procedure *FeasibleLayoutStep*$(V, E, C, C_h, C_v, \alpha)$
   $D \leftarrow ComputeDescentDirection(V, E, \alpha)$
   $d \leftarrow D - (V_x, V_y)$
   $\bar{D} \leftarrow ProjectDesiredPositions(C, D)$
   $C_h' \leftarrow C_h \cup GenerateHorizontalNonOverlapConstraints(V_x, V_y)$
   $x \leftarrow Project(C_h', \bar{D}_x)$
   $C_v' \leftarrow C_v \cup GenerateVerticalNonOverlapConstraints(x, V_y)$
   $y \leftarrow Project(C_v', \bar{D}_y)$
return $(x, y), |d|$

This procedure returns new positions (x, y) which improve the layout (depending on the quality of the result of *ComputeDescentDirection*), which are *strictly* feasible with respect to the separation constraints C_h, C_v and generated non-overlap constraints, and which are *close to feasible* with respect to the other constraints C. We discuss exactly what we mean by *close to feasible* in Section 5.1. We also return the size of the unconstrained gradient-descent step d. This is useful in heuristics for determining appropriate step-size α. We have had success using the adaptive *trust-region* step-size selection method proposed by Hu [15]. Though more costly, optimum step-size selection or Armijo Rules [3] could also be used to guarantee strict improvement as in [8].

The procedure *ComputeDescentDirection* returns updated positions for the nodes V after taking a gradient-related step, with size controlled by α to reduce a layout cost function. This could equally well be an unconstrained iteration of the *p-stress* minimization method described in [8] or an iteration of any "force"-based approach. In our experiments we use a Fast-Multipole method following Lauther [17] so that *ComputeDescentDirection* completes in $O(|V| \log |V| + |E|)$ time.

Fig. 4. A mesh graph with 576 nodes and eight circle constraints, used in timing and convergence tests

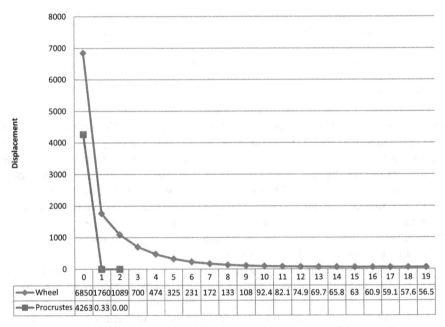

	0	1	2	3	4	5	6	7	8	9	10	11	12	13	14	15	16	17	18	19
◆—Wheel	6850	1760	1089	700	474	325	231	172	133	108	92.4	82.1	74.9	69.7	65.8	63	60.9	59.1	57.6	56.5
■—Procrustes	4263	0.33	0.00																	

Fig. 5. Total node displacement (the units are roughly screen pixels) versus iteration of constraint projection for the graph in Figure 4 using either Procrustes circle constraints or a wheel of Euclidean distance constraints

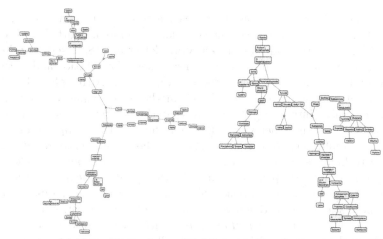

(a) Unconstrained

(b) Mixing a circular constraint with axis-aligned separation constraints to prevent overlap between nodes and to require directed edges to point downwards.

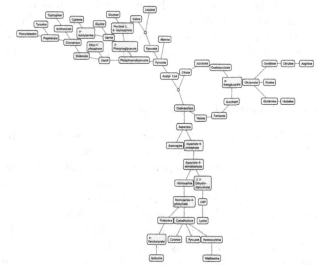

(c) In addition to the Procrustes circle constraint for the nodes involved in the cycle, this example shows the three subtrees constrained to layered configurations obtained with a Sugiyama algorithm, i.e. the local layout is used as the target configuration for Procrustes projection, which determines the optimal scale and rotation. Note, the subtrees could just as easily be DAGs or undirected subgraphs arranged with another algorithm, e.g. orthogonal layout.

Fig. 6. The citrate cycle metabolic pathway, arranged using various constraints

5.1   Gauss-Seidel Gradient Projection

For solving systems of linear equations, an iterative method of updating one variable at a time to satisfy one or more of the equations is commonly attributed to Gauss-Seidel. Jakobsen [16] and more recently Müller et al. [18] describe techniques using iterative constraint projection for rigid skeletal animation and cloth simulation in computer games, as "Gauss-Seidel" approaches. Although we know of no formal proof for the convergence of such methods our experiments with simple two-node constraints (see [5]) indicate that they work well in practical layout applications. In this paper we explore, for the first time, application of this approach to combining projection of different classes of constraints using different solver techniques. That is, whereas in [5] we considered only Euclidean distance constraints between pairs of nodes, in this paper we combine these with Procrustes projection and separation constraint projection giving us faster convergence, more stable interactive layout, and more flexible constraints.

Thus, the procedure *ProjectDesiredPositions* returns new positions for nodes by—starting from the desired positions D—cyclically projecting each constraint in C (note that C also conceptually includes any convex-hull non-overlap requirement, although the precise Euclidean projection operations are determined dynamically as per Section 3.2). Figure 5 shows a comparison of total displacement of nodes for the large example in Figure 4 with circular constraints using either wheel-like meshes of Euclidean distance constraints (see [5]) projected as described in Section 3.2 or Procrustes projection using the procedure *ProjectXonY* as in Section 4. Clearly, far fewer iterations are required for Procrustes projection.

5.2   Separation Constraint Projection

The final steps of *ComputeDescentDirection* apply axis-aligned separation constraints. The *GenerateHorizontalNonOverlapConstraints* uses the scan-line algorithm described in [9] to generate separation constraints to resolve horizontal overlap between rectangular node and cluster boundaries horizontally. Note that it uses the *starting* configuration V_x, V_y rather than the output of *ProjectDesiredPositions*. This is because, if the input is already feasible (i.e. not overlapping) the relative left-to-right arrangement of nodes should be preserved (unless nodes have, in the interim, moved vertically so that they can no longer potentially overlap horizontally). In practice we have found that this makes continuous layout while the user directly manipulates (drags) nodes much smoother and less surprising.

The next call to *Project* invokes the separation constraint solver [9] to place nodes horizontally as close as possible to the desired positions $\bar{D}$ subject to the generated and user-defined separation constraints. Next, vertical non-overlap constraints are generated based on the newly computed feasible horizontal positions x and the previous vertical positions V_y (again to preserve any applicable previous vertical ordering). Finally, *Project* is called again to determine feasible vertical positions y.

Applying separation constraints last means that they are always satisfied, while the Procrustes and Euclidean distance constraints projected cyclically by *ProjectDesiredPositions* may be slightly violated. This works well as any violation of the axis aligned and rectangular non-overlap constraints tends to be more noticeable than for the other types of constraints. Still, since the whole *FeasibleLayoutStep* procedure is applied many times inside a larger layout loop, all constraints tend to be resolved after a few iterations.

6 Discussion, Conclusion, Further Work

Figures 1 and 6 give practical examples of how the various types of constraints we have described can be applied in practice. The Procrustes constraints are very fast compared to the overall layout process: Figure 4 with 576 nodes, 1104 edges and 8 circle constraints took (on a 2.1Ghz PC) 1.86 seconds total layout time with about 0.01 seconds spent in projection operations due to the convergence criteria described in Section 5.1. Further work should be done to time much larger, pathological examples to really explore the convergence properties of cyclical constraint projection. Static layout of all the other smaller examples in this paper takes a fraction of a second. The real benefit of fast constraint layout, however, is in supporting incremental layout scenarios. All of the examples in this paper were produced in an interactive system where users can directly manipulate nodes and edit the constraints, getting immediate feedback from "rigid" constraint structures.

In addition to efficiency the Procrustes projection presented in this paper allows for variable radii circles enabling interlocking constraints as in Figure 3. Furthermore, they can be applied to obtain scaling and rigid rotation of any initial arrangement of nodes such as layout from a different algorithm, see Figure 6(c). The other contribution of this paper is to show that these and other types of constraints can be combined through cyclical projection as described in Section 5.1.

Detecting satisfiability of constraints, and where satisfiable, finding a feasible starting configuration require much more research. One imperfect strategy is to detect if error does not significantly decrease inside the cyclical constraint satisfaction loop. Once unsatisfiable constraints have been detected, communicating this to the user in a way allows them to resolve the unsatisfiability is also a challenge. In our rudimentary interactive test systems, where constraints are added incrementally by the user, we have found the most useful strategy has been a simple undo facility to remove the most recently added constraint. Of course users unfamiliar with constraint layout would need an intuitive interface that prevents unsatisfiable constraints from being created at all.

Acknowledgements. Thanks to Lev Nachmanson and Ted Hart for providing various pieces used in our layout software.

References

1. Baur, M., Brandes, U.: Multi-circular layout of micro/macro graphs. In: Hong, S.-H., Nishizeki, T., Quan, W. (eds.) GD 2007. LNCS, vol. 4875, pp. 255–267. Springer, Heidelberg (2008)

2. Becker, M.Y., Rojas, I.: A graph layout algorithm for drawing metabolic pathways. Bioinformatics 17(5), 461–467 (2001)
3. Bertsekas, D.P.: Nonlinear Programming. Athena Scientific, Belmont (1999)
4. Borg, I., Groenen, P.J.F.: Modern Multidimensional Scaling: Theory and Applications, 2nd edn. Springer, Heidelberg (2005)
5. Dwyer, T.: Scalable, versatile and simple constrained graph layout. In: Proc. Eurographics/IEEE-VGTC Symp. on Visualization (Eurovis 2009). IEEE, Los Alamitos (2009) (to appear)
6. Dwyer, T., Koren, Y., Marriott, K.: Drawing directed graphs using quadratic programming. IEEE Transactions on Visualization and Computer Graphics 12(4), 536–548 (2006)
7. Dwyer, T., Koren, Y., Marriott, K.: IPSep-CoLa: an incremental procedure for separation constraint layout of graphs. IEEE Transactions on Visualization and Computer Graphics 12(5), 821–828 (2006)
8. Dwyer, T., Marriott, K., Schreiber, F., Stuckey, P.J., Woodward, M., Wybrow, M.: Exploration of networks using overview+detail with constraint-based cooperative layout. IEEE Transactions on Visualization and Computer Graphics 14(6), 1293–1300 (2008)
9. Dwyer, T., Marriott, K., Stuckey, P.: Fast node overlap removal. In: Healy, P., Nikolov, N.S. (eds.) GD 2005. LNCS, vol. 3843, pp. 153–164. Springer, Heidelberg (2006)
10. Dwyer, T., Marriott, K., Wybrow, M.: Integrating edge routing into force-directed layout. In: Kaufmann, M., Wagner, D. (eds.) GD 2006. LNCS, vol. 4372, pp. 8–19. Springer, Heidelberg (2007)
11. Dwyer, T., Marriott, K., Wybrow, M.: Dunnart: A constraint-based network diagram authoring tool. In: Tollis, I.G., Patrignani, M. (eds.) GD 2008. LNCS, vol. 5417, pp. 420–431. Springer, Heidelberg (2009)
12. Dwyer, T., Marriott, K., Wybrow, M.: Topology preserving constrained graph layout. In: Tollis, I.G., Patrignani, M. (eds.) GD 2008. LNCS, vol. 5417, pp. 230–241. Springer, Heidelberg (2009)
13. Everson, R.: Orthogonal, but not orthonormal, procrustes problems. Advances in Computational Mathematics (submitted) (1998),
    http://secamlocal.ex.ac.uk/people/staff/reverson/uploads/Site/procrustes.pdf
14. Friedrich, C., Eades, P.: Graph drawing in motion. Graph Algorithms and Applications 6(3), 353–370 (2002)
15. Hu, Y.: Efficient and high quality force-directed graph drawing. The Mathematica Journal 10(1), 37–71 (2005)
16. Jakobsen, T.: Advanced character physics. In: San Jose Games Developers' Conference (2001),
    http://www.gamasutra.com/resource_guide/20030121/jacobson_01.shtml
17. Lauther, U.: Multipole-based force approximation revisited - a simple but fast implementation using a dynamized enclosing-circle-enhanced k-d-tree. In: Kaufmann, M., Wagner, D. (eds.) GD 2006. LNCS, vol. 4372, pp. 20–29. Springer, Heidelberg (2007)
18. Müller, M., Heidelberger, B., Hennix, M., Ratcliff, J.: Position based dynamics. In: Proc. of Virtual Reality Interactions and Physical Simulations (VRIPhys), pp. 71–80 (2006)
19. Six, J.M., Tollis, I.G.: A framework for user-grouped circular drawings. In: Liotta, G. (ed.) GD 2003. LNCS, vol. 2912, pp. 135–146. Springer, Heidelberg (2004)

GMap: Drawing Graphs as Maps

Emden R. Gansner, Yifan Hu, and Stephen G. Kobourov

AT&T Labs - Research, Florham Park, NJ USA
{erg,yifanhu,skobourov}@research.att.com

1 Introduction

In graph drawing, vertices are typically represented as points in two or three dimensional space and edges are represented as lines between the corresponding vertices. Other representations have also been considered. For example, treemaps use a recursive space filling approach to represent trees. There is also a large body of work on representing planar graphs as contact graphs of geometrical objects. GMap is an algorithm that represents general graphs as maps [2]. Our overall goal is to create a representation which makes the underlying data easy to understand and visually appealing. Our map representation is especially effective when the underlying graph contains structural information such as clusters and/or hierarchy. The traditional point-and-line representation of graphs can require considerable effort to comprehend, and often puts off general users. On the other hand, a map representation is more intuitive, as people are very familiar with maps and even enjoy carefully examining maps.

2 The Mapping and Coloring Algorithm

The first step in our GMap algorithm is to embed the graph in the plane. In our implementation we use a scalable force directed algorithm [3]. The second step is a cluster analysis of the underlying graph or the embedded pointset. Here we use modularity based clustering [4] as it is a good fit [5] for the force directed algorithm we employ. In the third step the embedding and the clustering are used to create the map. A Voronoi diagram of the vertices is generated. To create "European-style" borders we use the vertex sets in each cluster together with some random points and generate "form fitting" outer boundaries. Vertex weights are used to determine the font size of the vertex label, and the size of the label is used to create the area in the map that corresponds to the vertex. We then merge Vononoi cells that belong to the same cluster, thus forming regions of complicated shapes. The overall algorithm has complexity $O(|V| \log |V|)$ and easily scales to graphs with tens of thousands of vertices.

Because countries in GMap are not necessarily contiguous, we need as many colors as the total number of countries, in order to make each country uniquely identifiable by its color. We use a two-step heuristic color assignment algorithm to ensure that neighboring countries are colored with as distinctive colors as possible. In the first step we apply a spectral algorithm to the *country graph* that

D. Eppstein and E.R. Gansner (Eds.): GD 2009, LNCS 5849, pp. 405–407, 2010.

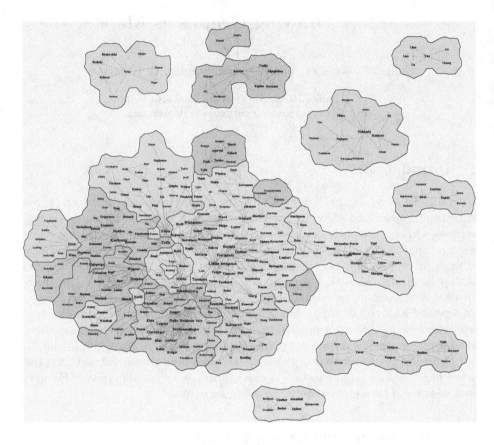

Fig. 1. Author collaboration map for the GD conference, 1994-2004

captures the neighboring structure of countries. In the second step we apply a greedy color-swapping algorithm, with the objective of maximizing the minimal color difference between any two neighboring countries. For further details see [2].

3   GMap Example

We consider the collaboration graph of authors with publications in the Symposium on Graph Drawing in the period 1994 to 2004. Authors are vertices and there is an edge between two authors if they have collaborated on at least one paper in this period. The font size of the vertices is proportional to the number of papers by that author; see Fig. 1. It is easy to see that European authors dominate the main continent. Several well-defined German groups can be seen on the west and southwest coasts. A largely Italian cluster occupies the center, with an adjacent Spanish peninsula in the east. The northwest contains a mostly Australasian cluster. Two North American clusters are to be found in the southeast and in the southwest, the latter one made up of three distinct components.

A combinatorial geometry cluster forms the northernmost point of the main continent. Most Canadian researchers can be found in the central Italian cluster and the Spanish peninsula. Northeast of the mainland lies a large Japanese island and southeast of the mainland there is a large Czech island. Northwest of the mainland is *Crossing Number* island.

GMap has been used to visualize data from several different areas [2, 1], including similarities between musicians (last.fm data) and books (Amazon.com data).

References

1. http://www.research.att.com/~yifanhu/gmap/
2. Gansner, E.R., Hu, Y.F., Kobourov, S.G.: Gmap: Drawing graphs as maps (2009), http://arxiv1.library.cornell.edu/abs/0907.2585v1
3. Hu, Y.F.: Efficient and high quality force-directed graph drawing. Mathematica Journal 10, 37–71 (2005)
4. Newman, M.E.J.: Modularity and community structure in networks. Proc. Natl. Acad. Sci. USA 103, 8577–8582 (2006)
5. Noack, A.: Modularity clustering is force-directed layout. Physical Review E (Statistical, Nonlinear, and Soft Matter Physics) 79 (2009)

Using High Dimensions to Compare Drawings of Graphs*

Stina Bridgeman

Department of Mathematics and Computer Science
Hobart and William Smith Colleges, Geneva, NY 14456
bridgeman@hws.edu

1 Introduction

We describe a simple visualization technique which allows a user to quickly assess the overall similarity of drawings of similar graphs, and to easily find regions of stability and of change. It can be used to simultaneously compare any number of drawings, and does not require that the layouts be adjusted to minimize changes.

The underlying idea — grouping vertices whose relative positions undergo little change — is shared by the animation technique of Friedrich and Houle [1], but both our approach to identifying such groups (visually via planes in high dimensions) and our application (comparing drawings) are quite different.

2 Algorithm

Let D_1 and D_2 be 2D drawings of some graph G, and let (x_{v_i}, y_{v_i}) be the co-ordinates assigned to vertex v in drawing i. Create a 4D drawing D of G in which each vertex is associated with the 4-tuple $(x_{v_1}, y_{v_1}, x_{v_2}, y_{v_2})$, then project that 4D drawing into 3D for viewing. Groups of vertices whose relative positions are maintained under translation, rotation, and/or scaling appear on the same plane in both the 4D drawing D and the 3D projection, with different planes indicating different combinations of transformations. See Figure 1.

The algorithm can be extended to include edges by evenly spacing a number of points along each edge and creating 4D "edge points" in the same manner as vertex points. Changes in the graph structure can be accommodated by assigning appropriate coordinates to the missing points. In addition, k drawings can be compared simultaneously by building a $2k$-tuple for each point from the coordinates in each drawing, then projecting to 3D for viewing.

3 Applications

In dynamic graph drawing and layout adjustment, the user must frequently orient herself to a new drawing of the same or nearly the same graph. The "high-dimensional comparison" technique of section 2 can be used to provide visual

---

* Some of this work was completed while the author was a visiting researcher at the National ICT Australia (ATP Sydney) in 2007.

D. Eppstein and E.R. Gansner (Eds.): GD 2009, LNCS 5849, pp. 408–410, 2010.

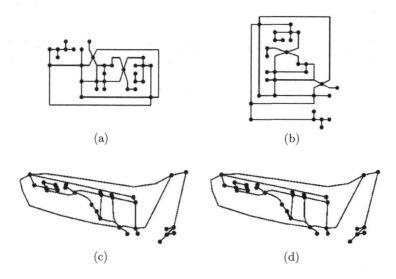

(a)   (b)

(c)   (d)

Fig. 1. D_1 and D_2 are shown in (a) and (b). Finding similar regions in these drawings requires repeatedly looking back and forth between the drawings and a lot of mental gymnastics. Figures (c) and (d) are a stereo pair showing a 3D projection of D — defocus your eyes until (c) and (d) merge into a third picture. (The visualization is most effective when the user can also manipulate the viewing angle to enhance the 3D sense.) Flat areas indicating similar regions can be identified immediately. Brushing can be used in an interactive viewer to match flat regions to the original drawings.

cues to aid the reorientation process, and is particularly useful when different parts of the layout have undergone different transformations (see Figure 1).

Time-varying graphs are often drawn by creating a separate layout for each time step, and either animating the transitions between time steps or stacking the layouts to create a 2.5D drawing. The success of these techniques requires adjusting consecutive layouts to reduce changes between time steps, which precludes the use of drawing algorithms designed to display particular structural elements of the graph (such as clusters). High-dimensional comparison provides a new way to visualize the change and stability between layers and over a series of time steps, and does not require adjustment of consecutive layouts.

In addition, the fit of the high-dimensional points to a single plane provides a potential similarity measure for comparing drawings and has the novel feature of being suitable for comparing more than two drawings at once.

Finally, several other applications are based on automatically clustering high-dimensional points which lie on or near the same plane (e.g. using the techniques of Friedrich and Houle [1] or Tseng [2]). Clustering information can be used to color-code similar groups in the D_i drawings, to produce a meta-visualization of how cluster position and membership changes over a series of drawings, or to define similarity measures based on common features rather than individual nodes.

References

1. Friedrich, C., Houle, M.E.: Graph drawing in motion II. In: Mutzel, P., Jünger, M., Leipert, S. (eds.) GD 2001. LNCS, vol. 2265, pp. 220–231. Springer, Heidelberg (2002)
2. Tseng, P.: Nearest q-flat to m points. J. Optim. Theory Appl. 105(1), 249–252 (2000)

On ρ-Constrained Upward Topological Book Embeddings

Tamara Mchedlidze and Antonios Symvonis

Dept. of Mathematics, National Technical University of Athens, Athens, Greece
{mchet,symvonis}@math.ntua.gr

1 Introduction

Giordano, Liotta and Whitesides [1] developed an algorithm that, given an embedded planar st-digraph and a topological numbering ρ of its vertices, computes in $O(n^2)$ time a ρ-constrained upward topological book embedding with at most $2n-4$ spine crossings per edge. The number of spine crossings per edge is asymptotically worst case optimal.

In this poster, we present improved results with respect to the number of spine crossings per edge and the time required to compute the book embedding. Firstly, for any embedded planar st-digraph G and any topological numbering ρ of its vertices, there exists a ρ-constrained upward topological book embedding with at most $n-3$ spine crossings per edge and, moreover, $n-3$ spine crossing per edge are required for some graphs. In this result, we allow edge (s,t) to be internal in the embedding of the graph. If edge (s,t) is always on the external face, the corresponding number of spine crossings reduces to at most $n-4$ and is worst case optimal. Secondly, a ρ-contrained upward topological book embedding with minimum number of spine crossings and at most $n-3$ spine crossings per edge can be computed by an output sensitive algorithm in $O(\alpha+n)$ time, where α is the total number of spine crossings.

2 Results

The problem of *Acyclic HP-completion with crossing minimization problem* (for short, *Acyclic-HPCCM*) was defined in [2] as follows: Given an embedded upward planar st-digraph $G = (V,E)$ and a non-negative integer c, the Acyclic-HPCCM problem asks whether there exists a *Hamiltonian Path Completion* set (for short, *HP-completion* set) E_c and a drawing $\Gamma(G')$ of graph $G' = (V, E \cup E_c)$ such that (i) G' remains acyclic and has a hamiltonian path from vertex s to vertex t, (ii) $\Gamma(G')$ preserves the drawing of the embedded planar graph G and, (iii) $\Gamma(G')$ has at most c edge crossings. In Theorem 2 of [2], the equivalence between the Acyclic-HPCCM problem and the problem of determining an upward topological book embedding with minimum number of spine crossings was established. Moreover, in the proof of the theorem it was shown that if we are given an acyclic HP-completion set E_c and an embedding of its edges on the original drawing of G causing c edge crossings, then there is an upward topological

D. Eppstein and E.R. Gansner (Eds.): GD 2009, LNCS 5849, pp. 411–412, 2010.

book embedding of G with c spine crossings and vice versa. In the construction supporting the proof, the crossings of the original edges of G with the edges of the HP-completion set are exactly the crossings of the original edges with the spine in the constructed upward topological book embedding. In the following theorems, we assume that the edges of E_c do not cross each other and that each pair of edges in the HP-completed drawing cross at most once.

Theorem 1. *Let G be an embedded planar st-digraph and E_c be an acyclic HP-completion set of G. Then, there exists a unique upward drawing $\Gamma(G')$ of $G' = (V, E \cup E_c)$ that respects the original embedding. If edge (s,t) is not on the external face of G, $\Gamma(G')$ has at most $n - 3$ crossings per edge, otherwise it has at most $n - 4$ crossings per edge. Moreover, there exist embedded graphs and acyclic HP-completions sets that require so many edge crossings.*

Theorem 2. *Let G be an embedded planar st-digraph, E_c be an acyclic HP-completion set of G and P be the implied Hamiltonian path. Then, the unique upward drawing $\Gamma(G')$ of $G' = (V, E \cup E_c)$ that respects the original embedding can be computed in $O(n+\alpha)$ time, where α is the total number of edge crossings.*

Let $\rho = (s = v_1, v_2, \ldots, v_n = t)$ be a topological ordering of the vertices of G. Then, observe that the edge set $E^\rho = \{(v_i, v_{i+1}) \mid (v_i, v_{i+1}) \notin E, 1 \leq i < n\}$ is an HP-completion set for G. Based on this fact, Theorems 1 and 2 and the equivalence between the Acyclic-HPCCM and the book embedding problems established in [2], we can state the following theorems:

Theorem 3. *Let G be an embedded planar st-digraph and ρ be a topological numbering of G. Then, G admits a unique ρ-constrained upward topological book embedding with at most $n-3$ spine crossings per edge for the case where edge (s,t) is not on the external face, otherwise it admits a embedding with at most $n - 4$ spine crossings per edge. Moreover, there exist embedded graphs and topological orderings that require so many spine crossings.*

Theorem 4. *Let G be an embedded planar st-digraph and ρ be a topological numbering of G. Then, the unique ρ-constrained upward topological book embedding of G (with a minimum number of spine crossings) can be computed in $O(n + \alpha)$ time, where α is the total number of spine crossings.*

We note that the improved results on ρ-constrained upward topological book embeddings can be used to also improve results presented in [1] regarding upward point set embeddability.

References

1. Giordano, F., Liotta, G., Whitesides, S.H.: Embeddability Problems for Upward Planar Digraphs. In: Tollis, I.G., Patrignani, M. (eds.) GD 2008. LNCS, vol. 5417, pp. 242–253. Springer, Heidelberg (2009)
2. Mchedlidze, T., Symvonis, A.: Crossing-optimal acyclic hamiltonian path completion and its application to upward topological book embeddings. In: Das, S., Uehara, R. (eds.) WALCOM 2009. LNCS, vol. 5431, pp. 250–261. Springer, Heidelberg (2009)

4-Labelings and Grid Embeddings of Plane Quadrangulations

Lali Barrière* and Clemens Huemer**

Departament de Matemàtica Aplicada IV, Universitat Politècnica de Catalunya
{lali,clemens}@ma4.upc.edu

Finding aesthetic drawings of planar graphs is a main issue in graph drawing. Of special interest are *rectangle of influence drawings*.The graphs considered here are quadrangulations, that is, planar graphs all whose faces have degree four. We show that each quadrangulation on n vertices has a closed rectangle of influence drawing on the $(n-2) \times (n-2)$ grid. Biedl, Bretscher and Meijer [2] proved that every planar graph on n vertices without separating triangle has a closed rectangle of influence drawing on the $(n-1) \times (n-1)$ grid.Our method, which is completely different from that of [2], is in analogy to Schnyder's algorithm for embedding triangulations on an integer grid [9] and gives a simple algorithm.

Schnyder [9] showed that labeling the angles of a triangulation T with 3 colors, with special rules, gives a 3-coloring and 2-orientation of the edges of T such that the edges of each color form a directed tree. For each interior vertex of T, the three colored paths to the sinks of the respective trees divide T into three regions. Counting the number of faces in each region gives the coordinates of the interior vertex in the grid drawing. Felsner [3] extended this result to the class of 3-connected plane graphs. In [8] it was studied to adapt this method to quadrangulations. In this case, the angles of a quadrangulation Q can be colored with 2 colors, which gives an analogous 2-coloring and 2-orientation of the edges of Q such that the edges of each color form a directed tree, and for each interior vertex the two colored paths to the respective sinks divide Q into two regions. In [5] it is shown that counting the number of faces in a region of an interior vertex v of Q gives the coordinate of v in a book embedding of Q with two pages. Each page in this book embedding for Q contains one of the two trees. Book embeddings of quadrangulations were also found in [6]. Whether this approach also gives a grid embedding for quadrangulations remained open.

We show here that labeling the angles of Q with 4 colors instead of 2 (which gives a 4-coloring and 2-orientation of the edges) allows to obtain a pair of book embeddings of Q such that the coordinates of a vertex v in the two book embeddings are the coordinates of v in the grid drawing of Q. It turns out that this embedding is a closed rectangle of influence drawing. It has the further property that edges of different colors are oriented in different directions (northeast, south-east, south-west, north-west). As a by-product of the rectangle of influence drawing, we also obtain a grid drawing of a quadrangulation on an

---

* Research supported by project MTM2008-06620-C03-01/MTM.
** Research supported by projects DGR 2009SGR-1040 and MEC MTM2009-07242.

D. Eppstein and E.R. Gansner (Eds.): GD 2009, LNCS 5849, pp. 413–414, 2010.

$\lceil \frac{n}{2} \rceil \times \lceil \frac{3n}{4} \rceil$ grid by simple scaling. This is not optimal, because quadrangulations on n vertices have a straight-line embedding on an $(\lceil \frac{n}{2} \rceil - 1) \times \lfloor \frac{n}{2} \rfloor$ grid. However, the known algorithms by Biedl and Brandenburg [1] and Fusy [7], both require to add edges to make the quadrangulation 4-connected. An advantage of our simple algorithm is that it does not need to add edges and also works for quadrangulations with connectivity 2.

Quadrangulations Q are known to admit a touching segment representation: de Fraysseix, de Mendez and Pach [6] showed that one can assign vertical segments and horizontal segments to the vertices of Q such that two segments touch if and only if the two corresponding vertices of Q are adjacent. A different proof of this result, based on book embeddings of Q, is by Felsner et al. [4], who provided a bijection between the two trees of book embeddings of quadrangulations and rectangulations of a diagonal point set. The 4-labeling of a quadrangulation Q gives two book embeddings and therefore two rectangulations by [4]. This pair of rectangulations has the further nice property that in each rectangulation the boxes correspond isomorphically to the faces of Q (that is, the dual graphs are isomorphic), both rectangulations have the same fixed outer face, and each segment intersects the line with slope 1 in one rectangulation and intersects the line with slope -1 in the other one.

This work builds upon previous results on binary labelings of quadrangulations from [5,8]. The novelty is the use of four colors instead of two and its application to the grid drawing. 4-labelings are in bijection with binary labelings from [5]. Using four colors allows us to get more insight into the combinatorial structure of quadrangulations.

References

1. Biedl, T., Brandenburg, F.: Drawing planar bipartite graphs with small area. In: Proceedings of the 17th Canadian Conference on Computational Geometry, Windsor, Canada, pp. 105–108 (2005)
2. Biedl, T., Bretscher, A., Meijer, H.: Rectangle of Influence Drawings of Graphs without Filled 3-Cycles. In: Kratochvíl, J. (ed.) GD 1999. LNCS, vol. 1731, pp. 359–368. Springer, Heidelberg (1999)
3. Felsner, S.: Convex Drawings of Planar Graphs and the Order Dimension of 3-Polytopes. Order 18, 19–37 (2001)
4. Felsner, S., Fusy, E., Noy, M., Orden, D.: Bijections for Baxter Families and Related Objects (2008), http://arxiv.org/abs/0803.1546
5. Felsner, S., Huemer, C., Kappes, S., Orden, D.: Binary Labelings for Plane Quadrangulations and their Relatives (2008), http://arxiv.org/abs/math/0612021v3
6. de Fraysseix, H., de Mendez, P.O., Pach, J.: A left-first search algorithm for planar graphs. Discrete and Computational Geometry 13, 459–468 (1995)
7. Fusy, E.: Straight-line drawing of quadrangulations. In: Kaufmann, M., Wagner, D. (eds.) GD 2006. LNCS, vol. 4372, pp. 234–239. Springer, Heidelberg (2007)
8. Huemer, C., Kappes, S.: A binary labelling for plane Laman graphs and quadrangulations. In: Proceedings of the 22nd European Workshop on Computational Geometry, Delphi, Greece, pp. 83–86 (2006)
9. Schnyder, W.: Embedding Planar Graphs on the Grid. In: Proceedings of the first annual ACM-SIAM Symposium on Discrete Algorithms, pp. 138–148 (1990)

IBM ILOG Graph Layout for Eclipse

Jerome Joubert, Stephane Lizeray, Romain Raugi, and Georg Sander

IBM ILOG Visualization Group
{j.joubert,lizeray,romain.raugi}@fr.ibm.com, georg.sander@de.ibm.com

1 Introduction

Eclipse is becoming increasingly popular within the Java developers' community, and with the availability of RCP (Rich Client Platform), Eclipse is also seen as a very attractive framework for building professional stand-alone applications. For creating visual interfaces and diagram displays, Eclipse provides the Graphical Modeling Framework (GMF) and the Graphical Editing Framework (GEF). Both frameworks provide only very simple support for automatic graph layout natively but lack professional quality layout capabilities. IBM ILOG JViews Graph Layout for Eclipse fills this gap. It brings a concrete solution to developers who need to produce professional diagram visualization and modeling on Eclipse. It contains a wide collection of layout algorithms and configuration services, but also interacts well with GEF and GMF. The loose coupling of the application architecture and the graph layout provides last-minute integration capabilities giving a chance to incorporate layout services very late in the development process.

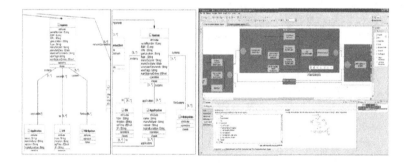

Fig. 1. Left: UML Diagram, right: Eclipse Application made with IBM ILOG Graph Layout for Eclipse

2 Highlights

IBM ILOG JViews Graph Layout for Eclipse provides advanced algorithms to automatically arrange diagrams so that they are readable by a human being. There are different types of algorithm. Some optimize node and link placement, some route links in order to minimize crossings, some compute text label placement with minimal overlap. The following families of algorithms are available:

D. Eppstein and E.R. Gansner (Eds.): GD 2009, LNCS 5849, pp. 415–416, 2010.

- Normal tree layouts, Org chart layouts, Radial tree layouts
- Hierarchical layouts
- Spring Embedder
- Bus layout
- Topological mesh layout
- Circular layout
- Recursive layout for nested graphs
- Label layout, annotation layout
- Link routing (several algorithms and variants)
- Improved display of link crossings

The layout services are available as different Eclipse plug-ins. The plug-in based on GEF delivers graph layout core capabilities via an adapted graph model that issues GEF requests and triggers GEF commands to position nodes and links during layout. This is the natural, nondestructive approach for GEF. This architecture ensures that layout is undoable and persistent. The ILOG JViews Graph Model for Eclipse hides the complexity and can easily be attached to any GEF Graph Editing Part. Similarly, the plug-in based on GMF utilizes all GEF services and additionally provides support for the modelling capabilities of EMF. An animation framework and various predefined graphic shapes (subgraph shapes, links with crossing jogs) complete the services offered by IBM ILOG Graph Layout for Eclipse.

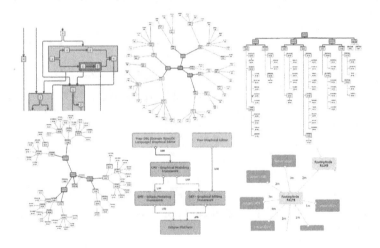

Fig. 2. Sample Layouts of IBM ILOG Graph Layout for Eclipse

Layout Techniques Coupled with Web2.0-Based Business Process Modeling

Philip Effinger[1] and Gero Decker[2]

[1] Wilhelm-Schickard-Institut, Eberhard-Karls-Universität Tübingen, Germany
[2] Hasso-Plattner-Institut, Universität Potsdam, Germany
Effinger@informatik.uni-tuebingen.de,
Gero.Decker@hpi.uni-potsdam.de

1 Introduction

The most wide-spread notation for process models is the Business Process Modeling Notation (BPMN). With Oryx, we have a open-source modeling tool at hand that supports collaborative and web-based modeling of BPMN diagrams. Here, we show how automatic layout of diagrams can support the designer when starting to model a process in BPMN. We provide an automatic layout integrated into Oryx that computes a new layout for a given diagram considering BPMN drawing conventions, e.g. orthogonal edges, hierarchical structures, partitions, etc.

2 Oryx – A Web2.0-Based Collaborative Graphical Editor

Oryx (http://oryx-project.org) is an extensible framework for graphical modeling in the web browser. Using JavaScript and Scalable Vector Graphics (SVG), Oryx uses modern web technologies that realize a similar user experience like a classical modeling tool that runs on the desktop. The application is loaded into the browser whenever a graphical model is opened for editing.

In Oryx, each artifact is identified by a URL, so that models can be shared by passing references, rather than by exchanging model documents as email attachments. Oryx follows the Representational State Transfer (REST) architectural style, using the HTTP verbs GET, PUT, POST and DELETE for reading and updating models. This enables a highly scalable architecture, allowing for caching mechanisms at the protocol level.

The Oryx source code is available under an Open Source license and has become a widely used technology platform, especially in the Business Process Management (BPM) community. Here, process modeling using languages such as the Business Process Modeling Notation (BPMN) is a central activity.

3 The Automatic Layout Algorithm and Integration into Oryx

Our layout approach for support of automatic layout in ORYX is developed by extending previous works on layout techniques adoptable for BPMN [1, 2]. The layout is computed on BPMN diagrams that are based on a graphs. An automatic layout approach for BPMN-graphs has to support the drawing conventions that represent specific layout requirements of the BPMN notation. In our case, specfic layout requirements are:

D. Eppstein and E.R. Gansner (Eds.): GD 2009, LNCS 5849, pp. 417–418, 2010.

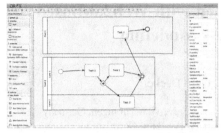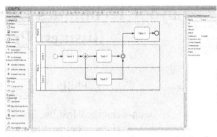

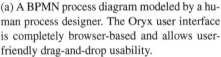

(a) A BPMN process diagram modeled by a human process designer. The Oryx user interface is completely browser-based and allows user-friendly drag-and-drop usability.

(b) The resulting process diagram of (a) after our approach is applied. The drawing conventions of BPMN are fulfilled.

- Elements have different sizes.
- We have to consider partitions, e.g. (collapsed/expanded) pools and swimlanes.
- Subprocesses may be nested and edges can start/end at a subprocess. This concept corresponds to graph clustering.
- Handle labels of pools, swimlanes, elements and edges.

Since BPMN-graphs are usually drawn using orthogonal routes for edges, we use an orthogonal layout approach for calculating the initial layout of a given BPMN-graph.

Our layout approach employs the implementation described in [3] that incorporates different constraints needed for the automatic layout of activity diagrams which are related to business process diagrams. The supported constraints include partitions (a generalization of swimlanes), clusters (subprocesses/groups) as well as a common workflow direction of edges which is especially important for such diagrams. Used techniques are based on Sugiyama's algorithm [4] and the *Topology-Shape-Metrics* (TSM) approach [5]. All layout requirements and drawing conventions demanded by BPMN models can be satisfied. Further details of this implementation can be found in [2].

For the integration of the layout implementations into Oryx, a wrapper was implemented in JAVA that offers interfaces for the connection to the JavaScript-based BPMN-editor Oryx and support of its diagram model.

References

[1] Effinger, P., Kaufmann, M., Siebenhaller, M.: Enhancing visualizations of business processes. In: Tollis, I.G., Patrignani, M. (eds.) GD 2008. LNCS, vol. 5417, pp. 437–438. Springer, Heidelberg (2009)
[2] Effinger, P., Siebenhaller, M., Kaufmann, M.: An Interactive Layout Tool for BPMN. In: BPMN 2009 - 1st International Workshop on BPMN, co-located with 11th IEEE Conference on Commerce and Enterprise Computing, CEC 2009 (2009)
[3] Siebenhaller, M., Kaufmann, M.: Drawing activity diagrams. In: Proceedings of ACM 2006 Symposium on Software Visualization, SoftVis 2006, pp. 159–160. ACM, New York (2006)
[4] Sugiyama, K., Tagawa, S., Toda, M.: Methods for visual understanding of hierarchical system structures. IEEE Transactions on Systems, Man, and Cybernetics SMC-11(2), 109–125 (1981)
[5] Tamassia, R.: On embedding a graph in the grid with the minimum number of bends. SIAM Journal on Computing 16(3), 421–444 (1987)

Proving or Disproving Planar Straight-Line Embeddability onto Given Rectangles

Michael Kaufmann and Stephan Kottler

Wilhelm–Schickard–Institute, University of Tübingen, Germany

Encoding as SAT Problem

Given a plane graph $G = (V, E)$ and a rectangle we ask whether there exists a planar straight-line embedding of G onto the grid-points of the rectangle. For this NP-hard problem [5] some powerful heuristics have been developed to minimise the area of an embedding of a given graph [5,4]. Moreover, for particular families of graphs upper and lower bounds on the area have been proven [2]. However, in the general case it is not possible to ensure whether there is an embedding that preserves a particular area restriction $A = h \cdot w$. We present an implementation[1] based on a translation into SAT to tackle this kind of problems for small graphs. We only describe the direct encoding into CNF[2] that turned out to be most suitable.

Matching Vertices to Grid Positions. We introduce boolean variables $x_{i,j}$ representing whether or not vertex $0 < i \leq N$ ($N = |V|$) is located at grid position $0 < j \leq A$. This causes $N \cdot A$ variables. Moreover, in case $N < A$ we further introduce one variable $x_{.,j}$ per grid position to represent the fact that position j is not used by any vertex. Another possibility would be to introduce $A - N =: d$ disconnected dummy vertices, but this causes $d \cdot A$ additional variables.

The following clauses ensure that each vertex is placed at (at least) one position: $(x_{i,1} \vee x_{i,2} \vee \ldots \vee x_{i,A})$ $\forall\, 0 < i \leq N$. Analogously we ensure that each grid position is either used by a vertex or, in case of $N < A$ may be free: $(x_{.,j} \vee x_{1,j} \vee x_{2,j} \vee \ldots \vee x_{N,j})$ $\forall\, 0 < j \leq A$ *[$x_{.,j}$ is omitted if $N = A$]* In case $N = A$ the conjunction of the above *position clauses* would be sufficient to guarantee that each vertex is placed at exactly one postion and vice versa. When $N < A$ a valid mapping of vertices to positions has to be enforced by additional constraints. One possibility is to introduce binary clauses $(\overline{x_{i,j}} \vee \overline{x_{k,l}})$ for each pair of literals within a clause. Note, that this is necessary for each of the above clauses. In practice it is important to have these – possibly redundant – constraints to guide the SAT-solver by enabling early recognition of conflicting assignments. We modified or SAT-solver *SApperloT*[3] to treat the position clauses as special constraints where **exactly one** literal has to be true. This simulates all $O(N \cdot A^2 + A \cdot N^2)$ binary clauses by simultaneously using a linear amount of memory.

---

[1] www-pr.informatik.uni-tuebingen.de/?site=forschung/sat/algo_engineering
[2] A formula in CNF (conjunctive normal form) is a conjunction of clauses; clauses are disjunctions of literals, whereas a literal is a boolean variable or its negation.

D. Eppstein and E.R. Gansner (Eds.): GD 2009, LNCS 5849, pp. 419–420, 2010.

Planar Embedding and Symmetry Breaking. To achieve a planar embedding of the graph crossings have to be prohibited explicitely. In order to avoid symmetric constraints this is done by introducing further A^2 variables: For each possible straight-line connection between any two grid positions we hold a variable $y_{k,l}$ $(0 < k, l \leq A)$ indicating whether or not the edge between grid position k and position l is present in an embedding of G.

With this, any placement of any two adjacent vertices onto grid positions causes a particular edge embedding to be drawn. If, for instance, two adjacent vertices i and j are placed at the positions k and l then the edge between these two positions is actually drawn. Hence, $x_{a,k} \wedge x_{j,l}$ implies variable $y_{k,l}$ to be true. This can be expressed by the clause $(\overline{x_{a,k}} \vee \overline{x_{j,l}} \vee y_{k,l})$. Note that there will be another clause $(\overline{x_{a,l}} \vee \overline{x_{j,k}} \vee y_{k,l})$ for the symmetric case. Given that the number of adjacent vertex pairs in a planar graph is bounded by $O(N)$ the number of introduced ternary clauses by this kind of constraints is bounded by $O(N \cdot A^2)$. It remains to prohibit the crossing of any two embedded edges. For this reason we disallow all combinations of crossing edge embeddings by introducing binary clauses of the form $(\overline{y_{k,l}} \vee \overline{y_{q,t}})$. At the same time we forbid any edge embedding that crosses a grid position k unless k is chosen to contain no vertex $(x_{.,k} = true)$. The number of binary clauses introduced by these constraints is bounded by A^4.

Experimental Results and Conclusion

The purpose of our approach is to realise both: either proving or disproving the existence of a graph embedding onto a specified rectangle for small graphs. For our experimental setup we chose seven different kinds of planar graphs (using [1]): biconnected, triconnected, not biconnected, nested triangles, nested triangles completely triangulated, trees and grids. We generated the graphs with $16 \leq N \leq 100$ vertices. For the first three types of graphs each N was combined with $|E| \in \{N, \frac{3}{2}N, 2N, \frac{5}{2}N, 3N - 6\}$ edges. For 393 graphs from a total of 576 test graphs our software was able to prove (269) or disprove (124) the existence of a planar straight-line embedding on a specified tight rectangle. We see strong potential to extend our tool to confirm conjectures for special kinds of graphs (as e.g. in [2]). For such cases additional constraints could be added to prune the search space of the solver.

References

1. Open Graph Drawing Framework, http://www.ogdf.net
2. Frati, F., Patrignani, M.: A note on minimum area straight-line drawings of planar graphs. In: 15th International Symposium on Graph Drawing (2007)
3. Kottler, S.: Solver descriptions for the SAT competition (2009), satcompetition.org
4. Krug, M.: Minimizing the area for planar straight-line grid drawings. Master's thesis, University of Karlsruhe (2007)
5. Krug, M., Wagner, D.: Minimizing the area for planar straight-line grid drawings. In: Hong, S.-H., Nishizeki, T., Quan, W. (eds.) GD 2007. LNCS, vol. 4875, pp. 207–212. Springer, Heidelberg (2008)

Visualization of Complex BPEL Models

Benjamin Albrecht[1], Philip Effinger[1], Markus Held[2], Michael Kaufmann[1],
and Stephan Kottler[1,2]

[1] Parallel Computing, Universität Tübingen, Germany
[2] Symbolic Computation Group, Universität Tübingen, Germany
{albrecht,effinger,mheld,mk,
kottlers}@informatik.uni-tuebingen.de

1 Introduction

In this work, we present our approach for producing layouts of complex workflows given in the *Business Process Execution Language* (BPEL) [1]. BPEL is a verbose, hierarchical workflow language containing nested, alternative and concurrent execution paths. Our approach enhances the Sugiyama algorithm [2] by introducing special paths, which are constrained to be drawn in parallel, and hence, orthogonally to the layers in the Sugiyama model. To prove the feasibility of our approach, we have developed an extension to the collaborative BPEL development system HOBBES [3] [4]. Collaboration enhances the need for visualizations of complex workflow models, as team members have to coordinate their activities.

2 Collaborative Workflow Development with HOBBES

The Business Process Execution Language The XML-based *Business Process Execution Language.* (BPEL) has become the *de-facto* standard for business workflows. It is a key element of the *Service Oriented Architecture* (SOA) [1]. BPEL control flow is a mixture of block-oriented and graph-oriented elements. Atomic tasks like service invocations or waiting commands are called *basic activities*. Control structures are expressed as *structured activities* (e.g. Sequence, If, While), which can contain child activities. Concurrency can be modeled using the structured activities Flow and ForEach. Flow allows the definition of Directed Acyclic Graphs of activities while ForEach loops may be marked as "parallel". Expressions, given in the XPath language, are used for conditionals, for triggering links between activities and for assignments.

The HOBBES system. HOBBES is a web-based BPEL development system, which enables synchronous collaboration sessions [3, 4]. A team leader may grant privileged access to workflow parts or assign tasks to team members, as well as inspect the workflow using BPEL-specific software metrics (*workflow metrics*). Team members have different views on the BPEL model edited in a HOBBES session, to enable parallel development activities. For communication purposes and to enable a better understanding of the entire model, the need for graphical visualizations arises. These have to present the control flow paths in a concise way and preserve the hierarchies of the process' structure.

D. Eppstein and E.R. Gansner (Eds.): GD 2009, LNCS 5849, pp. 421–423, 2010.

HOBBES has been implemented using the Adobe Flex framework, which enables rich user interfaces as well as server-to-client notification. Communication between clients is relayed via a Java-based server, which governs access to a central object model. A demo version of HOBBES can be accessed at:

`http://www-sr.informatik.uni-tuebingen.de/workflows`

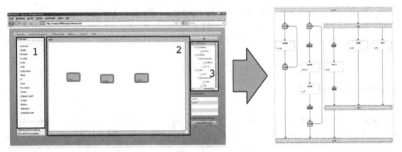

(a) Hobbes' user interface (left): The modeling frame (2) shows a single hierarchical level of a BPEL model, the tree of the whole current process is provided (3). Rectangle (1) depicts the palette of available BPEL elements. The result of our layout approach for this process is shown on the right: The hierarchies of the BPEL model are visualized unfolded. Start and Exit of a hierarchical element are depicted by green and red boxes, and surrounded by colored rectangles.

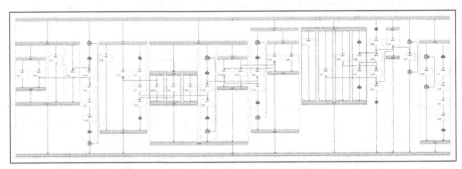

(b) Example of a layout for a complex BPEL-Process.

3   Realizing Layout Capabilities in Hobbes

A visualization of BPEL processes suggests a layered drawing technique. Thus, our layout approach is based on the Sugiyama algorithm. Since there is no unique method to derive paths from a BPEL model, we consider the number of descendant BPEL activites of structured activities for path construction and embedding. In order to draw each path in parallel, orthogonally to the layers in the Sugiyama model, three steps of the standard algorithm (Cycle Removal, Layer Assignment and Computation of the Horizontal Coordinates) have to be modified.

Each path consists of so-called *path-edges* and should preferably be drawn from top to bottom. Regarding cycles in the graph, we have to consider the special case of a cycle that consists only of path-edges. In this case, we perform a division of at least one of the paths contributing at least one edge to the cycle. Furthermore, additional constraints are added to the layer assignment phase in order to ensure that the source node of each path-edge is placed in a layer above the target node.

To increase the readability of the main activities in a workflow, paths should be drawn in a straight vertical manner. Thus, for the computation of the horizontal coordinates, the standard algorithm is applied as a first step followed by a postprocessing step: For each path p, the barycenter b_p of the x-coordinates of all nodes in p is computed. The horizontal coordinates of nodes contained in exactly one path p are set to b_p. For nodes contained in paths $p_1, \ldots, p_k$ with $k \geq 2$, the x-coordinate is set to the barycenter of $b_{p_1}, \ldots, b_{p_k}$.

Finally, in order to avoid overlapping nodes, non-path nodes are moved to the left and to the right until all nodes of the graph adhere to the minimal node distance. Final modifications are applied to the graphical representation of the elements of the graph: Each node is assigned its own shape according to its representative activity. All edges are drawn in a way that they respect the flow from the top to the bottom. To further improve the readability of the workflow, all hierarchical activities containing several subordinate activities are surrounded by colored rectangles.

References

[1] Alves, A., Arkin, A., Askary, S., Baretto, C., Bloch, B., Curbera, F., Ford, M., Goland, Y., Guizar, A., Kartha, N., Liu, C., Khalaf, R., König, D., Marin, M., Mehta, V., Thatte, S., van der Rijn, D., Yendluri, P., Yiu, A.: Web Services Business Process Execution Language Version 2.0. OASIS standard (April 2007)

[2] Sugiyama, K., Tagawa, S., Toda, M.: Methods for visual understanding of hierarchical system structures. IEEE Transactions on Systems, Man, and Cybernetics SMC-11(2), 109–125 (1981)

[3] Held, M., Blochinger, W.: Collaborative BPEL Design in a Rich Internet Application. In: CCGRID 2008: 8th International Symposium on Cluster Computing and the Grid. IEEE Computer Society Press, Los Alamitos (2008)

[4] Held, M., Blochinger, W.: Structured collaborative workflow design. Future Generation Computer Systems FCGS 25(6), 638–653 (2009)

DAGmaps and Dominance Relationships

Vassilis Tsiaras and Ioannis G. Tollis

Institute of Computer Science, Foundation for Research and Technology-Hellas
{tsiaras,tollis}@ics.forth.gr

1 Introduction

In [2] we use the term *DAGmap* to describe space filling visualizations of DAGs according to constraints that generalize treemaps and we show that deciding whether or not a DAG admits a DAGmap drawing is NP-complete. Let $G = (V, E)$ be a DAG with a single source s. A component st-graph $G_{u,v}$ of G is a subgraph of G with a single source u and a single sink v that contains at least two edges and that is connected with the rest of G through vertex u and/or vertex v. A vertex w *dominates* a vertex v if every path from s to v passes through w. The dominance relation in G can be represented in compact form as a tree T, called the dominator tree of G, in which the dominators of a vertex v are its ancestors. Vertex w is the *immediate dominator* of v if w is the parent of v in T. A simple and fast algorithm to compute T has been proposed by Cooper et al. [1]. The *post-dominators* of G are defined as the dominators in the graph obtained from G by reversing all directed edges and assuming that all vertices are reachable from a (possibly artificial) vertex t. Using the definition of DAGmaps, it is easy to prove that in a DAGmap of G the rectangle of a vertex u includes the rectangles of all vertices that are dominated (resp. post-dominated) by u. Therefore when vertex u dominates vertex v and vertex v post-dominates vertex u then the rectangles R_u and R_v of u and v coincide. Based on this observation, we propose a heuristic algorithm that transforms a DAG G into a DAG G' that admits a DAGmap. When G contains component st-graphs then our algorithm performs significantly fewer duplications than the transformation of G into a tree.

2 Transforming a DAG So That It Admits a DAGmap

A component st-graph $G_{u,v}$ of G behaves under vertex duplications as if it is encapsulated in a cluster vertex (c-vertex). When vertex u is duplicated then $G_{u,v}$ is duplicated as a whole. Additionally duplications that start at a vertex w of $G_{u,v}$, where $w \neq u, v$, stop at vertex v (i.e., they do not propagate to the rest of G since $R_v = R_u$). To identify the component st-graphs of G first we compute the dominator and the post-dominator trees of G. Then for each non-leaf node v of the post-dominator tree that has more than one incoming edges we find its immediate dominator u. The st-graph $G_{u,v}$ that is induced by all paths from u to v in G is a component st-graph of G. If the immediate post-dominator of u is v

D. Eppstein and E.R. Gansner (Eds.): GD 2009, LNCS 5849, pp. 424–425, 2010.

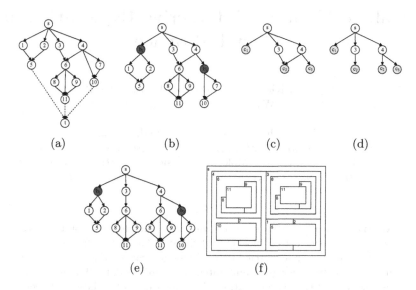

Fig. 1. Transforming DAG G into G' that admits a DAGmap. a) A DAG with an artificial sink vertex t. b) Introduction of artificial b-vertices. c) Folding of component st-graphs into c-vertices. d) Vertex duplications. e) Unfolding of c-vertices; subgraph $G_{6,11}$ appears twice in DAG G'. f) DAGmap drawing of G'.

then $G_{u,v}$ is clearly discernible, else it is recommended to introduce an artificial vertex b that is adjacent to u and collects all outgoing edges of u that belong to paths that lead to v. It is easy to show that a DAG G is converted into a DAG G' that admits a DAGmap if we first recursively fold the component st-graphs of G and then we repeat until termination the following: unfold a c-vertex and in case that the unfolded subgraph does not admit a DAGmap then perform duplications of its vertices that have more than one incoming edge. Note that if we perform duplications even when the unfolded st-graph admits a DAGmap then an initial st-graph G is converted into a TTSP digraph [2]. In Fig. 1 vertex s dominates vertex 5 but vertex 5 does not post-dominate vertex s. Therefore an artificial vertex b_1 is introduced. On the other hand the subgraph induced by vertices $s, 3, 4$ and 6 cannot be separated since vertex 4 has an outgoing edge to a vertex that does not belong to the subgraph. Finally c-vertex c_2 is duplicated since it has two incoming edges and this leads to duplication of the whole st-graph $G_{6,11}$.

References

1. Cooper, K.D., Harvey, T.J., Kennedy, K.: A simple, fast dominance algorithm (2001), http://www.cs.rice.edu/~keith/EMBED/dom.pdf
2. Tsiaras, V., Triantafilou, S., Tollis, I.G.: DAGmaps: Space Filling Visualization of Directed Acyclic Graphs. JGAA 13(3), 319–347 (2009)

Scaffold Hunter – Interactive Exploration of Chemical Space

Karsten Klein[1], Nils Kriege[1], Petra Mutzel[1],
Herbert Waldmann[2], and Stefan Wetzel[2]

[1] Technische Universität Dortmund, Germany
{karsten.klein,nils.kriege,petra.mutzel}@cs.tu-dortmund.de
[2] Max-Planck-Institute for Molecular Physiology, Dortmund, Germany
{herbert.waldmann,stefan.wetzel}@mpi-dortmund.mpg.de

Abstract. Scaffold Hunter is a Java-based software tool for the analysis of structure-related biochemical data. It facilitates the interactive
exploration of chemical space by enabling generation of and navigation
in a scaffold tree hierarchy annotated with various data. The graphical
visualization of structural relationships allows to analyze large data sets,
e.g., to correlate chemical structure and biochemical activity.

The search for small molecules that are biologically relevant, e.g., to design new
drugs and diagnostics, is an important task in chemical biology. Even though
high throughput methods are available to test large numbers of chemical compounds, the overwhelming size of chemical space—containing up to 10^{160} different molecules—renders any exhaustive search infeasible. Methods are needed
that allow to focus the search on the tiny fraction of chemical space that contains
the most promising candidates for biological activity.

Schuffenhauer et al. [1] developed a hierarchical scaffold classification strategy
to chart chemical spaces. A rule set is defined to reduce molecules to *scaffolds*,
on which a hierarchy is defined based on a substructure relation, cf. Fig. 1. This
way, the compounds are classified by structural similarity, reflecting the fact that
the biological relevance is closely coupled to the molecular structure.

We developed Scaffold Hunter, a software tool that admits the graphical representation of chemical compound databases based on this classification strategy.
The software visualizes the set of scaffold trees resulting from the hierarchy generation step using graph layouts, and allows interactive exploration of the data

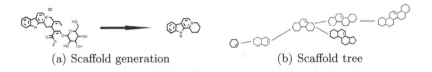

(a) Scaffold generation   (b) Scaffold tree

Fig. 1. Example for a reduction step that derives a scaffold structure from a given
molecule structure (a) and a partial scaffold tree (b)

D. Eppstein and E.R. Gansner (Eds.): GD 2009, LNCS 5849, pp. 426–427, 2010.
© Springer-Verlag Berlin Heidelberg 2010

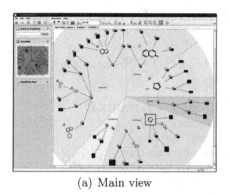
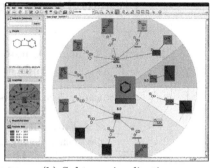

(a) Main view   (b) Subtree visualization

Fig. 2. Screenshots of the main view and the visualization of a subtree generated by a filtering step, both with property-based color shading

based on this representation, cf. Fig. 2(a). Scaffold Hunter provides several layout styles— adaptions of radial, tree, and balloon layout— and iterative scaffold property filtering can be used to reduce the number of scaffolds displayed. Color shading based on selected properties, both of scaffolds and specific segments of the dataset, facilitates orientation and navigation within the dataset. The set of molecules corresponding to scaffolds of interest and also subtrees derived by filtering and selection steps can be viewed in a separate window, cf. Fig. 2(b).

The use of Scaffold Hunter fits into the chemist's workflow for the analysis of large compound libraries and allows to intuitively explore the underlying chemical space. It helps to identify 'holes' in the structure space analyzed that may serve as promising starting points for compound library design. Scaffold Hunter identifies virtual scaffolds that do not represent molecules in the dataset and that should share bioactivity properties with their parents or child scaffolds. These virtual scaffolds may provide new opportunities for the identification of new biologically relevant scaffold classes. An initial proof-of-concept analysis demonstrated the usefulness of our approach [3].

Scaffold Hunter is implemented in Java and freely available under the GPL [2].

References

1. Schuffenhauer, A., Ertl, P., Roggo, S., Wetzel, S., Koch, M.A., Waldmann, H.: The scaffold tree - visualization of the scaffold universe by hierarchical scaffold classification. J. Chem. Inf. Model. 47(1), 47–58 (2007)
2. Scaffold Hunter Project Webpage,
   http://sourceforge.net/projects/scaffoldhunter/
3. Wetzel, S., Klein, K., Renner, S., Rauh, D., Oprea, T.I., Mutzel, P., Waldmann, H.: Interactive exploration of chemical space with scaffold hunter. Nat. Chem. Biol. 5(8), 581–583 (2009)

Graph Drawing Contest Report

Christian A. Duncan[1], Carsten Gutwenger[2], Lev Nachmanson[3],
and Georg Sander[4]

[1] Louisiana Tech University, Ruston, LA 71272, USA
duncan@latech.edu
[2] University of Dortmund, Germany
carsten.gutwenger@cs.uni-dortmund.de
[3] Microsoft, USA
levnach@microsoft.com
[4] IBM, Germany
georg.sander@de.ibm.com

Abstract. This report describes the 16th Annual Graph Drawing Contest, held in conjunction with the 2009 Graph Drawing Symposium in Chicago, USA. The purpose of the contest is to monitor and challenge the current state of graph-drawing technology.

1 Introduction

This year's Graph Drawing Contest had two topics: Partial Graph Drawing and the Graph Drawing Challenge. The partial graph drawing topic provided four data sets of different kinds of graphs, which were partially laid out. Some of the nodes had specified coordinates, and some of the edges had specified routing points. These nodes and edges were considered fixed. Other nodes and edges had no coordinates and had to be laid out. These nodes and edges were free to be moved. All nodes had specified sizes. The task was to find the nicest layout that integrates the free nodes and edges into the layout of the fixed nodes and edges without changing the positions of fixed nodes and edges. The data sets were a simple tree, an organization chart, a flow chart, and a mystery graph.

The Graph Drawing Challenge, which took place during the conference, focused on minimizing the number of crossings of upward grid drawings of graphs with edge bends. We received 27 submissions: 10 submissions for the Partial Graph Drawing topic and 17 submissions for the Graph Drawing Challenge.

2 Simple Tree

The simple tree had 56 nodes and 55 directed edges. 16 nodes were fixed. There were no special requirements concerning the edge shapes. The data was artificially created and the fixed nodes were roughly arranged in a top down manner. The challenge was to fit the free nodes into the dense area between the fixed nodes. We received 4 submissions for the tree data. Two submissions were based on a spring embedder approach, and two other submissions were based on a specialized tree layout algorithm.

D. Eppstein and E.R. Gansner (Eds.): GD 2009, LNCS 5849, pp. 428–433, 2010.

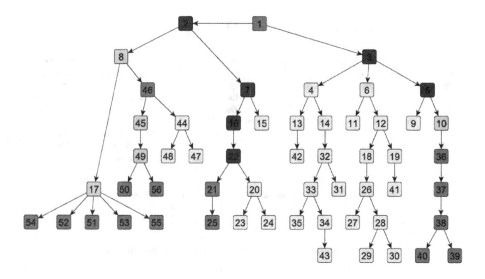

Fig. 1. First place, Simple Tree (original in color)

The winning submission by Melanie Badent and Michael Baur from the Universities of Konstanz and Karlsruhe (Figure 1) used a variation of the algorithm by Reingold and Tilford [5]. They determined the node spacing from the positions of the fixed nodes, so that the layout looks uniform. Then, they categorized the nodes into red nodes (in Figure 1 medium gray) that were fixed, orange nodes (in Figure 1 lighter gray) that have only fixed or orange nodes as successors, hence their position is determined by the layout principle as the center above their successors, blue nodes (in Figure 1 black) that have some red or orange nodes as successors and hence are partially constrained in their position, and green nodes (in Figure 1 white) that are completely free to be placed. For the green subtrees, they computed all possible permutations of successors to obtain the subtree with minimum width for the previously determined node spacing to fit into the space between the fixed nodes. For blue nodes, the ordering of their green child nodes was determined by a test routine that checked all insertion points between red and orange nodes. The algorithm was specially tailored for this input graph. They reported that for general trees a more sophisticated method would be advisable.

3   Organization Chart

The organization chart had 171 nodes and 170 directed edges. 101 nodes and 96 edges were fixed. The fixed nodes had a minimal spacing of approximately 40 units (border to border). The edges had to be laid out in an orthogonal fashion, which was only satisfied by one of the two submissions.

The winning submission came from Nicholas Jefferson from the University of Sydney (Figure 2; the dark nodes and edges were fixed). The layout was produced

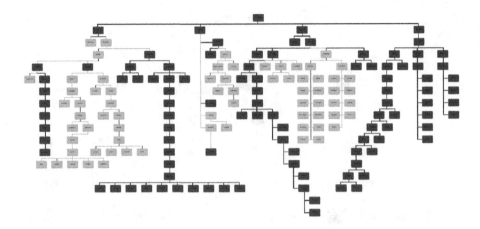

Fig. 2. First place, Organization Chart (original in color)

using a backtracking search algorithm that used a minimum-size layout for all descendants of a node whenever possible but otherwise recursed for each permutation of the children of the node, up to subtree isomorphism. The minimum-size layouts were generated using an approach similar to the Stockmeyer merge [3] and again the Reingold-Tilford algorithm [5]. The implementation used the programming language Ruby and the relational database PostgreSQL, using spatial indexes to detect occlusion and continuations and savepoints to restore state on backtracking.

4   Flow Chart

The flow chart had 57 nodes and 72 directed edges. 26 nodes were fixed and 24 edges were fixed. The graph had a single source and single target and was artificially created. The majority of the edges should point downwards. Orthogonal edges were preferred but not required.

The winning submission came from Hui Liu from the University of Sydney (Figure 3; the dark nodes and their connecting edges in the left picture were fixed). First, an automatic layout was produced using the Neato algorithm of GraphViz [1]. From this sketch, a drawing was created with Microsoft Visio that was manually laid out. Finally, a background picture was added showing a diamond pattern to highlight the fixed nodes (Figure 3, right).

5   Mystery Graph

The mystery graph had the 17 cities of the past Graph Drawing Conferences as nodes fixed on a circle. 374 city nodes were added to keep the graph connected. These nodes were free to be moved. The additional task was to determine the

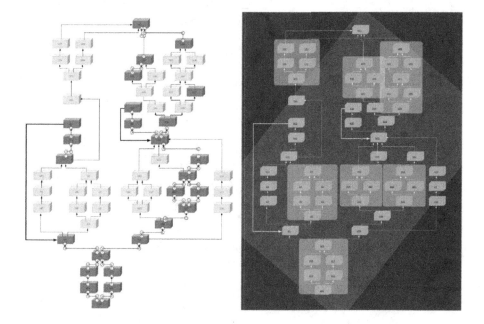

Fig. 3. First place, Flow Chart (original in color)

meaning of the 911 undirected edges, which depicted the twin towns or sister city relationships between the cities as collected from Wikipedia [7].

The winning team, again Melanie Badent and Michael Baur, found the correct answer. Since the fixed nodes formed a circle, they decided to arrange all nodes in circles as well. The fixed nodes formed the innermost circle, and the free nodes were placed at outer circles depending on their topological distance from fixed nodes. The team used a radial layout algorithm from the visone software project [4] and rotated the circles with free nodes to bring sister cities on different circles closer to each other.

6   Graph Drawing Challenge

Following the Graph Drawing Challenge tradition of having the same topic in two subsequent years, we repeated the same challenge as in year 2008, with new sample graphs however. The challenge, a subproblem of the popular layered layout technique by Sugiyama et al. [6] known to be NP-hard, dealt with minimizing the number of crossings of upward grid drawings of graphs allowing edge bends. This technique requires that all nodes be placed on grid positions, that nodes and edge bends don't overlap, and that all edge segments point strictly upwards. At the start of the one-hour on-site competition, the contestants were given six nonplanar directed acyclic graphs with a legal upward layout that however had a huge number of crossings. The goal was to rearrange the layout to reduce the

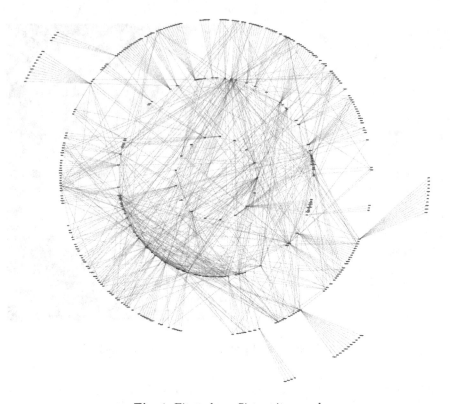

Fig. 4. First place, Sister city graph

number of crossings. Only the number of crossings was judged; other aesthetic criteria such as the number of edge bends or the area were ignored.

We partitioned the challenge into two subcategories: automated and manual. In the automated subcategory, contestants received graphs ranging in size from 26 nodes / 48 edges to 953 nodes / 1529 edges and were allowed to use their own sophisticated software tools with specialized algorithms. Only one team (TU Dortmund: Hoi-Ming Wong and Karsten Klein) submitted results and hence were the winner of this subcategory. They found the optimal solution of four of the graphs by using a modified version [2] of the same software as last year. For the largest graph, they found a solution with 57 crossings, which is not far from the optimal layout with 12 crossings.

The 16 manual teams solved the problems by hand using IBM's Simple Graph Editing Tool provided by the committee. They received graphs ranging in size from 25 nodes / 40 edges to 144 nodes / 365 edges. To determine the winner among the manual teams, the scores of each graph, determined by dividing the crossing number of the best submission by the crossing number of the current submission, were summed up. With a score of 4.87, the winner was J. Joseph Fowler from the University of Wisconsin-Milwaukee, who found the optimal result for the 3 smallest graphs and very good results for graphs 4 and 5. For the

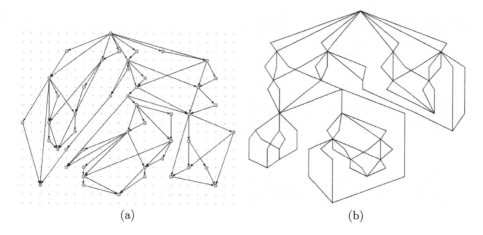

(a)   (b)

Fig. 5. Challenge graph with 48 nodes / 83 edges and 4 crossings: (a) the best manually obtained result by J. Joseph Fowler, (b) the best automated result by team Dortmund

largest graph in the manual category, which can be laid out optimally with only 4 crossings, no contestant found a good solution manually. The submissions for that graph ranged between 4287 and 5995 crossings. Figure 5 shows the results of a graph with 48 nodes and 83 edges. This graph was used both in the manual and automated subcategory and had 4 crossings in the best results, which is optimal.

Acknowledgments. We thank the generous sponsors of the symposium and all the contestants for their participation.

References

1. AT&T. Graphviz, http://www.graphviz.org
2. Chimani, M., Gutwenger, C., Mutzel, P., Wong, H.: Upward planarization layout. In: Eppstein, D., Gansner, E.R. (eds.) GD 2009. LNCS, vol. 5849, pp. 94–106. Springer, Heidelberg (2010)
3. Eades, P., Lin, T., Lin, X.: Minimum size h-v drawings. In: Advanced Visual Interfaces (Proc. AVI 1992). World Series in Computer Science, vol. 36, pp. 386–394 (1992)
4. University of Karlsruhe. visone, http://visone.info
5. Reingold, E., Tilford, J.: Tidier drawing of trees. IEEE Trans. on Software Engineering 7(2), 223–228 (1981)
6. Sugiyama, K., Tagawa, S., Toda, M.: Methods for visual understanding of hierarchical systems. IEEE Trans. Sys. Man, and Cyb. SMC11(2), 109–125 (1981)
7. Wikipedia. Sister cities, http://en.wikipedia.org/wiki/Sister_cities (accessed, May 2009)

Author Index